Aging and Age-Related Disorders

Aging and Age-Related Disorders

From Molecular Mechanisms to Therapies

Topical Collection Editor

Vladimir Titorenko

MDPI • Basel • Beijing • Wuhan • Barcelona • Belgrade

MDPI

Topical Collection Editor
Vladimir Titorenko
Concordia University
Canada

Editorial Office
MDPI
St. Alban-Anlage 66
4052 Basel, Switzerland

This is a reprint of articles from the Topical Collection published online in the open access journal *International Journal of Molecular Sciences* (ISSN 1422-0067) from 2017 to 2019 (available at: https://www.mdpi.com/journal/ijms/special_issues/aging_disorders and https://www.mdpi.com/journal/ijms/special_issues/aging_mech_therap).

For citation purposes, cite each article independently as indicated on the article page online and as indicated below:

LastName, A.A.; LastName, B.B.; LastName, C.C. Article Title. *Journal Name* **Year**, *Article Number, Page Range.*

ISBN 978-3-03921-355-9 (Pbk)
ISBN 978-3-03921-356-6 (PDF)

Contents

About the Topical Collection Editor

Vladimir Titorenko, Ph.D., Professor, University Research Fellow, Biology Department, Concordia University, Montreal, Quebec, Canada. Vladimir Titorenko received his bachelor's degree in biochemistry from Lviv University in Ukraine. He did graduate work in the laboratory of Dr. Andriy Sibirny and received his PhD in genetics from the Institute of Genetics of Industrial Microorganisms in Moscow, Russia. Following postdoctoral work with Dr. Marten Veenhuis in the Laboratory of Electron Microscopy at Groningen University in the Netherlands, he moved to the Department of Cell Biology at the University of Alberta in Edmonton where he was a research associate in the laboratory of Dr. Richard Rachubinski. Vladimir Titorenko is currently a full professor and a university research fellow in the Biology Department at Concordia University in Montreal. Using the budding yeast Saccharomyces cerevisiae as a model organism, his laboratory investigates molecular mechanisms of cellular aging and aging delay.

Preface to "Aging and Age-Related Disorders"

Aging of unicellular and multicellular eukaryotic organisms is a convoluted biological phenomenon, which is manifested as an age-related functional decline caused by progressive dysregulation of certain cellular and organismal processes. Many chronic diseases are associated with human aging. These aging-associated diseases include cardiovascular diseases, chronic obstructive pulmonary disease, chronic kidney disease, diabetes, osteoarthritis, osteoporosis, sarcopenia, stroke, neurodegenerative diseases (including Parkinson's, Alzheimer's, and Huntington's diseases), and many forms of cancer. Studies in yeast, roundworms, fruit flies, fishes, mice, primates, and humans have provided evidence that the major aspects and basic mechanisms of aging and aging-associated pathology are conserved across phyla. The focus of this *International Journal of Molecular Sciences* Special Issue is on molecular and cellular mechanisms, diagnostics, and therapies and diseases of aging. Fifteen original research and review articles in this Special Issue provide important insights into how various genetic, dietary, and pharmacological interventions can affect certain longevity-defining cellular and organismal processes to delay aging and postpone the onset of age-related pathologies in evolutionarily diverse organisms. These articles outline the most important unanswered questions and directions for future research in the vibrant and rapidly evolving fields of mechanisms of biological aging, aging-associated diseases, and aging-delaying therapies.

Vladimir Titorenko
Topical Collection Editor

International Journal of
Molecular Sciences

MDPI

Review

Articular Cartilage Aging-Potential Regenerative Capacities of Cell Manipulation and Stem Cell Therapy

Magdalena Krajewska-Włodarczyk [1,2,3,]*[ ], Agnieszka Owczarczyk-Saczonek [4][ ],
Waldemar Placek [4], Adam Osowski [3][ ] and Joanna Wojtkiewicz [3,5][ ]

[1]   Department of Rheumatology, Municipal Hospital in Olsztyn, 10-900 Olsztyn, Poland
[2]   Department of Internal Medicine, School of Medicine, Collegium Medicum, University of Warmia and
      Mazury, 10-900 Olsztyn, Poland
[3]   Department of Pathophysiology, School of Medicine, Collegium Medicum, University of Warmia and
      Mazury, 10-900 Olsztyn, Poland; adam.osowski@uwm.edu.pl (A.O.); joanna.wojtkiewicz@uwm.edu.pl (J.W.)
[4]   Department of Dermatology, Sexually Transmitted Diseases and Clinical Immunology, School of Medicine,
      Collegium Medicum, University of Warmia and Mazury, 10-900 Olsztyn, Poland; aganek@wp.pl (A.O.-S);
      w.placek@wp.pl (W.P.)
[5]   Laboratory for Regenerative Medicine, School of Medicine, Collegium Medicum, University of Warmia and
      Mazury, 10-900 Olsztyn, Poland
*   Correspondence: magdalenakw@op.pl; Tel.: +48-89-678-6650; Fax: +48-89-678-6668

Received: 28 January 2018; Accepted: 16 February 2018; Published: 22 February 2018

Abstract: Changes in articular cartilage during the aging process are a stage of natural changes in the human body. Old age is the major risk factor for osteoarthritis but the disease does not have to be an inevitable consequence of aging. Chondrocytes are particularly prone to developing age-related changes. Changes in articular cartilage that take place in the course of aging include the acquisition of the senescence-associated secretory phenotype by chondrocytes, a decrease in the sensitivity of chondrocytes to growth factors, a destructive effect of chronic production of reactive oxygen species and the accumulation of the glycation end products. All of these factors affect the mechanical properties of articular cartilage. A better understanding of the underlying mechanisms in the process of articular cartilage aging may help to create new therapies aimed at slowing or inhibiting age-related modifications of articular cartilage. This paper presents the causes and consequences of cellular aging of chondrocytes and the biological therapeutic outlook for the regeneration of age-related changes of articular cartilage.

Keywords: aging; articular cartilage; cell manipulation; stem cells

---

1. Introduction

The motor system is the basis for the motion of the human body and the processes that take place in it during aging result not only in limiting the self-sufficiency in performing everyday activities, changes in attitude and gait and increased risk of falling but they also affect the function of internal organs and the quality of life [1].

Attempts to elucidate the causes of age-related changes in articular cartilage and the development of osteoarthritis (OA) include several theories, including the wear and tear theory [2]. Causes of OA development are also sought in processes related to cellular aging, such as the acquisition of the so-called senescence-associated secretory phenotype (SASP), which derives from a change of gene expression in old cells [3], the weakening of the chondrocyte response to growth factors, including the insulin-like growth factor-1 (IGF-1) and transforming growth factor-β (TGF-β) [4], mitochondria function disorders and the effect of oxidative stress [5] as well as the accumulation of advanced glycation end products (AGEs) [6]. Therefore, the destruction of articular cartilage which progresses with age is not only a result of mechanical overload caused by obesity, posture, gait disorders or trauma. Depletion of cartilage that penetrates into the sub-cartilage bone is a gate to non-differentiated mesenchymal cells which—depending on numerous local factors—can differentiate towards osteoblasts which produce bone tissue, towards fibroblasts which contribute to the formation of fibrous connective tissue or towards chondroblasts, which can produce fibrous or hyaline-like cartilage [7]. Unfortunately, physiological regenerative capability of articular cartilage is highly limited and, in cases of significant damage, only external therapeutic intervention can improve the local condition.

2. Cellular Senescence

Cellular senescence affects all cells. After the number of divisions (ca. 30–60) determined by the replication limit (Hayflick's limit) in an in vitro culture, cells lose their replication potential and stop dividing while continuing to age but they do not die at once and can remain metabolically active for a long time [8]. The aging of cells and organs is probably caused by the accumulation of old cells, which—because of changed metabolism and secreted proteins—create their own microenvironment, affecting their own activity and adjacent cells. The low-grade inflammation which develops as a result of these processes accompanies a majority of old-age diseases [9]. Cellular senescence not only reflects the aging of the body but it also plays a significant role in tissue regeneration in young individuals and probably reduces the risk of neoplasm formation by inhibiting mitotic divisions of cells with damaged genetic material [10]. Unlike in apoptotic cells, the activity of senescence associated β-galactosidase (SA-β-gal) increases in the aging process [11]. As a result of DNA damage, aging cells acquire a specific secretory phenotype which leads to their elimination by immune system phages, while at the same time contributing to the development of age-related diseases. The essence of SASP lies in secretion to the environment of a range of cytokines (interleukin: IL-1, -6, -7, -13, -15), inflammatory chemokines (CCL2/MCP-1, CCL8/MCP-2, CCL26, CXCL8/IL-8, CXCL12/SDF-1), growth factors (amphiregulins, EGF, hFGF, HGF, heregulins, KGF, NGF, VEGF), metalloproteinases (MMP) -1, -3, -10, -12, -13, -14, other proteases and their modulators (TIMP-2, PAI-1,PAI-2, t-PA, u-PA) [12].

Two types of cellular senescence are distinguished: replicative and accelerated [13]. Replicative senescence is associated with exhaustion of the division limit. This is caused by shortening of telomeres, whose function is to protect the ends of chromosomes from joining and preserving the genome integrity. Human telomeres consist of thousands of repetitions of motifs made up of six pairs of bases TTAGGG and they constitute a specific counter of cell divisions. The telomere structure is supported by shelterins—proteins that ensure maintenance of their specific structure. Telomeres are multiplied with a specific reverse transcriptase (RT)—telomerase in a complex process, coordinated by genomic replication. Human telomerase is made up of two main subunits—the telomerase RNA template (hTERC) and the catalytic enzyme telomerase reverse transcriptase (hTERT), which are responsible for replication of base pairs in a specific sequence and for the length of telomeres, respectively [14]. Activity of telomerases in regular human tissues is not sufficient to keep the telomere length constant, which results in telomeres becoming shorter with each cellular division [15]. When telomeres are shortened down to half of their original length on five chromosomes, phenotypic cell senescence occurs [16]. Stress-induced premature senescence (SIPS), associated with DNA damage, is another type of cellular senescence that takes place regardless of physiological telomere shortening. Accelerated senescence can be triggered by oxidative stress, oncogenes, UV radiation or a chronic inflammation [17]. This process is faster than replicative senescence and it does not result directly from exhausting the division potential. The causes of each of these types of senescence is different but they are associated with activation of the same path of response to DNA damage. Such a signal in replicative senescence is generated by shortened telomeres or ones without shelterins. A similar response in accelerated senescence is triggered by a rupture of a double strand of DNA in telomere sections which are inaccessible to the repair systems due to their specific structure and protective proteins [17]. A response to DNA damage is controlled, inter alia, with protein p53, which is an inhibitor of cyclin-dependent kinases, which is responsible for inhibiting cellular divisions and hypophosphorylated Rb (retinoblastoma protein), which is responsible for recruiting enzymes associated with epigenetic chromatin modification [18].

3. Senescence-Related Changes of Articular Cartilage

Homeostasis of cartilage depends on the regularity of function of mature chondrocytes and progenitor cells. With age, the matrix of articular cartilage also undergoes molecular, structural and mechanical changes, there are some changes in the composition and structure of proteoglycans, collagen cross-linking increases and the elongation strength of cartilage decreases. The balance between anabolic activity of chondrocytes and destructive processes is disturbed. Articular cartilage thins slowly as the matrix reduces, cartilage hydration decreases and the chondrocyte count decreases (Figure 1). An age-related decrease in the number of chondrocytes in articular cartilage has been observed in people without clinical symptoms of arthritis, although more pronounced chondrocyte loss is present in patients with OA [19]. The number of chondrocytes in the hip joint cartilage in people aged 30–70 years was reduced by ca. 40% [20]; similar differences have been observed in animal studies [21]. However, in a study of human knee joint cartilage, no significant changes in the number of chondrocytes were observed [22]. The frequency of chondrocyte divisions observed in articular cartilage in adults is low, which—with a very small number of local progenitor cells—may suggest that chondrocytes in elderly people are the same cells as many years earlier but are considerably changed. It also appears that the number of apoptotic cells in human articular cartilage does not increase significantly with progressing age [22].

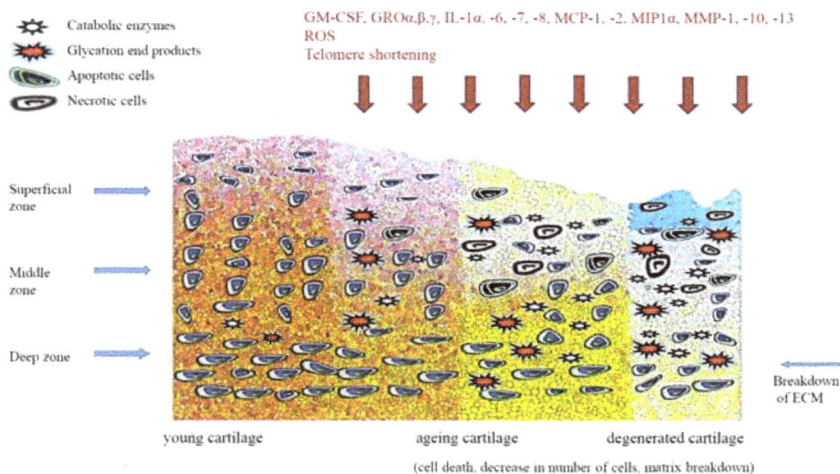

Figure 1. The age-related changes in cartilage.

Senescence of chondrocytes is accompanied by aging-related changes in extracellular cartilage matrix (ECM). The unique properties of extracellular matrix of cartilage have their source in collagenic and non-collagenic glycoproteins, proteoglycans and hyaluronic acid. ECM in articular cartilage plays a crucial role in regulating chondrocyte functions via cell-matrix interaction, organized cytoskeleton and integrin-mediated signalling [23]. A reduction in cartilage volume can be caused by a reduction of water content dependent largely on the content of aggrecan, which is the principal proteoglycan in an articular cartilage matrix. Sulphated, negatively-charged glycosaminoglycans (GAG), which make up aggrecan, are characterised by high hydrophilicity and are responsible for cartilage elasticity.

There have been reports describing age-related changes of size, structure and degree of sulphation of aggrecan which resulted in a decrease in hydration and elasticity of cartilage [24].

3.1. Telomere Shortening

Like in cells of other tissues, telomeres have been found to be significantly shortened in chondrocytes of articular cartilage which have passed through a larger number of cellular divisions [25]; larger numbers of shortened telomeres have also been reported in chromosomes of chondrocytes in elderly people [25]. Age-dependent shortening of telomeres has been observed recently in a study conducted with people with no symptoms of arthritis and in a group of OA patients [26]. The activity of telomerase is higher in chondrocytes in young individuals, which enables repair procedures in young chondrocytes; this activity is reduced considerably after puberty [27]. Attempts have been made to explain the importance of telomeres shortening for senescence of chondrocytes and their precursors—mesenchymal stem cells (MSCs). The length of telomeres varies depending on the donor's age. Embryonic or foetal cells have longer telomeres, telomerase activity in them is higher and they senesce later than cells collected from adult individuals [28]. The length of telomeres in chondrocytes has been reported as 9 to 11 kbp in donors over 55 years old and below 12 kbp in donors under 22 years old [3]. Guillot et al. report that telomeres in foetal stem cells were significantly longer (10 to 11 kbp) than in MSCs obtained from the bone marrow of adult individuals (under 7 kbp) [28], whereas Mareschi et al. found the length of telomeres in MSCs of young donors to be approx. 10 kbp [29]. In these in vitro studies, the telomere length in bone marrow MSCs decreased by 1.5–2 kbp per passage. Telomere length in bone marrow MSCs decreased by 17 bp a year, as found in an in vivo study [30]. Interesting findings were presented in a paper by Parsch et al. where telomeres in bone marrow MSC remained shorter than in chondrocytes even after they were chondrogenically differentiated and their shortening was inhibited at a length of approx. 10 kbp [31]. Apart from the length reduction caused by replicative senescence, telomeres shortening is caused by oxidative stress and destruction of DNA strands [32]. Telomeres in chondrocytes and in MSCs have been found to shorten considerably in cell culture subjected both to sublethal oxidative stress and to prolonged low-level stress [33,34]. A relationship has been described between a considerable shortening of telomeres in chondrocytes and intensified senescence of chondrocytes and the severity of OA [35], although no differences have been observed in another study in the length of telomeres in regular chondrocytes and in those collected from OA lesions [36]. The reason for the large differences between these studies may be the use of different telomere length estimation methods.

3.2. Oxidative Stress

Chondrocytes and MSCs are known to be exposed naturally to under-physiological oxygen concentrations. Reactive oxygen species (ROS) induce telomere shortening stimulated by DNA damage [37,38]. The amount of ROS in chondrocytes increases with age, excessive mechanical load and activity of inflammatory cytokines [39]. The addition of ROS to a chondrocyte culture resulted in developing aging-related phenotypic traits in chondrocytes [39]. The amount of proteins (which are shelterins) associated with telomerase TRF1, TRF2 (telomeric repeat binding factor 1, 2) increases considerably in chondrocytes under oxidative stress during early cell division. These proteins are responsible for the formation and maintaining the structure of telomeres, protein XRCC5 (X-ray repair complementing defective repair in Chinese hamster cells 5) participating in the repair of two-strand DNA and sirtulin 1 (SIRT1), which suppresses protein p53 and prevents inhibition of cell divisions. Secretion of these shelterins is much weaker in later divisions [37]. This study suggests a protective effect of proteins TRF1, TRF2, XRCC5 and SIRT1 on young chondrocytes against shortening of telomeres associated with damage to DNA strands under oxidative stress, whereas in chondrocytes which divide later, a decrease in the activity of these regulatory proteins results in decreased tolerance to ROS and in the accumulation of damaged DNA, which may induce aging-related processes. ROS appear to induce acceleration of chondrocyte senescence by intensifying expression of p53 and p21 and by activation of p38 MAPK (mitogen-activated protein kinase) and phosphatidylinositol 3-kinase/Akt (PI3K/Akt) signalling pathways [40]. Chondrocytes in areas of cartilage affected by OA have been found to contain an increased amount of nitrotyrosine—an oxidative damage marker—which was proportionate to intensification of histological changes [41]. Oxidative stress affecting chondrocytes participates in inducing apoptosis, decreases the cells sensitivity to growth factors, leads to mitochondria dysfunction, telomere-related genomic instability and loss of cartilage matrix [42–45]. ROS also contribute to senescence of MSCs, which are precursors of chondrocytes. The presence of antioxidants (such as *N*-acetylcysteine (NAC) and ascorbic acid) in cell cultures, increased the proliferative activity of MSCs [44,46] and the proliferative and differentiating capability of MSCs in low oxygen concentration (3–5%) was higher than in physiological concentrations (20%) [47,48].

With age, excessive formation of ROS takes place and the oxidative-antioxidative equilibrium is disturbed in cartilage matrix. In reaction with the core protein of proteoglycans, reactive oxygen species modify amino acid residues and cause ruptures of the polypeptide chain of the core protein and formation of proteoglycan fragmentation products in the form of glycosaminoglycan chains bound to core protein residues and free glycosaminoglycans [49].

3.3. Inflammatory Cytokines

Epidemiological studies indicate a relationship between a low-level systemic inflammation and an increase in a concentration of inflammatory cytokines, including C-reactive protein (CRP), IL-6 and TNF-α and the development of OA [50,51]. Inflammatory cytokines can be generated locally in chondrocytes, synovial membrane cells and infrapatellar fat pad-derived cells [50]. Articular cartilage aging may result from the acquisition of a specific secretory phenotype by chondrocytes, whose characteristic features include an increase in the production and secretion of interleukins, matrix metalloproteinases and growth factors, including epidermal growth factor (EGF) [38,52]. SASP-inducing factors include granulocyte macrophage colony stimulating factor (GM-CSF), growth regulated oncogene-α, -β, -γ (GRO-α, -β, -γ), IL-1α, IL-6, IL-7, IL-8, monocyte chemoatractant protein (MCP)-1, -2, IGF-1, macrophage inflammatory protein-1α (MIP-1α) as well as MMP-1, MMP-10 and MMP-13 [50]. Literature reports have described increased expression of metalloproteinases MMP-1 and MMP-13 in aging cartilage [53] and the accumulation of neoepitopes of collagens formed as a result of denaturation and fragmentation of collagen [54]. Other studies have found the capability for production and secretion of IL-1 [55] and IL-7 [56] by isolated chondrocytes to significantly increase with the donor's age and an increase in IL-7 secretion to be associated with increased production of MMP-13 [56]. Moreover, the inflammation process induced by the administration of IL-1β was associated with an increase in the expression of p16^{INK4a}, resulting in increased production of MMP-1 and MMP-13 [52].

The total amount of all proteoglycans in cartilage decreases with age. Age-related modifications of proteoglycans in cartilage matrix are related to synthesis disorders and enzymatic and non-enzymatic degradation. Degradation of proteoglycan macromolecules which intensifies with age is accompanied by overexpression of MMP, including MMP-1, MMP-8, MMP-13 with a concomitant decrease in their tissue inhibitors (TIMP) [57]. The activity of disintegrin and metalloproteinase with thrombospondin motifs (ADAMTS), including aggrecanases ADAMTS-4 and ADAMTS-5, which participate in the digestion of core proteins of aggrecans, increases with age [58]. Apart from a direct effect degrading the extracellular matrix of cartilage, these enzymes stimulate the secretion of inflammatory cytokines, e.g.,: IL-1, IL-6 and the tumour necrosis factor-α (TNF-α) [38]. IL-1, IL-6 and TNF-α induce chondrocytes to synthesise increased amounts of matrix metalloproteinases, at the same time inhibiting the production of natural inhibitors of these endopeptidases [58]. Additionally, IL-1 and TNF-α stimulate the production of insulin-like growth factor-binding protein-1 (IGFBP-1). IGFBP-1 binds IGF-1, thereby decreasing its binding capability with an appropriate receptor on chondrocytes, which leads to a decreased response of mature chondrocytes to IGF-1 [59]. An effect of IL-1 and TNF-α results in an increase in the activity of inducible nitric oxide synthase (iNOS), and—secondarily—in an increase in secretion to the matrix of catabolic metalloproteinases and prostaglandins [58].

3.4. Altered Responsiveness to Growth Factors

With time, the anabolic activity of chondrocytes in cartilage decreases and the chondrocyte metabolic equilibrium shifts towards catabolic mechanisms.

Chondrocyte response to IGF-1 decreases with age [4]. Similar disturbances also occur in isolated chondrocytes from cartilage with symptoms of OA [60]. IGF-1 stimulates the proliferation of cartilage cells, supports the synthesis of cartilage matrix cells and inhibits chondrocytes apoptosis through phosphoinositide 3-kinase (PI3K) and extracellular signal-regulated kinase (ERK) [61]. IGF-1 increases the synthesis of proteoglycans of cartilage matrix under in vitro conditions by activating the kinases PI3K/Akt/mTOR/p70S6 pathway [62]. Through PI3K, IGF-1 stimulates MSCs to chondrogenic differentiation [62].

Expression and amount in cartilage of osteogenic protein OP-1 (BMP-7), which is a member of the bone morphogenetic proteins superfamily, decreases with age. The addition of anabolic protein OP-1 to a culture of chondrocytes collected from mature individuals does not affect the activity of telomerase, whereas an addition of inflammatory cytokine IL-1α inhibits its activity [27].

The concentration of TGF-β2 and TGF-β3 (but not TGF-β1) in cartilage also decreases in an age-related manner, like the number of TGF-β receptors [21]. TGF-β is secreted by cells as a latent form (latent TGF-β, L-TGF-β). An active form is generated after dissociation of the non-covalently bonded latency-associated peptide (LAP). LAP dissociation is effected by the plasminogen/plasmin proteolytic system [63], thrombospondin-1 (TSP-1) [64], metalloproteases [65] as well as by mechanical stress [66]. After TGF-β binds to the II type receptor, an activin receptor-like kinase 5 (ALK5) recruiting complex is formed, which is a TGF-β type I receptor. Such a compound induces phosphorylation of serine and threonine residues of type I receptor by type II receptor. The primed receptor transmits the signal directly to cytoplasm, where R–SMAD proteins are phosphorylated and translocated to the cell nucleus [67]. TGF-β signal transmission can be mediated by an alternative ALK1 type I receptor, resulting in the final cellular differentiation and hypertrophy [68]. Binding of TGF-β to the ALK5 receptor leads to phosphorylation of proteins: SMAD2 and SMAD3, whereas binding of TGF-β to the ALK1 results in phosphorylation of SMAD1, SMAD5 and SMAD8. Activation of SMAD2/3 and SMAD1/5/8 pathways differs by the response. Signal transduction by SMAD2/3 is associated with a protective effect and by SMAD1/5/8—with terminal differentiation and hypertrophy [69]. Activation of signal pathways is necessary for in vitro development of MSCs population. Inhibition of TGF-β pathways in rat and human cultures prevented their differentiation [70]. TGF-β in MSCs cultures can be a factor used in the induction of chondrogenesis [71]; on the other hand, TGF-β can accelerate processes of cellular senescence by increasing the activity of senescence-associated-galactosidase (SA-Gal) and the production of mitochondrial reactive oxygen species (mROS) [72]. Philipot et al. found that—during chondrogenic differentiation of BM-MSCs caused by TGF-β3—expression of p16^{INK4a} accompanying the production of type IIB collagen and MMP13 took place, which was a sign of terminal cell differentiation [52].

3.5. Advanced Glycation End-Products

Modifications of proteoglycans of articular cartilage taking place with time are caused by the catabolic effect of enzymes and oxidative stress but also by accumulation of products of late glycation in cartilage (advanced glycation end-products, AGEs). The production of AGEs takes place as a result of spontaneous, non-enzymatic glycation of proteins which, in turn, is a result of reaction of reducing sugars, including sucrose, fructose and ribose with lysine and arginine residues. Due to its relatively slow metabolism, cartilage is particularly predisposed to form AGEs. The half-life of type II collagen, which is the most widespread protein of extracellular cartilage matrix, has been estimated to exceed 100 years [73]. Although products of late glycation decrease the sensitivity of proteoglycans to proteolytic effect of metalloproteinases, the total pool of proteoglycans in cartilage is decreased proportionally to the amount of AGEs [74]. The effect of AGEs on processes that take place in aging cartilage has not been fully elucidated. The appropriate receptors (receptor for advanced glycation end products, RAGE) on the chondrocyte cell membrane are probably responsible for inhibition of synthesis and secretion of proteoglycans to cartilage extracellular matrix and for inducing the synthesis of matrix metalloproteinases (MMP-1, MMP-3, MMP-13) and prostaglandine E2 [57].

3.6. Autophagy

Autophagy has gained interest in the past decade due to its role in regulation of the aging process. Autophagy is a naturally occurring catabolic process that removes unnecessary of dysfunctional cellular components in cytoplasm as aggregated proteins and redundant or damaged organelles [75]. Little is known about the role of autophagy in articular cartilage. In articular cartilage, which has a very low rate of cell turnover, autophagy appears to be protective process for maintaining cartilage homeostasis. Autophagy regulates maturation and promotes survival of terminally differentiated chondrocytes under stress and hypoxia conditions [76]. Transiently increased autophagy is a compensatory response to cellular stress. During the early degenerative phase, autophagy is increased in cartilage, with increased accumulation of autophagic proteins, such as ULK-1, LC3 and Beclin-1 messenger RNA in chondrocytes [77]. Reduced expression of ULK1, Beclin-1 and LC3 protein was observed in aging joints in humans and mice [78]. The reduction in these major autophagy regulators was accompanied by increased chondrocyte apoptosis [77,78]. Many studies identified correlations between autophagy and the mTOR signalling pathway. The cartilage-specific deletion of mTOR upregulates autophagy and results in increased autophagy signalling and a significant protection from the articular cartilage degradation, apoptosis and synovial fibrosis [79]. The intra-articular injection of rapamycin- an mTOR inhibitor [80], or intra-articular administration of gelatin hydrogels incorporating rapamycin-micelles [81], inhibited mTOR expression suppressed the development of the articular cartilage degeneration. Additionally, REDD1- an endogenous inhibitor of mTOR that regulates cellular stress responses is highly expressed in normal human articular cartilage and reduced with age [82]. Chondrocytes are adapted to hypoxic conditions. Two main HIF hypoxia-inducible factors isoforms (HIF-1α and HIF-2α) mediate the response of chondrocytes to hypoxia. HIF-1α supports metabolic adaptation to a hypoxic environment and by suppression of mTOR causes increased autophagy [83]. In contrast, HIF-2α has been shown to be a suppressor of autophagy under hypoxic conditions in vitro [84].

4. Possible Anti-Aging Strategies

Old age and, obviously, the aging of tissues and organs pose a challenge to contemporary medicine. Neither the treatment of pain as the main symptom of changes related to aging of the motor system, nor burdensome surgeries are fully satisfying to patients; hence, the need for alternative methods for slowing down the aging process and supporting regeneration of articular cartilage.

4.1. Cell Manipulation

4.1.1. Telomerase Activators

Several studies have been conducted to assess the effect of an increase in the hTERT activity on the lifespan of MSCs. Increased expression of hTERT by transduction in MSCs resulted in extended in vitro cell replication capability and in maintaining the potential for in vitro adipo-, chondro- and in vivo osteogenic differentiation [85,86]. Neither a tendency for neoplasm formation nor changes in the MSCs caryotype were observed in these studies. Recently, several telomerase-inducing factors have been discovered and described [87–89]. Astragaloside (AST) and its active metabolite cysloastragenole (CAG) activated telomerase and slowed down the aging process in human embryonic kidney HEK293 fibroblasts [90]. Cynomorium songaricum polysaccharide increased telomerase activity and led to elongation of telomeres in murine cells [91]. Tichon et al. found chemical telomerase activators: AGS-499 and AGS-500 to induce expression of TERT in MSCs, to increase the length of telomeres and to stimulate cell resistance to apoptosis induced by oxidative stress. No chromosomal aberrations or MSCs differentiation disorders were observed [92]. Telomerase differentiation in vitro was increased by curcuminoid derivatives with the core and at least one *n*-pentylpyridine side chain [93]. An extract of Astragalus membranaceus root (TA-65) used in another study extended the proliferative activity of T cells [94] and it significantly decreased the amount of extremely short telomeres in a study of mice [95].

4.1.2. Antioxidants and Hypoxia

Several antioxidants have been described with a protective effect on chondrocytes and MSC against cellular senescence and oxidative stress-induced apoptosis. NAC inhibited apoptosis in chondrocytes [40] and MSCs [44] subjected to oxidative stress. In the study by Liu et al., stimulation of chondrocytes with IL-1β caused a significant up-regulation of TLR4 and its downstream targets MyD88 and TRAF6 resulting in NF-κB activation associated with the synthesis of IL-1β and TNFα. These IL-1β-induced inflammatory responses were all effectively reversed by resveratrol- a polyphenol of plant origin. Furthermore, activation of NF-κB in chondrocytes treated with TLR4 siRNA was significantly attenuated but not abolished and exposure to resveratrol further reduced NF-κB translocation [96]. In another study, following resveratrol injection, the expression of collagen type II was retained but the expression of inducible nitric oxide synthase and matrix metalloproteinase-13 was reduced in OA cartilage. Moreover, the administration of resveratrol significantly induced the activation of SIRT1 expression in mouse OA cartilage and in IL-1β-treated human chondrocytes [97]. The effectiveness of antioxidants in preventing damage to chondrocytes seems to increase when biomaterial scaffolds are used [98]. A cell-free collagen/resveratrol (Col/Res) complex in the form of hydrogel significantly reduced the activity of IL-1β, MMP-13 and COX-2 in an animal OA model; additionally, it improved the condition of articular cartilage after only 12 weeks [99]. The physiological oxygen concentration was found to weaken the growth of human chondrocytes and MSCs in a culture and an increase in antioxidant production as a response to these oxygen concentrations resulted in cell destruction [100,101]. Hypoxia in culturing regular human chondrocytes enabled maintaining the chondrogenic cell potential [102] but this was not observed in a study with osteoarthritic human chondrocytes [103]. In the study by Choi et al. chondrogenic differentiation of ASCs was found to be increased under hypoxia as evidenced by the greater amount of proteoglycan formation and increased chondrogenic genes expression, ACAN, COL2 and SOX9 [104]. It appears that human MSCs acquire aging-related traits in cultures under hypoxic conditions later than under normoxic ones. Additionally, the frequency of MSC divisions increases at low oxygen concentrations with the intact genotype maintained [105]. An increase in the chondrogenic potential of MSCs caused by hypoxia has been reported [104,106], although—as with chondrocytes—the direction of MSCs differentiation depends on the oxygen concentration but also on other factors, such as the substrate used [107,108].

4.1.3. Mechanical Load

Mechanical load stimulates the metabolic activity of chondrocytes, which is responsible for keeping the metabolic balance of matrix proteins [109] and participates in processes which regulate MSCs differentiation towards chondrogenesis [110]. Dynamic compressive loading is the most common system of mechanical stimulation of MSC-dependent cartilage regeneration used in mechanobiological studies. The process of mechano-transduction is, at least partly, mediated by the TGF-β signalling pathway [111–113]. Synthesis of matrix proteins by MSCs stimulated by multiaxial pressure is more effective than with monoaxial stimulation [114]. Multidirectional forces support the production of type II collagen, aggrecan and GAG, with or without the participation of external growth factors, including TGF-β [113]; intensification of the production and activity of endogenous TGF-β caused by mechanical load has been reported [111]. In a human MSCs culture, dynamic compressive loading significantly inhibited the expression of hypertrophy markers, such as type X collagen X, MMP-13 and alkaline phosphatase (ALP) [114]. In another study, mechanical inhibition of chondrocyte hypertrophy was mediated by TGF-β/SMAD and integrin β1/focal adhesion kinase (FAK)/extracellular-signal-regulated kinase (ERK) pathways [115]. These findings suggest that applying a mechanical load to a joint following the implantation of cartilage obtained by bioengineering or applying a load to cellular structures before intra-articular implantation can support cartilage maturation and inhibit its hypertrophy, giving better results than immobilisation of the joint, which can have significant application in clinical practice.

4.2. Cell-Based Therapy

Attempts have been made for a long time to surgically stimulate local regeneration of weakened articular cartilage in such procedures as abrasion arthroplasty, drilling and microfracture. Alternatively, there are therapies based on autologous chondrocyte implantation (ACI), autologous matrix-induced chondrogenesis (AMIC) and intra-articular injection of mesenchymal stem cells [116–119].

4.2.1. Autologous Chondrocyte Implantation

ACI is an example of the application of tissue engineering in the treatment of small- and medium-size cartilage loss. This technique includes a surgical biopsy of cartilage from an affected joint, isolation of chondrocytes with collagenase and creation of a monolayer culture, which requires a series of cellular passages to achieve sufficient propagation of chondrocytes. At the next stage, cells acquired in vitro are implanted at the site of a loss and immobilised with a periosteal patch, which has a protective role but which also has chondrogenic capability, or they are protected with a collagenic layer [120]. The results of the ACI procedure observed in population studies have been encouraging—pain within the joint decreased and the formation of high-strength tissue was observed [121]. However, there are some limitations to the application of ACI. However, it seems that the proliferative capability and the chondrocytes' ability to propagate in a one-layer culture decrease with the age of a donor [122]. The study by Barbero et al. did not show any difference in the proliferative activity of chondrocytes collected from donors under 40 years of age, whereas proliferation of chondrocytes from donors over 40 years of age slowed down considerably [123]. A decrease in the proliferative capability of chondrocytes collected from older donors can obviously limit the use of ACI in treating age-related changes. Changes in the phenotype are also an issue in monolayer chondrocyte cultures. The culture conditions do not reflect fully the in vivo conditions. When cultured in vitro, cells can change shape from polygonal or round to flat after as few as four passages; at the time, they start synthesising type X collagen, which is a sign of a loss of the chondrocytal phenotype. With a growing number of passages, aging and dedifferentiation of chondrocytes in a monolayer culture is taking place, which is associated with a decrease in the production of type II collagen, proteoglycans and glycoproteins [40]. Transplantation of old and dedifferentiated cells results in an unwanted growth of fibrous cartilage at the site under treatment [124]. Apart from a conventional monolayer chondrocyte culture, there are methods currently in use using growth factors [62], pellet culture [125] and bioreactors [126]. De-differentiation can be inhibited and chondrogenicity can be regained successfully by functionalization of the dynamic culture surfaces with ECM [127], culture on a continuously expanding surface [128], surface coating with Col I [129] and by applying a mechanical load [130]. Although the methods are effective in maintaining the chondrocytal phenotype, they may not be able to achieve sufficient cell expansion [131]. Much attention has been recently devoted to supplying genes associated with cellular growth factors, including TGF-β1, BMP, IGF-1, fibroblast growth factor (FGF) and transcription factors, i.e., SRY (sex-determining region Y)-box 9 (SOX9), RUNX2 and SMAD; microRNAs and Col II promoter-binding proteins [132].

4.2.2. Mesenchymal Stem Cells

Mesenchymal stem cells are able to differentiate in multiple directions and to self-regenerate [133]. Since MSCs are present in many embryonic tissues (embryonic stem cells, ESC) and in adult individuals (adult stem cells, ASC), there are many methods of acquiring them. Embryonic stem cells can be collected from umbilical cord blood after delivery, from Wharton's jelly, from the placenta and amniotic fluid and as well as from the subamniotic membrane and perivascular area of the umbilical cord. MSCs have been identified in the following tissues in adult individuals: marrow, adipose tissue, skin, lungs, dental pulp, periosteum, skeletal muscles, tendons and synovial membrane [134] but clinical application of "adult" MSCs is limited mainly to bone marrow-derived mesenchymal stromal cells (BM-MSCs) and adipose-derived stem cells (ADSC, ASC) [135]. According to criteria issued by The International Society for Cellular Therapy (ISCT), characteristic features of mesenchymal cells include the ability to adhere to a plastic base, the presence of three surface antigens: CD105 (endoglin), CD90 (Thy-1), CD73 (ecto-5′-nucleotidase) and concomitant absence of antigens CD45, CD34, CD14 or CD11a, CD79a, or CD19 and class II HLA and the capability of in vitro differentiation towards three cellular lines: osteoblasts, adipocytes and chondroblasts (Figure 2) [133]. Additionally, a detailed description of stem cells includes information, such as the cell origin (tissue, organ, systemic), culture conditions, medium composition, the presence of other antigens of positive identification and absence of negative markers, potential for differentiation, cloning, proteomes, secretomes and transcriptone data [136]. MSCs have a clinically promising immunomodulatory and regenerative potential but it must be noted that MSC also age and die in cultures after several passages [137].

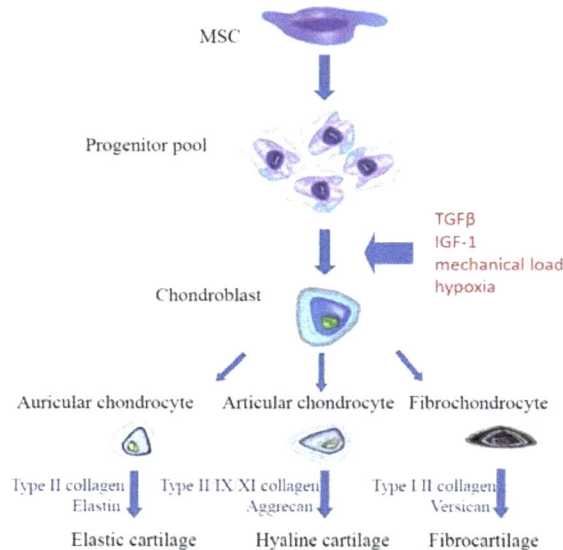

Figure 2. Chondrogenic differentiation of mesenchymal stem cells.

The results of the application of MSCs in regeneration of articular cartilage are promising. In a study by Davatchi et al., intra-articular administration of MSCs to the knee joint ($8–9 \times 10^6$ cells/patient) resulted in a significant, long term reduction of pain as assessed with the visual analogue scale (VAS), as well as in a reduction of flexion contracture in the joint, of cracking during movements and in an extension of the walking distance [138]. In a randomised, multi-centre clinical trial, a considerable reduction of pain assessed in the VAS and an improvement of the function assessed in the WOMAC (Western Ontario and McMaster Universities OA Index) scale and by measurement of the joint movement range was observed in patients following intra-articular administration of autologous BM-MSCs at 100×10^6 cells/patient in combination with hyaluronic acid. Conversely, an improvement was observed only initially in patients who received BM-MSCs in smaller amounts (10×10^6 cells/patient) but when measured after six months, the change in the parameters under assessment was not significant [139]. An improvement after the administration of even small doses of autologous BM-MSCs into the knee joint (2×10^6 cells/patient) was reported recently by Pers et al. [140]. In the study by Emadedin et al. BM-MSCs at 5×10^5 cells/kg/bw (body weight) were administered to patients with OA of knee joints but also to patients with OA of the ankle and of the hip joint. An improvement of the clinical parameters was observed in these patients for a year but it gradually decreased over the next 18 months [141]. A clinical improvement in all the studies corresponded to that observed in imaging examinations (MRI). Reports published later also confirmed the efficacy and safety of the use of autologous MSCs (Table 1) [138,140,142–156]. Use of allogeneic cells could be an alternative to therapy with autologous MSCs (Table 2) [157–160]. Administration to the knee joint of allogeneic BM-MSCs at 40×10^6 cells/patient [158], 50×10^6 cells/patient, 150×10^6 cells/patient [157] was safe and resulted in a clinical improvement, observed in MRI examinations. Similar results were reported in a recent study by de Windt et al. in which a mixture of allogeneic BM-MSCs and autologous chondrones (chondrocytes with their native pericellular matrix) was used [159,160]. However, a growing body of evidence has been emerging of the time of survival of MSCs at the implantation site being much shorter than expected [161]. Therefore, it cannot be ruled out that MSCs have a regenerative effect on articular cartilage, not only as a source of cells which differentiate towards chondrocytes but also through the secretome that they secrete, which covers a broad spectrum of paracrine factors [162]. After intra-articular administration, MSCs use cytokines to stimulate secretion of extracellular matrix proteases and growth factors, including TGF-β, IGF-1 and FGF [163]. After in vitro addition of articular fluid collected from patients in an early and late stage of OA, larger amounts of factors secreted by MSCs were observed, such as: chemokine (C–X–C motif) ligands, chemokine (C–C motif) ligands and IL-6 in the fluid taken from patients in the initial phase of OA [164]. These findings are indicative of different reactions of MSCs following an intra-articular administration, depending on the degree of local cartilage damage, which can determine the therapeutic indications.

However, there are a number of unknowns associated with the use of MSCs. When administered intra-articularly, MSCs after long culturing may lose their chondrogenic phenotype and die quickly [165]. There have been many attempts at maintaining or restoring the chondrogenic differentiation capability of MSCs, which includes the addition of growth factors, modification of culture conditions and specifying the sources of MSCs; however, there are currently no recommendations in this regard [166–168]. Most of the studies conducted to date have been based on MSCs obtained from a patient's own bone marrow or adipose tissue, which was supposed to limit the immune response; allogeneic MSCs were used only in a few studies. The follow-up period in the published studies conducted to assess the efficacy and safety of intra-articular administration of cultured MSCs (except in studies by Davatchi et al. [138] and Wakitani et al. [145]) was short: from six months to two years. A clinical issue is the tendency of implanted MSCs to differentiate towards fibro-like tissue instead of towards hyaline cartilage [169], which considerably affects the properties of the newly-formed tissue. New sources of MSCs from tissues are sought with a high potential of differentiation towards chondrocytes. Synovial membrane was thought to be such a place. Intra-articular administration of autologous synovium stem cells (SSC) in studies with animal models [170] and in human studies [154,171] proved to cause improvement in imaging and histological parameters of articular cartilage. There are some ongoing studies into maintaining the chondroidal phenotype with advanced methods of genetic modifications and epigenetic regulations.

Table 1. List of publications of autologous mesenchymal stem cells application for cartilage repair.

Author Year Study Type	Number of Patients	Stem Cell Origin	Number of Cells	Control Group	Follow-up Time	Co-Treatment	Delivery System	Assessments
Wakitani et al. Case and control study [142]	12	Autologous BM-MSCs from iliac crest	1.3×10^7	12 cell free	28–95 weeks	High tibial osteotomy; gel-cell composite; autologous periosteum	Implantation of collagen cell sheets	Arthroscopic photography; Histological samples; Clinical evaluation
Niejadnik et al. Cohort study [143]	36	Autologous BM-MSCs from iliac crest	$1.0–1.5 \times 10^7$	36 autologous chondrocyte implantation	24 months	Fibrin glue	Implantation upon autologous periosteal patch	ICRS; Lysholm score; Tegner activity scale X-ray; Clinical evaluation
Buda et al. Case series study [144]	20	Autologous BM-MSCs from iliac crest	N/A (2 mL of bone marrow concentrate)	None	24 months	Arthroscopic debridement; Hyaluronic acid membrane scaffold	Transplantation of Arthroscopic Bone-Marrow-Derived Mesenchymal Stem Cells	Clinical evaluation (IKDC; KOOS); MRI; Histochemical analysis
Wakitani et al. Case series [145]	41	Autologous BM-MSCs from iliac crest	5.0×10^6 /mL	None	5–137 months	Gel-cell composite	Implantation cell sheets	X-ray; Clinical evaluation
Saw et al. Case series study [146]	5	Autologous peripheral blood progenitor cells	N/A (7–8 mL of peripheral blood progenitor cells	None	10–26 months	Microfracture; Hyaluronic acid intra-articular injections	Intra-articular injection	Second-look arthroscopy; Histological samples;
Davatchi et al. Case series study [147]	4	Autologous BM-MSCs from iliac crest	$8.0–9.0 \times 10^6$	None	12 months	Saline with 2% human serum albumine	Intra-articular injection	X-ray; Clinical evaluation; VAS
Koh et al. Case and control study [148]	25	Autologous A-MSCs from infrapatellar fat pad	$1.2–2.3 \times 10^6$	25 cell free	12–18 months	Arthroscopic debridement; Synovectomy	Intra-articular injection	Lysholm score; Tegner activity scale; VAS
Koh et al. Case series study [149]	18	Autologous A-MSCs from infrapatellar fat pad	$0.3–2.7 \times 10^6$	None	24–26 months	Arthroscopic debridement; Synovectomy; Platelet-rich plasma	Intra-articular injection	Lysholm score; Tegner activity scale; VAS; MRI
Saw et al. Randomised control trial [150]	49	Autologous peripheral blood progenitor cells	N/A (7–8 mL of PBPC)	25 cell free	24 months	Microfracture; Hyaluronic acid intra-articular injections	Intra-articular injection	IKDC; MRI; Second-look arthroscopy; Histological samples
Orozko et al. Case series study [151]	12	Autologous BM-MSCs from iliac crest	40×10^6	None	12 months	Ringer lactate, human albumin, glucose	Intra-articular injection	VAS; Clinical evaluation; WOMAC; MRI

Table 1. *Cont.*

Author Year Study Type	Number of Patients	Stem Cell Origin	Number of Cells	Control Group	Follow-up Time	Co-Treatment	Delivery System	Assessments
Wong et al. Randomized control trial [152]	28	Autologous BM-MSCs from iliac crest	1.46×10^6	28 cell free	24 months	Microfracture and high tibial osteotomy; Hyaluronic acid;	Intra-articular injection	IKDC; Lysholm score; Tegner activity scale; MOCART
Jo et al. Cohort study [153]	18	Autologous A-MSCs from abdominal subcutaneous fats	1.0×10^7, 5.0×10^7, 1.0×10^8	None	6 months	None	Intra-articular injection	WOMAC; VAS; KSS; X-ray, MRI; Second-look arthroscopy; Histological samples
Akgun et al. Prospective, single-site, randomized, single-blind pilot study. [154]	7	Autologous synovium-derived MSCs	4.0×10^6	7 matrix-induced autologous chondrocyte implantation	24 months	Type I/III collagen membrane (2 cm × 2 cm)	Implantation of MSC preloaded collagen membrane	Clinical evaluation: VAS, KOOS; MRI
Davatchi et al. Case series study [38]	4	Autologous BM-MSCs	8.0–9.0×10^6	None	60 months	Glucosamine was permitted	Intra-articular injection	X-ray; Clinical evaluation: VAS, walking time to pain
Fodor et al. Case series study [155]	6	Adipose-derived stromal vascular cells (the stromal vascular fraction of adipose tissue)	14.0×10^6	None	12 months	None	Intra-articular injection	Clinical evaluation: WOMAC, VAS, ROM, TUG; MRI
Koh et al. Randomized control trial [156]	40	Autologous A-MSCs	5.0×10^6	40 Microfracture treatment	24 months	Debridement; Microfracture, Fibrin glue	Arthroscopic implantation of MSC loaded in fibrin glue	Second-look arthroscopy; the Lysholm score, KOOS; VAS; MRI; Histological samples
Pers et al. Cohort study [140]	18	Autologous A-MSCs	2.0×10^6, 10.0×10^6, 50.0×10^6	None	6 months	None	Intra-articular injection	WOMAC, VAS, PGA, SAS, KOOS; Histological samples

A-MSCs: adipose-derived mesenchymal stem cells; BM-MSCs: bone marrow-derived mesenchymal stem cells; ICRS: International Cartilage Repair Society score; IKDC: International Knee Documentation Committee; KOOS: the Knee Injury and Osteoarthritis Outcome; KSS: Knee Society Clinical Rating System score; MOCART: Magnetic Resonance Observation of Cartilage Repair Tissue score; MRI: magnetic resonance imaging; MSCs: mesenchymal stem cells; N/A: not available; PGA: Patient Global Assessment; ROM: range of motion; SAS: Short Arthritis Assessment Scale; TUG: timed up-and-go; VAS: visual analogue pain scale; WOMAC: the Western Ontario and McMaster Universities Arthritis Index.

Table 2. List of publications of allogenic mesenchymal stem cells application for cartilage repair.

Author Year Study Type	Number of Patients	Stem Cell Origin	Number of Cells	Control Group	Follow-up Time	Co-Treatment	Delivery System	Assessments
Vangsness et al. Randomized, double-blind, controlled study [157]	36	Allogenic BM-MSCs from 18–30-year old donors	5.0×10^7, 1.5×10^8	20 cell free vehicle control	24 months	Partial medial meniscectomy	Intra-articular injection	Measurement of immune cell markers; MRI; VAS; Lysholm score
Vega et al. Randomized control trial [158]	15	Allogenic BM-MSCs	40×10^6	15 Hyaluronic acid (60 mg, single dose)	12 months	None	Intra-articular injection	VAS; WOMAC, SF-12; MRI
De Windt et al. Case series study [159]	10	Allogenic BM-MSCs from the iliac crest of 2 healthy donors (age 2 and 5)	N/A. Mixed cells in the fibrinogen component of fibrin glue at $1.5–2 \times 10^6$ cells/mL	10% or 20% autologous chondrons (a standard or a high yield mixture)	12 months	None	Defect site-specific implantation	KOOS, VAS; Second-look arthroscopy; Histological analysis;
De Windt et al. Case series study [160]	35	Allogenic cryopreserved BM-MSCs	N/A. Mixed cells in the fibrinogen component of fibrin glue at $1.5–2 \times 10^6$ cells/mL. Approximately 0.9 mL cell product/cm^2 defect	10% or 20% autologous chondrons (a standard or a high yield mixture)	18 months	None	Defect site-specific implantation	KOOS, VAS; Second-look arthroscopy; Histological analysis;

BM-MSCs: bone marrow-derived mesenchymal stem cells; KOOS: the Knee Injury and Osteoarthritis Outcome; MRI: magnetic resonance imaging; N/A: not available; SF-12: short form-12 life quality questionnaire; VAS: visual analogue pain scale; WOMAC: the Western Ontario and McMaster Universities Arthritis Index.

4.3. Cell-Free Procedures

Currently numerous, single-phase scaffold-based cartilage repair techniques exist. Chondral scaffolds used in cartilage repair can be based on components of cartilaginous matrix, such as collagen or hyaluronate [172,173], proteins and natural polymers (such as fibrin, agarose, alginate and chitosan) [174] or synthetic polyesters (such as polylactic acid, polyglycolic acid, polylactide glycolide, polyethylene oxide and polypropylene oxide) [175]. Chondral scaffolds consist of a monolayer material (monophasic scaffold) or they are multi-layer structures, which copy the architecture of cartilage more effectively. Currently, matrices consisting of collagen are used more commonly in regeneration of damaged articular cartilage [118].

As an alternative to cell seeded scaffolds, another treatment approach involves the implantation of acellular biomaterials for cartilage regeneration by stimulating bone marrow stem-cell recruitment and differentiation induced by the scaffold. One of these cell-free procedures is autologous matrix-induced chondrogenesis (AMIC). The method of autologous matrix-induced chondrogenesis is often used in treatment of small, mainly post-traumatic, loss of articular cartilage. AMIC combines stimulation of bone marrow through microfractures with the use of a no-cell membrane made up of type I/III collagen. A collagen membrane stabilises the clot, creates a biological chamber, protects progenitor cells and stimulates the cells to differentiate towards chondrocytes. Collagen membranes are fixed with partial autologous fibrin glue or surgical sutures [176]. Gille et al. showed a significant improvement in all clinical parameters in up to 87% of patients. An analysis of MRI examinations revealed in most cases the total or moderate filling of the cartilage loss with a normal-to-incidentally hyperintense signal. However, an improvement in joint mobility seems to depend on the age of the patients [118]. Moreover, the application of the AMIC technique is recommended in the treatment of local, small cartilage loss and it is contraindicated in cases of multiple cartilage loss, rheumatic diseases and a considerable restriction of joint mobility [176], which considerably limits AMIC applicability in the treatment of aging-related changes.

5. Summary

Aging-related changes in articular cartilage lower the possibility of maintaining the properties and regeneration of cartilage. Obviously, time-related factors, such as chondrocytes secretory phenotype, weakening of chondrocytes' response to growth factors, oxidative stress and accumulation of final glycation products in cartilage can also lead to osteoarthritis, which is frequent in the population of the elderly. Aging of cartilage with time is inevitable and the changes it causes are irreversible. Because of the aging of the global population, age-related diseases of the motor system are a significant social issue. Studies aimed at elucidating the mechanisms of intensified aging of articular cartilage may help to introduce a therapy to slow down the aging-related changes or to support local processes of regeneration of articular cartilage. Due to a growing number of reports published in recent years, increasing attention has been attracted by cell therapies, including the possibility of using MSCs in the regeneration of cartilage. Therapies which employ MSCs offer great potential. However, use of MSCs in the regeneration of aging cartilage raises a lot of doubt. There are no detailed recommendations regarding the culture conditions, the tissue origin and the dosage of stem cells. The degree of changes and the presence of inflammation are the issues which occur in elderly people. Additionally, autologous MSCs in elderly people exhibit lower proliferative activity and there is some difficulty regarding the production of cells with a chondrogenic phenotype and a specific biological activity. However, due to the importance of the issue, further studies are needed of the methods for improving the quality of aging cartilage.

Acknowledgments: The study was supported by the School of Medicine, University of Warmia and Mazury in Olsztyn, Poland.

Conflicts of Interest: The authors declare no conflict of interest.

References

1. Bijlsma, J.W.; Berenbaum, F.; Lafeber, F.P. Osteoarthritis: An update with relevance for clinical practice. *Lancet* **2011**, *377*, 2115–2126. [CrossRef]
2. Aigner, T.; Rose, J.; Martin, J.; Buckwalter, J. Aging theories of primary osteoarthritis: From epidemiology to molecular biology. *Rejuvenation Res.* **2004**, *7*, 134–145. [CrossRef] [PubMed]
3. Martin, J.A.; Buckwalter, J.A. The role of chondrocyte senescence in the pathogenesis of osteoarthritis and in limiting cartilage repair. *J. Bone Jt. Surg. Am.* **2003**, *85*, 106–110. [CrossRef]
4. Messai, H.; Duchossoy, Y.; Khatib, A.M.; Mitrovic, D.R. Articular chondrocytes from aging rats respond poorly to insulin-like growth factor-1: An altered signalling pathway. *Mech. Ageing Dev.* **2000**, *115*, 21–37. [CrossRef]
5. Blanco, F.J.; Rego, I.; Ruiz-Romero, C. The role of mitochondria in osteoarthritis. *Nat. Rev. Rheumatol.* **2011**, *7*, 161–169. [CrossRef] [PubMed]
6. Huang, C.Y.; Lai, K.Y.; Hung, L.F.; Wu, W.L.; Liu, F.C.; Ho, L.J. Advanced glycation end products cause collagen II reduction by activating Janus kinase/signal transducer and activator of transcription 3 pathway in porcine chondrocytes. *Rheumatology (Oxford)* **2011**, *50*, 1379–1389. [CrossRef] [PubMed]
7. Galle, J.; Bader, A.; Hepp, P.; Grill, W.; Fuchs, B.; Käs, J.A.; Krinner, A.; Marquass, B.; Müller, K.; Schiller, J.; et al. Mesenchymal Stem Cells in Cartilage Repair: State of the Art and Methods to monitor Cell Growth, Differentiation and Cartilage Regeneration. *Curr. Med. Chem.* **2010**, *17*, 2274–2291. [CrossRef] [PubMed]
8. Hayflick, L. Intracellular determinants of cell aging. *Mech. Ageing Dev.* **1984**, *28*, 177–185. [CrossRef]
9. Sikora, E.; Arendt, T.; Bennett, M.; Narita, M. Impact of cellular senescence signature on ageing research. *Ageing Res. Rev.* **2011**, *10*, 146–152. [CrossRef] [PubMed]
10. López-Otín, C.; Blasco, M.A.; Partridge, L.; Serrano, M.; Kroemer, G. The hallmarks of aging. *Cell* **2013**, *153*, 1194–1217. [CrossRef] [PubMed]
11. Hunt, A.; Betts, D.; King, W.A.; Madan, P. Senescence or apoptosis? The choice bovine fibroblasts make in the presence of increasing concentrations of extracellular H_2O_2. *SURG J.* **2010**, *3*, 64–68.
12. Coppe, J.P.; Desprez, P.Y.; Krtolica, A.; Campisi, J. The senescence-associated secretory phenotype: The dark side of tumor suppression. *Annu. Rev. Pathol.* **2010**, *5*, 99–118. [CrossRef] [PubMed]
13. Von Zglinicki, T.; Petrie, J.; Kirkwood, T.B. Telomere-driven replicative senescence is a stress response. *Nat. Biotechnol.* **2003**, *21*, 229–230. [CrossRef] [PubMed]
14. Lu, W.; Zhang, Y.; Liu, D.; Songyang, Z.; Wan, M. Telomeres-structure, function and regulation. *Exp. Cell Res.* **2013**, *319*, 133–141. [CrossRef] [PubMed]
15. Gilley, D.; Herbert, B.S.; Huda, N.; Tanaka, H.; Reed, T. Factors impacting human telomere homeostasis and age-related disease. *Mech. Ageing Dev.* **2008**, *129*, 27–34. [CrossRef] [PubMed]
16. Kaul, Z.; Cesare, A.J.; Huschtscha, L.I.; Neumann, A.A.; Reddel, R.R. Five dysfunctional telomeres predict onset of senescence in human cells. *EMBO Rep.* **2011**, *13*, 52–59. [CrossRef] [PubMed]
17. Bielak-Zmijewska, A.; Wnuk, M.; Przybylska, D.; Grabowska, W.; Lewinska, A.; Alster, O.; Korwek, Z.; Cmoch, A.; Myszka, A.; Pikula, S.; et al. A comparison of replicative senescence and doxorubicin-induced premature senescence of vascular smooth muscle cells isolated from human aorta. *Biogerontology* **2014**, *15*, 47–64. [CrossRef] [PubMed]
18. D'Adda di Fagagna, F. Living on a break: Cellular senescence as a DNA-damage response. *Nat. Rev. Cancer* **2008**, *8*, 512–522. [CrossRef] [PubMed]
19. Kühn, K.; D'Lima, D.D.; Hashimoto, S.; Lotz, M. Cell death in cartilage. *Osteoarthr. Cartil.* **2004**, *12*, 1–16. [CrossRef] [PubMed]
20. Vignon, E.; Arlot, M.; Patricot, L.M.; Vignon, G. The cell density of human femoral head cartilage. *Clin. Orthop. Relat. Res.* **1976**, *121*, 303–308. [CrossRef]
21. Blaney Davidson, E.N.; Scharstuhl, A.; Vitters, E.L.; van der Kraan, P.M.; van den Berg, W.B. Reduced transforming growth factor-beta signalling in cartilage of old mice: Role in impaired repair capacity. *Arthritis Res. Ther.* **2005**, *7*, 1338–1347. [CrossRef] [PubMed]
22. Aigner, T.; Hemmel, M.; Neureiter, D.; Gebhard, P.M.; Zeiler, G.; Kirchner, T.; McKenna, L. Apoptotic Cell Death Is Not a Widespread Phenomenon in Normal Aging and Osteoarthritis Human Articular Knee Cartilage: A Study of Proliferation, Programmed Cell Death (Apoptosis) and Viability of Chondrocytes in Normal and Osteoarthritic Human Knee Cartilage. *Arthritis Rheum.* **2001**, *44*, 1304–1312. [PubMed]

23. Ding, C.; Cicuttini, F.; Blizzard, L.; Scott, F.; Jones, G. A longitudinal study of the effect of sex and age on rate of change in knee cartilage volume in adults. *Rheumatology (Oxford)* **2007**, *46*, 273–279. [CrossRef] [PubMed]

24. Wells, T.; Davidson, C.; Morgelin, M.; Bird, J.L.E.; Bayliss, M.T.; Dudhia, J. Age-related changes in the composition, the molecular stoichiometry and the stability of proteoglycan aggregates extracted from human articular cartilage. *Biochem. J.* **2003**, *370*, 69–79. [CrossRef] [PubMed]

25. Parsch, D.; Brümmendorf, T.H.; Richter, W.; Fellenberg, J. Replicative aging of human articular chondrocytes during ex vivo expansion. *Arthritis Rheum.* **2002**, *46*, 2911–2916. [CrossRef] [PubMed]

26. Harbo, M.; Delaisse, J.M.; Kjaersgaard-Andersen, P.; Soerensen, F.B.; Koelvraa, S.; Bendix, L. The relationship between ultra-short telomeres, aging of articular cartilage and the development of human hip osteoarthritis. *Mech. Ageing Dev.* **2013**, *134*, 367–372. [CrossRef] [PubMed]

27. Wilson, B.; Novakofski, K.D.; Donocoff, R.S.; Liang, Y.X.; Fortier, L.A. Telomerase Activity in Articular Chondrocytes Is Lost after Puberty. *Cartilage* **2014**, *5*, 215–220. [CrossRef] [PubMed]

28. Guillot, P.V.; Gotherstrom, C.; Chan, J.; Kurata, H.; Fisk, N.M. Human first-trimester fetal MSC express pluripotency markers and grow faster and have longer telomeres than adult MSC. *Stem Cells* **2007**, *25*, 646–654. [CrossRef] [PubMed]

29. Mareschi, K.; Ferrero, I.; Rustichelli, D.; Aschero, S.; Gammaitoni, L.; Aglietta, M.; Madon, E.; Fagioli, F. Expansion of mesenchymal stem cells isolated from pediatric and adult donor bone marrow. *J. Cell. Biochem.* **2006**, *97*, 744–754. [CrossRef] [PubMed]

30. Baxter, M.A.; Wynn, R.F.; Jowitt, S.N.; Wraith, J.E.; Fairbairn, L.J.; Bellantuono, I. Study of telomere length reveals rapid aging of human marrow stromal cells following in vitro expansion. *Stem Cells* **2004**, *22*, 675–682. [CrossRef] [PubMed]

31. Parsch, D.; Fellenberg, J.; Brümmendorf, T.; Eschlbeck, A.M.; Richter, W. Telomere length and telomerase activity during expansion and differentiation of human mesenchymal stem cells and chondrocytes. *J. Mol. Med.* **2004**, *82*, 49–55. [PubMed]

32. Dai, S.M.; Shan, Z.Z.; Nakamura, H.; Masuko-Hongo, K.; Kato, T.; Nishioka, K.; Yudoh, K. Catabolic stress induces features of chondrocyte senescence through overexpression of caveolin 1: Possible involvement of caveolin 1-induced down-regulation of articular chondrocytes in the pathogenesis of osteoarthritis. *Arthritis Rheum.* **2006**, *54*, 818–831. [CrossRef] [PubMed]

33. Jallali, N.; Ridha, H.; Thrasivoulou, C.; Butler, P.; Cowen, T. Modulation of intracellular reactive oxygen species level in chondrocytes by IGF-1, FGF and TGF-beta1. *Connect. Tissue Res.* **2007**, *48*, 149–158. [CrossRef] [PubMed]

34. Brandl, A.; Meyer, M.; Bechmann, V.; Nerlich, M.; Angele, P. Oxidative stress induces senescence in human mesenchymal stem cells. *Exp. Cell Res.* **2011**, *317*, 1541–1547. [CrossRef] [PubMed]

35. Harbo, M.; Bendix, L.; Bay-Jensen, A.C.; Graakjaer, J.; Soe, K.; Andersen, T.L.; Kjaersgaard-Andersen, P.; Koelvraa, S.; Delaisse, J.M. The distribution pattern of critically short telomeres in human osteoarthritic knees. *Arthritis Res. Ther.* **2012**, *14*, R12. [CrossRef] [PubMed]

36. Rose, J.; Söder, S.; Skhirtladze, C.; Schmitz, N.; Gebhard, P.M.; Sesselmann, S.; Aigner, T. DNA damage, discoordinated gene expression and cellular senescence in osteoarthritic chondrocytes. *Osteoarthr. Cartil.* **2012**, *20*, 1020–1028. [CrossRef] [PubMed]

37. Brandl, A.; Hartmann, A.; Bechmann, V.; Graf, B.; Nerlich, M.; Angele, P. Oxidative stress induces senescence in chondrocytes. *J. Orthop. Res.* **2011**, *29*, 1114–1120. [CrossRef] [PubMed]

38. Loeser, R.F. Aging and osteoarthritis. *Curr. Opin. Rheumatol.* **2011**, *23*, 492–496. [CrossRef] [PubMed]

39. Jallali, N.; Ridha, H.; Thrasivoulou, C.; Underwood, C.; Butler, P.E.; Cowen, T. Vulnerability to ROS-induced cell death in ageing articular cartilage: The role of antioxidant enzyme activity. *Osteoarthr. Cartil.* **2005**, *13*, 614–622. [CrossRef] [PubMed]

40. Yu, S.M.; Kim, S.J. Thymoquinone-induced reactive oxygen species causes apoptosis of chondrocytes via PI3K/Akt and p38kinase pathway. *Exp. Biol. Med.* **2013**, *238*, 811–820. [CrossRef] [PubMed]

41. Yudoh, K.; van Trieu, N.; Nakamura, H.; Hongo-Masuko, K.; Kato, T.; Nishioka, K. Potential involvement of oxidative stress in cartilage senescence and development of osteoarthritis: Oxidative stress induces chondrocyte telomere instability and downregulation of chondrocyte function. *Arthritis Res. Ther.* **2005**, *7*, R380–R391. [CrossRef] [PubMed]

42. Ashraf, S.; Cha, B.H.; Kim, J.S.; Ahn, J.; Han, I.; Park, H.; Lee, S.H. Regulation of senescence associated signalling mechanisms in chondrocytes for cartilage tissue regeneration. *Osteoarthr. Cartil.* **2016**, *24*, 196–205. [CrossRef] [PubMed]

43. Loeser, R.F.; Gandhi, U.; Long, D.L.; Yin, W.; Chubinskaya, S. Aging and oxidative stress reduce the response of human articular chondrocytes to insulin-like growth factor 1 and osteogenic protein 1. *Arthritis Rheumatol.* **2014**, *66*, 2201–2209. [CrossRef] [PubMed]

44. Li, C.J.; Sun, L.Y.; Pang, C.Y. Synergistic Protection of N-acetylcysteine and ascorbic acid 2-phosphate on human mesenchymal stem cells against mitoptosis, necroptosis and apoptosis. *Sci. Rep.* **2015**, *5*, 9819. [CrossRef] [PubMed]

45. Sakata, S.; Hayashi, S.; Fujishiro, T.; Kawakita, K.; Kanzaki, N.; Hashimoto, S.; Iwasa, K.; Chinzei, N.; Kihara, S.; Haneda, M.; et al. Oxidative stress-induced apoptosis and matrix loss of chondrocytes is inhibited by eicosapentaenoic acid. *J. Orthop. Res.* **2015**, *33*, 359–365. [CrossRef] [PubMed]

46. Lin, T.M.; Tsai, J.L.; Lin, S.D.; Lai, C.S.; Chang, C.C. Accelerated growth and prolonged lifespan of adipose tissue-derived human mesenchymal stem cells in a medium using reduced calcium and antioxidants. *Stem Cells Dev.* **2005**, *14*, 92–102. [CrossRef] [PubMed]

47. Estrada, J.C.; Torres, Y.; Benguria, A.; Dopazo, A.; Roche, E.; Carrera-Quintanar, L.; Pérez, R.A.; Enríquez, J.A.; Torres, R.; Ramírez, J.C.; et al. Human mesenchymal stem cell-replicative senescence and oxidative stress are closely linked to aneuploidy. *Cell Death Dis.* **2013**, *4*, e691. [CrossRef] [PubMed]

48. Munir, S.; Foldager, C.; Lind, M.; Zachar, V.; Søballe, K.; Koch, T. Hypoxia enhances chondrogenic differentiation of human adipose tissue-derived stromal cells in scaffold-free and scaffold systems. *Cell. Tissue Res.* **2014**, *355*, 89–102. [CrossRef] [PubMed]

49. Lomri, A. Role of reactive oxygen species and superoxide dismutase in cartilage aging and pathology. *Future Reumatol.* **2008**, *3*, 381–392. [CrossRef]

50. Greene, M.A.; Loeser, R.F. Aging-related inflammation in osteoarthritis. *Osteoarthr. Cartil.* **2015**, *23*, 1966–1971. [CrossRef] [PubMed]

51. Livshits, G.; Zhai, G.; Hart, D.J.; Kato, B.S.; Wang, H.; Williams, F.M.; Spector, T.D. Interleukin-6 is a significant predictor of radiographic knee osteoarthritis: The Chingford study. *Arthritis Rheum.* **2009**, *60*, 2037–2045. [CrossRef] [PubMed]

52. Philipot, D.; Guerit, D.; Platano, D.; Chuchana, P.; Olivotto, E.; Espinoza, F.; Dorandeu, A.; Pers, Y.M.; Piette, J.; Borziet, R.M.; et al. p16INK4a and its regulator miR-24 link senescence and chondrocyte terminal differentiation-associated matrix remodeling in osteoarthritis. *Arthritis Res. Ther.* **2014**, *16*, R58. [CrossRef] [PubMed]

53. Wu, W.; Billinghurst, R.C.; Pidoux, I.; Antoniou, J.; Zukor, D.; Tanzer, M.; Poole, A.R. Sites of collagenase cleavage and denaturation of type II collagen in aging and osteoarthritic articular cartilage and their relationship to the distribution of matrix metalloproteinase 1 and matrix metalloproteinase 13. *Arthritis Rheum.* **2002**, *46*, 2087–2094. [CrossRef] [PubMed]

54. Aurich, M.; Poole, A.R.; Reiner, A.; Mollenhauer, C.; Margulis, A.; Kuettner, K.E.; Cole, A.A. Matrix homeostasis in aging normal human ankle cartilage. *Arthritis Rheum.* **2002**, *46*, 2903–2910. [CrossRef] [PubMed]

55. Forsyth, C.B.; Cole, A.; Murphy, G.; Bienias, J.L.; Im, H.J.; Loeser, R.F. Increased matrix metalloproteinase-13 production with aging by human articular chondrocytes in response to catabolic stimuli. *J. Gerontol. A Biol. Sci. Med. Sci.* **2005**, *60*, 1118–1124. [CrossRef] [PubMed]

56. Long, D.; Blake, S.; Song, X.Y.; Lark, M.; Loeser, R.F. Human articular chondrocytes produce IL-7 and respond to IL-7 with increased production of matrix metalloproteinase-13. *Arthritis Res. Ther.* **2008**, *10*, R23. [CrossRef] [PubMed]

57. Loeser, R.F. Aging and osteoarthritis: The role of chondrocyte senescence and aging changes in the cartilage matrix. *Osteoarthr. Cartil.* **2009**, *17*, 971–979. [CrossRef] [PubMed]

58. Hashimoto, M.; Nakasa, T.; Hikata, T.; Asahara, H. Molecular network of cartilage homeostasis and osteoarthritis. *Med. Res. Rev.* **2008**, *28*, 464–481. [CrossRef] [PubMed]

59. Goldberg, A. Effects of growth factors on articular cartilage. *Ortop. Traumatol. Rehab.* **2001**, *3*, 190–193.

60. De Ceuninck, F.; Caliez, A.; Dassencourt, L.; Anract, P.; Renard, P. Pharmacological disruption of insulin-like growth factor 1 binding to IGF-binding proteins restores anabolic responses in human osteoarthritic chondrocytes. *Arthritis Res. Ther.* **2004**, *6*, 393–403. [CrossRef] [PubMed]

61. Yin, W.; Park, J.I.; Loeser, R.F. Oxidative stress inhibits insulin-like growth factor-I induction of chondrocyte proteoglycan synthesis through differential regulation of phosphatidylinositol 3-Kinase-Akt and MEK-ERK MAPK signaling pathways. *J. Biol. Chem.* **2009**, *284*, 31972–31981. [CrossRef] [PubMed]

62. Starkman, B.G.; Cravero, J.D.; Delcarlo, M.; Loeser, R.F. IGF-I stimulation of proteoglycan synthesis by chondrocytes requires activation of the PI 3-kinase pathway but not ERK MAPK. *Biochem. J.* **2005**, *389*, 723–729. [CrossRef] [PubMed]

63. Allan, E.H.; Martin, T.J. The plasminogen activator inhibitor system in bone cell function. *Clin. Orthop. Relat. Res.* **1995**, *313*, 54–63.

64. Murphy-Ullrich, J.E.; Poczatek, M. Activation of latent TGF-beta by thrombospondin-1: Mechanisms and physiology. *Cytokine Growth Factor Rev.* **2000**, *11*, 59–69. [CrossRef]

65. Dangelo, M.; Sarment, D.P.; Billings, P.C.; Pacifici, M. Activation of transforming growth factor beta in chondrocytes undergoing endochondral ossification. *J. Bone Miner. Res.* **2001**, *16*, 2339–2347. [CrossRef] [PubMed]

66. Albro, M.B.; Cigan, A.D.; Nims, R.J.; Yeroushalmi, K.J.; Oungoulian, S.R.; Hung, C.T.; Ateshian, G.A. Shearing of synovial fluid activates latent TGF-β. *Osteoarthr. Cartil.* **2012**, *20*, 1374–1382. [CrossRef] [PubMed]

67. Gordon, K.J.; Blobe, G.C. Role of transforming growth factor-beta superfamily signaling pathways in human disease. *Biochim. Biophys. Acta* **2008**, *1782*, 197–228. [CrossRef] [PubMed]

68. Blaney Davidson, E.N.; Remst, D.F.; Vitters, E.L.; van Beuningen, H.M.; Blom, A.B.; Goumans, M.J.; van den Berg, W.B.; van der Kraan, P.M. Increase in ALK1/ALK5 ratio as a cause for elevated MMP-13 expression in osteoarthritis in humans and mice. *J. Immunol.* **2009**, *182*, 7937–7945. [CrossRef] [PubMed]

69. Finnson, K.W.; Parker, W.L.; ten Dijke, P.; Thorikay, M.; Philip, A. ALK1 opposes ALK5/Smad3 signaling and expression of extracellular matrix components in human chondrocytes. *J. Bone Miner. Res.* **2008**, *23*, 896–906. [CrossRef] [PubMed]

70. Li, W.; Ding, S. Generation of novel rat and human pluripotent stem cells by reprogramming and chemical approaches. *Methods Mol. Biol.* **2010**, *636*, 293–300. [CrossRef] [PubMed]

71. Johnstone, B.; Hering, T.M.; Caplan, A.I.; Goldberg, V.M.; Yoo, J.U. In vitro chondrogenesis of bone marrow-derived mesenchymal progenitor cells. *Exp. Cell Res.* **1998**, *238*, 265–272. [CrossRef] [PubMed]

72. Wu, J.; Niu, J.; Li, X.; Wang, X.; Guo, Z.; Zhang, F. TGF-β1 induces senescence of bone marrow mesenchymal stem cells via increase of mitochondrial ROS production. *BMC Dev. Biol.* **2014**, *14*, 21. [CrossRef] [PubMed]

73. Ahmed, U.; Anwar, A.; Savage, R.S.; Thornalley, P.J.; Rabbani, N. Protein oxidation, nitration and glycation biomarkers for early-stage diagnosis of osteoarthritis of the knee and typing and progression of arthritic disease. *Arthritis Res. Ther.* **2016**, *18*, 250. [CrossRef] [PubMed]

74. Saudek, D.M.; Kay, J. Advanced glycation endproducts and osteoarthritis. *Curr. Reumatol. Rep.* **2003**, *5*, 33–40. [CrossRef]

75. Mizumura, K.; Choi, A.M.; Ryster, S.W. Emerging role of selective autophagy in human diseases. *Front. Pharmacol.* **2014**, *5*, 244. [CrossRef] [PubMed]

76. Srinivas, V.; Bohensky, J.; Shapiro, I.M. Autophagy: A new phase in the maturation of growth plate chondrocytes is regulated by HIF, mTOR and AMP kinase. *Cells Tissues Organs* **2009**, *189*, 88–92. [CrossRef] [PubMed]

77. Shapiro, I.M.; Layfield, R.; Lotz, M.; Settembre, C.; Whitehouse, C. Boning up on autophagy: The role of autophagy in skeletal biology. *Autophagy* **2014**, *10*, 7–19. [CrossRef] [PubMed]

78. Carames, B.; Taniguchi, N.; Otsuki, S.; Blanco, F.J.; Lotz, M. Autophagy is a protective mechanism in normal cartilage and its aging-related loss is linked with cell death and osteoarthritis. *Arthritis Rheum.* **2010**, *62*, 791–801. [CrossRef] [PubMed]

79. Zhang, Y.; Vasheghani, F.; Li, Y.H.; Blati, M.; Simeone, K.; Fahmi, H.; Lussier, B.; Roughley, P.; Lagares, D.; Pelletier, J.P.; et al. Cartilage-specific deletion of mTOR upregulates autophagy and protects mice from osteoarthritis. *Ann. Rheum. Dis.* **2015**, *74*, 1432–1440. [CrossRef] [PubMed]

80. Takayama, K.; Kawakami, Y.; Kobayashi, M.; Greco, N.; Cummins, J.H.; Matsushita, T.; Kuroda, R.; Kurosaka, M.; Fu, F.H.; Huard, J. Local intra-articular injection of rapamycin delays articular cartilage degeneration in a murine model of osteoarthritis. *Arthritis Res. Ther.* **2014**, *16*, 482–485. [CrossRef] [PubMed]

81. Matsuzaki, T.; Matsushita, T.; Tabata, Y.; Saito, T.; Matsumoto, T.; Nagai, K.; Kuroda, R.; Kurosaka, M. Intra-articular administration of gelatin hydrogels incorporating rapamycin-micelles reduces the development of experimental osteoarthritis in a murine model. *Biomaterials* **2014**, *35*, 9904–9911. [CrossRef] [PubMed]

82. Alvarez-Garcia, O.; Olmer, M.; Akagi, R.; Akasaki, Y.; Fisch, K.M.; Shen, T.; Su, A.I.; Lotz, M.K. Suppression of REDD1 in osteoarthritis cartilage, a novel mechanism for dysregulated mTOR signaling and defective autophagy. *Osteoarthr. Cartil.* **2016**, *24*, 1639–1647. [CrossRef] [PubMed]

83. Zhang, F.J.; Luo, W.; Lei, G.H. Role of HIF-1alpha and HIF-2alpha in osteoarthritis. *Jt. Bone Spine* **2015**, *82*, 144–147. [CrossRef] [PubMed]

84. Bohensky, J.; Terkhorn, S.P.; Freeman, T.A.; Adams, C.S.; Garcia, J.; Shapiro, I.M.; Srinivas, V. Regulation of autophagy in humanand murine cartilage: Hypoxia-inducible factor 2 suppresses chondrocyteautophagy. *Arthritis Rheum.* **2009**, *60*, 1406–1415. [CrossRef] [PubMed]

85. Simonsen, J.L.; Rosada, C.; Serakinci, N.; Justesen, J.; Stenderup, K.; Rattan, S.I.; Jensen, T.G.; Kassem, M. Telomerase expression extends the proliferative life-span and maintains the osteogenic potential of human bone marrow stromal cells. *Nat. Biotechnol.* **2002**, *20*, 592–596. [CrossRef] [PubMed]

86. Shi, S.; Gronthos, S.; Chen, S.; Reddi, A.; Counter, C.M.; Robey, P.G.; Wang, C.Y. Bone formation by human postnatal bone marrow stromal stem cells is enhanced by telomerase expression. *Nat. Biotechnol.* **2002**, *20*, 587–591. [CrossRef] [PubMed]

87. Pearce, V.P.; Sherrell, J.; Lou, Z.; Kopelovich, L.; Wright, W.E.; Shay, J.W. Immortalization of epithelial progenitor cells mediated by resveratrol. *Oncogene* **2008**, *27*, 2365–2374. [CrossRef] [PubMed]

88. Sprouse, A.A.; Steding, C.E.; Herbert, B.S. Pharmaceutical regulation of telomerase and its clinical potential. *J. Cell. Mol. Med.* **2012**, *16*, 1–7. [CrossRef] [PubMed]

89. Shen, C.Y.; Jiang, J.G.; Yang, L.; Wang, D.W.; Zhu, W. Anti-aging active ingredients from herbs and nutraceuticals used in TCM: Pharmacological mechanisms and implications for drug discovery. *Br. J. Pharmacol.* **2017**, *174*, 1395–1425. [CrossRef] [PubMed]

90. Yung, L.Y.; Lam, W.S.; Ho, M.K.; Hu, Y.; Ip, F.C.; Pang, H.; Chin, A.C.; Harley, C.B.; Ip, N.Y.; Wong, Y.H. Astragaloside IV and cycloastragenol stimulate the phosphorylation of extracellular signal-regulated protein kinase in multiple cell types. *Planta Med.* **2012**, *78*, 115–121. [CrossRef] [PubMed]

91. Meng, H.C.; Wang, S.; Li, Y.; Kuang, Y.Y.; Ma, C.M. Chemical constituents and pharmacologic actions of Cynomorium plants. *Chin. J. Nat. Med.* **2013**, *11*, 321–329. [CrossRef] [PubMed]

92. Tichon, A.; Eitan, E.; Kurkalli, B.G.; Braiman, A.; Gazit, A.; Slavin, S.; Beith-Yannai, E.; Priel, E. Oxidative stress protection by novel telomerase activators in mesenchymal stem cells derived from healthy and diseased individuals. *Curr. Mol. Med.* **2013**, *13*, 1010–1022. [CrossRef] [PubMed]

93. Taka, T.; Changtam, C.; Thaichana, P.; Kaewtunjai, N.; Suksamrarn, A.; Lee, T.R.; Tuntiwechapikul, W. Curcuminoid derivatives enhance telomerase activity in an in vitro TRAP assay. *Bioorg. Med. Chem. Lett.* **2014**, *24*, 5242–5246. [CrossRef] [PubMed]

94. Molgora, B.; Bateman, R.; Sweeney, G.; Finger, D.; Dimler, T.; Effros, R.B.; Valenzuela, H.F. Functional assessment of pharmacological telomerase activators in human T cells. *Cells* **2013**, *2*, 57–66. [CrossRef] [PubMed]

95. De Jesus, B.B.; Schneeberger, K.; Vera, E.; Tejera, A.; Harley, C.B.; Blasco, M.A. The telomerase activator TA-65 elongates short telomeres and increases health span of adult/old mice without increasing cancer incidence. *Aging Cell.* **2011**, *10*, 604–621. [CrossRef] [PubMed]

96. Liu, L.; Gu, H.; Liu, H.; Jiao, Y.; Li, K.; Zhao, Y.; An, L.; Yang, J. Protective effect of resveratrol against IL-1β-induced inflammatory response on human osteoarthritic chondrocytes partly via the TLR4/MyD88/NF-κB signaling pathway: An "in vitro study". *Int. J. Mol. Sci.* **2014**, *15*, 6925–6940. [CrossRef] [PubMed]

97. Li, W.; Cai, L.; Zhang, Y.; Cui, L.; Shen, G. Intra-articular resveratrol injection prevents osteoarthritis progression in a mouse model by activating SIRT1 and thereby silencing HIF-2α. *J. Orthop. Res.* **2015**, *33*, 1061–1070. [CrossRef] [PubMed]

98. Toh, W.S.; Loh, X.J. Advances in hydrogel delivery systems for tissue regeneration. *Mater. Sci. Eng. C. Mater. Biol. Appl.* **2014**, *45*, 690–697. [CrossRef] [PubMed]

99. Wang, W.; Sun, L.; Zhang, P.; Song, J.; Liu, W. An anti-inflammatory cell-free collagen/resveratrol scaffold for repairing osteochondral defects in rabbits. *Acta Biomater.* **2014**, *10*, 4983–4995. [CrossRef] [PubMed]

100. Moussavi-Harami, F.; Duwayri, Y.; Martin, J.A.; Moussavi-Harami, F.; Buckwalter, J.A. Oxygen effects on senescence in chondrocytes and mesenchymal stem cells: Consequences for tissue engineering. *Iowa Orthop. J.* **2004**, *24*, 15–20. [PubMed]

101. Martin, J.A.; Klingelhutz, A.J.; Moussavi-Harami, F.; Buckwalter, J.A. Effects of oxidative damage and telomerase activity on human articular cartilage chondrocyte senescence. *J. Gerontol. A. Biol. Sci. Med. Sci.* **2004**, *59*, B324–B336. [CrossRef]

102. Egli, R.J.; Bastian, J.D.; Ganz, R.; Hofstetter, W.; Leunig, M. Hypoxic expansion promotes the chondrogenic potential of articular chondrocytes. *J. Orthop. Res.* **2008**, *26*, 977–985. [CrossRef] [PubMed]

103. Schrobback, K.; Klein, T.J.; Crawford, R.; Upton, Z.; Malda, J.; Leavesley, D.I. Effects of oxygen and culture system on in vitro propagation and redifferentiation of osteoarthritic human articular chondrocytes. *Cell Tissue Res.* **2012**, *347*, 649–663. [CrossRef] [PubMed]

104. Choi, J.R.; Pingguan-Murphy, B.; Wan Abas, W.A.B.; Noor Azmi, M.A.; Omar, S.Z.; Chua, K.H.; Wan Safwani, W.K. Impact of low oxygen tension on stemness, proliferation and differentiation potential of human adipose-derived stem cells. *Biochem. Biophys. Res. Commun.* **2014**, *448*, 218–224. [CrossRef] [PubMed]

105. Tsai, C.C.; Chen, Y.J.; Yew, T.L.; Chen, L.L.; Wang, J.Y.; Chiu, C.H.; Hung, S.C. Hypoxia inhibits senescence and maintains mesenchymal stem cell properties through down-regulation of E2A-p21 by HIF-TWIST. *Blood* **2011**, *117*, 459–469. [CrossRef] [PubMed]

106. Wan Safwani, W.K.Z.; Choi, J.R.; Yong, K.W.; Ting, I.; Mat Adenan, N.A.; Pingguan-Murphy, B. Hypoxia enhances the viability, growth and chondrogenic potential of cryopreserved human adipose-derived stem cells. *Cryobiology* **2017**, *75*, 91–99. [CrossRef] [PubMed]

107. Wan Safwani, W.K.Z.; Wong, C.W.; Yong, K.W.; Choi, J.R.; Mat Adenan, N.A.; Omar, S.Z.; Wan Abas, W.A.B.; Pingguan-Murphy, B. The effects of hypoxia and serum-free conditions on the stemness properties of human adipose-derived stem cells. *Cytotechnology* **2013**, *68*, 1859–1872. [CrossRef] [PubMed]

108. Öztürk, E.; Hobiger, S.; Despot-Slade, E.; Pichler, M.; Zenobi-Wong, M. Hypoxia regulates RhoA and Wnt/β-catenin signaling in a context-dependent way to control re-differentiation of chondrocytes. *Sci. Rep.* **2017**, *7*, 9032. [CrossRef] [PubMed]

109. Grodzinsky, A.J.; Levenston, M.E.; Jin, M.; Frank, E.H. Cartilage tissue remodeling in response to mechanical forces. *Annu. Rev. Biomed Eng.* **2000**, *2*, 691–713. [CrossRef] [PubMed]

110. Schätti, O.; Grad, S.; Goldhahn, J.; Salzmann, G.; Li, Z.; Alini, M.; Stoddart, M.J. A combination of shear and dynamic compression leads to mechanically induced chondrogenesis of human mesenchymal stem cells. *Eur. Cell Mater.* **2011**, *22*, 214–225. [CrossRef] [PubMed]

111. Gardner, O.F.; Fahy, N.; Alini, M.; Stoddart, M.J. Differences in human mesenchymal stell cell secretomes during chondrogenic induction. *Eur. Cell Mater.* **2016**, *31*, 221–235. [CrossRef] [PubMed]

112. Li, Z.; Kupcsik, L.; Yao, S.J.; Alini, M.; Stoddart, M.J. Mechanical Load Modulates Chondrogenesis of Human Mesenchymal Stem Cells through the TGFβ Pathway. *J. Cell Mol. Med.* **2010**, *14*, 1338–1346. [CrossRef] [PubMed]

113. Guo, T.; Yu, L.; Lim, C.G.; Goodley, A.S.; Xiao, X.; Placone, J.K.; Ferlin, K.M.; Nguyen, B.N.; Hsieh, A.H.; Fisher, J.P. Effect of Dynamic Culture and Periodic Compression on Human Mesenchymal Stem Cell Proliferation and Chondrogenesis. *Ann. Biomed. Eng.* **2016**, *44*, 2103–2113. [CrossRef] [PubMed]

114. Bian, L.; Zhai, D.Y.; Zhang, E.C.; Mauck, R.L.; Burdick, J.A. Dynamic compressive loading enhances cartilage matrix synthesis and distribution and suppresses hypertrophy in hMSC-laden hyaluronic acid hydrogels. *Tissue Eng. Part A* **2012**, *18*, 715–724. [CrossRef] [PubMed]

115. Zhang, T.; Wen, F.; Wu, Y.; Goh, G.S.; Ge, Z.; Tan, L.P.; Hui, J.H.; Yang, Z. Cross-talk between TGF-beta/SMAD and integrin signaling pathways in regulating hypertrophy of mesenchymal stem cell chondrogenesis under deferral dynamic compression. *Biomaterials* **2015**, *38*, 72–85. [CrossRef] [PubMed]

116. Simon, T.M.; Jackson, D.W. Articular Cartilage: Injury Pathways and Treatment Options. *Sports Med. Arthrosc. Rev.* **2018**, *26*, 31–39. [CrossRef] [PubMed]

117. Al-Najar, M.; Khalil, H.; Al-Ajlouni, J.; Al-Antary, E.; Hamdan, M.; Rahmeh, R.; Alhattab, D.; Samara, O.; Yasin, M.; Abdullah, A.A.; et al. Intra-articular injection of expanded autologous bone marrow mesenchymal cells in moderate and severe knee osteoarthritis is safe: A phase I/II study. *J. Orthop. Surg. Res.* **2017**, *12*, 190. [CrossRef] [PubMed]

118. Gille, J.; Behrens, P.; Volpi, P.; de Girolamo, L.; Reiss, E.; Zoch, W.; Anders, S. Outcome of Autologous Matrix Induced Chondrogenesis (AMIC) in cartilage knee surgery: Data of the AMIC Registry. *Arch. Orthop. Trauma Surg.* **2013**, *133*, 87–93. [CrossRef] [PubMed]

119. Jo, C.H.; Chai, J.W.; Jeong, E.C.; Oh, S.; Shin, J.S.; Shim, H.; Yoon, K.S. Intra-articular Injection of Mesenchymal Stem Cells for the Treatment of Osteoarthritis of the Knee: A 2-Year Follow-up Study. *Am. J. Sports Med.* **2017**, *45*, 2774–2783. [CrossRef] [PubMed]

120. Brittberg, M.; Lindahl, A.; Nilsson, A.; Ohlsson, C.; Isaksson, O.; Peterson, L. Treatment of deep cartilage defects in the knee with autologous chondrocyte transplantation. *N. Engl. J. Med.* **1994**, *331*, 889–895. [CrossRef] [PubMed]

121. Peterson, L.; Brittberg, M.; Kiviranta, I.; Akerlund, E.L.; Lindahl, A. Autologous chon-drocyte transplantation. Biomechanics and long-term durability. *Am. J. Sports Med.* **2002**, *30*, 2–12. [CrossRef] [PubMed]

122. Li, Y.; Wei, X.; Zhou, J.; Wei, L. The age-related changes in cartilage and osteoarthritis. *BioMed Res. Int.* **2013**, *2013*, 916530. [CrossRef] [PubMed]

123. Barbero, A.; Grogan, S.; Schafer, D.; Heberer, M.; Mainil-Varlet, P.; Martin, I. Age related changes in human articular chondrocyte yield, proliferation and post-expansion chondrogenic capacity. *Osteoarthr. Cartil.* **2004**, *12*, 476–484. [CrossRef] [PubMed]

124. Schulze-Tanzil, G. Activation and dedifferentiation of chondrocytes: Implications in cartilage injury and repair. *Ann. Anat.* **2009**, *191*, 325–338. [CrossRef] [PubMed]

125. Caron, M.M.; Emans, P.J.; Coolsen, M.M.; Voss, L.; Surtel, D.A.; Cremers, A.; van Rhijn, L.W.; Welting, T.J. Redifferentiation of dedifferentiated human articular chondrocytes: Comparison of 2D and 3D cultures. *Osteoarthr. Cartil.* **2012**, *20*, 1170–1178. [CrossRef] [PubMed]

126. Meinert, C.; Schrobback, K.; Hutmacher, D.W.; Klein, T.J. A novel bioreactor system for biaxial mechanical loading enhances the properties of tissue-engineered human cartilage. *Sci. Rep.* **2017**, *7*, 16997. [CrossRef] [PubMed]

127. Rosenzweig, D.H.; Solar-Cafaggi, S.; Quinn, T.M. Functionalization of dynamic culture surfaces with a cartilage extracellular matrix extract enhances chondrocyte phenotype against dedifferentiation. *Acta Biomater.* **2012**, *8*, 3333–3341. [CrossRef] [PubMed]

128. Rosenzweig, D.H.; Matmati, M.; Khayat, G.; Chaudhry, S.; Hinz, B.; Quinn, T.M. Culture of primary bovine chondrocytes on a continuously expanding surface inhibits dedifferentiation. *Tissue Eng. Part A* **2012**, *18*, 2466–2476. [CrossRef] [PubMed]

129. Kino-Oka, M.; Yashiki, S.; Ota, Y.; Mushiaki, Y.; Sugawara, K.; Yamamoto, T.; Takezawa, T.; Taya, M. Subculture of chondrocytes on a collagen type I-coated substrate with suppressed cellular dedifferentiation. *Tissue Eng.* **2005**, *11*, 597–608. [CrossRef] [PubMed]

130. Das, R.H.; Jahr, H.; Verhaar, J.A.; van der Linden, J.C.; van Osch, G.J.; Weinans, H. In vitro expansion affects the response of chondrocytes to mechanical stimulation. *Osteoarthr. Cartil.* **2008**, *16*, 385–391. [CrossRef] [PubMed]

131. Barbero, A.; Martin, I. Human articular chondrocytes culture. *Methods Mol. Med.* **2007**, *140*, 237–247. [PubMed]

132. Gurusinghe, S.; Strappe, P. Gene modification of mesenchymal stem cells and articular chondrocytes to enhance chondrogenesis. *BioMed Res. Int.* **2014**, *2014*, 369528. [CrossRef] [PubMed]

133. Dominici, M.; Le Blanc, K.; Mueller, I.; Slaper-Cortenbach, I.; Marini, F.; Krause, D.; Deans, R.; Keating, A.; Prockop, D.J.; Horwitz, E. Minimal criteria for defining multipotent mesenchymal stromal cells. The International Society for Cellular Therapy position statement. *Cytotherapy* **2006**, *8*, 315–317. [CrossRef] [PubMed]

134. Girlovanu, M.; Susman, S.; Soritau, O.; Rus-Ciuca, D.; Melincovici, C.; Constantin, A.M.; Mihu, C.M. Stem cells—Biological update and cell therapy progress. *Clujul. Med.* **2015**, *88*, 265–271. [CrossRef] [PubMed]

135. Im, G.I. Bone marrow-derived stem/stromal cells and adipose tissue-derived stem/stromal cells: Their comparative efficacies and synergistic effects. *J. Biomed. Mater. Res. A* **2017**, *105*, 2640–2648. [CrossRef] [PubMed]

136. Keating, A. Mesenchymal stromal cells: New directions. *Cell Stem Cell* **2012**, *10*, 709–716. [CrossRef] [PubMed]

137. Ho, A.D.; Wagner, W.; Franke, W. Heterogeneity of mesenchymal stromal cell preparations. *Cytotherapy* **2008**, *10*, 320–330. [CrossRef] [PubMed]

138. Davatchi, F.; Sadeghi Abdollahi, B.; Mohyeddin, M.; Nikbin, B. Mesenchymal stem cell therapy for knee osteoarthritis: 5 years follow-up of three patients. *Int. J. Rheum. Dis.* **2016**, *19*, 219–225. [CrossRef] [PubMed]

139. Lamo-Espinosa, J.M.; Mora, G.; Blanco, J.F.; Granero-Moltó, F.; Nuñez-Córdoba, J.M.; Sánchez-Echenique, C.; Bondía, J.M.; Aquerreta, J.D.; Andreu, E.J.; Ornilla, E.; et al. Intra-articular injection of two different doses of autologous bone marrow mesenchymal stem cells versus hyaluronic acid in the treatment of knee osteoarthritis: Multicenter randomized controlled clinical trial (phase I/II). *J. Transl. Med.* **2016**, *14*, 246. [CrossRef] [PubMed]

140. Pers, Y.M.; Rackwitz, L.; Ferreira, R.; Pullig, O.; Delfour, C.; Barry, F.; Sensebe, L.; Casteilla, L.; Fleury, S.; Bourin, P.; et al. Adipose Mesenchymal Stromal Cell-Based Therapy for Severe Osteoarthritis of the Knee: A Phase I Dose-Escalation Trial. *Stem Cells Transl. Med.* **2016**, *5*, 847–856. [CrossRef] [PubMed]

141. Emadedin, M.; Ghorbani Liastani, M.; Fazeli, R.; Mohseni, F.; Moghadasali, R.; Mardpour, S.; Hosseini, S.E.; Niknejadi, M.; Moeininia, F.; Aghahossein Fanni, A.; et al. Long-Term Follow-up of Intra-articular Injection of Autologous Mesenchymal Stem Cells in Patients with Knee, Ankle, or Hip Osteoarthritis. *Arch. Iran. Med.* **2015**, *18*, 336–344. [PubMed]

142. Wakitani, S.; Imoto, K.; Yamamoto, T.; Saito, M.; Murata, N.; Yoneda, M. Human autologous culture expanded bone marrow mesenchymal cell transplantation for repair of cartilage defects in osteoarthritic knees. *Osteoarthr. Cartil.* **2002**, *10*, 199–206. [CrossRef] [PubMed]

143. Nejadnik, H.; Hui, J.H.; Feng Choong, E.P.; Tai, B.C.; Lee, E.H. Autologous bone marrow-derived mesenchymal stem cells versus autologous chondrocyte implantation: An observational cohort study. *Am. J. Sports Med.* **2010**, *38*, 1110–1116. [CrossRef] [PubMed]

144. Buda, R.; Vannini, F.; Cavallo, M.; Grigolo, B.; Cenacchi, A.; Giannini, S. Osteochondral lesions of the knee: A new one-step repair technique with bone-marrow-derived cells. *J. Bone Jt. Surg. Am.* **2010**, *92*, 2–11. [CrossRef] [PubMed]

145. Wakitani, S.; Okabe, T.; Horibe, S.; Mitsuoka, T.; Saito, M.; Koyama, T.; Nawata, M.; Tensho, K.; Kato, H.; Uematsu, K.; et al. Safety of autologous bone marrow-derived mesenchymal stem cell transplantation for cartilage repair in 41 patients with 45 joints followed for up to 11 years and 5 months. *J. Tissue Eng. Regen. Med.* **2011**, *5*, 146–150. [CrossRef] [PubMed]

146. Saw, K.Y.; Anz, A.; Merican, S.; Tay, Y.G.; Ragavanaidu, K.; Jee, C.S.; McGuire, D.A. Articular cartilage regeneration with autologous peripheral blood progenitor cells and hyaluronic acid after arthroscopic subchondral drilling: A report of 5 cases with histology. *Arthroscopy* **2011**, *27*, 493–506. [CrossRef] [PubMed]

147. Davatchi, F.; Abdollahi, B.S.; Mohyeddin, M.; Shahram, F.; Nikbin, B. Mesenchymal stem cell therapy for knee osteoarthritis. Preliminary report of four patients. *Int. J. Rheum. Dis.* **2011**, *14*, 211–215. [CrossRef] [PubMed]

148. Koh, Y.G.; Choi, Y.J. Infrapatellar fat pad-derived mesenchymal stem cell therapy for knee osteoarthritis. *Knee* **2012**, *19*, 902–907. [CrossRef] [PubMed]

149. Koh, Y.G.; Jo, S.B.; Kwon, O.R.; Suh, D.S.; Lee, S.W.; Park, S.H.; Choi, Y.J. Mesenchymal stem cell injections improve symptoms of knee osteoarthritis. *Arthroscopy* **2013**, *29*, 748–755. [CrossRef] [PubMed]

150. Saw, K.Y.; Anz, A.; Siew-Yoke Jee, C.; Merican, S.; Ching-Soong Ng, R.; Roohi, S.A.; Ragavanaidu, K. Articular cartilage regeneration with autologous peripheral blood stem cells versus hyaluronic acid: A randomized controlled trial. *Arthroscopy* **2013**, *29*, 684–694. [CrossRef] [PubMed]

151. Orozco, L.; Munar, A.; Soler, R.; Alberca, M.; Soler, F.; Huguet, M.; Sentís, J.; Sánchez, A.; García-Sancho, J. Treatment of knee osteoarthritis with autologous mesenchymal stem cells: A pilot study. *Transplantation* **2013**, *95*, 1535–1541. [CrossRef] [PubMed]

152. Wong, K.L.; Lee, K.B.; Tai, B.C.; Law, P.; Lee, E.H.; Hui, J.H. Injectable cultured bone marrow-derived mesenchymal stem cells in varus knees with cartilage defects undergoing high tibial osteotomy: A prospective, randomized controlled clinical trial with 2 years' follow-up. *Arthroscopy* **2013**, *29*, 2020–2028. [CrossRef] [PubMed]

153. Jo, C.H.; Lee, Y.G.; Shin, W.H.; Kim, H.; Chai, J.W.; Jeong, E.C.; Kim, J.E.; Shim, H.; Shin, J.S.; Shin, I.S.; et al. Intra-articular injection of mesenchymal stem cells for the treatment of osteoarthritis of the knee: A proof-of-concept clinical trial. *Stem Cells* **2014**, *32*, 1254–1266. [CrossRef] [PubMed]

154. Akgun, I.; Unlu, M.C.; Erdal, O.A.; Ogut, T.; Erturk, M.; Ovali, E.; Kantarci, F.; Caliskan, G.; Akgun, Y. Matrix-induced autologous mesenchymal stem cell implantation versus matrix-induced autologous chondrocyte implantation in the treatment of chondral defects of the knee: A 2-year randomized study. *Arch. Orthop. Trauma Surg.* **2015**, *135*, 251–263. [CrossRef] [PubMed]

155. Fodor, P.B.; Paulseth, S.G. Adipose derived stromal cell (ADSC) injections for pain management of osteoarthritis in the human knee joint. *Aesthet. Surg. J.* **2016**, *36*, 229–236. [CrossRef] [PubMed]

156. Koh, Y.G.; Kwon, O.R.; Kim, Y.S.; Choi, Y.J.; Tak, D.H. Adipose-Derived mesenchymal stem cells with microfracture versus microfracture alone: 2-year follow-up of a prospective randomized trial. *Arthroscopy* **2016**, *32*, 97–109. [CrossRef] [PubMed]

157. Vangsness, C.T.; Farr, J.; Boyd, J.; Dellaero, D.T.; Mills, C.R.; LeRoux-Williams, M. Adult human mesenchymal stem cells delivered via intra-articular injection to the knee following partial medial meniscectomy: A randomized, double-blind, controlled study. *J. Bone Jt. Surg. Am.* **2014**, *96*, 90–98. [CrossRef] [PubMed]

158. Vega, A.; Martin-Ferrero, M.A.; del Canto, F.; Alberca, M.; Garcia, V.; Munar, A.; Orozco, L.; Soler, R.; Fuertes, J.J.; Huguet, M.; et al. Treatment of knee osteoarthritis with allogeneic bone marrow mesenchymal stem cells: A randomized controlled trial. *Transplantation* **2015**, *99*, 1681–1690. [CrossRef] [PubMed]

159. De Windt, T.S.; Vonk, L.A.; Slaper-Cortenbach, I.C.; van den Broek, M.P.; Nizak, R.; van Rijen, M.H.; de Weger, R.A.; Dhert, W.J.; Saris, D.B. Allogeneic Mesenchymal Stem Cells Stimulate Cartilage Regeneration and Are Safe for Single-Stage Cartilage Repair in Humans upon Mixture with Recycled Autologous Chondrons. *Stem Cells* **2017**, *35*, 256–264. [CrossRef] [PubMed]

160. De Windt, T.S.; Vonk, L.A.; Slaper-Cortenbach, I.C.M.; Nizak, R.; van Rijen, M.H.P.; Saris, D.B.F. Allogeneic MSCs and Recycled Autologous Chondrons Mixed in a One-Stage Cartilage Cell Transplantion: A First-in-Man Trial in 35 Patients. *Stem Cells* **2017**, *35*, 1984–1993. [CrossRef] [PubMed]

161. Baldari, S.; Di Rocco, G.; Piccoli, M.; Pozzobon, M.; Muraca, M.; Toietta, G. Challenges and Strategies for Improving the Regenerative Effects of Mesenchymal Stromal Cell-Based Therapies. *Int. J. Mol. Sci.* **2017**, *18*, 2087. [CrossRef] [PubMed]

162. Liang, X.; Ding, Y.; Zhang, Y.; Tse, H.F.; Lian, Q. Paracrine mechanisms of mesenchymal stem cell-based therapy: Current status and perspectives. *Cell Transpl.* **2014**, *23*, 1045–1059. [CrossRef] [PubMed]

163. Steinert, A.F.; Rackwitz, L.; Gilbert, F.; Noth, U.; Tuan, R.S. Concise review: The clinical application of mesenchymal stem cells for musculoskeletal regeneration: Current status and perspectives. *Stem Cells Transl. Med.* **2012**, *1*, 237–247. [CrossRef] [PubMed]

164. Gomez-Aristizabal, A.; Sharma, A.; Bakooshli, M.A.; Kapoor, M.; Gilbert, P.M.; Viswanathan, S.; Gandhi, R. Stage-specific differences in secretory profile of mesenchymal stromal cells (MSCs) subjected to early- vs. late-stage OA synovial fluid. *Osteoarthr. Cartil.* **2017**, *25*, 737–741. [CrossRef] [PubMed]

165. Farrell, M.J.; Fisher, M.B.; Huang, A.H.; Shin, J.I.; Farrell, K.M.; Mauck, R.L. Functional properties of bone marrow-derived MSC-based engineered cartilage are unstable with very longterm in vitro culture. *J. Biomech.* **2014**, *47*, 2173–2182. [CrossRef] [PubMed]

166. Perez-Silos, V.; Camacho-Morales, A.; Fuentes-Mera, L. Mesenchymal stem cells subpopulations: Application for orthopedic regenerative medicine. *Stem Cells Int.* **2016**, 3187491. [CrossRef] [PubMed]

167. Clause, K.C.; Liu, L.J.; Tobita, K. Directed stem cell differentiation: The role of physical forces. *Cell Commun. Adhes.* **2010**, *17*, 48–54. [CrossRef] [PubMed]

168. Gugjoo, M.B.; Amarpal; Sharma, G.T.; Aithal, H.P.; Kinjavdekar, P. Cartilage tissue engineering: Role of mesenchymal stem cells along with growth factors & scaffolds. *Indian J. Med. Res.* **2016**, *144*, 339–347. [CrossRef] [PubMed]

169. Vinardell, T.; Sheehy, E.J.; Buckley, C.T.; Kelly, D.J. A comparison of the functionality and in vivo phenotypic stability of cartilaginous tissues engineered from different stem cell sources. *Tissue Eng. Part A* **2012**, *18*, 1161–1170. [CrossRef] [PubMed]

170. Mak, J.; Jablonski, C.L.; Leonard, C.A.; Dunn, J.F.; Raharjo, E.; Matyas, J.R.; Biernaskie, J.; Krawetz, R.J. Intra-articular injection of synovial mesenchymal stem cells improves cartilage repair in a mouse injury model. *Sci. Rep.* **2016**, *6*, 23076. [CrossRef] [PubMed]

171. Sekiya, I.; Muneta, T.; Horie, M.; Koga, H. Arthroscopic Transplantation of Synovial Stem Cells Improves Clinical Outcomes in Knees with Cartilage Defects. *Clin. Orthop. Relat. Res.* **2015**, *473*, 2316–2326. [CrossRef] [PubMed]

Int. J. Mol. Sci. **2018**, *19*, 623

172. Van Osch, G.J.; Brittberg, M.; Dennis, J.E.; Bastiaansen-Jenniskens, Y.M.; Erben, R.G.; Konttinen, Y.T.; Luyten, F.P. Cartilage repair: Past and future—Lessons for regenerative medicine. *J. Cell. Mol. Med.* **2009**, *13*, 792–810. [CrossRef] [PubMed]

173. Gomoll, A.H. Microfracture and augments. *J. Knee Surg.* **2012**, *25*, 9–15. [CrossRef] [PubMed]

174. Bonzani, I.C.; George, J.H.; Stevens, M.M. Novel materials for bone and cartilage regeneration. *Curr. Opin. Chem. Biol.* **2006**, *10*, 568–575. [CrossRef] [PubMed]

175. Kon, E.; Roffi, A.; Filardo, G.; Tesei, G.; Marcacci, M. Scaffold-based cartilage treatments: With or without cells? A systematic review of preclinical and clinical evidence. *Arthroscopy* **2015**, *31*, 767–775. [CrossRef] [PubMed]

176. Benthien, J.P.; Behrens, P. Autologous Matrix-Induced Chondrogenesis (AMIC): Combining Microfracturing and a Collagen I/III Matrix for Articular Cartilage Resurfacing. *Cartilage* **2010**, *1*, 65–68. [CrossRef] [PubMed]

International Journal of
Molecular Sciences

MDPI

Article

Aged Mouse Cortical Microglia Display an Activation Profile Suggesting Immunotolerogenic Functions

Tanja Zöller [1], Abdelraheim Attaai [1,2,3], Phani Sankar Potru [4], Tamara Ruß [4] and Björn Spittau [1,4,*]

[1]   Institute for Anatomy and Cell Biology, Department of Molecular Embryology, Faculty of Medicine, University of Freiburg, Freiburg 79104, Germany; tanja.zoeller@googlemail.com (T.Z.); abdelraheim.attaai@anat.uni-freiburg.de (A.A.)
[2]   Faculty of Biology, University of Freiburg, Freiburg 79104, Germany
[3]   Department of Anatomy and Histology, Faculty of Veterinary Medicine, Assiut University, Assiut 71526, Egypt
[4]   Institute of Anatomy, University of Rostock, Rostock 18057, Germany; PhaniSankar.Potru@med.uni-rostock.de (P.S.P.); TamaraHeike.Russ@med.uni-rostock.de (T.R.)
*   Correspondence: bjoern.spittau@med.uni-rostock.de; Tel.: +49-381-494-8409; Fax: +49-381-494-8402

Received: 31 January 2018; Accepted: 25 February 2018; Published: 1 March 2018

Abstract: Microglia are the resident immune cells of the central nervous system (CNS) and participate in physiological and pathological processes. Their unique developmental nature suggests age-dependent structural and functional impairments that might contribute to neurodegenerative diseases. In the present study, we addressed the age-dependent changes in cortical microglia gene expression patterns and the expression of M1- and M2-like activation markers. Iba1 immunohistochemistry, isolation of cortical microglia followed by fluorescence-activated cell sorting and RNA isolation to analyze transcriptional changes in aged cortical microglia was performed. We provide evidence that aging is associated with decreased numbers of cortical microglia and the establishment of a distinct microglia activation profile including upregulation of *Ifi204*, *Lilrb4*, *Arhgap*, *Oas1a*, *Cd244* and *Ildr2*. Moreover, flow cytometry revealed that aged cortical microglia express increased levels of Cd206 and Cd36. The data presented in the current study indicate that aged mouse cortical microglia adopt a distinct activation profile, which suggests immunosuppressive and immuno-tolerogenic functions.

Keywords: microglia; aging; cerebral cortex; neuroinflammation

---

1. Introduction

Aging has been described as one of the major risk factors for development, onset and progression of several diseases, including cancer, cardiovascular pathologies, as well as neurodegenerative disorders [1]. In the central nervous system (CNS), age-dependent changes are believed to foster the development of neuropathologies including Alzheimer's disease (AD) and Parkinson's disease (PD). The high incidence of neurodegenerative diseases in elderly individuals has been linked to dysregulated functions of innate immune responses mediated by CNS-resident microglia [2]. Microglia develop from yolk sac progenitors in a PU.1- and interferon regulatory factor 8 (Irf8)-dependent manner [3] and further colonize the embryonic CNS parenchyma by chemotactic attraction driven by neuron-derived IL34 [4]. Sensing of neuronal IL-34 is mediated by colony stimulating factor-1 receptor (CSF1R), which is essential for microglia homing and migration towards the developing CNS [5]. After induction of a microglia-specific gene expression pattern [6] in a transforming growth factor β1 (TGFβ1)-dependent mechanism [7] and the establishment of the blood-brain barrier (BBB), microglia turnover in the adult CNS involves microglia proliferation and apoptosis [8], but not replacement by bone-marrow-derived progenitor cells [9,10]. This self-renewal capacity causing high proliferative activity of microglia has been hypothesized to result in telomere shortening and subsequent senescence of aged microglia [11]. However, it appears that microglia in aged mice do not develop telomere shortening, and only transgenic approaches, such as those described in the DNA repair-deficient *Ercc1* mutant mice, display accelerated aging associated with microglial senescence, dystrophy and functional impairments [12,13]. However, it has been reported that aged microglia exhibit enhanced expression of inflammatory markers tumor necrosis factor α (Tnfα), Il1β and Il6 after challenge with lipopolysaccharide (LPS), indicating an increased response of aged microglia [14], which is caused by age-dependent priming of microglia. Toll-like receptors Tl2, Tlr3, Tlr4 [15], as well as high-mobility group box 1 (Hmgb1) [16], have been described to be important mediators of microglia priming. Holtman et al. [17] recently identified a highly conserved gene expression profile of primed microglia, which significantly differed from gene expression patterns observed after LPS-driven acute microglia activation. Interestingly, RNA sequencing from aged total brain microglia revealed increased expression of genes involved in microglia-mediated neuroprotective effects [18].

In the present study, we addressed the age-dependent changes of cortical microglia in 24 months old C57BL/6 mice and observed decreased microglia numbers, enhanced expression of genes related to innate immune responses, and increased numbers of Cd206$^+$ and Cd36$^+$ microglia. Our data indicate that aging is associated with changes in microglia gene expression, which points towards activation of alternative markers and genes involved in immunosuppressive functions and immune toleration.

2. Results

2.1. Decreased Numbers of Cortical Iba1$^+$ Microglia in Aged Mice

In order to address age-dependent changes in cortical thickness and total microglia numbers in the frontal cortex of C57BL/6JRj mice, 50 μm vibratome sections were stained against the microglia marker ionized calcium binding adaptor molecular 1 (Iba1). Figure 1A displays the area used for analysis of cortical thickness. No obvious differences in cortical thickness were observed between young (Figure 1B) and aged (Figure 1C) mice. Moreover, quantification of thicknesses and statistical analyses also revealed no significant differences between young and aged mice (Figure 1D). Interestingly, quantification of cortical microglia (Iba1$^+$) numbers revealed significant reduction of Iba1$^+$ microglia in aged mice. Figure 1F displays microglia from young mice, showing normal ramifications with fine processes and homogenous distributions throughout the cortex (Figure 1F). However, aged cortical microglia presented a reduced branching pattern and an uneven distribution compared to microglia in young mice (Figure 1G). These data indicate that aging is associated with a reduction of cortical microglia numbers, which tend to cluster and thereby show an uneven distribution pattern.

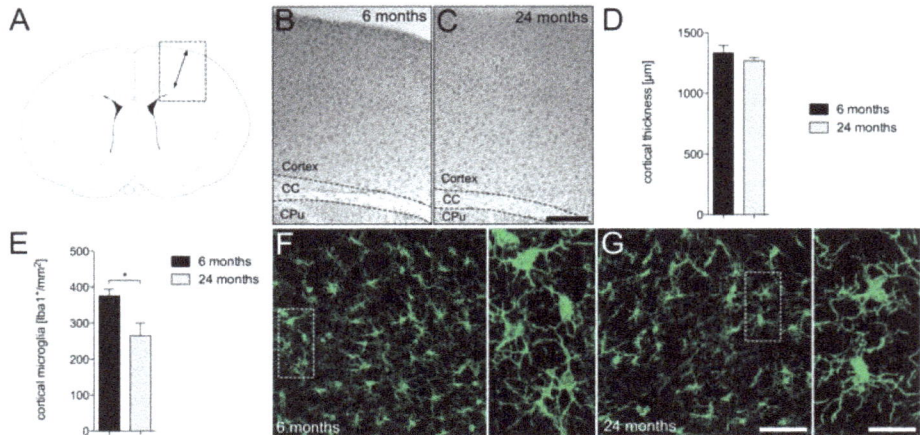

Figure 1. Age-dependent changes in cortical microglia. (**A**) Orientation scheme monitoring cortical areas for evaluation. Arrows mark the area for cortical thickness measurements. Representative images of Iba1+ microglia in cortices from 6-month-old (**B**) and 24-month-old mice (**C**). Scale bar represents 300 μm. Quantifications of cortical thickness (**D**) and cortical Iba1+ microglia numbers (**E**) are shown. Data are given as means ± SEM from three animals per age. *p*-value derived from Student's *t*-test is * $p < 0.05$. Differences in morphology and distribution of Iba1+ microglia between 6-month-old (**F**) and 24-month-old (**G**) mice. Scale bars indicate 20 μm in overview images and 7 μm in high magnification detail images. CC = corpus callsosum, CPu = caudatoputamen.

2.2. Detection of a Distinct Gene Expression Profile in Aged Cortical Microglia

As a next step, we addressed the transcriptional changes in aged cortical microglia. Therefore, frontal cortices from young and aged mice were dissected and microglia were isolated and stained for fluorescence activated cell sorting (FACS). Cd11b+/Cd45low-positive microglia were collected and RNA was isolated for cDNA microarrays (Figure 2A). Expression data were used for prediction of the biological processes (Figure 2B) and molecular functions (Figure 2C) of the top-regulated genes using the DAVID gene ontology (GO term) enrichment analysis. As shown in Figure 2B, immunological functions and cholesterol/steroid metabolic processes are predicted to be upregulated in aged cortical microglia. Activated molecular functions of aged microglia include ATP-binding, 2′-5′-oligoadenylate synthetase activity, double-stranded RNA binding and nucleotidyltransferase activity (Figure 2C). Data from cDNA microarray analysis revealed the upregulation of genes involved in immune responses including *Ifi204*, *Lilrb4*, *Arhgap15* and *Cd244*. Taken together, the data indicate activated immune responses and increased lipid metabolism in aged cortical microglia.

A

B

Aged cortical microglia

GO term_BP

- cholesterol metabolic process
- steroid metabolic process
- purine nucleotide biosynthetic process
- innate immune response
- immune system response
- defense response to virus
- response to virus
- immune response

number of genes

C

Aged cortical microglia

GO term_MF

- nucleotidyltransferase activity
- double-stranded RNA binding
- 2′-5′-oligoadenylate synthetase activity
- ATP binding

number of genes

D

gene	1	2	p-value
AI607873			0.037
Ifi204			0.012
Lilrb4			0.033
Mir872			0.020
Gm5970			0.003
Arhgap15			0.005
Gm16118			0.018
Sp100; n-R5s215			0.013
Oas1a			0.032
Cd244			0.012
Ildr2			0.019
B430306N03Rik			0.032
Gm6654			0.038
Gm22079			0.038
Mir6897			0.004
Olfr111			0.040
Gm25336			0.007
Soat2			0.005
Gm24400			0.004
Prkaa2			0.024
H2-Ab1			0.028
Oasl2			0.006
Tnfsf13b			0.046
Oas1g			0.035
Cxcl13			0.007
Slc7a11			0.043
Snora7a			0.018
Cdk6			0.001
Snord71			0.022
Gm1966			0.012
Snord53			0.015
Gm26797			0.020
St14			0.016
Gm23319			0.001
Hsph1			0.027
Arrdc3			0.048
Gm16128			0.034
Mir3108			0.001

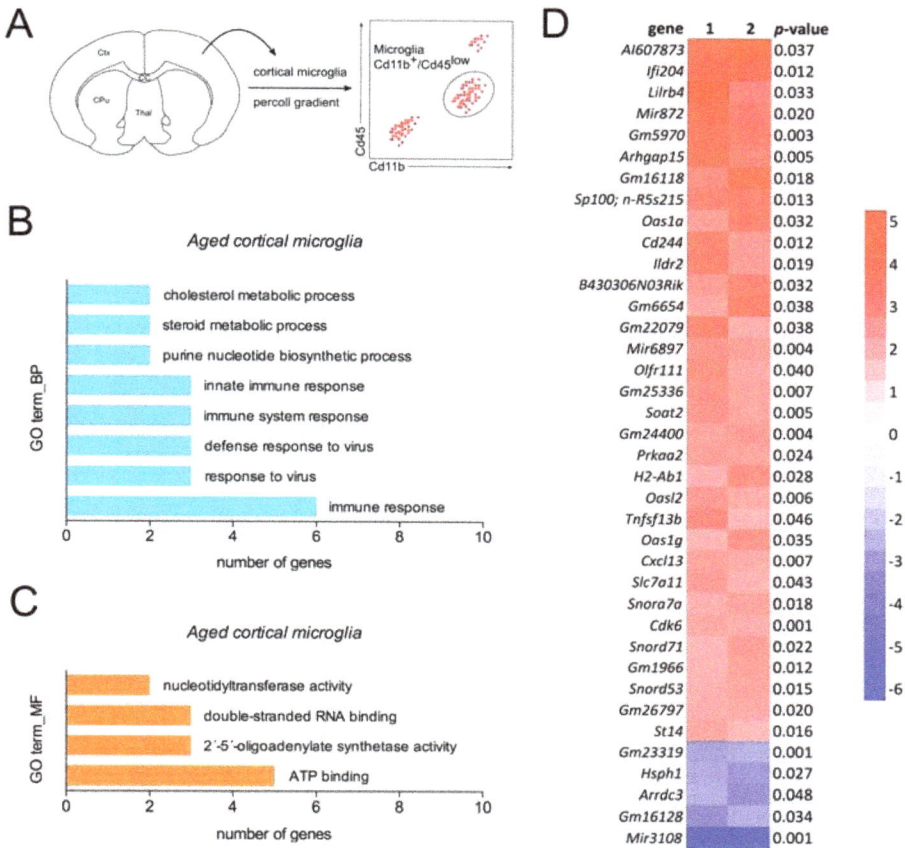

Figure 2. Gene expression pattern of aged cortical microglia. (**A**) Workflow scheme depicting microglia isolation and sorting strategy. (**B,C**) GO term enrichment analysis of biological processes (**B**) and molecular functions (**C**) as performed using DAVID Bioinformatics Resources 6.8. (**D**) Heatmap summarizing transcriptional changes in aged cortical microglia. Expression data from aged microglia are presented as $\log_2$-fold changes ($n = 2$) and compared to young (6-month-old) microglia. CC = corpus callosum, CPu = caudatoputamen, Ctx = cortex, Thal = thalamus.

2.3. Increased Expression of Cd206 and Cd36 in Aged Cortical Microglia

Using flow cytometry, the expression of microglia activation markers Cd206, Cd36 and Cd86 was determined. Cortical microglia were isolated using the percoll gradient method and subsequently stained against F4/80 as a microglia marker in combination with either Cd206, Cd36 or Cd86. Flow cytometry revealed that total numbers of F4/80$^+$ cortical microglia were significantly reduced in aged mice (Figure 3A,B), which confirms microglia quantifications depicted in Figure 1E. Moreover, we observed significant increases in Cd206$^+$ (Figure 3A,C), as well as Cd36$^+$ (Figure 3A,D) microglia. As given in Figure 3E, quantifications of F4/80$^+$/Cd86$^+$ microglia revealed that the percentages of Cd86^{high} microglia increase in the frontal cortices of aged mice; however, this increase did not reach statistical significance ($p = 0.069$). The results presented here demonstrate that aged cortical microglia display an activation pattern characterized by increased expression of M1-like as well as M2-like microglia activation markers. Based on the percentages of marker-positive microglia, it is highly likely that distinct cortical microglia subsets with specialized functions exist in aged mice. It further remains to be established to what extent these subsets contribute to anti-inflammatory and restorative functions in the aged cerebral cortex.

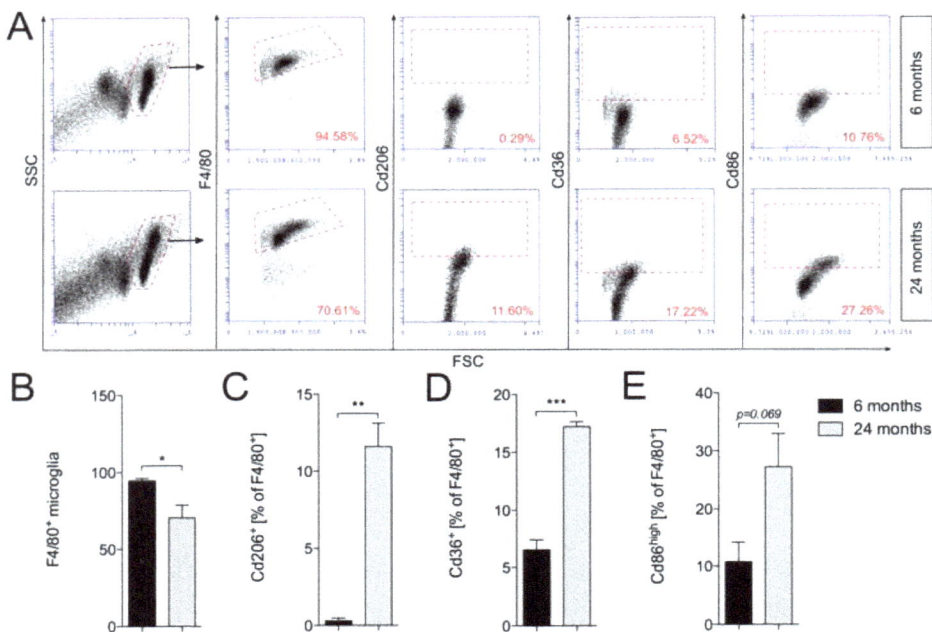

Figure 3. Expression of Cd206, Cd36 and Cd86 in F4/80$^+$ cortical microglia. (**A**) Gating strategy and representative dot plots of Cd206$^+$, Cd36$^+$ and Cd86^{high} cortical microglia from young (6 months) and aged (24 months) mice. Quantification of (**B**) F4/80$^+$, (**C**) Cd206$^+$, (**D**) Cd36$^+$ and (**E**) Cd86^{high} microglia. Data are given as percentages of F4/80$^+$ microglia $\pm$ SEM from three animals per age. *p*-values derived from student's *t*-test are * $p < 0.05$, ** $p < 0.01$ and *** $p < 0.001$.

3. Discussion

In the present study, we demonstrate that aging is associated with distinct changes in cortical microglia. We describe that numbers of cortical microglia significantly decreased in 24-month-old male mice and that the remaining microglia in aged mice showed slight morphological changes, as well as an uneven distribution pattern. This observation is in congruence with a recent report showing that microglia numbers decreased in the nigrostriatal system and cortices of 18-month- and 24-month-old mice, respectively [19]. However, a different mouse strain was used in the present study, thus resulting in the necessity of ruling out differences in age-dependent changes of microglia numbers between mouse strains. Morphological changes of aged microglia, including retracted processes and reduced process branching, have also been reported in aged human neocortex samples [20], indicating functional changes described during aging [21]. However, morphological changes of microglia do not necessarily reflect their functional states and, thus, analysis of microglial gene expression has been widely used to gain insights into their functional changes during aging and in several disease models [17,18]. In order to understand the impact of aging on cortical microglia, we have isolated aged (24-month-old) Cd11b$^+$/Cd45^{low} microglia and analyzed their gene expression profiles compared to young (6-month-old) microglia. Using cDNA microarray analysis, we demonstrated the upregulation of immune function related genes, including *Ifi204*, *Lilrb4*, *Arhgap15* and *Cd244*, and DAVID gene ontology (GO term) enrichment analysis further predicted the activation of innate immune responses in aged cortical microglia. However, detailed analysis of recent reports addressing the function of the above-mentioned genes suggests an immunoregulatory and/or anti-inflammatory phenotype of aged cortical microglia. However, the molecular functions of most of the observed upregulated genes are not understood in microglia and, thus, the prediction of potential microglia functions in the aged cerebral cortex can only be made based on recent reports addressing the role of these genes in other cell types. For instance, *Ifi204* expression has been described to increase in macrophages to inhibit proliferation and foster their differentiation [22]. Moreover, macrophage autophagy activation and IFN-β (interferon-β) release after bacterial infections has been linked to *Ifi204* expression [23]. Interestingly, the type I interferon IFN-β has been reported to limit CNS damage by abrogating chronic cytokine release. This functional feature of IFN-β might further explain its therapeutic potential in chronic autoimmune CNS pathologies such as multiple sclerosis [24]. The myeloid inhibitory receptor Lilrb4 (leukocyte immunoglobulin-like receptor b4), also referred to as Ilt3 (immunoglobulin-like transcript 3), Lir-5 (immunoglobulin-like receptor 5) or Cd85k, is a member of the leukocyte immunoglobulin-like receptor family (LILRs/LIRs), and represents an important mediator of immune tolerance [25]. Although the functions of LILRB4 in microglia are unknown and, thus, remain elusive, *Lilrb4*-deficiency has been demonstrated to result in increased Nf-κB (nuclear factor κB) signaling in atherosclerotic plaque-associated macrophages [26] and exaggerated LPS-induced cytokine/chemokine release from neutrophils [27]. In the CNS, increased expression of *Lilrb4* has been described in Cd11c⁺ microglia, which are believed to counteract amyloid deposition by increased amyloid β-uptake and degradation in a mouse model for AD [28]. Another interesting upregulated gene in aged cortical microglia is the Rho GTPase activating protein 15 (*Arhgap15*), which serves as a potent inhibitor of Rac1 (rac family small GTPase 1) [29]. The small GTPase Rac1 has been implicated in Nox2 (NADPH oxidase 2 or gp91^{phox})-mediated generation of reactive oxygen species (ROS) in microglia [30]. Notably, ROS generation via the Rac1-Nox2 axis has been demonstrated to be responsible for microglia-mediated neurotoxicity associated by Tnfα release and Nf-κB activation [31]. The SLAM (signaling lymphocytic activation molecule) receptor Cd244 was described to be essential in inhibiting the LPS-induced release of proinflammatory cytokines from splenic dendritic cells [32]. However, these functions in microglia have not been analyzed, so far. Taken together, the functional properties of the above-mentioned upregulated genes suggest a distinct phenotype of aged cortical microglia associated with regulation/inhibition of innate immune functions and anti-inflammatory and neuroprotective functions. This notion is further supported by increased surface expression of Cd206, also known as mannose receptor 1 (Mrc1) in aged cortical microglia. Cd206/Mrc1 expression was reported to be increased in alternatively activated microglia facilitating neuroprotective effects [33,34]. Furthermore, we provide evidence that the number of Cd36⁺ microglia was significantly increased in frontal cortices of aged mice. Recently, increased expression of the scavenger receptor Cd36 in microglia was linked to triggering receptor expressed on myeloid cells 2 (Trem2)-mediated alleviations of AD symptoms by enhanced uptake of Aβ and abrogation of memory loss [35]. Moreover, lack of Cd36 exacerbated injury in cerebral ischemia models associated with reduced engulfment of apoptotic neurons and enhanced Nf-κB signaling [36,37]. It remains elusive how microglia might promote neuroprotection in the aged brain, but expression and release of neuroprotective factors is likely to be one major mechanism. Release of insulin like growth factor 1 (Igf1) from microglia has recently been shown to be essential for survival of cortical layer V neurons during postnatal development [38] and could also be a candidate in the aged cerebral cortex. However, sophisticated proteome-based studies are necessary to understand which microglia-derived factors contribute to neuroprotective effects.

It has been described that microglia in different brain regions show distinct expression patterns of immunoregulatory markers, suggesting the existence of a huge immunological diversity between microglia from different functional CNS regions [39]. In the present study, we clearly demonstrate that cortical microglia in aged mice show a distinct activation profile, which is characterized by immunoregulatory and anti-inflammatory functions. These results suggest that aged microglia might support neuron survival rather than promoting detrimental effects on cortical neuron populations. However, upregulation of Cd86 further indicates M1-like microglia activation in aged mice suggesting a microglia activation phenotype fostering neurodegeneration. It remains to be established to what extent different functional microglia subsets are present in the aged murine cerebral cortex, and whether these cortical microglia phenotypes are region-specific or can be detected in other functional systems of the aged CNS. Finally, data from aged mice need to be compared with recently published reports addressing age-dependent changes in aged human microglia [40,41]. Together, the results presented here further broaden our understanding of age-dependent changes in cortical microglia functions and activation states, and indicate that aging alone—without any additional inflammatory trigger—does not necessarily result in a pro-inflammatory microglia activation.

4. Materials and Methods

4.1. Animals

All animal experiments in this study were approved by the Federal Ministry for Nature, Environment and Consumer's Protection of the federal state of Baden-Württemberg (X-15/01A (9 February 2015), X-15/06A (1 December 2015), G-15/111 (3 December 2015)) and were conducted in accordance with the respective national, federal and institutional regulations. 6-month- and 24-month-old male C57BL/6JRj mice used for cortical microglia isolations (flow cytometry and RNA isolation) and immunohistochemistry were obtained from Janvier (Le Genest Saint Isle, France).

4.2. Microglia Isolation

Mice were deeply anesthetized by intraperitoneal injections of Ketamin (75 mg/kg)/Rompun (5.8 mg/kg) and transcardially perfused with ice-cold phosphate buffer solution (PBS). Brains were dissected and meninges were removed on absorbent paper. Brains were collected in cold buffer (1× HBSS (Hanks' balanced salt solution) containing 1% BSA (bovine serum albumin) and 1 mM EDTA (ethylenediaminetetraacetic acid)), homogenized using a glass homogenizer and filtered through 75 µm cell strainers (Falcon). Samples were centrifuged 12 min at 300× g and 10 °C. For density gradient centrifugation, the pellet was resuspended in 5 mL 37% Percoll (P1644, Sigma Aldrich, St. Louis, MO, USA) in PBS, underlayed with 4 mL 70% Percoll and overlayed with 4 mL 30% Percoll in a 15 mL Falcon Tube (Corning, New York, NY, USA). Gradients were centrifuged for 40 min, 600× g, 10 °C, without breaks. Finally, the cell layer was collected from the 70% and 37% interface and transferred to PBS containing 1% FCS and centrifuged for 5 min, 200× g, 4 °C.

4.3. Flow Cytometry

Cells were stained with primary antibodies directed against Cd11b (1:20, 53-0112-82, eBiosciene, Thermo Fisher Scientific, Waltham, MA, USA), Cd206 (5 µL, FAB2535C, R&D Systems, Minneapolis, MN, USA), Cd36 (5 µL, MCA2748A647, AbD Serotech, BIO-Rad, Puchheim, Germany), Cd45 (1:20, 17-0451-82, eBiosciene), Cd86 (5 µL, MCA2463PE, AbD Serotech) and F4/80 (5 µL, MCA497A488, AbD Serotech) at 4 °C for 15 min. Fc (fragment crystallizable region) receptor blocking (TrueStain fcX, 101319, Biolegend, San Diego, CA, USA) was used to avoid unspecific antibody binding. Cells were washed and analyzed using the BD Accuri C6 flow cytometer (BD, Heidelberg, Germany).

4.4. Histology and Immunohistochemistry

Anesthetized mice were transcardially perfused using PBS followed by 4% paraformaldehyde (PFA). Brains were extracted and post-fixed in 4% PFA overnight. Free-floating 50 µm vibratome (Leica, Wetzlar, Germany) sections were incubated overnight with anti-IBA1 (1:1000, 019-19741, Wako, Japan). Alexa Flour 488-conjugated secondary antibodies (Cell Signaling Technology, Danvers, MA, USA) were used at 1:200 for 2 h at room temperature. Nuclei were counterstained using 4′6-diamidino-2-phenylindole (DAPI, Roche, Basel, Switzerland) and sections were mounted on

Int. J. Mol. Sci. **2018**, *19*, 706

glass cover slips. Imaging was performed using the Leica TCS SP8 confocal laser scanning microscope (Leica, Wetzlar, Germany) and LAS AF image analysis software (Leica, Wetzlar, Germany). For DAB (3,3′-diaminobenzidine) staining, peroxidase-conjugated secondary antibodies (Dianova, Hamburg, Germany) were used. Images were captured using A Zeiss AxioImager I (Zeiss, Göttingen, Germany) and the Stereoinvestigator Software 8.0 (MicroBrightField, Magdeburg, Germany).

4.5. Determination of Cortical Thickness and Cortical Microglia Numbers

Cortical thicknesses and microglia numbers were determined after immunohistochemical stainings (Iba1) of 50 μm coronal vibratome sections. After image acquisition using a Zeiss AxioImager I (Zeiss, Göttingen, Germany) and the Stereoinvestigator Software 8.0 (MicroBrightField, Magdeburg, Germany). Four serial sections were opened with ImageJ (National Institutes of Health, Bethesda, MD, USA), and the cortical thicknesses were analyzed after setting the scale for pixel/micron conversation using the ImageJ-integrated measurement function. Means were calculated from four serial sections, and cortical thicknesses are given in μm. Cortical microglia numbers were obtained using the automatic cell counting function of the ImageJ toolbox. Four serial sections were used and $Iba1^{+}$ microglia were determined in a 0.67 mm^2 rectangle. Microglia numbers were calculated for 1 mm^2 and means from four sections per animals were plotted.

4.6. cDNA Microarray

Total RNA was extracted from $Cd11b^{+}/Cd45^{low}$ microglia after sorting with BD Cell Sorter FACS Aria Fusion or BD Cell Sorter FACS Aria III using the Picopure RNA extraction kit (Life technologies, Carlsbad, CA, USA) according to the manufacturer's instructions. Cortices of three mice were pooled to obtain sufficient numbers of $Cd11b^{+}/Cd45^{low}$ microglia for RNA isolation. RNA quality was controlled using RNA pico chips on a Bioanalyzer 2100 (Agilent, Santa Clara, CA, USA). 2 ng total RNA was labeled using the GeneChip WT Pico Reagent kit (catalog number 902623, Thermo Fisher Scientific, Waltham, MA, USA) and hybridized to Affymetrix Mouse Transcriptome Array, MTA 1.0 (Affymetrix, Inc., Santa Clara, CA, USA). In this procedure, first strand cDNA was synthesized with a combination of a Poly-dT and random primers containing a 5′-adaptor sequence. A 3′-adaptor was added to the single-stranded cDNA followed by low-cycle PCR amplification. Next, the cDNA was used as a template for in vitro transcription (IVT), which produces amplified amounts of antisense mRNA (cRNA). The cRNA was then used as input for a second round of first-strand cDNA synthesis, producing single-stranded sense cDNA. Finally, the cDNA was fragmented using uracil-DNA glycosylase (UDG) and end-labeled with biotin and terminal deoxynucleotidyl transferase (TdT). The labeled targets were hybridized to GeneChip MTA 1.0 cartridge arrays, which were stained on a GeneChip Fluidics Station 450 and scanned on a GeneChip scanner 3000 7G (Thermo Fisher Scientific, Waltham, MA, USA).

4.7. Statistics

Data are given as means ± standard error of the mean (SEM). Two-group analysis (young vs. aged) was performed using Student's *t*-tests. All statistical analyses were performed using GraphPad Prism 6 (GraphPad Software Inc., La Jolla, CA, USA) and *p*-values < 0.05 were considered as being statistically significant. To determine differential expression of genes in young and aged cortical microglia, the two-sample Bayesian *t*-test [42] was used.

Acknowledgments: The authors thank Ludmila Butenko for her excellent technical support. The work was funded by grants from the Deutsche Forschungsgemeinschaft (DFG, SP 1555/2-1). Funds from the University of Rostock were received to cover the costs of publishing in open access.

Author Contributions: Björn Spittau and Tanja Zöller conceived and designed the experiments; Tanja Zöller, Abdelraheim Attaai and Phani Sankar Potru performed the experiments; Tanja Zöller and Björn Spittau analyzed the data; Björn Spittau, Phani Sankar Potru and Tamara Ruß wrote the paper.

Conflicts of Interest: The authors declare no conflict of interest.

References

1. McHugh, D.; Gil, J. Senescence and aging: Causes, consequences, and therapeutic avenues. *J. Cell Biol.* **2018**, *217*, 65–77. [CrossRef] [PubMed]
2. Molteni, M.; Rossetti, C. Neurodegenerative diseases: The immunological perspective. *J. Neuroimmunol.* **2017**, *313*, 109–115. [CrossRef] [PubMed]
3. Kierdorf, K.; Erny, D.; Goldmann, T.; Sander, V.; Schulz, C.; Perdiguero, E.G.; Wieghofer, P.; Heinrich, A.; Riemke, P.; Hölscher, C.; et al. Microglia emerge from erythromyeloid precursors via Pu.1- and Irf8-dependent pathways. *Nat. Neurosci.* **2013**, *16*, 273–280. [CrossRef] [PubMed]
4. Greter, M.; Lelios, I.; Pelczar, P.; Hoeffel, G.; Price, J.; Leboeuf, M.; Kündig, T.M.; Frei, K.; Ginhoux, F.; Merad, M.; et al. Stroma-derived interleukin-34 controls the development and maintenance of Langerhans cells and the maintenance of microglia. *Immunity* **2012**, *37*, 1050–1060. [CrossRef] [PubMed]
5. Ginhoux, F.; Greter, M.; Leboeuf, M.; Nandi, S.; See, P.; Gokhan, S.; Mehler, M.F.; Conway, S.J.; Ng, L.G.; Stanley, E.R.; et al. Fate mapping analysis reveals that adult microglia derive from primitive macrophages. *Science* **2010**, *330*, 841–845. [CrossRef] [PubMed]
6. Bennett, M.L.; Bennett, F.C.; Liddelow, S.A.; Ajami, B.; Zamanian, J.L.; Fernhoff, N.B.; Mulinyawe, S.B.; Bohlen, C.J.; Adil, A.; Tucker, A.; et al. New tools for studying microglia in the mouse and human CNS. *Proc. Natl. Acad. Sci. USA* **2016**, *113*, E1738–E1746. [CrossRef] [PubMed]
7. Butovsky, O.; Jedrychowski, M.P.; Moore, C.S.; Cialic, R.; Lanser, A.J.; Gabriely, G.; Koeglsperger, T.; Dake, B.; Wu, P.M.; Doykan, C.E.; et al. Identification of a unique TGF-β-dependent molecular and functional signature in microglia. *Nat. Neurosci.* **2014**, *17*, 131–143. [CrossRef] [PubMed]
8. Askew, K.; Li, K.; Olmos-Alonso, A.; Garcia-Moreno, F.; Liang, Y.; Richardson, P.; Tipton, T.; Chapman, M.A.; Riecken, K.; Beccari, S.; et al. Coupled Proliferation and Apoptosis Maintain the Rapid Turnover of Microglia in the Adult Brain. *Cell Rep.* **2017**, *18*, 391–405. [CrossRef] [PubMed]
9. Ajami, B.; Bennett, J.L.; Krieger, C.; Tetzlaff, W.; Rossi, F.M.V. Local self-renewal can sustain CNS microglia maintenance and function throughout adult life. *Nat. Neurosci.* **2007**, *10*, 1538–1543. [CrossRef] [PubMed]
10. Goldmann, T.; Wieghofer, P.; Jordão, M.J.C.; Prutek, F.; Hagemeyer, N.; Frenzel, K.; Amann, L.; Staszewski, O.; Kierdorf, K.; Krueger, M.; et al. Origin, fate and dynamics of macrophages at central nervous system interfaces. *Nat. Immunol.* **2016**, *17*, 797–805. [CrossRef] [PubMed]
11. Spittau, B. Aging Microglia-Phenotypes, Functions and Implications for Age-Related Neurodegenerative Diseases. *Front. Aging Neurosci.* **2017**, *9*, 194. [CrossRef] [PubMed]
12. Raj, D.D.A.; Jaarsma, D.; Holtman, I.R.; Olah, M.; Ferreira, F.M.; Schaafsma, W.; Brouwer, N.; Meijer, M.M.; de Waard, M.C.; van der Pluijm, I.; et al. Priming of microglia in a DNA-repair deficient model of accelerated aging. *Neurobiol. Aging* **2014**. [CrossRef] [PubMed]
13. Raj, D.D.A.; Moser, J.; van der Pol, S.M.A.; van Os, R.P.; Holtman, I.R.; Brouwer, N.; Oeseburg, H.; Schaafsma, W.; Wesseling, E.M.; den Dunnen, W.; et al. Enhanced microglial pro-inflammatory response to lipopolysaccharide correlates with brain infiltration and blood-brain barrier dysregulation in a mouse model of telomere shortening. *Aging Cell* **2015**, *14*, 1003–1013. [CrossRef] [PubMed]
14. Sierra, A.; Gottfried-Blackmore, A.C.; McEwen, B.S.; Bulloch, K. Microglia derived from aging mice exhibit an altered inflammatory profile. *Glia* **2007**, *55*, 412–424. [CrossRef] [PubMed]
15. Facci, L.; Barbierato, M.; Marinelli, C.; Argentini, C.; Skaper, S.D.; Giusti, P. Toll-like receptors 2, -3 and -4 prime microglia but not astrocytes across central nervous system regions for ATP-dependent interleukin-1β release. *Sci. Rep.* **2014**, *4*, 6824. [CrossRef] [PubMed]
16. Fonken, L.K.; Frank, M.G.; Kitt, M.M.; D'Angelo, H.M.; Norden, D.M.; Weber, M.D.; Barrientos, R.M.; Godbout, J.P.; Watkins, L.R.; Maier, S.F. The Alarmin HMGB1 Mediates Age-Induced Neuroinflammatory Priming. *J. Neurosci.* **2016**, *36*, 7946–7956. [CrossRef] [PubMed]
17. Holtman, I.R.; Raj, D.D.; Miller, J.A.; Schaafsma, W.; Yin, Z.; Brouwer, N.; Wes, P.D.; Möller, T.; Orre, M.; Kamphuis, W.; et al. Induction of a common microglia gene expression signature by aging and neurodegenerative conditions: A co-expression meta-analysis. *Acta Neuropathol. Commun.* **2015**, *3*, 31. [CrossRef] [PubMed]
18. Hickman, S.E.; Kingery, N.D.; Ohsumi, T.K.; Borowsky, M.L.; Wang, L.-C.; Means, T.K.; El Khoury, J. The microglial sensome revealed by direct RNA sequencing. *Nat. Neurosci.* **2013**, *16*, 1896–1905. [CrossRef] [PubMed]

19. Sharaf, A.; Krieglstein, K.; Spittau, B. Distribution of microglia in the postnatal murine nigrostriatal system. *Cell Tissue Res.* **2013**, *351*, 373–382. [CrossRef] [PubMed]

20. Davies, D.S.; Ma, J.; Jegathees, T.; Goldsbury, C. Microglia show altered morphology and reduced arborization in human brain during aging and Alzheimer's disease. *Brain Pathol.* **2017**, *27*, 795–808. [CrossRef] [PubMed]

21. Koellhoffer, E.C.; McCullough, L.D.; Ritzel, R.M. Old Maids: Aging and Its Impact on Microglia Function. *Int. J. Mol. Sci.* **2017**, *18*. [CrossRef] [PubMed]

22. Dauffy, J.; Mouchiroud, G.; Bourette, R.P. The interferon-inducible gene, *Ifi204*, is transcriptionally activated in response to M-CSF, and its expression favors macrophage differentiation in myeloid progenitor cells. *J. Leukoc. Biol.* **2006**, *79*, 173–183. [CrossRef] [PubMed]

23. Chunfa, L.; Xin, S.; Qiang, L.; Sreevatsan, S.; Yang, L.; Zhao, D.; Zhou, X. The Central Role of IFI204 in IFN-β Release and Autophagy Activation during *Mycobacterium bovis* Infection. *Front. Cell. Infect. Microbiol.* **2017**, *7*, 169. [CrossRef] [PubMed]

24. Blank, T.; Prinz, M. Type I interferon pathway in CNS homeostasis and neurological disorders. *Glia* **2017**, *65*, 1397–1406. [CrossRef] [PubMed]

25. Cheng, H.; Mohammed, F.; Nam, G.; Chen, Y.; Qi, J.; Garner, L.I.; Allen, R.L.; Yan, J.; Willcox, B.E.; Gao, G.F. Crystal structure of leukocyte Ig-like receptor LILRB4 (ILT3/LIR-5/CD85k): A myeloid inhibitory receptor involved in immune tolerance. *J. Biol. Chem.* **2011**, *286*, 18013–18025. [CrossRef] [PubMed]

26. Jiang, Z.; Qin, J.-J.; Zhang, Y.; Cheng, W.-L.; Ji, Y.-X.; Gong, F.-H.; Zhu, X.-Y.; Zhang, Y.; She, Z.-G.; Huang, Z.; et al. LILRB4 deficiency aggravates the development of atherosclerosis and plaque instability by increasing the macrophage inflammatory response via NF-κB signaling. *Clin. Sci.* **2017**. [CrossRef] [PubMed]

27. Zhou, J.S.; Friend, D.S.; Lee, D.M.; Li, L.; Austen, K.F.; Katz, H.R. gp49B1 deficiency is associated with increases in cytokine and chemokine production and severity of proliferative synovitis induced by anti-type II collagen mAb. *Eur. J. Immunol.* **2005**, *35*, 1530–1538. [CrossRef] [PubMed]

28. Kamphuis, W.; Kooijman, L.; Schetters, S.; Orre, M.; Hol, E.M. Transcriptional profiling of CD11c-positive microglia accumulating around amyloid plaques in a mouse model for Alzheimer's disease. *Biochim. Biophys. Acta* **2016**, *1862*, 1847–1860. [CrossRef] [PubMed]

29. Zamboni, V.; Armentano, M.; Sarò, G.; Ciraolo, E.; Ghigo, A.; Germena, G.; Umbach, A.; Valnegri, P.; Passafaro, M.; Carabelli, V.; et al. Disruption of ArhGAP15 results in hyperactive Rac1, affects the architecture and function of hippocampal inhibitory neurons and causes cognitive deficits. *Sci. Rep.* **2016**, *6*, 34877. [CrossRef] [PubMed]

30. Bedard, K.; Krause, K.-H. The NOX family of ROS-generating NADPH oxidases: Physiology and pathophysiology. *Physiol. Rev.* **2007**, *87*, 245–313. [CrossRef] [PubMed]

31. Li, Q.; Spencer, N.Y.; Pantazis, N.J.; Engelhardt, J.F. Alsin and SOD1(G93A) proteins regulate endosomal reactive oxygen species production by glial cells and proinflammatory pathways responsible for neurotoxicity. *J. Biol. Chem.* **2011**, *286*, 40151–40162. [CrossRef] [PubMed]

32. Georgoudaki, A.-M.; Khodabandeh, S.; Puiac, S.; Persson, C.M.; Larsson, M.K.; Lind, M.; Hammarfjord, O.; Nabatti, T.H.; Wallin, R.P.A.; Yrlid, U.; et al. CD244 is expressed on dendritic cells and regulates their functions. *Immunol. Cell Biol.* **2015**, *93*, 581–590. [CrossRef] [PubMed]

33. Porrini, V.; Lanzillotta, A.; Branca, C.; Benarese, M.; Parrella, E.; Lorenzini, L.; Calzà, L.; Flaibani, R.; Spano, P.F.; Imbimbo, B.P.; et al. CHF5074 (CSP-1103) induces microglia alternative activation in plaque-free Tg2576 mice and primary glial cultures exposed to beta-amyloid. *Neuroscience* **2015**, *302*, 112–120. [CrossRef] [PubMed]

34. Xiao, Q.; Yu, W.; Tian, Q.; Fu, X.; Wang, X.; Gu, M.; Lü, Y. Chitinase1 contributed to a potential protection via microglia polarization and Aβ oligomer reduction in D-galactose and aluminum-induced rat model with cognitive impairments. *Neuroscience* **2017**, *355*, 61–70. [CrossRef] [PubMed]

35. Kim, S.-M.; Mun, B.-R.; Lee, S.-J.; Joh, Y.; Lee, H.-Y.; Ji, K.-Y.; Choi, H.-R.; Lee, E.-H.; Kim, E.-M.; Jang, J.-H.; et al. TREM2 promotes Aβ phagocytosis by upregulating C/EBPα-dependent CD36 expression in microglia. *Sci. Rep.* **2017**, *7*, 11118. [CrossRef] [PubMed]

36. Li, F.; Faustino, J.; Woo, M.-S.; Derugin, N.; Vexler, Z.S. Lack of the scavenger receptor CD36 alters microglial phenotypes after neonatal stroke. *J. Neurochem.* **2015**, *135*, 445–452. [CrossRef] [PubMed]

37. Woo, M.-S.; Wang, X.; Faustino, J.V.; Derugin, N.; Wendland, M.F.; Zhou, P.; Iadecola, C.; Vexler, Z.S. Genetic deletion of CD36 enhances injury after acute neonatal stroke. *Ann. Neurol.* **2012**, *72*, 961–970. [CrossRef] [PubMed]

38. Ueno, M.; Fujita, Y.; Tanaka, T.; Nakamura, Y.; Kikuta, J.; Ishii, M.; Yamashita, T. Layer V cortical neurons require microglial support for survival during postnatal development. *Nat. Neurosci.* **2013**, *16*, 543–551. [CrossRef] [PubMed]

39. De Haas, A.H.; Boddeke, H.W.G.M.; Biber, K. Region-specific expression of immunoregulatory proteins on microglia in the healthy CNS. *Glia* **2008**, *56*, 888–894. [CrossRef] [PubMed]

40. Olah, M.; Patrick, E.; Villani, A.-C.; Xu, J.; White, C.C.; Ryan, K.J.; Piehowski, P.; Kapasi, A.; Nejad, P.; Cimpean, M.; et al. A transcriptomic atlas of aged human microglia. *Nat. Commun.* **2018**, *9*, 539. [CrossRef] [PubMed]

41. Mrdjen, D.; Pavlovic, A.; Hartmann, F.J.; Schreiner, B.; Utz, S.G.; Leung, B.P.; Lelios, I.; Heppner, F.L.; Kipnis, J.; Merkler, D.; et al. High-Dimensional Single-Cell Mapping of Central Nervous System Immune Cells Reveals Distinct Myeloid Subsets in Health, Aging, and Disease. *Immunity* **2018**. [CrossRef] [PubMed]

42. Fox, R.J.; Dimmic, M.W. A two-sample Bayesian *t*-test for microarray data. *BMC Bioinform.* **2006**, *7*, 126. [CrossRef] [PubMed]

International Journal of
Molecular Sciences

MDPI

Article

Characterization of Ageing- and Diet-Related Swine Models of Sarcopenia and Sarcopenic Obesity

Consolacion Garcia-Contreras [1,†], Marta Vazquez-Gomez [2,†] , Laura Torres-Rovira [1], Jorge Gonzalez [3], Esteban Porrini [4], Magali Gonzalez-Colaço [5], Beatriz Isabel [2], Susana Astiz [1] and Antonio Gonzalez-Bulnes [1,2,*]

[1]   Comparative Physiology Group, SGIT-INIA, 28040 Madrid, Spain; garcia.consolacion@inia.es (C.G.-C.); torres.laura@inia.es (L.T.-R.); astiz.susana@inia.es (S.A.)
[2]   Faculty of Veterinary Sciences, Universidad Complutense de Madrid, 28040 Madrid, Spain; mvgomez@ucm.es (M.V.-G.); bisabelr@ucm.es (B.I.)
[3]   Micros Veterinaria, Campus de Vegazana, 24007 Leon, Spain; info@microsvet.com
[4]   Institute of Biomedical Technology (ITB), Universidad de La Laguna, 38200 Tenerife, Spain; esteban.l.porrini@gmail.com
[5]   Central Unit of Clinical Research and Clinical Assays (UCICEC), Universitary Hospital of Canary Island, 28010 Tenerife, Spain; magaligch@hotmail.com
*   Correspondence: bulnes@inia.es; Tel.: +34-91-347-4022; Fax: +34-91-347-4014
†   These authors contributed equally to this work.

Received: 23 January 2018; Accepted: 7 March 2018; Published: 12 March 2018

Abstract: Sarcopenia and sarcopenic obesity are currently considered major global threats for health and well-being. However, there is a lack of adequate preclinical models for their study. The present trial evaluated the suitability of aged swine by determining changes in adiposity, fatty acids composition, antioxidant status and lipid peroxidation, development of metabolic disturbances and structural changes in tissues and organs. Iberian sows with clinical evidence of aging-related sarcopenia were fed a standard diet fulfilling their maintenance requirements or an obesogenic diet for 100 days. Aging and sarcopenia were related to increased lipid accumulation and cellular dysfunction at both adipose tissue and non-adipose ectopic tissues (liver and pancreas). Obesity concomitant to sarcopenia aggravates the condition by increasing visceral adiposity and causing dyslipidemia, insulin resistance and lipotoxicity in non-adipose tissues. These results support that the Iberian swine model represents certain features of sarcopenia and sarcopenic obesity in humans, paving the way for future research on physiopathology of these conditions and possible therapeutic targets.

Keywords: animal-model; insulin-resistance; lipotoxicity; obesity; sarcopenia; swine

1. Introduction

Sarcopenia is a syndrome characterized by loss of skeletal muscle mass and function [1], causing negative health outcomes like functional decline and disability [2–4] and increased mortality [5]. This may be further aggravated in case of obesity and increases in fat mass (sarcopenic obesity; [6]) which increases mobility and disability problems [7]. Currently, around 20 and 30% of elderly women and men, respectively, may have sarcopenic obesity [8]. Sarcopenic obesity is also related to metabolic disturbances, like metabolic syndrome and type-2 diabetes [9]; it is, therefore, considered a major global threat to health and well-being [10].

There are currently no drugs for the treatment of sarcopenia [7,11] and the recommended therapeutic interventions are based on lifestyle changes [9], with limited effects and with difficult implementation in elderly individuals with chronic diseases or different disabilities. Hence, there is a strong need for research efforts on pathogenesis and possible treatments. The results will contribute to relevant improvements of individual life-quality and savings in health-care resources.

Almost all the available studies are observational (with their inherent constraints), and interventional research is, therefore, necessary. Such research should be based on adequate translational animal models as elegantly reviewed by Palus in 2017 [12]. Most of the interventional trials on sarcopenia have been performed in rodents and have been based on induced acute muscle atrophy. These models do not represent sarcopenia by ageing [9], which needs long time to be achieved (around 20–24 months [13]). Moreover, the study of metabolic implications of sarcopenic obesity in rodents is constrained by their marked differences in metabolic and endocrine pathways with humans [14]. There is, consequently, a need for complementary preclinical animal models. The use of non-rodent mammals is essential to develop more relevant and predictive models of human disease. In this way, large animals have close similarities to humans in size and in many anatomical, physiological and pathological features. Moreover, large animal models of human diseases can be studied employing the same clinical approach used for human patients. In this sense, the pig has more similar patho-physiological features to humans than rodents [15].

Pigs used for biomedical research can be either random- or purpose-bred. Pigs from random-bred are typically farm animals (e.g., Landrace, Large White, Pietrain), where genetic selection is focused on animal production parameters. Purpose-bred pigs for biomedical research originate from closed colonies selected for a specific phenotype. The most commonly used swine breeds for research on obesity and metabolic diseases are Göttingen, Yucatan and Ossabaw Island pigs. However, all these breeds have a common ancestor: the Iberian pig.

The Iberian pig has an ancient origin, being traced back approximately to year 1000 Before Christ, and has been traditionally reared under extensive free-range conditions, using natural resources like pastures and acorns. Maintenance in semi-feral conditions has induced the pigs to develop an adaptive mechanism to the environment, which is known as *thrifty genotype* and which was firstly described in humans [16]. The *thrifty genotype* facilitates accommodation to seasonal cycles of feasting and famine because the ability to store fat in excess during food abundance enables survival during periods of scarcity. However, these individuals become obese in case of food-excess. Iberian pigs arise therefore as good model for studies on obesity and associated morbidities, since they develop cardiovascular and metabolic disorders in case of obesity, either during juvenile development [17] or adulthood [18].

The Iberian pig may be also useful for studies focused on sarcopenia, since the breed is reared under extensive traditional systems which facilitates the finding of older animals than in the case of breeds used for meat production under intensive management conditions (e.g., Landrace, Large-White, Pietrain). Intensive animal production systems are associated to intense genetic and productive selection and sows identified as having lower performance than the herd average are early culled, as early as age as 3–4 year-old [19]. Conversely, genetic and productive demands are not so strong in extensive traditional systems and culling is more related to either reproductive failures or musculoskeletal and locomotive disorders associated to age (which are similar to those associated with human sarcopenia).

Hence, the present study aimed to characterize a large-animal model for preclinical studies on sarcopenia and sarcopenic obesity, based on the use of old Iberian sows with aging-related sarcopenia and exposed or not to obesogenic diets. The validity of the models was assessed by determining possible links among adiposity, fatty acids composition, antioxidant status and lipid peroxidation, development of metabolic disturbances and structural changes in tissues and organs.

2. Results

2.1. Body-Weight and Adiposity

All the animals fed with the obesogenic diet showed a significant increase in body-weight (123 ± 6.1 to 202.3 ± 4.9 kg, $p < 0.0001$) and adiposity (16.1 ± 2.4 to 45.6 ± 2.9 mm of back-fat depth,

$p < 0.0001$) from Day 0 to 100 (Figure 1). The values in control sows were similar to the initial values in obese sows (127.9 ± 4.0 kg and 19.3 ± 1.3 mm, respectively). The macroscopic findings during necropsy indicated a severe intraabdominal fattening in obese sows, with large amounts of mesenteric fat covering all viscerae and mesenteric tissues. However, intramuscular fat content was similar in obese and non-obese sows (12.9 ± 1.1% vs. 12.8 ± 0.7%).

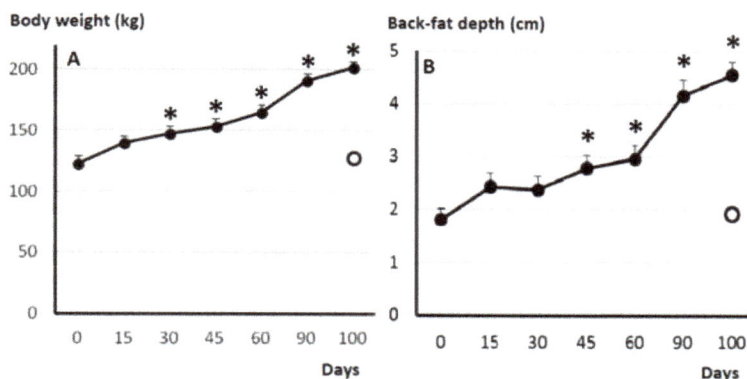

Figure 1. Mean (±S.E.M.) values in body-weight (**A**) and back-fat depth (**B**) in control sows (white dot; $n = 17$) and changes in such values over time after starting differential feeding with a diet enriched with saturated fat in treated sows (black line and dots; $n = 11$). Asterisks (*) denote significant differences between obese and non-obese sows ($p < 0.05$).

2.2. Fatty Acid Composition of Subcutaneous and Visceral Fat and Muscle

The composition of fatty acids (FA) was similar at the outer and inner layers of subcutaneous fat (Supplementary Tables S1 and S2). Overall, when compared to non-obese sows, obese females showed a higher monounsaturated FA (MUFA) content ($p < 0.05$) and lower contents of polyunsaturated FA (PUFA), n-3 and n-6 PUFA ($p < 0.005$ for all). Visceral fat in obese sows also evidenced a higher MUFA content ($p < 0.0001$; Supplementary Table S3) and a lower content of PUFA, n-3 and n-6 PUFA ($p < 0.0001$ for all) than non-obese females. There were no differences at subcutaneous fat, but visceral fat of obese females showed higher ratios of MUFA/saturated FA (SFA) and C18:1/C18:0 ($p < 0.05$ for both).

The analysis of the fatty acid composition in the *longissimus dorsi* showed major effects at the polar lipid fraction (Supplementary Table S4). Obese sows showed a higher MUFA content and a lower content of SFA and PUFA than the controls and, as a consequence, higher MUFA/SFA and C18:1/C18:0 ratios ($p < 0.005$ for all). The content of n-3 and n-6 PUFA were lower, with a higher Σn-6/Σn-3 ratio, in obese than in control sows ($p < 0.005$ for all). The effects from obesity were minor at the neutral fraction (Supplementary Table S5), with MUFA content being lower in obese than in control females ($p < 0.01$).

2.3. Hepatic Architecture, Fatty Acid Composition and Function

The total fat content in the liver was similar in control and obese groups (15.3 ± 0.2% vs. 14.5 ± 0.4%). Histological evaluation of the liver (Figure 2) indicated lipid accumulation in all the obese and in 82.4% of the control sows, with degrees varying from mild (63.6% and 76.5%, respectively) to moderate (36.4% and 23.5%, respectively). Increased presence of lipocytes (also known as perisinusoidal fat-storing cells, stellate cells or Ito cells) was found in all the obese and in 88.2% of the control sows. There were no evidences of inflammation or fibrosis, but hydropic degeneration was observed in 23.5% of the controls and all the obese animals ($p < 0.05$). All the controls had mild degeneration, whilst the obese sows showed mild (27.3%), moderate (63.7%, $p < 0.05$) or even severe degeneration (9.0%, $p < 0.05$).

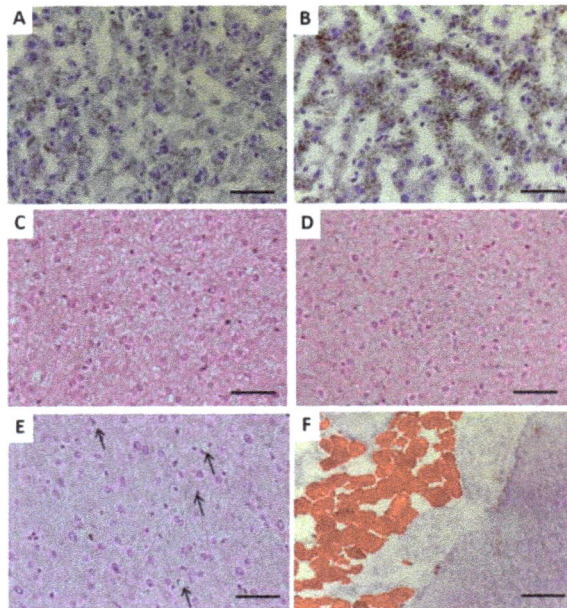

Figure 2. Histological images of liver and pancreas in obese sows. Upper pictures represent mild (**A**) and moderate (**B**) infiltration of lipids, stained in red, in the liver (oil red-O, 400×) Bar 50 μm. Pictures (**C**) and (**D**) of the liver represent mild and severe hydropic degeneration, respectively, whilst picture **E** exemplifies increased presence of lipocytes or Ito cells indicated with arrows (hematoxylin-eosin HE, 400×) Bar 50 μm. Picture **F** represents infiltration of adipocytes with intracytoplasmic accumulation of lipids, stained in red, in the pancreas (oil red-O, 100×) Bar 200 μm.

The major effects of obesity on fatty acids composition of the liver, like in the muscle, were found at the polar fraction (Supplementary Table S6), with higher MUFA/SFA, C18:1/C18:0 and Σn-6/Σn-3 ratios in the obese than in the control females ($p < 0.0001$ for all). The neutral fraction (Supplementary Table S7) was characterized by a higher SFA content ($p < 0.01$) and a higher Σn-6/Σn-3 ratio ($p < 0.0001$) in the obese animals.

The obese sows showed no major changes in the plasma biomarkers of liver function throughout the study, excepting a significant increase in the concentrations of alkaline-phosphatase (56.5 ± 8.7 to 76.8 ± 10.1 IU/L, $p < 0.05$) and leucine-aminopeptidase (16.5 ± 1.9 to 21.4 ± 0.7 IU/L, $p < 0.01$). Conversely, albumin concentration decreased over time (4.6 ± 0.1 to 4.1 ± 0.1 g/dL; $p < 0.05$), from initial values similar to controls (4.7 ± 0.1 g/dL).

2.4. Plasma Lipid Profile

Plasma concentrations of triglycerides and cholesterol increased with time of treatment in the obese group ($p < 0.05$; Figure 3), starting from values similar to those of control sows at Day 0. The increase in total cholesterol was accompanied by a similar increase in low-density lipoproteins cholesterol (LDL-c) levels ($p < 0.01$) but not in high-density lipoproteins cholesterol (HDL-c). Hence, total cholesterol/HDL-c, LDL-c/HDL-c and atherogenic dyslipidemia ratios increased with time ($p < 0.05$).

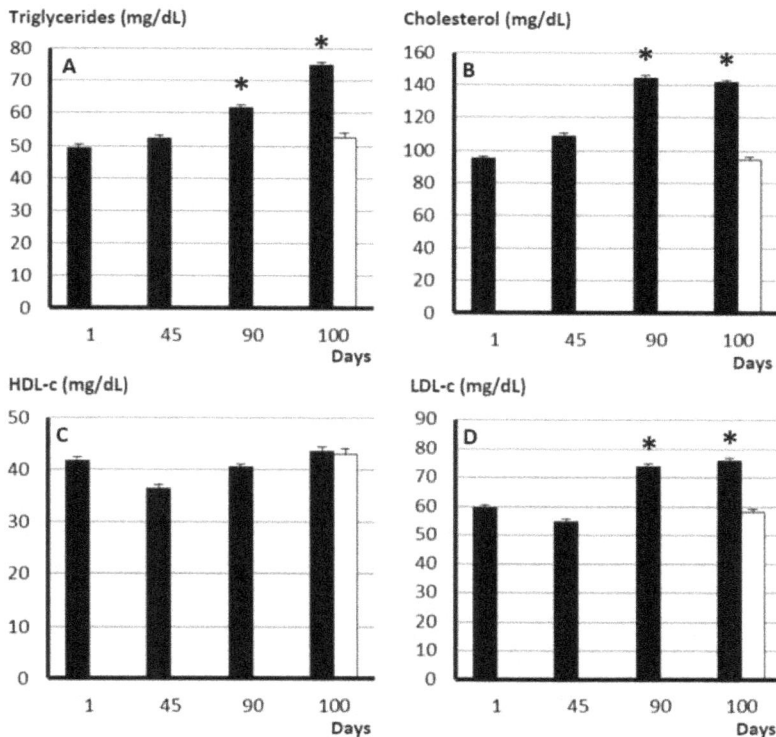

Figure 3. Mean (±S.E.M.) plasma concentrations of triglycerides (**A**), cholesterol (**B**), HDL-c (**C**) and LDL-c (**D**) in control sows (white bar; $n = 17$) and changes in such values over time after starting differential feeding with a diet enriched with saturated fat in treated sows (black bars; $n = 11$). Asterisks (*) denote significant differences between obese and non-obese sows ($p < 0.05$).

2.5. Plasma Antioxidant Capacity and Lipid Oxidation

The plasma antioxidant capacity (measured by FRAP; ferric reducing antioxidant power assay) decreased in the obesogenic-diet group (from 23.6 ± 2.7 at Day 0 to 16.9 ± 2.6 μmol/mL at Day 100, $p < 0.05$), starting from values similar to those of control animals (24.6 ± 4.0 μmol/mL). Concomitantly, the values for total lipid oxidation (assessed as MDA; malondialdehyde) increased with time of treatment in the obese group (from 10.7 ± 0.6 at Day 0 to 12.3 ± 0.3 μmol/mL at Day 100, $p < 0.05$), starting from values similar to those of controls (10.6 ± 0.4 μmol/mL).

2.6. Pancreatic Architecture and Exocrine Function

Microscopic evaluation of the pancreas showed increased lipid accumulation in all the obese sows. Around half of them showed severe infiltration (>66% of the cells; 45.5%), whilst the remaining obese females showed moderate (34–66%; 27.3%) or mild infiltration (5–33%; 27.3%). Conversely, lipid accumulation was found in 12 of the 17 controls (70.6%, $p < 0.05$), with degrees varying from mild (47.1%) to moderate (23.5%), but not severe ($p < 0.01$ with obese sows). On the other hand, no inflammation, fibrosis or cellular injury were observed in any of the animals. Concomitantly, the assessment of plasma biomarkers of exocrine pancreatic function (lipase and amylase) showed no major changes during the period of study.

2.7. Pancreatic Endocrine Function and Insulin Resistance

Mean plasma concentrations of glucose and insulin, and HOMA-IR (Homeostatic Model Assessment for Insulin Resistance) and HOMA-β (Homeostatic Model Assessment for β-cell function) values were similar in controls and obese sows at Day 0. Afterwards, plasma glucose concentrations in obese females were maintained by changes in insulin secretion throughout the period of study (Figure 4). Insulin and HOMA-β index increased at Day 45 ($p < 0.05$ and $p < 0.01$, respectively) and decreased again at Day 90 ($p < 0.05$ and $p < 0.01$, respectively). On the other hand, HOMA-IR increased throughout the study ($p < 0.05$), evidencing insulin resistance (IR). Changes in insulin secretion and glucose elimination were confirmed when analyzing the OGTTs (Oral Glucose Tolerance Tests) performed during the study (Figure 5). In this way, the areas under the curve (AUCs) for glucose and insulin were similar in control and treated sows at Day 0. Afterwards, at 45 and 90 days, AUCs for glucose remained almost stable whilst AUCs for insulin increased 55.7% at Day 45 and 28.3% at Day 90 when compared to Day 0 ($p < 0.01$ and $p < 0.05$, respectively).

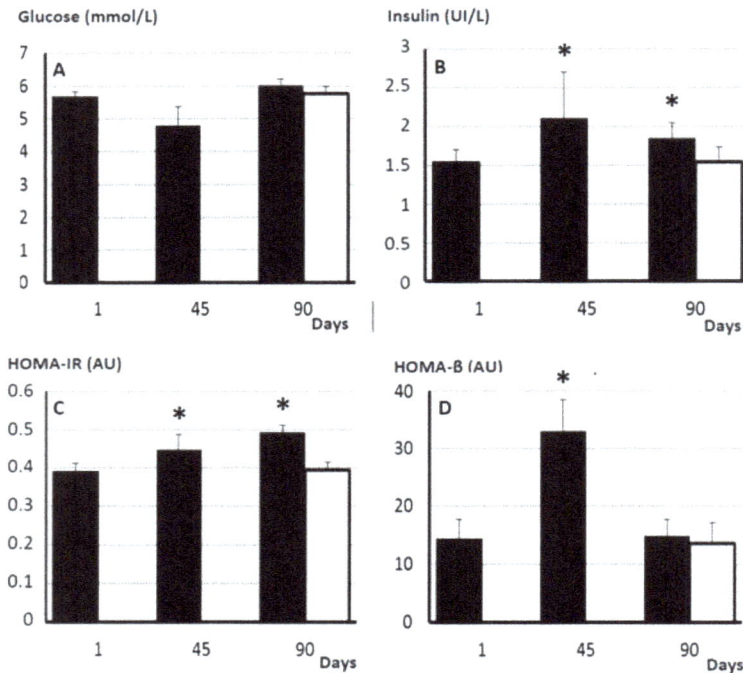

Figure 4. Mean (±S.E.M.) plasma concentrations of glucose (**A**) and insulin (**B**) and values for HOMA-IR and HOMA-β indexes (**C** and **D**, respectively) in control sows (white bar; $n = 17$) and changes in such values over time after starting differential feeding with a diet enriched with saturated fat in treated sows (black bars; $n = 11$). Asterisks (*) denote significant differences between obese and non-obese sows ($p < 0.05$).

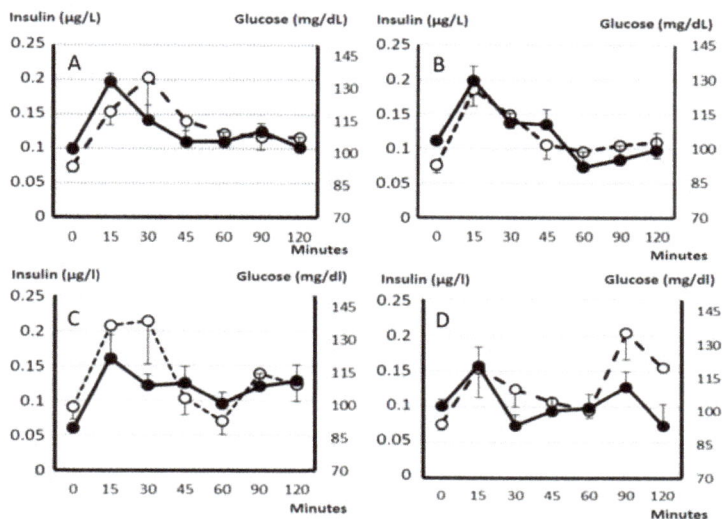

Figure 5. Changes in plasma concentration of glucose (continuous line with black circles) and insulin (discontinuous line with white circles) over time after oral administration of 2 g/kg live weight of D-glucose in control sows (**A**; *n* = 17) and at 0 (**B**), 45 (**C**) and 90 days (**D**) after starting differential feeding with a diet enriched with saturated fat in treated sows (*n* = 11). Areas under the Curve (AUCs) for glucose and insulin were similar in control and treated sows at Day 0. AUCs for glucose remained almost stable at 45 and 90 days but AUCs for insulin increased and were higher at both Day 45 and 90 than at Day 0 ($p < 0.01$ and $p < 0.05$, respectively).

3. Discussion

The present study characterizes aged Iberian pigs as a robust, amenable, and reliable translational model for studies on sarcopenia and sarcopenic obesity. The model represents certain characteristics of the human disease, since aging and sarcopenia were related to increased lipids accumulation and cellular dysfunction, not only at the adipose tissue but also at non-adipose ectopic tissues (liver and pancreas). The food-intake excess leading to sarcopenic obesity worsened the condition by increasing visceral adiposity, dyslipidemia, insulin resistance and lipotoxicity in non-adipose tissues.

3.1. Similarities of the Model with Human Sarcopenia

In the pigs of the current study, aging was characterized, like in humans, by a significant reduction in muscle mass; even 50% when compared to values obtained in younger adults. This sarcopenic state was related to systemic dyslipidemia, insulin resistance (IR) and lipotoxicity in adipose tissue, liver and pancreas (i.e., increased lipids accumulation and cellular dysfunction).

The comparison of lipid profiles in the current trial with data obtained in younger adult pigs in previous studies [18] shows that plasma concentrations of triglycerides and total- and LDL-cholesterol are significantly higher in aged than in young adult sows (around 50 vs. 25 mg/dL for triglycerides, around 95 vs. 55 mg/dL for total cholesterol and around 55 vs. 25 mg/dL for LDL-c). Conversely, increases in HDL-c were smaller (around 35 vs. 25 mg/dL). These data indicate that in our model, ageing and sarcopenia induce a dyslipidemic profile similar to that described in humans [20], resembling the so-called "lipid triad" (elevated triglycerides and LDL-c with normal HDL-c; [21]). In consequence, aged sows have an increased atherogenic dyslipidemia ratio evidencing cardiometabolic risk like human beings [22].

In human medicine, elevated triglycerides in blood and tissues have been largely linked to establishment of glucose intolerance and IR [23,24] and there is a well-known connection between aging, sarcopenia and IR [20]. Concomitantly, in our sows, assessment of the glycemic index showed that HOMA-IR and HOMA-β indexes are higher in aged and sarcopenic individuals than in younger individuals.

The states of dyslipidemia and IR were accompanied by histological evidences of lipotoxicity in the liver, with morphological alterations similar to those found in the early nonalcoholic fatty liver disease (NAFLD) in humans. This finding indicates the main roles of IR and abnormal lipoprotein metabolism in the pathogenesis of NAFLD in swine, like in humans [25]. Around 80–90% of the non-obese sarcopenic sows evidenced lipids infiltration and increased presence of lipocytes (mostly known as Ito cells). This finding supports the relationship between ageing and proliferation of Ito cells previously described in human patients [26]. In 24% of our sows, the presence of ballooning injury, if extrapolated to humans, would be indicative of predisposition to a later development of nonalcoholic steatohepatitis (NASH) [25,27], a condition also known as lipotoxic liver disease [28]. The primary role of Ito cells is the intracytoplasmic storage of fat and vitamin A [29]. However, damage of the liver tissue is associated with proliferation of Ito cells, which plays an important role in the pathogenesis of chronic liver disease [30] since the cells are subsequently transformed in myofibroblasts. Such a process results in fibrogenesis and, in consequence, the normal hepatic tissue is replaced by fibrotic tissue in advanced stages. There was no evidence of hepatic fibrosis in the current study, which indicates early stages of liver disease. This assumption is reinforced by the lack of significant changes in most of the enzymes used as markers of hepatic function that remained within physiological ranges during the whole study [31,32].

Hence, the state of sarcopenia induced by aging in the Iberian sows of the present study represents certain clinical and histological evidences of dyslipidemia, IR, lipotoxicity, NAFLD, and ultimately prodrome of NASH, in human medicine.

3.2. Similarities of the Model with Human Sarcopenic Obesity

The obesogenic diet offered to sarcopenic sows affected body weight and adiposity very early after starting the differentiated feeding, like previously described for younger pigs [18].

In the obese sows of the current study, increases in fat accretion were mainly observed at subcutaneous and visceral compartments. In human medicine, it is well-known that aging-related increases in visceral fat favor the development of dyslipidemia and IR [33]. In our model, increases in visceral fat were similarly related to significant changes in the plasma lipids profile, which supports that the lipidomic profile of swine is mainly determined by the visceral fat, as found in humans [33]. The increase in total cholesterol and LDL-c (indicating hyperlipidemia type-2) concurrently with maintained low HDL-c levels found in our obese sarcopenic sows, if extrapolated to human medicine, would be considered as indicative of type 2 of diabetes mellitus (T2DM; [34]. In our pigs, such diagnosis was confirmed by the evidence of impaired glucose regulation throughout the study, with altered β-cell function and IR. However, there were no changes in plasma glucose concentrations, like in the first stages of the human disease [35], and our sows were able to counterbalance the prodrome of T2DM at Day 90. These findings may be related to the short duration of the experiment and/or to the high plasticity of the swine pancreas, in agreement with previous studies [36]. Hence, the data obtained so far pave the way for further studies assessing IR in longer experimental trials.

An elegant revision of Mlinar and Marc [37] analyzes the different factors and steps in the link between adiposity and IR. In brief, a triglyceride overload in non-adipose ectopic tissues occurs when the capacity of the adipose tissue for storing lipids is exceeded. At the same time, the failure of the vasculature to expand together with the adipocyte hypertrophy causes hypoxia of the adipose tissue and, therefore, oxidative stress; in turn, oxidative stress is reinforced by an altered metabolism of the non-adipose tissue. Systemic oxidative stress and steatosis in non-adipose tissues induce low-grade inflammation. Finally, the concurrence of hypertriglyceridemia and ectopic lipid deposition, increased oxidative stress and inflammation trigger IR. All these factors described in humans were also found in the obese sarcopenic sows of the current study.

Moreover, in our study, the assessment of fatty acid composition of subcutaneous and visceral fat, where more than 95% were neutral lipids (triglycerides), offers outstanding information on its changes during sarcopenic obesity. In brief, obesity increases MUFA content and decreases PUFA content (specifically n-3 and n-6 PUFA) at both compartments, but also increases the ratios of MUFA/SFA and C18:1/C18:0 in visceral fat (i.e., increases desaturation index and Stearoyl-CoA desaturase 1 activity; SCD1). In humans, increased desaturation index and SCD1 activity have been related to metabolic disorders, like alterations in lipogenesis and insulin regulation [23,38]. Similar results have been found in the Iberian pig in previous studies [39,40].

In the present study, we observed major effects of obesity on fatty acid composition of the polar fraction of cell membranes at non-adipose ectopic tissues (muscle and liver). Specifically, obese sows showed higher MUFA/SFA, C18:1/C18:0 and Σn-6/Σn-3 ratios than control sows. The implications of the two first ratios have been previously considered, but the changes in the Σn-6/Σn-3ratio also indicate important metabolic impairments. In the skeletal muscle, the n-3 PUFA content of cellular membranes plays a main role favoring the action of insulin; a high Σn-6/Σn-3ratio seems to be deleterious to insulin sensitivity [41] and ultimately results in IR [42]. In the liver, the increase of Σn-6/Σn-3ratio indicates also a pro-inflammatory state related to increased peripheral lipolysis and, therefore, enlarged flux of fatty acids [43].

The histological evidence of lipotoxicity at the liver previously described in a high percentage of the non-obese sarcopenic sows of the current study were also observed in the obese sows. All of them (100%) evidenced lipid infiltration, increased presence of Ito cells and presence of ballooning injury indicating nonalcoholic fatty liver disease (NAFLD) and later development of NASH. These data again mimic results found in aged humans. The increasing prevalence of NAFLD with age was highlighted in the Rotterdam study [44] while its relationship with sarcopenia has been reported, two years ago, in the Korean Sarcopenic Obesity Study [45]. Further analysis of this last database highlighted that this association is independent of obesity [46]. However, these studies are precluded by the inherent limitation in conducting invasive experimentation in humans; hence, animal models for NAFLD are necessary to better understand this pathogenesis. A recent comprehensive review of studies with rodent models of NASH [47] states that the ideal model should encompass all the defining features of the human condition (obesity, IR, steatohepatitis, and ultimately fibrosis). However, no single murine model currently represents all subsets of human NASH. In this scenario, swine models have an outstanding translational value.

In humans, the fatty liver is also closely associated with increased pancreatic fat content (a disorder also known as nonalcoholic fatty pancreas disease, NAFPD) [48]. NAFPD is associated with visceral obesity, dyslipidemia, IR and ultimately T2DM [49,50], which is similar to data from swine models [51]. A similar relationship was found in our current study, where pancreatic steatosis was severe in half of the obese sows and moderate or mild in the remaining animals; conversely, only 70% of the controls showed steatosis and most of them showed it to a mild degree. Moreover, all the obese sows showed pancreatic lipomatosis, defined as fat accumulation around and within the pancreas. This is a condition associated with aging and IR and known to worsen pancreatic dysfunction caused by steatosis [52]. Hence, these results also reinforce the translational value of our model for studies on the effect of aging, obesity and IR on the development of NAFPD.

4. Materials and Methods

4.1. Ethics Statement

The experiment was performed in agreement with the Spanish Policy for Animal Protection RD1201/05, which meets the European Union Directive 86/609. The experiment was specifically assessed and approved by the Committee of Ethics in Animal Research of the National Institute of Agricultural and Food Research and Technology (INIA) (report CEEA 2012/012, 28 January 2012). The sows were housed at the animal facilities of INIA, which meet local, national and European requirements.

4.2. Animals and Management

The experiment involved 28 Iberian sows (8–10 years-old) with established sarcopenia. Presence of sarcopenia was assessed by ultrasonographic evaluation of muscle mass (cross-sectional diameter of the *longissimus dorsi*), with a 5–8 MHz lineal-array probe (SonoSite Inc., Bothell, WA, USA), since ultrasonography is accepted as a reliable method for evaluating muscle mass and sarcopenia [53]. Muscle diameter had a mean value of 21.9 ± 3.2 mm, which is 50% lower than values obtained in adults (2 years-old; around 38–40 mm) and similar to data from young individuals (6 months-old; around 20 mm) of the same breed reared at our facilities.

All the sows were fed, prior to the experimental procedure, with a standard diet (2.8% of polyunsaturated fat and 3.08 Mcal/kg of metabolizable energy) fulfilling their maintenance requirements (2.5 kg/animal/day). For 100 days, seventeen sows (control group) continued being fed with the same diet and amount whilst 11 sows randomly chosen (obese group) were fed with an obesogenic diet (6.8% of saturated fat and 3.36 Mcal/kg of metabolizable energy) for inducing obesity [18]. Meal intake in the obese group was 4 kg/animal/day for the first 45 days and 5 kg/animal/day for the following 55 days.

4.3. Evaluation of Body Weight and Subcutaneous Adiposity

Changes in body-weight and subcutaneous fat-depth in the obese group were measured every 15 days, from Day 0 to 90, and finally at Day 100 together with the control group. Fat depth was evaluated with the ultrasound probe previously described, at the right-side at 4 cm from the midline and the head of the last rib.

4.4. Blood and Tissue Sampling

Plasma samples for assessment of glucose and lipids metabolism were drawn, after fasting, at Days 0, 45, 90 and 100 in the obese group and at Day 100 in the control group and immediately biobanked at -80 °C. Samples obtained at Days 0 and 100 in the obese group and at Day 100 in the control group were also used to determine redox status. In addition, oral glucose tolerance tests (OGTTs) were performed at Days 0, 45 and 90 in the obese and at Day 90 in the control sows by serial blood samplings (0, 15, 30, 60, 90 and 120 min) after giving 2 g/kg live-weight of D-glucose by gavage through a gastric tube.

At the end of the study (Day 100), a systematic necropsy was performed in all the animals, assessing the macroscopic appearance of the organs. Immediately, two portions of subcutaneous and visceral (perirenal) fat, muscle (*longissimus dorsi*), pancreas and left lobe of the liver were collected in 2-mL cryotubes and biobanked at -80 °C for assessment of fatty acids and/or oil red-O staining. A third portion of each tissue was fixed in 10% neutral-buffered-formalin and processed for hematoxylin-eosin (HE).

4.5. Assessment of Plasma Lipid Profiles

Plasma concentrations of triglycerides, total cholesterol and high-density and low-density lipoproteins cholesterol (HDL-c and LDL-c, respectively) were measured with a clinical chemistry analyzer (Saturno 300-plus; Crony Instruments s.r.l., Rome, Italy). Plasma HDL-c ratio and LDL-c ratio were calculated by dividing total cholesterol by HDL-c and LDL-c concentrations, respectively; plasma LDL-c/HDL-c ratio was obtained by dividing LDL-c by HDL-c concentrations. Finally, the atherogenic dyslipidemia ratio was calculated as log(triglycerides)/HDL-c.

4.6. Assessment of Plasma Antioxidant Capacity and Lipid Peroxidation

Values for total antioxidant capacity were determined by FRAP (ferric reducing antioxidant power assay) [54] whilst lipids oxidative damage was assessed by MDA (malondialdehyde) [55].

4.7. Assessment of Hepatic and Exocrine Pancreatic Function

Plasma concentrations of alanine-transaminase, albumin, alkaline-phosphatase, amylase, aspartate-aminotransferase, total and direct bilirubin, creatinine, γ-glutamyl-transpeptidase, leucine-aminopeptidase, lipase and urea were measured with the Saturno 300-plus analyzer (Crony Instruments s.r.l., Rome, Italy).

4.8. Assessment of Glucose Metabolism and Endocrine Pancreatic Function

Plasma concentrations of glucose and insulin were measured, respectively, with the Saturno 300-plus analyzer and with a Porcine Insulin ELISA kit (Mercodia AB, Uppsala, Sweden; 0.26 IU/L of assay sensitivity and 3.5% of intra-assay variation coefficient). Insulin sensitivity/resistance were determined, concomitantly with data from OGTTs, throughout the study of the HOMA-IR index [(FINS × FGLU)/22.5], whilst possible changes in β-cell function were assessed by the HOMA-β index [(20 × FINS)/(FGLU − 3.5)]. FINS accounts for fasting plasma insulin concentration in U/L and FGLU for fasting plasma glucose in mmol/L.

4.9. Histological Assessment of Liver and Pancreas

The stains of liver and pancreas tissue were examined by light microscopy and blindly scored for the presence and severity of histological features indicating fat infiltration, steatosis, inflammation, ballooning injury, fibrosis and vacuolization. Evaluation was based on the scoring developed by Kleiner et al. [56], considering none (<5%), mild (5–33%), moderate (34–66%) and severe degree (>66%).

4.10. Evaluation of Fatty Acids Composition

Fatty acid composition was analysed in subcutaneous (after differentiation of outer and inner layers) and visceral fat, *longissimus dorsi* muscle and liver. Fat was extracted using the *Ball-mill procedure* [57] and gas-chromatography was used for identification and quantification of fatty acids (FA) in total lipid extracts at subcutaneous and visceral fat, and separately for neutral (triacylglycerols or triglycerides) and polar (phospolipids) lipids at muscle and liver [58]. Total saturated FA (SFA), monounsaturated FA (MUFA) and polyunsaturated FA (PUFA) and total content of n-3 and n-6 PUFA and their ratio (Σn-6/Σn-3) were calculated from individual FA percentages. Finally, the desaturation index was obtained from the ratio MUFA/SFA and the activity of the stearoyl-CoA desaturase enzyme 1 (SCD1) was inferred from the proportions of oleic and stearic acids (C18:1n-9 and C18:0; product and precursor of SCD1 activity, respectively).

4.11. Statistical Analysis

Effects of diet on body-weight, adiposity, fatty-acid composition, metabolic and antioxidant/oxidative status were assessed by ANOVA for repeated measures (split-plot ANOVA), whilst changes over time were determined by Pearson correlation procedures. Histological features were assessed by one-way ANOVA or by a Kruskall–Wallis test if a Levene's test showed non-homogeneous variables; Duncan's post-hoc test was performed to contrast the differences among groups. Statistical analysis of results expressed as percentages was performed after arc-sine transformation of the values for each individual percentage, while the response to OGTTs was compared after calculating the Area under the Curve (AUC). All results were expressed as mean ± SEM and statistical significance was accepted from $p < 0.05$.

5. Conclusions

The present study shows that the Iberian swine model represents features of sarcopenia and sarcopenic obesity in humans, paving the way for future research on physiopathology of the condition and possible therapeutic targets. The strength of the model is increased by the availability and low-price of aged individuals of this breed and the short time necessary for inducing obesity. The present study was performed on female individuals, since availability of aged sows is higher than aged boars and because management is easier in females, but sex-related effects may be studied on males in further studies.

Supplementary Materials: Supplementary materials can be found at www.mdpi.com/1422-0067/19/3/823/s1.

Acknowledgments: The authors thank the INIA animal staff for his assistance with animal care and P. Cuesta and I. Cano (Department of Research Support, Universidad Complutense de Madrid) for statistical analyses. Esteban Porrini is aresearcher of the program Ramón y Cajal (RYC-2014-16573). Consolación García-Contreras, Marta Vázquez-Gómez, Laura Torres-Rovira, Susana Astiz and Antonio González-Bulnes are members of the EU COST-Action BM1308 "Sharing Advances on Large Animal Models (SALAAM)".

Author Contributions: Consolacion Garcia-Contreras, Marta Vazquez-Gomez, Esteban Porrini, Magali Gonzalez-Colaço, Susana Astiz and Antonio Gonzalez-Bulnes designed the experiments; Consolacion Garcia-Contreras, Marta Vazquez-Gomez, Laura Torres-Rovira, Jorge Gonzalez, Beatriz Isabel, Susana Astiz and Antonio Gonzalez-Bulnes performed the experiments and analyzed samples; Consolacion Garcia-Contreras, Marta Vazquez-Gomez, Jorge Gonzalez, Susana Astiz, Beatriz Isabel and Antonio Gonzalez-Bulnes analysed the data; Consolacion Garcia-Contreras, Marta Vazquez-Gomez, Esteban Porrini, Magali Gonzalez-Colaço and Antonio Gonzalez-Bulnes wrote the original draft; Susana Astiz, Laura Torres-Rovira, Jorge Gonzalez and Beatriz Isabel revised the manuscript.

Conflicts of Interest: The authors declare no conflicts of interest.

References

1. Fielding, R.A.; Vellas, B.; Evans, W.J.; Bhasin, S.; Morley, J.E.; Newman, A.B.; Abellan van Kan, G.; Andrieu, S.; Bauer, J.; Breuille, D.; et al. Sarcopenia: An undiagnosed condition in older adults. Current consensus definition: Prevalence, etiology, and consequences. International working group on sarcopenia. *J. Am. Med. Dir. Assoc.* **2011**, *12*, 249–256. [CrossRef] [PubMed]
2. Delmonico, M.J.; Harris, T.B.; Lee, J.S.; Visser, M.; Nevitt, M.; Kritchevsky, S.B.; Tylavsky, F.A.; Newman, A.B. Alternative definitions of sarcopenia, lower extremity performance, and functional impairment with aging in older men and women. *J. Am. Geriatr. Soc.* **2007**, *55*, 769–774. [CrossRef] [PubMed]
3. Morley, J.E. Anorexia, sarcopenia, and aging. *Nutrition* **2001**, *17*, 660–663. [CrossRef]
4. Castillo, E.M.; Goodman-Gruen, D.; Kritz-Silverstein, D.; Morton, D.J.; Wingard, D.L.; Barrett-Connor, E. Sarcopenia in elderly men and women: The Rancho Bernardo study. *Am. J. Prev. Med.* **2003**, *25*, 226–231. [CrossRef]
5. Metter, E.J.; Talbot, L.A.; Schrager, M.; Conwit, R. Skeletal muscle strength as a predictor of all-cause mortality in healthy men. *J. Gerontol. A Biol. Sci. Med. Sci.* **2002**, *57*, B359–B365. [CrossRef] [PubMed]
6. Malafarina, V.; Úriz-Otano, F.; Iniesta, R.; Gil-Guerrero, L. Sarcopenia in the elderly: Diagnosis, physiopathology and treatment. *Maturitas* **2012**, *71*, 109–114. [CrossRef]
7. Burton, L.A.; Sumukadas, D. Optimal management of sarcopenia. *Clin. Interv. Aging* **2010**, *5*, 217–228. [PubMed]
8. Batsis, J.A.; Mackenzie, T.A.; Lopez-Jimenez, F.; Bartels, S.J. Sarcopenia, sarcopenic obesity, and functional impairments in older adults: National Health and Nutrition Examination Surveys 1999–2004. *Nutr. Res.* **2015**, *35*, 1031–1039. [CrossRef] [PubMed]
9. Cleasby, M.E.; Jamieson, P.M.; Atherton, P.J. Insulin resistance and sarcopenia: Mechanistic links between common co-morbidities. *J. Endocrinol.* **2016**, *229*, R67–R81. [CrossRef] [PubMed]
10. Batsis, J.A.; Mackenzie, T.A.; Barre, L.K.; Lopez-Jimenez, F.; Bartels, S.J. Sarcopenia, sarcopenic obesity and mortality in older adults: Results from the National Health and Nutrition Examination Survey III. *Eur. J. Clin. Nutr.* **2014**, *68*, 1001–1007. [CrossRef] [PubMed]
11. Bouchonville, M.F.; Villareal, D.T. Sarcopenic obesity: How do we treat it? *Curr. Opin. Endocrinol. Diabetes Obes.* **2013**, *20*, 412–419. [CrossRef] [PubMed]

12. Palus, S.; Springer, J.I.; Doehner, W.; von Haehling, S.; Anker, M.; Anker, S.D.; Springer, J. Models of sarcopenia: Short review. *Int. J. Cardiol.* **2017**, *238*, 19–21. [CrossRef] [PubMed]
13. Bollheimer, L.C.; Buettner, R.; Pongratz, G.; Brunner-Ploss, R.; Hechtl, C.; Banas, M.; Singler, K.; Hamer, O.W.; Stroszczynski, C.; Sieber, C.C. Sarcopenia in the aging high-fat fed rat: A pilot study for modeling sarcopenic obesity in rodents. *Biogerontology* **2012**, *13*, 609–620. [CrossRef] [PubMed]
14. Russell, J.C.; Proctor, S.D. Small animal models of cardiovascular disease: Tools for the study of the roles of metabolic syndrome, dyslipidemia, and atherosclerosis. *Cardiovasc. Pathol.* **2006**, *15*, 318–330. [CrossRef] [PubMed]
15. Roura, E.; Koopmans, S.J.; Lalles, J.P.; Le Huerou-Luron, I.; de Jager, N.; Schuurman, T.; Val-Laillet, D. Critical review evaluating the pig as a model for human nutritional physiology. *Nutr. Res. Rev.* **2016**, *29*, 60–90. [CrossRef] [PubMed]
16. Neel, J.V. Diabetes mellitus: A "thrifty" genotype rendered detrimental by "progress"? *Am. J. Hum. Genet.* **1962**, *14*, 353–362. [PubMed]
17. Torres-Rovira, L.; Gonzalez-Anover, P.; Astiz, S.; Caro, A.; Lopez-Bote, C.; Ovilo, C.; Pallares, P.; Perez-Solana, M.L.; Sanchez-Sanchez, R.; Gonzalez-Bulnes, A. Effect of an obesogenic diet during the juvenile period on growth pattern, fatness and metabolic, cardiovascular and reproductive features of Swine with obesity/leptin resistance. *Endocr. Metab. Immune Disord. Drug Targets* **2013**, *13*, 143–151. [CrossRef] [PubMed]
18. Torres-Rovira, L.; Astiz, S.; Caro, A.; Lopez-Bote, C.; Ovilo, C.; Pallares, P.; Perez-Solana, M.L.; Sanchez-Sanchez, R.; Gonzalez-Bulnes, A. Diet-induced swine model with obesity/leptin resistance for the study of metabolic syndrome and type 2 diabetes. *Sci. World J.* **2012**, *2012*, 510149. [CrossRef] [PubMed]
19. Friendship, R.M.; Wilson, M.R.; Almond, G.W.; McMillan, I.; Hacker, R.R.; Pieper, R.; Swaminathan, S.S. Sow wastage: Reasons for and effect on productivity. *Can. J. Vet. Res.* **1986**, *50*, 205–208. [PubMed]
20. Karakelides, H.; Nair, K.S. Sarcopenia of aging and its metabolic impact. *Curr. Top. Dev. Biol.* **2005**, *68*, 123–148. [PubMed]
21. Temelkova-Kurktschiev, T.; Hanefeld, M. The lipid triad in type 2 diabetes—Prevalence and relevance of hypertriglyceridaemia/low high-density lipoprotein syndrome in type 2 diabetes. *Exp. Clin. Endocrinol. Diabetes* **2004**, *112*, 75–79. [CrossRef] [PubMed]
22. Hermans, M.P.; Ahn, S.A.; Rousseau, M.F. The atherogenic dyslipidemia ratio [log(TG)/HDL-C] is associated with residual vascular risk, beta-cell function loss and microangiopathy in type 2 diabetes females. *Lipids Health Dis.* **2012**, *11*, 132. [CrossRef] [PubMed]
23. Roden, M.; Price, T.B.; Perseghin, G.; Petersen, K.F.; Rothman, D.L.; Cline, G.W.; Shulman, G.I. Mechanism of free fatty acid-induced insulin resistance in humans. *J. Clin. Investig.* **1996**, *97*, 2859–2865. [CrossRef] [PubMed]
24. Koyama, K.; Chen, G.; Lee, Y.; Unger, R.H. Tissue triglycerides, insulin resistance, and insulin production: Implications for hyperinsulinemia of obesity. *Am. J. Physiol.* **1997**, *273*, E708–E713. [CrossRef] [PubMed]
25. Angulo, P. Nonalcoholic fatty liver disease. *N. Engl. J. Med.* **2002**, *346*, 1221–1231. [CrossRef] [PubMed]
26. Kim, I.H.; Kisseleva, T.; Brenner, D.A. Aging and liver disease. *Curr. Opin. Gastroenterol.* **2015**, *31*, 184–191. [CrossRef] [PubMed]
27. Chitturi, S.; Farrell, G.C. Etiopathogenesis of nonalcoholic steatohepatitis. *Semin. Liver Dis.* **2001**, *21*, 27–41. [CrossRef] [PubMed]
28. Cusi, K. Role of obesity and lipotoxicity in the development of nonalcoholic steatohepatitis: Pathophysiology and clinical implications. *Gastroenterology* **2012**, *142*, 711–725. [CrossRef] [PubMed]
29. Sankin, A. Discovering Biomarkers within the Genomic Landscape of Renal Cell Carcinoma. *J. Kidney* **2016**, *2*. [CrossRef]
30. Flisiak, R. Role of Ito cells in the liver function. *Pol. J. Pathol.* **1997**, *48*, 139–145. [PubMed]
31. Rückert, I.-M.; Heier, M.; Rathmann, W.; Baumeister, S.E.; Döring, A.; Meisinger, C. Association between markers of fatty liver disease and impaired glucose regulation in men and women from the general population: The KORA-F4-study. *PLoS ONE* **2011**, *6*, e22932. [CrossRef] [PubMed]
32. Sanyal, D.; Mukherjee, P.; Raychaudhuri, M.; Ghosh, S.; Mukherjee, S.; Chowdhury, S. Profile of liver enzymes in non-alcoholic fatty liver disease in patients with impaired glucose tolerance and newly detected untreated type 2 diabetes. *Indian J. Endocrinol. Metab.* **2015**, *19*, 597–601. [CrossRef] [PubMed]
33. Despres, J.P. Dyslipidaemia and obesity. *Baillieres Clin. Endocrinol. Metab.* **1994**, *8*, 629–660. [CrossRef]

34. Goldberg, R.B. Lipid disorders in diabetes. *Diabetes Care* **1981**, *4*, 561–572. [CrossRef] [PubMed]
35. Kahn, S.E.; Prigeon, R.L.; McCulloch, D.K.; Boyko, E.J.; Bergman, R.N.; Schwartz, M.W.; Neifing, J.L.; Ward, W.K.; Beard, J.C.; Palmer, J.P.; et al. Quantification of the relationship between insulin sensitivity and beta-cell function in human subjects. Evidence for a hyperbolic function. *Diabetes* **1993**, *42*, 1663–1672. [CrossRef] [PubMed]
36. Koopmans, S.J.; Schuurman, T. Considerations on pig models for appetite, metabolic syndrome and obese type 2 diabetes: From food intake to metabolic disease. *Eur. J. Pharmacol.* **2015**, *759*, 231–239. [CrossRef] [PubMed]
37. Mlinar, B.; Marc, J. New insights into adipose tissue dysfunction in insulin resistance. *Clin. Chem. Lab. Med.* **2011**, *49*, 1925–1935. [CrossRef] [PubMed]
38. Poudyal, H.; Brown, L. Stearoyl-CoA desaturase: A vital checkpoint in the development and progression of obesity. *Endocr. Metab. Immune Disord. Drug Targets* **2011**, *11*, 217–231. [CrossRef] [PubMed]
39. Barbero, A.; Astiz, S.; Lopez-Bote, C.J.; Perez-Solana, M.L.; Ayuso, M.; Garcia-Real, I.; Gonzalez-Bulnes, A. Maternal malnutrition and offspring sex determine juvenile obesity and metabolic disorders in a swine model of leptin resistance. *PLoS ONE* **2013**, *8*, e78424. [CrossRef] [PubMed]
40. Gonzalez-Bulnes, A.; Astiz, S.; Ovilo, C.; Lopez-Bote, C.J.; Sanchez-Sanchez, R.; Perez-Solana, M.L.; Torres-Rovira, L.; Ayuso, M.; Gonzalez, J. Early-postnatal changes in adiposity and lipids profile by transgenerational developmental programming in swine with obesity/leptin resistance. *J. Endocrinol.* **2014**, *223*, M17–M29. [CrossRef] [PubMed]
41. Corcoran, M.P.; Lamon-Fava, S.; Fielding, R.A. Skeletal muscle lipid deposition and insulin resistance: Effect of dietary fatty acids and exercise. *Am. J. Clin. Nutr.* **2007**, *85*, 662–677. [PubMed]
42. Li, Y.; Xu, S.; Zhang, X.; Yi, Z.; Cichello, S. Skeletal intramyocellular lipid metabolism and insulin resistance. *Biophys. Rep.* **2015**, *1*, 90–98. [CrossRef] [PubMed]
43. Valenzuela, R.; Videla, L.A. The importance of the long-chain polyunsaturated fatty acid n-6/n-3 ratio in development of non-alcoholic fatty liver associated with obesity. *Food Funct.* **2011**, *2*, 644–648. [CrossRef] [PubMed]
44. Koehler, E.M.; Schouten, J.N.; Hansen, B.E.; van Rooij, F.J.; Hofman, A.; Stricker, B.H.; Janssen, H.L. Prevalence and risk factors of non-alcoholic fatty liver disease in the elderly: Results from the Rotterdam study. *J. Hepatol.* **2012**, *57*, 1305–1311. [CrossRef] [PubMed]
45. Hong, H.C.; Hwang, S.Y.; Choi, H.Y.; Yoo, H.J.; Seo, J.A.; Kim, S.G.; Kim, N.H.; Baik, S.H.; Choi, D.S.; Choi, K.M. Relationship between sarcopenia and nonalcoholic fatty liver disease: The Korean Sarcopenic Obesity Study. *Hepatology* **2014**, *59*, 1772–1778. [CrossRef] [PubMed]
46. Lee, Y.H.; Jung, K.S.; Kim, S.U.; Yoon, H.J.; Yun, Y.J.; Lee, B.W.; Kang, E.S.; Han, K.H.; Lee, H.C.; Cha, B.S. Sarcopaenia is associated with NAFLD independently of obesity and insulin resistance: Nationwide surveys (KNHANES 2008–2011). *J. Hepatol.* **2015**, *63*, 486–493. [CrossRef] [PubMed]
47. Ibrahim, S.H.; Hirsova, P.; Malhi, H.; Gores, G.J. Animal Models of Nonalcoholic Steatohepatitis: Eat, Delete, and Inflame. *Dig. Dis. Sci.* **2016**, *61*, 1325–1336. [CrossRef] [PubMed]
48. Van Geenen, E.J.; Smits, M.M.; Schreuder, T.C.; van der Peet, D.L.; Bloemena, E.; Mulder, C.J. Nonalcoholic fatty liver disease is related to nonalcoholic fatty pancreas disease. *Pancreas* **2010**, *39*, 1185–1190. [CrossRef] [PubMed]
49. Tushuizen, M.E.; Bunck, M.C.; Pouwels, P.J.; Bontemps, S.; van Waesberghe, J.H.; Schindhelm, R.K.; Mari, A.; Heine, R.J.; Diamant, M. Pancreatic fat content and beta-cell function in men with and without type 2 diabetes. *Diabetes Care* **2007**, *30*, 2916–2921. [CrossRef] [PubMed]
50. Lee, J.S.; Kim, S.H.; Jun, D.W.; Han, J.H.; Jang, E.C.; Park, J.Y.; Son, B.K.; Kim, S.H.; Jo, Y.J.; Park, Y.S.; et al. Clinical implications of fatty pancreas: Correlations between fatty pancreas and metabolic syndrome. *World J. Gastroenterol.* **2009**, *15*, 1869–1875. [CrossRef] [PubMed]
51. Yin, W.; Liao, D.; Kusunoki, M.; Xi, S.; Tsutsumi, K.; Wang, Z.; Lian, X.; Koike, T.; Fan, J.; Yang, Y.; et al. NO-1886 decreases ectopic lipid deposition and protects pancreatic beta cells in diet-induced diabetic swine. *J. Endocrinol.* **2004**, *180*, 399–408. [CrossRef] [PubMed]
52. Noel, P.; Patel, K.; Durgampudi, C.; Trivedi, R.N.; de Oliveira, C.; Crowell, M.D.; Pannala, R.; Lee, K.; Brand, R.; Chennat, J.; et al. Peripancreatic fat necrosis worsens acute pancreatitis independent of pancreatic necrosis via unsaturated fatty acids increased in human pancreatic necrosis collections. *Gut* **2016**, *65*, 100–111. [CrossRef] [PubMed]

53. Minetto, M.A.; Caresio, C.; Menapace, T.; Hajdarevic, A.; Marchini, A.; Molinari, F.; Maffiuletti, N.A. Ultrasound-Based Detection of Low Muscle Mass for Diagnosis of Sarcopenia in Older Adults. *PM R* **2016**, *8*, 453–462. [CrossRef] [PubMed]

54. Benzie, I.F.; Strain, J.J. The ferric reducing ability of plasma (FRAP) as a measure of "antioxidant power": The FRAP assay. *Anal. Biochem.* **1996**, *239*, 70–76. [CrossRef] [PubMed]

55. Larstad, M.; Ljungkvist, G.; Olin, A.C.; Toren, K. Determination of malondialdehyde in breath condensate by high-performance liquid chromatography with fluorescence detection. *J. Chromatogr. B Anal. Technol. Biomed. Life Sci.* **2002**, *766*, 107–114. [CrossRef]

56. Kleiner, D.E.; Brunt, E.M.; Van Natta, M.; Behling, C.; Contos, M.J.; Cummings, O.W.; Ferrell, L.D.; Liu, Y.C.; Torbenson, M.S.; Unalp-Arida, A.; et al. Design and validation of a histological scoring system for nonalcoholic fatty liver disease. *Hepatology* **2005**, *41*, 1313–1321. [CrossRef] [PubMed]

57. Segura, J.; Lopez-Bote, C.J. A laboratory efficient method for intramuscular fat analysis. *Food Chem.* **2014**, *145*, 821–825. [CrossRef] [PubMed]

58. Olivares, A.; Rey, A.I.; Daza, A.; Lopez-Bote, C.J. High dietary vitamin A interferes with tissue alpha-tocopherol concentrations in fattening pigs: A study that examines administration and withdrawal times. *Animal* **2009**, *3*, 1264–1270. [CrossRef] [PubMed]

International Journal of
Molecular Sciences

MDPI

Review

Some Metabolites Act as Second Messengers in Yeast Chronological Aging

Karamat Mohammad, Paméla Dakik, Younes Medkour, Mélissa McAuley, Darya Mitrofanova and Vladimir I. Titorenko *

Department of Biology, Concordia University, 7141 Sherbrooke Street, West, SP Building, Room 501-13, Montreal, QC H4B 1R6, Canada; karamat.mohammad@concordia.ca (K.M.); pameladakik@gmail.com (P.D.); writetoyounes@gmail.com (Y.M.); melissa.mcauley@concordia.ca (M.M.); mitrofanova_darya@hotmail.com (D.M.)
*   Correspondence: vladimir.titorenko@concordia.ca; Tel.: +1-514-848-2424 (ext. 3424)

Received: 8 February 2018; Accepted: 13 March 2018; Published: 15 March 2018

Abstract: The concentrations of some key metabolic intermediates play essential roles in regulating the longevity of the chronologically aging yeast *Saccharomyces cerevisiae*. These key metabolites are detected by certain ligand-specific protein sensors that respond to concentration changes of the key metabolites by altering the efficiencies of longevity-defining cellular processes. The concentrations of the key metabolites that affect yeast chronological aging are controlled spatially and temporally. Here, we analyze mechanisms through which the spatiotemporal dynamics of changes in the concentrations of the key metabolites influence yeast chronological lifespan. Our analysis indicates that a distinct set of metabolites can act as second messengers that define the pace of yeast chronological aging. Molecules that can operate both as intermediates of yeast metabolism and as second messengers of yeast chronological aging include reduced nicotinamide adenine dinucleotide phosphate (NADPH), glycerol, trehalose, hydrogen peroxide, amino acids, sphingolipids, spermidine, hydrogen sulfide, acetic acid, ethanol, free fatty acids, and diacylglycerol. We discuss several properties that these second messengers of yeast chronological aging have in common with second messengers of signal transduction. We outline how these second messengers of yeast chronological aging elicit changes in cell functionality and viability in response to changes in the nutrient, energy, stress, and proliferation status of the cell.

Keywords: yeast; chronological aging; mechanisms of longevity regulation; metabolism; cell signaling; second messengers; mitochondria; interorganellar communications; proteostasis; regulated cell death

1. Introduction

Studies of the budding yeast *Saccharomyces cerevisiae* have led to the discovery of genes, signaling pathways, and small molecules that define the rate of cellular aging in this unicellular eukaryote [1–3]. Some of the genes, signaling pathways, and small molecules discovered in *S. cerevisiae* have been later found to play essential roles in cellular and organismal aging in diverse species of multicellular eukaryotes; it is believed, therefore, that the mechanisms underlying aging and longevity assurance have been conserved throughout the course of evolution [4–6]. Thus, the use of *S. cerevisiae* as a model organism is fundamental for the progress in aging research [2–4].

There are two different ways of studying aging in *S. cerevisiae*; each of these ways investigates a different mode of yeast aging.

A so-called replicative mode of yeast aging is monitored by counting the total number of asymmetric mitotic divisions—each producing a small daughter cell—that a mother cell could undergo on the surface of a solid nutrient-rich medium before it becomes senescent [2,7,8]. Yeast replicative aging has long been considered to resemble the aging of those cells in humans and other mammals that are able to divide mitotically; among these mitotically active cells are lymphocytes, monocytes, granulocytes, fibroblasts, and some stem cell types [2,4,6–8]. Recent studies have revealed that (1) many genes that modulate aging of post-mitotic cells in adults of the nematode *Caenorhabditis elegans* also influence yeast replicative aging [9–11]; and (2) many hallmarks of aging characteristic of post-mitotic cells in humans and other mammals are also cellular hallmarks of yeast replicative aging [12]. Hence, it is conceivable that the replicative mode of yeast aging may also mirror the aging of post-mitotic cells and even the aging of the entire organism in nematodes, humans, and other mammals [9–12].

A so-called chronological mode of yeast aging is monitored by determining how long a yeast cell cultured in a liquid medium can retain viability after it undergoes cell cycle arrest and enters a state of quiescence [2,13,14]. Yeast chronological aging is believed to model the aging of human and mammalian cells that lose the ability to divide mitotically; these post-mitotic cells include adipocytes, mature muscle cells, and mature neurons [2,14,15]. The chronological mode of yeast aging is also considered to be a simple model of organismal aging in multicellular eukaryotes [14,16].

Although the replicative and chronological modes of aging in yeast are usually examined separately from each other, recent evidence indicates that these two modes of yeast aging most likely converge into a single aging process [17–19].

Here, we review mechanisms through which the spatiotemporal dynamics of changes in the concentrations of some metabolites regulate the longevity of chronologically aging yeast. Based on the important advance in our understanding of these mechanisms, we conclude that a distinct group of metabolites act as second messengers that define the pace of yeast chronological aging.

2. Concentrations of Some Metabolites Define the Rate of Chronological Aging in Yeast

Recent studies have demonstrated that the intracellular and extracellular concentrations of some key metabolites play essential roles in regulating the longevity of chronologically aging *S. cerevisiae* [2–5,15,16,20–32]. These key metabolites are detected by certain protein sensors, which respond to concentration changes of the metabolites by altering the efficiencies of cellular processes known to define yeast chronological lifespan (CLS) [2–4,16,20–32]. In this section, we describe the metabolites whose concentration changes affect the pace of yeast chronological aging and discuss mechanisms through which these key metabolites influence yeast CLS.

2.1. NADPH

NADPH is generated in the Zwf1- and Gnd1-dependent reactions of the pentose phosphate pathway operating in the cytosol of *S. cerevisiae*, as well as in the Ald4-, Pos5-, Mae1-, and Idp1-driven reactions confined to yeast mitochondria [3,20,33]. NADPH is indispensable for the growth and viability of yeast cells because it serves as the major source of reducing equivalents for amino acid, fatty acid, and sterol synthesis [20,33]. In addition, NADPH has a specific role in longevity assurance of chronologically aging yeast because it provides electrons for thioredoxin and glutathione reductase systems (TRR and GTR, respectively) [34,35]. Both these NADPH-dependent reductase systems play essential roles in the maintenance of intracellular redox homeostasis, thereby decreasing the extent of oxidative damage to thiol-containing proteins that reside in the cytosol, nucleus, and mitochondria of yeast cells [34,35]. Such NADPH-driven, TRR- and GTR-dependent protection of many thiol-containing proteins from oxidative damage slows down yeast chronological aging (Figure 1A) [35].

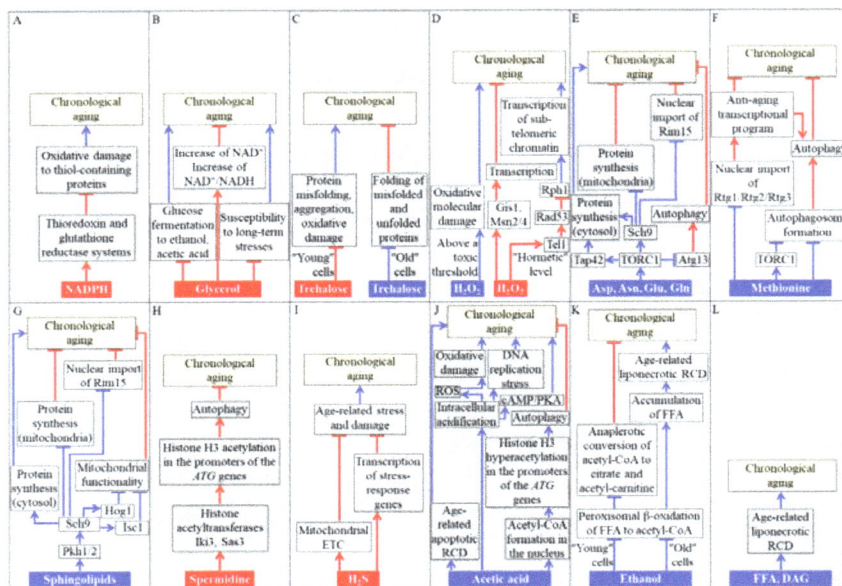

Figure 1. Mechanisms through which the concentrations of some key metabolites define the rate of chronological aging in the yeast *Saccharomyces cerevisiae*. These key metabolites include NADPH (**A**), glycerol (**B**), trehalose (**C**), H$_2$O$_2$ (**D**), the amino acids aspartate, asparagine, glutamate and glutamine (**E**), methionine (**F**), sphingolipids (**G**), spermidine (**H**), H$_2$S (**I**), acetic acid (**J**), ethanol (**K**), as well as FFA and DAG (**L**). These key metabolites are detected by a distinct set of ligand-specific protein sensors that respond to the concentration changes of the metabolites by altering the rates and efficiencies of longevity-defining cellular processes, thus creating a pro- or anti-aging cellular pattern and affecting the pace of yeast chronological aging. See text for more details. Activation arrows and inhibition bars denote pro-aging processes (displayed in blue color) or anti-aging processes (displayed in red color). Pro-aging or anti-aging metabolites are displayed in blue color or red color, respectively. Abbreviations: Asp, aspartate; Asn, asparagine; *ATG*, autophagy-related genes; ETC, electron transport chain; FFA, free (non-esterified) fatty acids; DAG, diacylglycerol; Glu, glutamate; Gln, glutamine; PKA, protein kinase A; RCD, regulated cell death; ROS, reactive oxygen species.

2.2. Glycerol

Glycerol is one of the products of glucose fermentation in the cytosol of *S. cerevisiae* cells [33]. An increase in the intracellular and extracellular concentrations of glycerol has been shown to decelerate yeast chronological aging [15,36]. Three mechanisms have been proposed to underlie such aging-delaying action of glycerol. These mechanisms are depicted in Figure 1B and outlined below. First mechanism: an increase in glucose fermentation to glycerol lowers metabolite flow into glucose fermentation to ethanol and acetic acid, both of which accelerate yeast chronological aging (Figure 1B) [3,15,36,37]. Second mechanism: glycerol decreases the susceptibility of yeast cells to long-term oxidative, thermal, and osmotic stresses; an age-related intensification of all these stresses is a potent pro-aging factor in chronologically aging yeast (Figure 1B) [3,36]. It is presently unknown if this second mechanism involves some protein sensors that respond to an increase in glycerol concentration by stimulating certain stress response processes in yeast cells. Third mechanism: an increase in glucose fermentation to glycerol allows an increase in both the intracellular concentration of NAD$^+$ and the intracellular NAD$^+$/NADH ratio, thereby setting up a pro-longevity cellular pattern in chronologically aging *S. cerevisiae* (Figure 1B) [3,36].

2.3. Trehalose

Trehalose, a non-reducing disaccharide synthesized from glucose, has long been considered only as a reserve carbohydrate in *S. cerevisiae* cells [38]. However, recent evidence indicates that trehalose is also essential for regulating the longevity of chronologically aging yeast [37,39–45]. Depending on the chronological age of *S. cerevisiae*, trehalose exhibits either an anti-aging or pro-aging effect. Both these effects are due to the ability of trehalose to directly bind unfolded domains of proteins, thereby modulating various aspects of cellular proteostasis throughout CLS. In chronologically "young" yeast cells that proliferate, trehalose assists in sustaining an anti-aging pattern of cellular proteostasis. This is because trehalose binding to newly synthesized proteins in these cells helps to decrease the misfolding, aggregation, and oxidative damage of such proteins (Figure 1C) [37,41,46,47]. In chronologically "old" yeast cells that enter a non-proliferative state, trehalose contributes to the establishment of a pro-aging pattern of cellular proteostasis. This is because trehalose binding to hydrophobic amino acid side chains of misfolded, partially folded, and unfolded proteins in these cells shields extended patches of hydrophobic amino acid residues from the molecular chaperones needed for the folding of such proteins (Figure 1C) [37,41,46].

2.4. Hydrogen Peroxide (H_2O_2)

H_2O_2 is the major molecule of reactive oxygen species (ROS) [48]. In chronologically aging yeast, H_2O_2 and other ROS are initially generated in mitochondria and peroxisomes as by-products of mitochondrial respiration and peroxisomal oxidative metabolism, respectively [21–23]. After being made in mitochondria and peroxisomes, H_2O_2 is released from these two organelles and may play a dual role in regulating yeast CLS [21–23]. If the intracellular concentration of H_2O_2 exceeds a toxic threshold, it accelerates the process of chronological aging by eliciting oxidative damage to proteins, lipids, and nucleic acids in various locations within the yeast cell [21,48]. If the intracellular concentration of H_2O_2 is maintained at a "hormetic" level (i.e., a sub-lethal concentration of H_2O_2 that is insufficient to damage cellular macromolecules), this ROS can activate at least two signaling networks that extend yeast CLS by creating an anti-aging cellular pattern [2,3,27]. One of these hormetic H_2O_2-responsive signaling networks activates the Gis1, Msn2, and Msn4 transcription factors; after being activated, Gis1, Msn2, and Msn4 establish an anti-aging cellular pattern by stimulating the transcription of many genes involved in carbohydrate and lipid metabolism, nutrient sensing, cell cycle progression, autophagy, stress response and protection, and stationary phase survival (Figure 1D) [2,49,50]. Another hormetic H_2O_2-responsive signaling network is initiated when the DNA damage response (DDR) kinase Tel1 phosphorylates and activates the DDR kinase Rad53; Rad53 then phosphorylates and inactivates the histone demethylase Rph1, thus creating an anti-aging cellular pattern by suppressing the Rph1-driven transcription of sub-telomeric chromatin regions (Figure 1D) [51,52]. Mechanisms though which the Rad53-dependent suppression of sub-telomeric transcription prolongs yeast CLS remain unknown [51].

2.5. Amino Acids

The synthesis of the amino acids aspartate, asparagine, glutamate, and glutamine from tricarboxylic acid (TCA) cycle intermediates occurs in mitochondria of yeast cells [20,33]. These amino acids then exit mitochondria and activate protein kinase activity of the target of rapamycin complex 1 (TORC1), a pro-aging regulator confined to the surface of yeast vacuoles (Figure 1E) [53–56]. Active TORC1 orchestrates the establishment and maintenance of a pro-aging cellular pattern by phosphorylating at least three downstream protein targets and altering their activities. As outlined below, each of these downstream protein targets of TORC1 defines yeast CLS because it regulates one or more longevity-defining cellular processes.

One of the proteins phosphorylated by TORC1 is the nutrient-sensing protein kinase Sch9. Once phosphorylated by TORC1, Sch9 influences some longevity-defining cellular processes as follows: (1) it promotes ribosome assembly and stimulates translation initiation, thus activating the pro-aging process of protein synthesis in the cytosol [57–60]; (2) it slows down the anti-aging process of protein synthesis in mitochondria [55,56,61,62]; and (3) it inhibits nuclear import of Rim15, a nutrient-sensing protein kinase that elicits an anti-aging cellular pattern by stimulating the Gis1, Msn2, and Msn4 transcription factors in the nucleus [63–65]; this pro-aging effect of Sch9 is reinforced by the nutrient-sensing protein kinase A (PKA), which impedes nuclear import of Msn2 and Msn4 (Figure 1E) [65–67].

Another downstream phosphorylation target of active TORC1 is Tap42, a protein that (akin to Sch9) stimulates the pro-aging process of protein synthesis in the cytosol by promoting ribosome assembly and activating translation initiation (Figure 1E) [58,68].

The autophagy-initiating protein Atg13 is also a downstream phosphorylation target of active TORC1 [69,70]. The TORC1-driven phosphorylation of Atg13 suppresses the anti-aging process of autophagy by inhibiting autophagosome formation in the cytosol (Figure 1E) [71–73]. Atg13 can also be phosphorylated by PKA, which strengthens the pro-aging effect of TORC1 by further suppressing the formation of autophagosomes [69–73].

Like aspartate, asparagine, glutamate, and glutamine, the amino acid methionine is a pro-aging metabolite in yeast that ages chronologically (Figure 1F). A likely pro-aging effect of methionine consists in stimulating the methylation of tRNAs in the cytosol [74]. The resulting decline in the concentration of non-methylated tRNAs weakens nuclear import of the Rtg1/Rtg2/Rtg3 heterotrimeric transcription factor needed for driving an anti-aging transcriptional program (Figure 1F) [74]. Another pro-aging effect of methionine is manifested in its ability to inhibit autophagy by activating TORC1 and/or by suppressing autophagosome formation in a TORC1-independent manner (Figure 1F) [75]. This shortens yeast CLS by decreasing the acidity of the vacuole and by eliciting the accumulation of extracellular acetic acid, a potent pro-aging metabolite [74–77].

2.6. Sphingolipids

Sphingolipids are pro-aging metabolites in chronologically aging yeast because some genetic and pharmaceutical interventions that change the intracellular concentrations of certain sphingolipids have been shown to prolong yeast CLS [24,28,29,32]. Mechanisms underlying the essential role of sphingolipids in defining yeast CLS have begun to emerge; these mechanisms are outlined below.

After being synthesized in the endoplasmic reticulum (ER), the sphingolipid backbone base phytosphingosine stimulates protein kinase activities of Pkh1 and Pkh2 (Pkb-activating kinase homolog proteins 1 and 2) [78,79]. Pkh1 and Pkh2 then phosphorylate and activate the nutrient-sensing protein kinase Sch9 [24,80,81]. Such Pkh1/2-dependent phosphorylation of Sch9 establishes a Pkh1/2-Sch9 branch of a network that integrates nutrient and sphingolipid signaling (Figure 1G) [24,29,32,81]. As mentioned in Section 2.5, the phosphorylated and therefore active form of Sch9 in this signaling branch accelerates yeast chronological aging by stimulating the pro-aging process of protein synthesis in the cytosol [57–60], decelerating the anti-aging process of protein synthesis in mitochondria [55,56,61,62] and inhibiting nuclear import of the nutrient-sensing protein kinase Rim15 (Figure 1G) [63–65]. The Pkh1/2-dependent phosphorylation of Sch9 also elicits the Sch9-driven phosphorylation and activation of: (1) Hog1, a mitogen-activated protein kinase that attenuates mitochondrial functionality; and (2) Isc1, an inositol phosphosphingolipid phospholipase C that is involved in the formation of ceramides from complex sphingolipids in yeast mitochondria (Figure 1G) [24,28,29,32]. This ability of the Pkh1/2-Sch9 signaling branch to promote the phosphorylation of Hog1 and Isc1 allows it to coordinate mitochondrial functionality and sphingolipid metabolism, thus making an essential contribution to longevity regulation in chronologically aging yeast [24,28,29,32].

2.7. Spermidine

The natural polyamine spermidine is an aging-delaying metabolite that extends yeast CLS [82–84]. In yeast cells, spermidine is synthesized from the amino acids arginine and methionine in a series of reactions confined to mitochondria, the cytosol, and peroxisomes [84]. Spermidine delays yeast chronological aging because it inhibits the histone acetyltransferases Iki3 and Sas3 (Figure 1H) [82]. Although such spermidine-dependent inhibition of the two acetyltransferases attenuates transcription of many nuclear genes because it impairs histone H3 acetylation at their promoter regions, the extent of histone H3 acetylation at the promoter regions of several autophagy-related (*ATG*) genes is increased in yeast cells that exhibit high spermidine concentrations [82]. This selectively activates the transcription of the *ATG* genes, thus enhancing the cytoprotective process of autophagy and delaying yeast chronological aging (Figure 1H) [82,83].

2.8. Hydrogen Sulfide (H_2S)

H_2S is a metabolite that plays an essential role in the delay of yeast chronological aging by caloric restriction (CR) [85], a dietary regimen that delays aging, increases lifespan, and improves healthspan in evolutionarily distant eukaryotes [1,86,87]. In yeast, this water- and fat-soluble gas can be generated endogenously via two different metabolic pathways. One pathway of H_2S synthesis involves a unique yeast assimilation of exogenous inorganic sulfate [88]. Another pathway of H_2S synthesis is an evolutionarily conserved trans-sulfuration pathway (TSP) of transfer from methionine to cysteine [88]. In yeast cultured in a liquid synthetic medium, only H_2S that is endogenously synthesized via the TSP pathway and then released to the culture medium is responsible for yeast CLS extension under CR conditions [85]. Mechanisms through which an exogenous (extracellular) pool of H_2S delays chronological aging of yeast limited in calorie supply remain to be determined. It has been suggested that low, hormetic concentrations of H_2S may protect chronologically aging yeast from age-related stress and damage by accelerating the electron transport chain in mitochondria and/or by activating the transcription of many stress-response genes in the nucleus (Figure 1I) [85,88,89].

2.9. Acetic Acid

Acetic acid is one of the products of glucose fermentation by yeast cultured in a liquid medium under high oxygenation [2,15,33]. It is formed in the Ald6-dependent reaction that occurs in the cytosol and in the Ald4-driven reaction confined to mitochondria [3]. Acetic acid is a pro-aging metabolite that shortens yeast CLS [15,90].

Three mechanisms through which extracellular and/or intracellular pools of acetic acid may accelerate yeast chronological aging have been suggested, all based on known effects of these acetic acid pools. These mechanisms are outlined below.

First mechanism: acetic acid (and/or the acidification of the liquid culture medium it causes) may, directly or indirectly, elicit an age-related apoptotic mode of regulated cell death (Figure 1J) [15,90–94].

Second mechanism: extracellular acidification caused by the buildup of acetic acid in the culture medium may trigger intracellular acidification, which may then accelerate chronological aging by: (1) stimulating the pro-aging cyclic adenosine monophosphate (cAMP)/PKA signaling pathway; (2) increasing the concentration of ROS, thus eliciting oxidative damage to cellular macromolecules; and/or (3) enhancing DNA replication stress (Figure 1J) [2,90,92].

Third mechanism: nuclear transport of acetic acid and its subsequent Acs2-dependent conversion into acetyl-Coenzyme A (CoA) may suppress the transcription of nuclear *ATG* genes by causing histone H3 hyperacetylation at their promoter regions, thus slowing down the cytoprotective process of autophagy and accelerating yeast chronological aging (Figure 1J) [25,26,95,96].

It needs to be emphasized that exogenous acetic acid accelerates yeast chronological aging not only because it creates an acidic extracellular environment; indeed, neither other organic acids present in culture medium of aged yeast nor hydrochloric acid can shorten yeast CLS if added exogenously [15].

Moreover, only exogenous acetic acid at acidic pH, but not its conjugate base at neutral pH, exhibits the aging-accelerating effect in yeast [15]. It is conceivable, therefore, that only a combinatory action of acetic acid and an acidic extracellular environment is responsible for the acceleration of yeast chronological aging via the above mechanisms [15].

2.10. Ethanol

Yeast cells cultured in a liquid medium under low oxygenation produce ethanol as the main product of glucose fermentation in the Adh1-dependent reaction confined to the cytosol [2,3,13,14,33]. Ethanol is a pro-aging metabolite that shortens yeast CLS [37,97]. This pro-aging action of ethanol can be enhanced by the sirtuin deacetylase Sir2, which inhibits the Adh2-driven conversion of ethanol to acetaldehyde [97].

Ethanol accelerates yeast chronological aging via two different mechanisms. Both of these mechanisms are elicited in response to ethanol-dependent suppression of the Fox1-, Fox2-, and Fox3-driven peroxisomal β-oxidation of fatty acids to acetyl-CoA [21,37]. Each of these two mechanisms is specific to the chronological age of a yeast cell.

In chronologically "young" yeast, the ethanol-dependent suppression of acetyl-CoA formation in peroxisomes decreases its availability for anaplerotic conversion to citrate and acetyl-carnitine in these organelles [3,21,37]. This accelerates yeast chronological aging by weakening the longevity assurance process of supplying citrate and acetyl-carnitine to mitochondria for the replenishment of these two TCA cycle intermediates (Figure 1K) [3,21,37].

In chronologically "old" yeast, the ethanol-dependent suppression of peroxisomal β-oxidation of fatty acids leads to the excessive accumulation of free (non-esterified) fatty acids (FFA) [3,21,37]. Such build-up of FFA accelerates yeast chronological aging by increasing the risk of an age-related mode of regulated cell death (RCD) called "liponecrosis" (Figure 1K) [3,98–100].

2.11. FFA and Diacylglycerol (DAG)

The ethanol-driven excessive accumulation of unoxidized FFA in yeast peroxisomes (see Section 2.10) elicits several negative-feedback loops whose action causes a buildup of FFA and DAG in the ER and lipid droplets (LD) [3,100,101]. This buildup of FFA and DAG accelerates the onset of the liponecrotic mode of RCD, thereby increasing the risk of death and accelerating yeast chronological aging (Figure 1L) [3,21,98–100]. Thus, both FFA and DAG are pro-aging metabolites that shorten yeast CLS.

3. The Spatiotemporal Dynamics of Changes in Concentrations of Some Metabolites Define Yeast CLS

It is becoming increasingly evident that the concentrations of the key metabolites that influence the pace of yeast chronological aging are controlled spatially and temporally at different levels.

One level of the spatial control over the concentrations of these metabolites consists in regulating their intracellular and extracellular concentrations [3,102]. Some of the key metabolites can influence longevity-defining processes only within the cell they were produced; these metabolites: (1) define the pace of yeast chronological aging only via cell-autonomous mechanisms; (2) cannot act as low molecular weight transmissible longevity factors; and (3) include NADPH, trehalose, sphingolipids, FFA, and DAG [3,24,28,32,35,37,41,102]. In contrast, some of the key metabolites can influence longevity-defining processes both within the cell they were generated as well as within other cells in the yeast population; these metabolites: (1) define the pace of yeast chronological aging via both cell-autonomous and cell-non-autonomous mechanisms; (2) can act as low molecular weight transmissible longevity factors; and (3) include glycerol, H_2O_2, amino acids, spermidine, H_2S, acetic acid, and ethanol [3,15,36,50–52,54,71–73,82,85,93,94,97,102]. It is noteworthy that some of those metabolites that cannot act as low molecular weight transmissible longevity factors (and thus define the pace of yeast chronological aging only via cell-autonomous mechanisms) may change metabolism within the "host" cell so that this cell may respond by altering the production of acetic acid and/or other metabolites that can act via both cell-autonomous and cell-non-autonomous mechanisms.

Another level at which the concentrations of these key metabolites are controlled spatially consists in regulating their abundance in different organellar compartments within the cell. The effective concentrations of the metabolites that influence the pace of yeast chronological aging are established and maintained within an intricate network that involves the unidirectional and/or bidirectional flow of certain metabolites between mitochondria, the nucleus, vacuoles, peroxisomes, the ER, the plasma membrane, LD, and the cytosol [22,23,27].

In addition, the concentrations of the key metabolites that define yeast CLS are controlled temporally. Indeed, it has been found that: (1) these metabolite concentrations undergo age-related changes; and (2) the effects of these metabolites on their protein sensors are also subjected to age-related changes; these protein sensors respond to concentration changes of the key metabolites by altering the efficiencies of certain longevity-defining cellular processes [3,27]. It has been therefore proposed that the concentration changes of different key metabolites are temporally limited to certain longevity-defining periods called "checkpoints" [2,3,27,41,103]. Many of these checkpoints exist early in the life of chronologically aging yeast cultured in glucose-containing liquid media, during diauxic and post-diauxic growth phases [3,27]. NADPH, glycerol, H_2O_2, amino acids, sphingolipids, and spermidine are the key metabolites whose concentrations only at these "early" checkpoints play essential roles in defining the pace of yeast chronological aging [24,27,28,32,35,36,42,51–54,71–73,82]. Some of these checkpoints are the "late" checkpoints that occur after chronologically aging yeast cultured in glucose-containing liquid media undergoes cell cycle arrest and enters the non-proliferative stationary phase of culturing [3,27]. H_2S, acetic acid, FFA, and DAG are the key metabolites whose concentrations only at these "late" checkpoints make essential contributions to defining the pace of yeast chronological aging [15,27,37,85,93,94]. It is noteworthy that trehalose and ethanol are the two key metabolites whose concentrations affect the longevity of chronologically aging yeast at both the "early" and the "late" checkpoints; for each of these two metabolites, the longevity-defining effect at the "early" checkpoint differs from that at the "late" checkpoint (see Sections 2.3 and 2.10) [3,27,41]. It has been also proposed that at each of these "early" and "late" checkpoints, the concentration changes of the key metabolites are detected by a distinct set of checkpoint-specific protein sensors. These protein sensors have been called "master regulators" because their suggested common role consists in responding to the concentration changes of different key metabolites by altering the rates and efficiencies of various longevity-defining cellular processes throughout the chronological lifespan, thus establishing a pro- or anti-aging cellular pattern and defining yeast CLS [3,27].

4. Conclusions

Based on our analysis of how the spatiotemporal dynamics of changes in the concentrations of some metabolites influence yeast CLS, we conclude that a distinct group of metabolites can act as second messengers that define the pace of yeast chronological aging. Molecules that can operate both as intermediates of yeast metabolism and as second messengers of yeast chronological aging include NADPH, glycerol, trehalose, H_2O_2, amino acids, sphingolipids, spermidine, H_2S, acetic acid, ethanol, FFA, and DAG. Akin to second messengers of signal transduction [104,105], these second messengers of yeast chronological aging: (1) are generated and can undergo substantial concentration changes within a yeast cell in response to some extracellular stimuli (i.e., nutrients and hormetic stresses); these extracellular stimuli are detected by certain hierarchically organized proteins and protein complexes (i.e., the TORC1 and PKA signaling pathways) that then transduce the signal to some metabolic reactions involved in the generation of the intracellular second messengers; (2) differ from each other in their chemical properties and solubilities; these differences allow the second messengers to transduce the signal in the polar chemical environment of the cytosol or organelle interior, be distributed within the hydrophobic chemical environment of the cellular or organellar membrane, or be translocated across such membranes as a gas or free radical; (3) are subjected to stringent spatial control; indeed, the movement of these second messengers within the cell and their exit out of the cell are regulated in response to changes in the nutrient, energy, stress, and/or proliferation status of the cell; (4) are also controlled temporally, as their intracellular and extracellular concentrations at different periods of chronological lifespan vary significantly and are regulated by alterations in nutrient availability, stress intensity, and cell proliferation rate; and (5) are detected by a distinct set of ligand-specific protein sensors that respond to the concentration changes of different second messengers by altering the rates and efficiencies of longevity-defining cellular processes, thereby creating a pro- or anti-aging cellular pattern and affecting the pace of yeast chronological aging.

Of note, Pietrocola et al. [26] have recently concluded that acetyl-CoA can operate both as a metabolic intermediate and as a second messenger in regulating the delicate balance between cellular catabolism and anabolism in eukaryotic organisms across phyla. In conjunction with the demonstration that acetyl-CoA plays essential roles in regulating the longevity of chronologically aging yeast (see Sections 2.9 and 2.10; [21,25,37,95,96]), this important conclusion further supports the notion that intermediates of some metabolic pathways in eukaryotic cells can function as second messengers that orchestrate global changes in cell functionality in response to changes in the nutrient, energy, stress, and proliferation status of the cell.

Acknowledgments: We are grateful to current and former members of the Titorenko laboratory for discussions. This research was supported by grants from the NSERC of Canada and Concordia University Chair Fund to Vladimir I. Titorenko. Karamat Mohammad and Paméla Dakik were supported by the Concordia University Graduate Fellowship Awards. Younes Medkour was supported by the Concordia University Public Scholars Program Award. Vladimir I. Titorenko is a Concordia University Research Chair in Genomics, Cell Biology, and Aging.

Conflicts of Interest: The authors declare no conflict of interest.

Abbreviations

Asp	aspartate
Asn	asparagine
ATG	autophagy-related genes
ER	endoplasmic reticulum
ETC	electron transport chain
FFA	free (non-esterified) fatty acids
CLS	chronological lifespan
CR	caloric restriction
DAG	diacylglycerol
DDR	DNA damage response
Glu	glutamate
Gln	glutamine
GTR	glutathione reductase
LD	lipid droplets
PKA	protein kinase A
Pkh1	Pkb-activating kinase homolog protein 1
Pkh2	Pkb-activating kinase homolog protein 2
RCD	regulated cell death
ROS	reactive oxygen species
TCA	tricarboxylic acid
TORC1	target of rapamycin complex 1
TRR	thioredoxin reductase
TSP	transsulfuration pathway

References

1. Fontana, L.; Partridge, L.; Longo, V.D. Extending healthy life span—From yeast to humans. *Science* **2010**, *328*, 321–326. [CrossRef] [PubMed]
2. Longo, V.D.; Shadel, G.S.; Kaeberlein, M.; Kennedy, B. Replicative and chronological aging in *Saccharomyces cerevisiae*. *Cell Metab.* **2012**, *16*, 18–31. [CrossRef] [PubMed]
3. Arlia-Ciommo, A.; Leonov, A.; Piano, A.; Svistkova, V.; Titorenko, V.I. Cell-autonomous mechanisms of chronological aging in the yeast *Saccharomyces cerevisiae*. *Microb. Cell* **2014**, *1*, 163–178. [CrossRef] [PubMed]
4. Kaeberlein, M. Lessons on longevity from budding yeast. *Nature* **2010**, *464*, 513–519. [CrossRef] [PubMed]
5. Váchová, L.; Cáp, M.; Palková, Z. Yeast colonies: A model for studies of aging, environmental adaptation, and longevity. *Oxid. Med. Cell. Longev.* **2012**, *2012*. [CrossRef] [PubMed]
6. Denoth Lippuner, A.; Julou, T.; Barral, Y. Budding yeast as a model organism to study the effects of age. *FEMS Microbiol. Rev.* **2014**, *38*, 300–325. [CrossRef] [PubMed]
7. Steinkraus, K.A.; Kaeberlein, M.; Kennedy, B.K. Replicative aging in yeast: The means to the end. *Annu. Rev. Cell Dev. Biol.* **2008**, *24*, 29–54. [CrossRef] [PubMed]
8. Steffen, K.K.; Kennedy, B.K.; Kaeberlein, M. Measuring replicative life span in the budding yeast. *J. Vis. Exp.* **2009**, *28*, 1209. [CrossRef] [PubMed]
9. Smith, E.D.; Tsuchiya, M.; Fox, L.A.; Dang, N.; Hu, D.; Kerr, E.O.; Johnston, E.D.; Tchao, B.N.; Pak, D.N.; Welton, K.L.; et al. Quantitative evidence for conserved longevity pathways between divergent eukaryotic species. *Genome Res.* **2008**, *18*, 564–570. [CrossRef] [PubMed]
10. Ghavidel, A.; Baxi, K.; Ignatchenko, V.; Prusinkiewicz, M.; Arnason, T.G.; Kislinger, T.; Carvalho, C.E.; Harkness, T.A. A genome scale screen for mutants with delayed exit from mitosis: Ire1-independent induction of autophagy integrates ER homeostasis into mitotic lifespan. *PLoS Genet.* **2015**, *11*, e1005429. [CrossRef] [PubMed]
11. McCormick, M.A.; Delaney, J.R.; Tsuchiya, M.; Tsuchiyama, S.; Shemorry, A.; Sim, S.; Chou, A.C.; Ahmed, U.; Carr, D.; Murakami, C.J.; et al. A comprehensive analysis of replicative lifespan in 4698 single-gene deletion strains uncovers conserved mechanisms of aging. *Cell Metab.* **2015**, *22*, 895–906. [CrossRef] [PubMed]
12. Janssens, G.E.; Veenhoff, L.M. Evidence for the hallmarks of human aging in replicatively aging yeast. *Microb. Cell* **2016**, *3*, 263–274. [CrossRef] [PubMed]
13. Fabrizio, P.; Longo, V.D. The chronological life span of *Saccharomyces cerevisiae*. *Methods Mol. Biol.* **2007**, *371*, 89–95. [PubMed]
14. Longo, V.D.; Fabrizio, P. Chronological aging in *Saccharomyces cerevisiae*. *Subcell. Biochem.* **2012**, *57*, 101–121. [PubMed]
15. Burtner, C.R.; Murakami, C.J.; Kennedy, B.K.; Kaeberlein, M. A molecular mechanism of chronological aging in yeast. *Cell Cycle* **2009**, *8*, 1256–1270. [CrossRef] [PubMed]
16. Longo, V.D.; Kennedy, B.K. Sirtuins in aging and age-related disease. *Cell* **2006**, *126*, 257–268. [CrossRef] [PubMed]
17. Murakami, C.; Delaney, J.R.; Chou, A.; Carr, D.; Schleit, J.; Sutphin, G.L.; An, E.H.; Castanza, A.S.; Fletcher, M.; Goswami, S.; et al. pH neutralization protects against reduction in replicative lifespan following chronological aging in yeast. *Cell Cycle* **2012**, *11*, 3087–3096. [CrossRef] [PubMed]
18. Delaney, J.R.; Murakami, C.; Chou, A.; Carr, D.; Schleit, J.; Sutphin, G.L.; An, E.H.; Castanza, A.S.; Fletcher, M.; Goswami, S.; et al. Dietary restriction and mitochondrial function link replicative and chronological aging in *Saccharomyces cerevisiae*. *Exp. Gerontol.* **2013**, *48*, 1006–1013. [CrossRef] [PubMed]
19. Molon, M.; Zadrag-Tecza, R.; Bilinski, T. The longevity in the yeast *Saccharomyces cerevisiae*: A comparison of two approaches for assessment the lifespan. *Biochem. Biophys. Res. Commun.* **2015**, *460*, 651–656. [CrossRef] [PubMed]
20. Cai, L.; Tu, B.P. Driving the cell cycle through metabolism. *Annu. Rev. Cell Dev. Biol.* **2012**, *28*, 59–87. [CrossRef] [PubMed]
21. Titorenko, V.I.; Terlecky, S.R. Peroxisome metabolism and cellular aging. *Traffic* **2011**, *12*, 252–259. [CrossRef] [PubMed]
22. Beach, A.; Burstein, M.T.; Richard, V.R.; Leonov, A.; Levy, S.; Titorenko, V.I. Integration of peroxisomes into an endomembrane system that governs cellular aging. *Front. Physiol.* **2012**, *3*, 283. [CrossRef] [PubMed]

23. Leonov, A.; Titorenko, V.I. A network of interorganellar communications underlies cellular aging. *IUBMB Life* **2013**, *65*, 665–674. [CrossRef] [PubMed]

24. Huang, X.; Withers, B.R.; Dickson, R.C. Sphingolipids and lifespan regulation. *Biochim. Biophys. Acta* **2014**, *1841*, 657–664. [CrossRef] [PubMed]

25. Schroeder, S.; Zimmermann, A.; Carmona-Gutierrez, D.; Eisenberg, T.; Ruckenstuhl, C.; Andryushkova, A.; Pendl, T.; Harger, A.; Madeo, F. Metabolites in aging and autophagy. *Microb. Cell* **2014**, *1*, 110–114. [CrossRef] [PubMed]

26. Pietrocola, F.; Galluzzi, L.; Bravo-San Pedro, J.M.; Madeo, F.; Kroemer, G. Acetyl coenzyme A: A central metabolite and second messenger. *Cell Metab.* **2015**, *21*, 805–821. [CrossRef] [PubMed]

27. Dakik, P.; Titorenko, V.I. Communications between mitochondria, the nucleus, vacuoles, peroxisomes, the endoplasmic reticulum, the plasma membrane, lipid droplets, and the cytosol during yeast chronological aging. *Front. Genet.* **2016**, *7*, 177. [CrossRef] [PubMed]

28. Eltschinger, S.; Loewith, R. TOR complexes and the maintenance of cellular homeostasis. *Trends Cell Biol.* **2016**, *26*, 148–159. [CrossRef] [PubMed]

29. Teixeira, V.; Costa, V. Unraveling the role of the target of rapamycin signaling in sphingolipid metabolism. *Prog. Lipid Res.* **2016**, *61*, 109–133. [CrossRef] [PubMed]

30. Laxman, S. Conceptualizing eukaryotic metabolic sensing and signaling. *J. Indian Inst. Sci.* **2017**, *97*, 59–77. [CrossRef] [PubMed]

31. Li, X.; Handee, W.; Kuo, M.H. The slim, the fat, and the obese: Guess who lives the longest? *Curr. Genet.* **2017**, *63*, 43–49. [CrossRef] [PubMed]

32. Mitrofanova, D.; Dakik, P.; McAuley, M.; Medkour, Y.; Mohammad, K.; Titorenko, V.I. Lipid metabolism and transport define longevity of the yeast *Saccharomyces cerevisiae*. *Front. Biosci.* **2018**, *23*, 1166–1194.

33. Fraenkel, D.G. *Yeast Intermediary Metabolism*; Cold Spring Harbor Laboratory Press: Cold Spring Harbor, NY, USA, 2011; ISBN 978-0-87969-797-6.

34. Grant, C.M. Role of the glutathione/glutaredoxin and thioredoxin systems in yeast growth and response to stress conditions. *Mol. Microbiol.* **2001**, *39*, 533–541. [CrossRef] [PubMed]

35. Brandes, N.; Tienson, H.; Lindemann, A.; Vitvitsky, V.; Reichmann, D.; Banerjee, R.; Jakob, U. Time line of redox events in aging postmitotic cells. *Elife* **2013**, *2*, e00306. [CrossRef] [PubMed]

36. Wei, M.; Fabrizio, P.; Madia, F.; Hu, J.; Ge, H.; Li, L.M.; Longo, V.D. Tor1/Sch9-regulated carbon source substitution is as effective as calorie restriction in life span extension. *PLoS Genet.* **2009**, *5*, e1000467. [CrossRef] [PubMed]

37. Goldberg, A.A.; Bourque, S.D.; Kyryakov, P.; Gregg, C.; Boukh-Viner, T.; Beach, A.; Burstein, M.T.; Machkalyan, G.; Richard, V.; Rampersad, S.; et al. Effect of calorie restriction on the metabolic history of chronologically aging yeast. *Exp. Gerontol.* **2009**, *44*, 555–571. [CrossRef] [PubMed]

38. François, J.; Parrou, J.L. Reserve carbohydrates metabolism in the yeast *Saccharomyces cerevisiae*. *FEMS Microbiol. Rev.* **2001**, *25*, 125–145. [CrossRef] [PubMed]

39. Samokhvalov, V.; Ignatov, V.; Kondrashova, M. Reserve carbohydrates maintain the viability of *Saccharomyces cerevisiae* cells during chronological aging. *Mech. Ageing Dev.* **2004**, *125*, 229–235. [CrossRef] [PubMed]

40. Wang, J.; Jiang, J.C.; Jazwinski, S.M. Gene regulatory changes in yeast during life extension by nutrient limitation. *Exp. Gerontol.* **2010**, *45*, 621–631. [CrossRef] [PubMed]

41. Kyryakov, P.; Beach, A.; Richard, V.R.; Burstein, M.T.; Leonov, A.; Levy, S.; Titorenko, V.I. Caloric restriction extends yeast chronological lifespan by altering a pattern of age-related changes in trehalose concentration. *Front. Physiol.* **2012**, *3*, 256. [CrossRef] [PubMed]

42. Ocampo, A.; Liu, J.; Schroeder, E.A.; Shadel, G.S.; Barrientos, A. Mitochondrial respiratory thresholds regulate yeast chronological life span and its extension by caloric restriction. *Cell Metab.* **2012**, *16*, 55–67. [CrossRef] [PubMed]

43. Cao, L.; Tang, Y.; Quan, Z.; Zhang, Z.; Oliver, S.G.; Zhang, N. Chronological lifespan in yeast is dependent on the accumulation of storage carbohydrates mediated by Yak1, Mck1 and Rim15 kinases. *PLoS Genet.* **2016**, *12*, e1006458. [CrossRef] [PubMed]

44. Svenkrtova, A.; Belicova, L.; Volejnikova, A.; Sigler, K.; Jazwinski, S.M.; Pichova, A. Stratification of yeast cells during chronological aging by size points to the role of trehalose in cell vitality. *Biogerontology* **2016**, *17*, 395–408. [CrossRef] [PubMed]

45. Leonov, A.; Feldman, R.; Piano, A.; Arlia-Ciommo, A.; Lutchman, V.; Ahmadi, M.; Elsaser, S.; Fakim, H.; Heshmati-Moghaddam, M.; Hussain, A.; et al. Caloric restriction extends yeast chronological lifespan via a mechanism linking cellular aging to cell cycle regulation, maintenance of a quiescent state, entry into a non-quiescent state and survival in the non-quiescent state. *Oncotarget* **2017**, *8*, 69328–69350. [CrossRef] [PubMed]

46. Singer, M.A.; Lindquist, S. Multiple effects of trehalose on protein folding in vitro and in vivo. *Mol. Cell* **1998**, *1*, 639–648. [CrossRef]

47. Jain, N.K.; Roy, I. Effect of trehalose on protein structure. *Protein Sci.* **2009**, *18*, 24–36. [CrossRef] [PubMed]

48. Giorgio, M.; Trinei, M.; Migliaccio, E.; Pelicci, P.G. Hydrogen peroxide: A metabolic by-product or a common mediator of ageing signals? *Nat. Rev. Mol. Cell Biol.* **2007**, *8*, 722–728. [CrossRef] [PubMed]

49. Causton, H.C.; Ren, B.; Koh, S.S.; Harbison, C.T.; Kanin, E.; Jennings, E.G.; Lee, T.I.; True, H.L.; Lander, E.S.; Young, R.A. Remodeling of yeast genome expression in response to environmental changes. *Mol. Biol. Cell* **2001**, *12*, 323–337. [CrossRef] [PubMed]

50. Fabrizio, P.; Pozza, F.; Pletcher, S.D.; Gendron, C.M.; Longo, V.D. Regulation of longevity and stress resistance by Sch9 in yeast. *Science* **2001**, *292*, 288–290. [CrossRef] [PubMed]

51. Schroeder, E.A.; Raimundo, N.; Shadel, G.S. Epigenetic silencing mediates mitochondria stress-induced longevity. *Cell Metab.* **2013**, *17*, 954–964. [CrossRef] [PubMed]

52. Schroeder, E.A.; Shadel, G.S. Crosstalk between mitochondrial stress signals regulates yeast chronological lifespan. *Mech. Ageing Dev.* **2014**, *135*, 41–49. [CrossRef] [PubMed]

53. Crespo, J.L.; Powers, T.; Fowler, B.; Hall, M.N. The TOR-controlled transcription activators GLN3, RTG1, and RTG3 are regulated in response to intracellular levels of glutamine. *Proc. Natl. Acad. Sci. USA* **2002**, *99*, 6784–6789. [CrossRef] [PubMed]

54. Powers, R.W., 3rd; Kaeberlein, M.; Caldwell, S.D.; Kennedy, B.K.; Fields, S. Extension of chronological life span in yeast by decreased TOR pathway signaling. *Genes Dev.* **2006**, *20*, 174–184. [CrossRef] [PubMed]

55. Conrad, M.; Schothorst, J.; Kankipati, H.N.; Van Zeebroeck, G.; Rubio-Texeira, M.; Thevelein, J.M. Nutrient sensing and signaling in the yeast *Saccharomyces cerevisiae*. *FEMS Microbiol. Rev.* **2014**, *38*, 254–299. [CrossRef] [PubMed]

56. Swinnen, E.; Ghillebert, R.; Wilms, T.; Winderickx, J. Molecular mechanisms linking the evolutionary conserved TORC1-Sch9 nutrient signalling branch to lifespan regulation in *Saccharomyces cerevisiae*. *FEMS Yeast Res.* **2014**, *14*, 17–32. [CrossRef] [PubMed]

57. Urban, J.; Soulard, A.; Huber, A.; Lippman, S.; Mukhopadhyay, D.; Deloche, O.; Wanke, V.; Anrather, D.; Ammerer, G.; Riezman, H.; et al. Sch9 is a major target of TORC1 in *Saccharomyces cerevisiae*. *Mol. Cell* **2007**, *26*, 663–674. [CrossRef] [PubMed]

58. Huber, A.; Bodenmiller, B.; Uotila, A.; Stahl, M.; Wanka, S.; Gerrits, B.; Aebersold, R.; Loewith, R. Characterization of the rapamycin-sensitive phosphoproteome reveals that Sch9 is a central coordinator of protein synthesis. *Genes Dev.* **2009**, *23*, 1929–1943. [CrossRef] [PubMed]

59. Lee, J.; Moir, R.D.; Willis, I.M. Regulation of RNA polymerase III transcription involves SCH9-dependent and SCH9-independent branches of the target of rapamycin (TOR) pathway. *J. Biol. Chem.* **2009**, *284*, 12604–12608. [CrossRef] [PubMed]

60. Wei, Y.; Zheng, X.F. Sch9 partially mediates TORC1 signaling to control ribosomal RNA synthesis. *Cell Cycle* **2009**, *8*, 4085–4090. [CrossRef] [PubMed]

61. Bonawitz, N.D.; Chatenay-Lapointe, M.; Pan, Y.; Shadel, G.S. Reduced TOR signaling extends chronological life span via increased respiration and upregulation of mitochondrial gene expression. *Cell Metab.* **2007**, *5*, 265–277. [CrossRef] [PubMed]

62. Pan, Y.; Shadel, G.S. Extension of chronological life span by reduced TOR signaling requires down-regulation of Sch9p and involves increased mitochondrial OXPHOS complex density. *Aging* **2009**, *1*, 131–145. [CrossRef] [PubMed]

63. Roosen, J.; Engelen, K.; Marchal, K.; Mathys, J.; Griffioen, G.; Cameroni, E.; Thevelein, J.M.; de Virgilio, C.; de Moor, B.; Winderickx, J. PKA and Sch9 control a molecular switch important for the proper adaptation to nutrient availability. *Mol. Microbiol.* **2005**, *55*, 862–880. [CrossRef] [PubMed]

64. Wanke, V.; Cameroni, E.; Uotila, A.; Piccolis, M.; Urban, J.; Loewith, R.; de Virgilio, C. Caffeine extends yeast lifespan by targeting TORC1. *Mol. Microbiol.* **2008**, *69*, 277–285. [CrossRef] [PubMed]

65. Smets, B.; Ghillebert, R.; de Snijder, P.; Binda, M.; Swinnen, E.; de Virgilio, C.; Winderickx, J. Life in the midst of scarcity: Adaptations to nutrient availability in *Saccharomyces cerevisiae. Curr. Genet.* **2010**, *56*, 1–32. [CrossRef] [PubMed]

66. Medvedik, O.; Lamming, D.W.; Kim, K.D.; Sinclair, D.A. MSN2 and MSN4 link calorie restriction and TOR to sirtuin-mediated lifespan extension in *Saccharomyces cerevisiae. PLoS Biol.* **2007**, *5*, e261. [CrossRef] [PubMed]

67. Lee, P.; Cho, B.R.; Joo, H.S.; Hahn, J.S. Yeast Yak1 kinase, a bridge between PKA and stress-responsive transcription factors, Hsf1 and Msn2/Msn4. *Mol. Microbiol.* **2008**, *70*, 882–895. [CrossRef] [PubMed]

68. Broach, J.R. Nutritional control of growth and development in yeast. *Genetics* **2012**, *192*, 73–105. [CrossRef] [PubMed]

69. Laplante, M.; Sabatini, D.M. mTOR signaling in growth control and disease. *Cell* **2012**, *149*, 274–293. [CrossRef] [PubMed]

70. Alers, S.; Wesselborg, S.; Stork, B. ATG13: Just a companion, or an executor of the autophagic program? *Autophagy* **2014**, *10*, 944–956. [CrossRef] [PubMed]

71. Yorimitsu, T.; Zaman, S.; Broach, J.R.; Klionsky, D.J. Protein kinase A and Sch9 cooperatively regulate induction of autophagy in *Saccharomyces cerevisiae. Mol. Biol. Cell* **2007**, *18*, 4180–4189. [CrossRef] [PubMed]

72. Stephan, J.S.; Yeh, Y.Y.; Ramachandran, V.; Deminoff, S.J.; Herman, P.K. The Tor and PKA signaling pathways independently target the Atg1/Atg13 protein kinase complex to control autophagy. *Proc. Natl. Acad. Sci. USA* **2009**, *106*, 17049–17054. [CrossRef] [PubMed]

73. Stephan, J.S.; Yeh, Y.Y.; Ramachandran, V.; Deminoff, S.J.; Herman, P.K. The Tor and cAMP-dependent protein kinase signaling pathways coordinately control autophagy in *Saccharomyces cerevisiae. Autophagy* **2010**, *6*, 294–295. [CrossRef] [PubMed]

74. Johnson, J.E.; Johnson, F.B. Methionine restriction activates the retrograde response and confers both stress tolerance and lifespan extension to yeast, mouse and human cells. *PLoS ONE* **2014**, *9*, e97729. [CrossRef] [PubMed]

75. Ruckenstuhl, C.; Netzberger, C.; Entfellner, I.; Carmona-Gutierrez, D.; Kickenweiz, T.; Stekovic, S.; Gleixner, C.; Schmid, C.; Klug, L.; Sorgo, A.G.; et al. Lifespan extension by methionine restriction requires autophagy-dependent vacuolar acidification. *PLoS Genet.* **2014**, *10*, e1004347. [CrossRef] [PubMed]

76. Wu, Z.; Song, L.; Liu, S.Q.; Huang, D. Independent and additive effects of glutamic acid and methionine on yeast longevity. *PLoS ONE* **2013**, *8*, e79319. [CrossRef] [PubMed]

77. Ruckenstuhl, C.; Netzberger, C.; Entfellner, I.; Carmona-Gutierrez, D.; Kickenweiz, T.; Stekovic, S.; Gleixner, C.; Schmid, C.; Klug, L.; Hajnal, I.; et al. Autophagy extends lifespan via vacuolar acidification. *Microb. Cell* **2014**, *1*, 160–162. [CrossRef] [PubMed]

78. Liu, K.; Zhang, X.; Lester, R.L.; Dickson, R.C. The sphingoid long chain base phytosphingosine activates AGC-type protein kinases in *Saccharomyces cerevisiae* including Ypk1, Ypk2, and Sch9. *J. Biol. Chem.* **2005**, *280*, 22679–22687. [CrossRef] [PubMed]

79. Liu, K.; Zhang, X.; Sumanasekera, C.; Lester, R.L.; Dickson, R.C. Signalling functions for sphingolipid long-chain bases in *Saccharomyces cerevisiae. Biochem. Soc. Trans.* **2005**, *33*, 1170–1173. [CrossRef] [PubMed]

80. Huang, X.; Liu, J.; Dickson, R.C. Down-regulating sphingolipid synthesis increases yeast lifespan. *PLoS Genet.* **2012**, *8*, e1002493. [CrossRef] [PubMed]

81. Liu, J.; Huang, X.; Withers, B.R.; Blalock, E.; Liu, K.; Dickson, R.C. Reducing sphingolipid synthesis orchestrates global changes to extend yeast lifespan. *Aging Cell* **2013**, *12*, 833–841. [CrossRef] [PubMed]

82. Eisenberg, T.; Knauer, H.; Schauer, A.; Büttner, S.; Ruckenstuhl, C.; Carmona-Gutierrez, D.; Ring, J.; Schroeder, S.; Magnes, C.; Antonacci, L.; et al. Induction of autophagy by spermidine promotes longevity. *Nat. Cell Biol.* **2009**, *11*, 1305–1314. [CrossRef] [PubMed]

83. Morselli, E.; Galluzzi, L.; Kepp, O.; Criollo, A.; Maiuri, M.C.; Tavernarakis, N.; Madeo, F.; Kroemer, G. Autophagy mediates pharmacological lifespan extension by spermidine and resveratrol. *Aging* **2009**, *1*, 961–970. [CrossRef] [PubMed]

84. Minois, N.; Carmona-Gutierrez, D.; Madeo, F. Polyamines in aging and disease. *Aging* **2011**, *3*, 716–732. [CrossRef] [PubMed]

85. Hine, C.; Harputlugil, E.; Zhang, Y.; Ruckenstuhl, C.; Lee, B.C.; Brace, L.; Longchamp, A.; Treviño-Villarreal, J.H.; Mejia, P.; Ozaki, C.K.; et al. Endogenous hydrogen sulfide production is essential for dietary restriction benefits. *Cell* **2015**, *160*, 132–144. [CrossRef] [PubMed]

86. De Cabo, R.; Carmona-Gutierrez, D.; Bernier, M.; Hall, M.N.; Madeo, F. The search for antiaging interventions: From elixirs to fasting regimens. *Cell* **2014**, *157*, 1515–1526. [CrossRef] [PubMed]

87. Lee, C.; Longo, V. Dietary restriction with and without caloric restriction for healthy aging. *F1000Res.* **2016**, *5*. [CrossRef] [PubMed]

88. Hine, C.; Mitchell, J.R. Calorie restriction and methionine restriction in control of endogenous hydrogen sulfide production by the transsulfuration pathway. *Exp. Gerontol.* **2015**, *68*, 26–32. [CrossRef] [PubMed]

89. Shim, H.S.; Longo, V.D. A protein restriction-dependent sulfur code for longevity. *Cell* **2015**, *160*, 15–17. [CrossRef] [PubMed]

90. Burhans, W.C.; Weinberger, M. Acetic acid effects on aging in budding yeast: Are they relevant to aging in higher eukaryotes? *Cell Cycle* **2009**, *8*, 2300–2302. [CrossRef] [PubMed]

91. Burtner, C.R.; Murakami, C.J.; Olsen, B.; Kennedy, B.K.; Kaeberlein, M. A genomic analysis of chronological longevity factors in budding yeast. *Cell Cycle* **2011**, *10*, 1385–1396. [CrossRef] [PubMed]

92. Mirisola, M.G.; Longo, V.D. Acetic acid and acidification accelerate chronological and replicative aging in yeast. *Cell Cycle* **2012**, *11*, 3532–3533. [CrossRef] [PubMed]

93. Giannattasio, S.; Guaragnella, N.; Zdralević, M.; Marra, E. Molecular mechanisms of *Saccharomyces cerevisiae* stress adaptation and programmed cell death in response to acetic acid. *Front. Microbiol.* **2013**, *4*, 33. [CrossRef] [PubMed]

94. Falcone, C.; Mazzoni, C. External and internal triggers of cell death in yeast. *Cell. Mol. Life Sci.* **2016**, *73*, 2237–2250. [CrossRef] [PubMed]

95. Eisenberg, T.; Schroeder, S.; Andryushkova, A.; Pendl, T.; Küttner, V.; Bhukel, A.; Mariño, G.; Pietrocola, F.; Harger, A.; Zimmermann, A.; et al. Nucleocytosolic depletion of the energy metabolite acetyl-coenzyme A stimulates autophagy and prolongs lifespan. *Cell Metab.* **2014**, *19*, 431–444. [CrossRef] [PubMed]

96. Mariño, G.; Pietrocola, F.; Eisenberg, T.; Kong, Y.; Malik, S.A.; Andryushkova, A.; Schroeder, S.; Pendl, T.; Harger, A.; Niso-Santano, M.; et al. Regulation of autophagy by cytosolic acetyl-coenzyme A. *Mol. Cell* **2014**, *53*, 710–725. [CrossRef] [PubMed]

97. Fabrizio, P.; Gattazzo, C.; Battistella, L.; Wei, M.; Cheng, C.; McGrew, K.; Longo, V.D. Sir2 blocks extreme life-span extension. *Cell* **2005**, *123*, 655–667. [CrossRef] [PubMed]

98. Richard, V.R.; Beach, A.; Piano, A.; Leonov, A.; Feldman, R.; Burstein, M.T.; Kyryakov, P.; Gomez-Perez, A.; Arlia-Ciommo, A.; Baptista, S.; et al. Mechanism of liponecrosis, a distinct mode of programmed cell death. *Cell Cycle* **2014**, *13*, 3707–3726. [CrossRef] [PubMed]

99. Sheibani, S.; Richard, V.R.; Beach, A.; Leonov, A.; Feldman, R.; Mattie, S.; Khelghatybana, L.; Piano, A.; Greenwood, M.; Vali, H.; et al. Macromitophagy, neutral lipids synthesis, and peroxisomal fatty acid oxidation protect yeast from "liponecrosis", a previously unknown form of programmed cell death. *Cell Cycle* **2014**, *13*, 138–147. [CrossRef] [PubMed]

100. Mohammad, K.; Dakik, P.; Medkour, Y.; McAuley, M.; Mitrofanova, D.; Titorenko, V.I. Yeast cells exposed to exogenous palmitoleic acid either adapt to stress and survive or commit to regulated liponecrosis and die. *Oxid. Med. Cell. Longev.* **2018**, *2018*. [CrossRef]

101. Beach, A.; Titorenko, V.I. In search of housekeeping pathways that regulate longevity. *Cell Cycle* **2011**, *10*, 3042–3044. [CrossRef] [PubMed]

102. Medkour, Y.; Svistkova, V.; Titorenko, V.I. Cell-nonautonomous mechanisms underlying cellular and organismal aging. *Int. Rev. Cell Mol. Biol.* **2016**, *321*, 259–297. [PubMed]

103. Burstein, M.T.; Kyryakov, P.; Beach, A.; Richard, V.R.; Koupaki, O.; Gomez-Perez, A.; Leonov, A.; Levy, S.; Noohi, F.; Titorenko, V.I. Lithocholic acid extends longevity of chronologically aging yeast only if added at certain critical periods of their lifespan. *Cell Cycle* **2012**, *11*, 3443–3462. [CrossRef] [PubMed]

104. Cantley, L.; Hunter, T.; Sever, R.; Thorner, J. *Signal Transduction: Principles, Pathways, and Processes*; Cold Spring Harbor Laboratory Press: Cold Spring Harbor, NY, USA, 2014; ISBN 978-0-87969-901-7.

105. Lim, W.; Mayer, B.; Pawson, T. *Cell Signaling: Principles and Mechanisms*; Garland Science: New York, NY, USA, 2015; ISBN 978-0-8153-4244-1.

International Journal of
Molecular Sciences

MDPI

Article

Aging and Intermittent Fasting Impact on Transcriptional Regulation and Physiological Responses of Adult *Drosophila* Neuronal and Muscle Tissues

Sharon Zhang [1,2,3], Eric P. Ratliff [1,3], Brandon Molina [1,3], Nadja El-Mecharrafie [1,3], Jessica Mastroianni [1,3], Roxanne W. Kotzebue [1,3], Madhulika Achal [1], Ruth E. Mauntz [1], Arysa Gonzalez [1,3], Ayeh Barekat [1,3], William A. Bray [1,3,4], Andrew M. Macias [1,3], Daniel Daugherty [5], Greg L. Harris [3], Robert A. Edwards [2,3] and Kim D. Finley [1,3,*]

[1]   Donald P. Shiley BioScience Center, San Diego State University, San Diego, CA 92182, USA;
       sharonzh10@yahoo.com (S.Z.); epratliff@hotmail.com (E.P.R.); Brandonrm.16@gmail.com (B.M.);
       nadjaelm@outlook.com (N.E.-M.); jmfmastroianni@yahoo.com (J.M.); rkotzebue@gmail.com (R.W.K.);
       madhulika.achal@gmail.com (M.A.); rmauntz@gmail.com (R.E.M.); gonzalez.arysa@gmail.com (A.G.);
       ayeh.sdsu@gmail.com (A.B.); yossarianassyrian@gmail.com (W.A.B.);
       maciasandrew01@gmail.com (A.M.M.)
[2]   Biological and Medical Informatics Research Center, San Diego State University, San Diego, CA 92182, USA;
       redwards@sdsu.edu
[3]   Department of Biology, San Diego State University, San Diego, CA 92182, USA; gharris@sdsu.edu
[4]   Sanford Consortium for Regenerative Medicine, 2880 Torrey Pines Scenic Dr., La Jolla, CA 92037, USA
[5]   The Salk Institute for Biological Studies, Cellular Neurobiology, 10010 North Torrey Pines Road,
       La Jolla, CA 92037, USA; ddaugherty@salk.edu
*   Correspondence: kfinley@sdsu.edu

Received: 5 March 2018; Accepted: 30 March 2018; Published: 10 April 2018

Abstract: The progressive decline of the nervous system, including protein aggregate formation, reflects the subtle dysregulation of multiple functional pathways. Our previous work has shown intermittent fasting (IF) enhances longevity, maintains adult behaviors and reduces aggregates, in part, by promoting autophagic function in the aging *Drosophila* brain. To clarify the impact that IF-treatment has upon aging, we used high throughput RNA-sequencing technology to examine the changing transcriptome in adult *Drosophila* tissues. Principle component analysis (PCA) and other analyses showed ~1200 age-related transcriptional differences in head and muscle tissues, with few genes having matching expression patterns. Pathway components showing age-dependent expression differences were involved with stress response, metabolic, neural and chromatin remodeling functions. Middle-aged tissues also showed a significant increase in transcriptional drift-variance (TD), which in the CNS included multiple proteolytic pathway components. Overall, IF-treatment had a demonstrably positive impact on aged transcriptomes, partly ameliorating both fold and variance changes. Consistent with these findings, aged IF-treated flies displayed more youthful metabolic, behavioral and basal proteolytic profiles that closely correlated with transcriptional alterations to key components. These results indicate that even modest dietary changes can have therapeutic consequences, slowing the progressive decline of multiple cellular systems, including proteostasis in the aging nervous system.

Keywords: aging; aging-delaying interventions; metabolism; cellular proteostasis; neural degeneration; intermittent fasting; RNA-sequencing; *Drosophila*

1. Introduction

Both human and model organism studies have shown even modest dietary modifications can have a significant positive influence on longevity and the health of aging individuals [1–7]. Existing research into the molecular mechanisms that underpin such improved health and longevity metrics indicates that the long-term function of multiple cellular systems is essential [3,5,8–10]. Indeed, numerous genetic studies have shown that a diverse range of regulatory, metabolic and clearance pathways are required for the healthy aging of the nervous system [8,11–13]. In addition, a wide range of environmental factors, especially modified diets, have consistently shown a positive influence on the age-related decline of select tissues as well as the overall functional health and performance of older individuals [2,10,14,15].

We have recently characterized a mild intermittent fasting (IF) protocol that promotes longevity and neural function of middle-aged adult *Drosophila* strains from divergent genetic backgrounds [1,16,17]. The beneficial effects of this fasting protocol included a delay in the progressive decline of locomotor behaviors and a reduction in neural aggregate levels [1,16,17]. This was associated with improvements in the acute and long-term functions of the autophagy pathway within middle-aged fly neurons [1,16,17]. Multiple studies have established that this pathway plays a central role in stress responses and in promoting healthy aging, partly through facilitating the clearance of protein aggregates and damaged cellular components [13,18,19]. At the mechanistic level, autophagy represents a downstream effector pathway, the activity of which is tightly regulated by multiple upstream factors that control and coordinate complex cellular responses [1,16,17,19,20]. The impact of IF-treatment on the aging CNS could reflect the alteration to multiple signaling and metabolic pathways that in turn influence pathway function, and as a result, the long-term maintenance of the aging nervous system [1,16,17,19,20].

Most organisms show well-defined aging phenotypes, which can reflect tissue-specific alterations to numerous molecular and metabolic pathways [15,21–23]. Taken together, these changes precede and eventually lead, to the functional decline (behavior, metabolism, stress responses), and the eventual death of an individual. Both genetic and drug-based studies using in vivo model systems have shown that the different genetic backgrounds and treatment regimens, which promote health and longevity, are often linked to basal changes to gene expression profiles [2,14,15,23,24]. Of note, a recent RNA-sequencing (RNA-seq) study that examined changes to the acult *C. elegans* transcriptome profiles, characterized a novel age-related phenomenon termed "transcriptional drift-variance" (TD), which reflects the progressive dysregulation of mRNA expression patterns [23]. In older worms, TD was significantly increased between replicate transcriptomes, which was partially suppressed in matching cohorts following exposure to the anti-aging compound, mianserin [23,25]. The analysis of transcriptome profiles from a range of mouse and human frontal cortex tissues samples also indicated that TD was an evolutionary conserved feature of aging [23]. A recent report examining the neural physiology of middle-age senescence-accelerated prone (SAMP8) mice also showed elevated TD levels in neural tissues. Following treatment with the neural protective compound, J147, TD levels were significantly reduced in mice and the average lifespan of adult *Drosophila* increased [26]. The implications from these studies are that environmental interventions that promote longevity, including IF-treatment, could facilitate more youthful mRNA expression patterns and variance profiles [1,2,5,26–28].

In this report, we examined the impact that aging and IF-treatment have upon the transcriptome profiles (non-fasted) in the adult *Drosophila* neural (head) and muscle (thorax) tissues [1,2,23,26]. Our analyses revealed that, by 4-week of age, both fly tissues showed significant basal changes to expression levels, with IF-treatment promoting more youthful global expression profiles. Both tissues exhibited the age-dependent phenomenon of transcriptional drift-variance, providing further evidence that the progressive disruption of transcriptional regulation also occurs in middle-aged flies [23,25,26]. Transcripts demonstrating fold changes and TD differences involved a wide array of genes representing a number of functional groups and pathways. In several cases, the age and IF-dependent changes were consistent with unique tissue-specific phenotypes and functional changes that included metabolic, behavioral and epigenetic systems [26]. Of particular interest were the dynamic variance changes, in neural tissues that occurred to multiple proteolytic pathway components, including members of the ubiquitin-proteasome system (UPS) and autophagy–lysosomal pathway. Globally these results indicate that even modest dietary manipulations could improve basal transcriptomic and phenotypic trends, which are themselves associated with partial suppression of several cellular processes linked to impaired cellular aging [23,26,29].

2. Results

2.1. Global Expression Differences Due to Age or IF-Treatment

To elaborate upon our original investigation into the impact of intermittent fasting (IF) on neuronal autophagy and aging, high throughput RNA-sequencing was used to examine global changes to the gene expression profiles in adult male *Drosophila* tissues. Examined were fly body segments that represented regions highly enriched with either neuronal (head) or skeletal muscle (thorax) tissues. Triplicate mRNA pools from adult heads were isolated from 1-week (1W), 4-week (4W) to 4-week IF-treated (4W-IF) male flies [1,20]. Matching thoracic mRNA pools (duplicates) were also isolated and used as reference non-neuronal control tissue set. Individual cDNA libraries were generated for each tissue and condition and then sequenced using Illumina HiSeq2000 technology [2,24,27,30]. Sequencing read mapping and alignments were done using the *Drosophila* reference genome (UCSC) and Genome Browser (genome.ucsc.edu, Santa Cruz, CA, USA), with sequence alignments and internal read mapping generated using TopHat and Bowtie (see Methods and Materials) [2,23,29]. Comparisons between annotated library sequences found that most reads (~80%) aligned with the reference transcriptome and represented ~94% exon-only sequences (Table S1).

As part of the predictive modeling of global transcriptome profiles, the high-dimensional variable analysis provided by the AltAnalyze analytical pipeline (Available online: https://www.altanalyze. org), was used to characterize the heterogeneity between individual RNA-seq datasets. This included performing principal component analyses (PCA) on total reads per kilobase of transcript, per million mapped reads (RPKM) values, which reduced the dimensionality of each expression dataset into two principal components (Figure 1) [23,29]. The PCA comparison of head transcriptomes demonstrated the close alignment between replicate samples as well as the expression differences that could largely be attributed to aging or IF-treatment (X axis, 98.9%, Figure 1A). PC analysis of head and thoracic datasets found distinct non-overlapping transcriptome profiles between the two tissues (PCA1 89.7%, Figure S1A), though both showed comparable shifts in global expression due to age or IF-treatment (PCA2 3.9%, Figure S1A). Interestingly, the head 4W-IF transcriptomes showed a significant tread toward more youthful patterns and had reduced TD variability when compared to matching non-fasted 4W controls (Figure 1A). Hierarchical clustering plots further highlighted the similarities between replicate samples as well as the impact that aging (4W) and IF-treatment (4W-IF) have upon global transcriptome trends (Figure 1B and Fgure S1B) [29,30]. The average RPKM values for transcripts typically expressed in muscle, neural or glial cell types were examined for both sets of tissues. In general, muscle specific genes were enriched in all thoracic samples ($n = 6$), while genes typically expressed in neural or glial cell types were preferentially detected across all RNA-seq samples from adult heads ($n = 9$, Table S2). This indicated that the transcriptome patterns from each body segment in large part reflected the expected tissue-specific expression profiles [2,29].

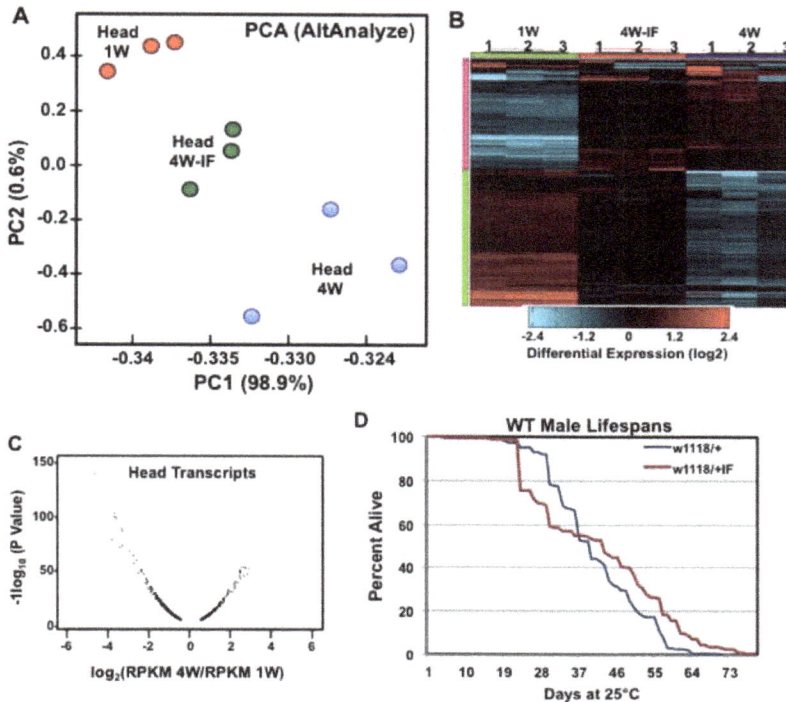

Figure 1. Principal component analysis (PCA) and expression clustering profiles of head transcriptomes. (**A**) The AltAnalyze software was used to compare PCA values for individual RNA-seq transcriptomes isolated from 1-week (1W), 4-week (4W) to IF-treated (4W-IF) male flies ($n = 3$). (**B**) AltAnalyzer was also used to establish the individual expression clustering profiles for head ($n = 9$) transcriptomes at different ages and treatment conditions. (**C**) Volcano plot showing significant message fold directional ($\pm$) changes in the CNS that occurs as a function of age (4W/1W, negative log10 of *p*-values as a function of log2). (**D**) Kaplan–Meier survival curves of male flies ($w^{1118}/+$) maintained on *ad libitum* or IF conditions starting at 1-week of age (25 °C) [1].

2.2. Age and IF-Dependent Changes in Transcriptome Profiles

The global assessment of neural profiles indicated that significant basal expression changes were occurring both as a consequence of aging and the IF-treatment regimen. Anticipating relatively modest changes to basal gene expression, the average fold change to each mRNA cohort was then examined, with significant cutoffs for aging set at >1.4 ± fold (4W/1W), and for IF-dependent changes, at >1.3 ± fold (4W/4W-IF) [2,29]. By 4-weeks, a similar number of genes from head (1197) and thoracic (1347) tissues showed substantial expression differences (Table 1). Age-related transcriptional changes in the fly head tended toward a broader deregulation of mRNA profiles (Figure 1C; Table 1), though the relative amplitude of such changes were more pronounced in 4W thoracic tissues (Figure S2A) [2,23]. In contrast, a larger number of neural transcripts showed a significant global response to IF-exposure when heads were compared to thoracic samples (294 versus 102 genes, Table 1). Further analysis of dynamic expression differences determined there was minimal gene overlap between fly head and thoracic tissues due to aging (Figure S2B; Table 1). Tissue-specific differences were also reflected in the IF-dependent responses, with neural tissues showing a significant shift toward more youthful expression patterns in 4W-IF cohorts, relative to those in the thorax (221 versus 58 genes, Table 1). The positive impact that

IF-treatment has on progressive transcriptional changes in both tissues was consistent with the more general phenotype of enhanced longevity, which was demonstrated in our previous studies for male flies from several genotypes (Figure 1D) [1].

Table 1. Genes with Fold Changes due to Aging and Intermittent Fasting.

Tissue	Change with Age	Age Down	Age Up	Change with IF	Age Down IF Up	Age Up IF Down	IF More Youthful
Thorax	1347	790	557	102	22	36	58
Head	1197	806	391	294	178	43	221

2.3. Gene Ontology (GO) or DAVID Analysis of Functional Pathway Changes

To assess the functional classification of genes or pathways influenced by age or IF-treatment, the DAVID bioinformatics program was used to identify functional clusters of genes with tissue-specific expression differences (Figure 2B; Table 1) [2,29]. For each condition, the transcriptome profiles from neural tissues showed a greater number of annotated functional groups with altered patterns of expression relative to thoracic tissues (Table S3). The largest functional gene clusters that showed dynamic age and youthful IF-dependent expression trends in the adult fly CNS were metabolic-related (Figure 2). Generally, these components are associated with longevity and lipid, carbohydrate (arrows) and mitochondrial related functions (Figure 2A). To confirm RNA-seq expression trends, quantitative RT-PCR (qRT-PCR) analysis of the *Tobi*, *Lps2* and *Sodh-1* genes was undertaken in neural tissues [31,32]. This indicated that the relative expression profiles of each gene matched the age and IF-dependent fluctuations detected by RNA-seq (Figure 2B–D). All three proteins have well-established metabolic functions and expression patterns that are influenced by aging, diet or alterations to the insulin-signaling pathway [31,32]. In thoracic tissues, the *Tobi* transcript also showed pronounced changes in expression due to aging and IF-treatment, albeit in the opposing directions (arrow, Figure S2C).

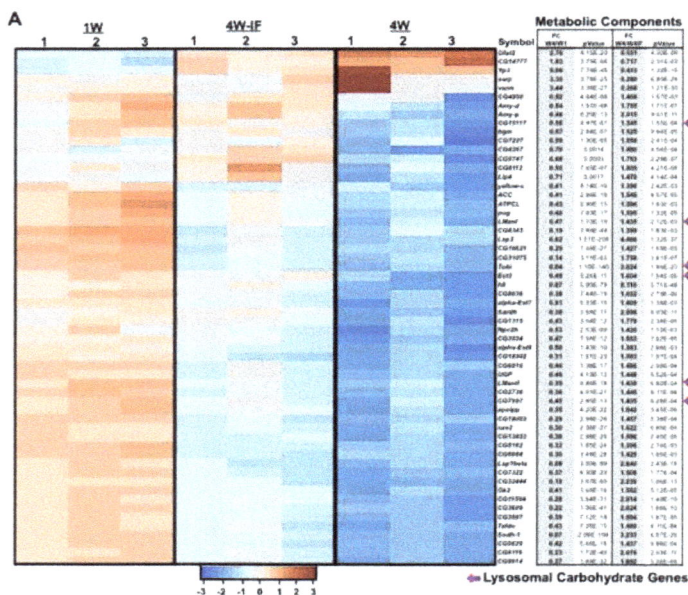

Figure 2. *Cont.*

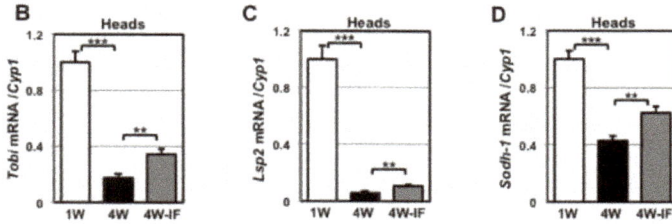

Figure 2. Age and IF-dependent fold changes (FC) to the expression profiles of metabolic pathway components in adult *Drosophila* heads. Quantitative RNA-sequencing and DAVID analysis identified metabolic genes that showed age and IF-dependent differences in expression levels. (**A**) Heatmap and corresponding table represents hierarchical clustering of scaled gene expression profiles and the respective fold change (FC) and *p* values between 4W/1W and 4W/4W-IF cohorts. Scaled expression values (z-score) were plotted in red–blue color scale with red indicating high expression and blue indicating low expression levels. Arrows highlight lysosomal proteins involved with carbohydrate metabolism. (**B–D**) qRT-PCR of expression changes to the *Tobi*, *Lsp2* and *Sodh-1* genes in neural tissues. ** $p \leq 0.01$, *** $p \leq 0.001$.

2.4. Age and IF-Dependent Neuronal and Behavioral Changes

Previously, we determined that IF-treated flies have lower insoluble protein aggregate profiles in neural tissues, which was linked to the preservation of adult climbing behaviors [1]. This was consistent with other modified diet studies that also detect a positive impact on the long-term function of neural tissues [1,33–37]. DAVID analysis revealed genes involved with a wide range of neural functions, including olfaction, had dynamic changes in expression patterns (Figure 3A). We had previously demonstrated that older male flies develop a marked increase in nighttime activity that was linked with the dysregulation of odorant binding protein (*Obp*) gene expression profiles and olfactory-based courtship behaviors [16]. In this study, elevated expression of *Atg8a* in neuronal tissues maintained both *Obp* expression patterns and suppressed the changes to courtship behaviors [16]. To determine whether IF-treatment could similarly alter nighttime activity profiles, 1W, 4W and 4W-IF treated male flies (w^{1118}/+) were placed in group-housed conditions (10 flies/tube) and examined using 12-h light:dark (LD) cycling conditions for 48-h (LAM system). The 24-h activity profiles showed normal behavior patterns for young males (1W), including typical morning and evening peaks of activity, with extended mid-day and mid-dark rest periods (ZT15-21, Figure 3B,C) [16]. Middle-aged males (4W) demonstrated the normal increase in nighttime activity, which was largely suppressed in IF-treated cohorts (Figure 3B,C). Similar to transgenic enhancement of neuronal *Atg8a* levels, these results indicated that even modest dietary manipulations can protect the progressive dysregulation of this and other behaviors, further indicating the preservation of neural function in aged animals [1,16,17,19].

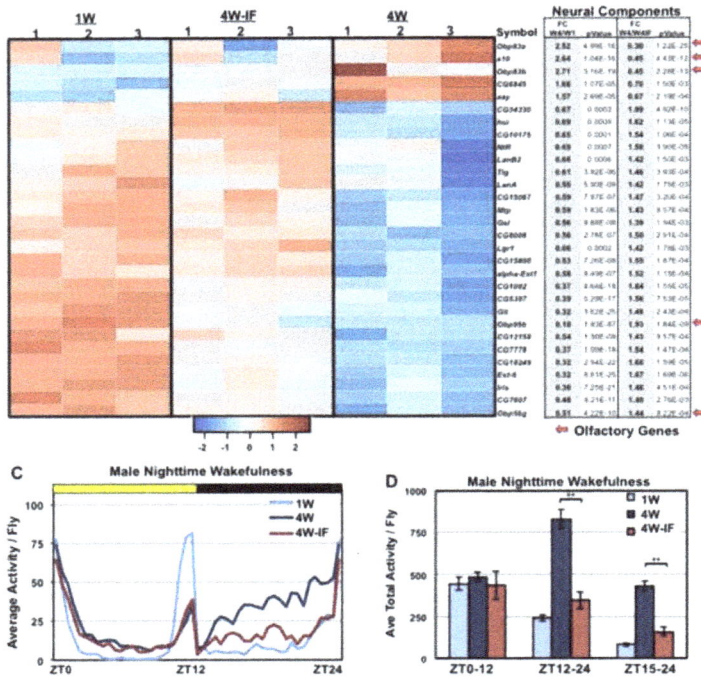

Figure 3. Age and IF-dependent expression changes to adult neuronal pathway components and adult olfactory-based behaviors. DAVID analysis identified genes with neuronal functions that demonstrated age and IF-dependent fluctuations in expression levels. (**A**) Heatmap and corresponding table represents hierarchical clustering of scaled gene expression profiles and fold change (FC) and *p* values between 4W/1W and 4W/4W-IF RNA-seq cohorts. Scaled expression values (Z-score) of RPKM expression levels were plotted in red (high) versus blue (low) color scale. Arrows indicate olfactory binding genes that are linked to behavioral defects. (**B,C**) Average 24 h activity profiles of grouped housed 1W, 4W and 4W IF male fly cohorts (*n* = 16 groups of 10 flies) and (**D**) Activity levels during light (ZT0 to 12) dark (ZT12 to 24) and mid-dark (ZT15 to 24) time periods. ** *p* ≤ 0.01.

2.5. Altered Expression Patterns of Stress Response and Epigenetic Pathway Components

DAVID analysis of thoracic samples identified significant fluctuations to the expression profiles of regulatory, proteolytic, vesicle transport and structural pathway components (Figure S3A–C). The fly CNS also demonstrated significant age and IF dependent fold changes to the expression levels of several stress response pathway components (Figure 4A). This included the *Drosophila Hsp22* transcript (mitochondrial heat shock protein), which in both head and thoracic tissues showed an age-dependent increase and IF-related decrease in expression levels (Figure 4B). Changes to *Hsp22* mRNA values were confirmed using quantitative (qRT-PCR) analyses (*n* = 16, Figure 4C), with expression patterns becoming highly variable in aged thorax samples (4W, Figure 4B,C). The age-related increase in expression variance for this gene suggested that transcription drift variance (TD) was occurring in older *Drosophila* tissues [23,26]. It also suggested that upstream transcription factors that regulated downstream expression and variance profiles could be influenced by protective dietary conditions [23,26]. Comparing the two tissues identified significant FC to the expression pattern of epigenetic pathway components, although there was minimal specific overlap between head and muscle samples (Figure 5A,B). For example the *CoRest* and *Df31* genes showed opposing response to aging IF exposure (blue arrows, Figure 5A,B). Taken together, our data suggests that both age and modified diets can produce unique tissue-specific changes to epigenetic pathway components in concert with the broader regulatory differences seen in the adult *Drosophila* transcriptome [38,39].

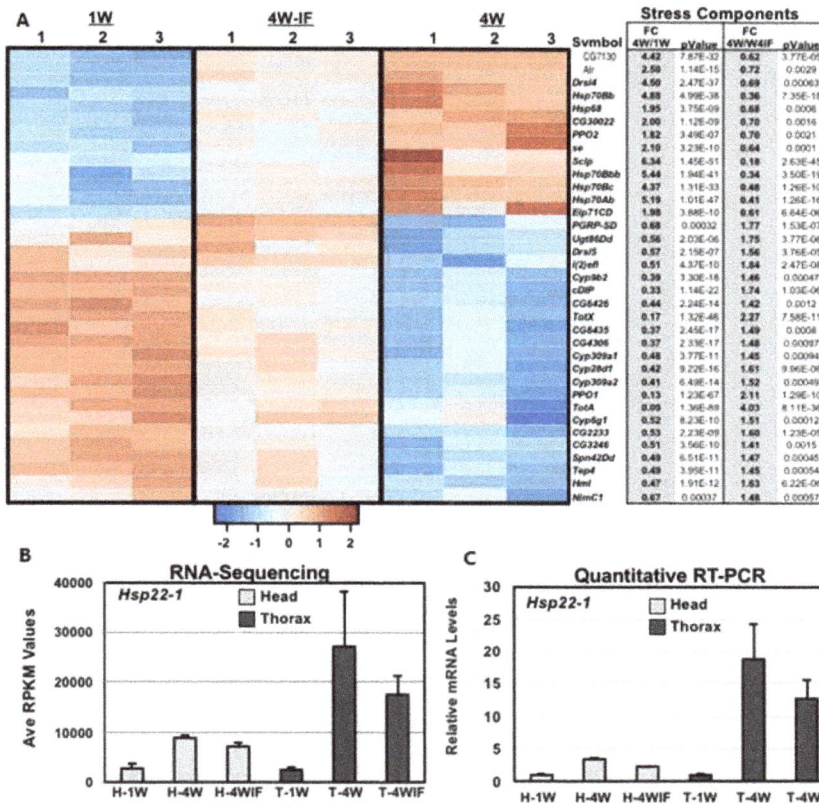

Symbol	FC 4W/1W	pValue	FC 4W/W4IF	pValue
CG7130	4.42	7.87E-32	0.62	3.77E-05
Atr	2.50	1.14E-15	0.72	0.0029
Drsl4	4.50	2.47E-37	0.69	0.00063
Hsp70Bb	4.88	4.99E-38	0.36	7.35E-18
Hsp68	1.95	3.75E-09	0.68	0.0006
CG30022	2.00	1.12E-09	0.70	0.0016
PPO2	1.82	3.49E-07	0.70	0.0021
se	2.19	3.23E-10	0.64	0.0001
Sclp	6.34	1.45E-51	0.18	2.63E-45
Hsp70Bbb	5.44	1.94E-41	0.34	3.50E-19
Hsp70Bc	4.37	1.31E-33	0.46	1.26E-10
Hsp70Ab	5.19	1.01E-47	0.41	1.26E-16
Eip71CD	1.98	3.88E-10	0.61	6.64E-06
PGRP-SD	0.68	0.00032	1.77	1.53E-07
Ugt86Dd	0.56	2.03E-06	1.75	3.77E-06
Drsl5	0.57	2.15E-07	1.56	3.76E-06
l(2)efl	0.51	4.37E-10	1.84	2.47E-08
Cyp9b2	0.39	3.30E-18	1.46	0.00047
cDIP	0.33	1.14E-22	1.74	1.03E-06
CG6426	0.44	2.24E-14	1.42	0.0012
TotX	0.17	1.32E-46	2.27	7.58E-11
CG6435	0.37	2.45E-17	1.49	0.0008
CG4306	0.37	2.33E-17	1.48	0.00097
Cyp309a1	0.48	3.77E-11	1.45	0.00094
Cyp28d1	0.42	9.22E-16	1.61	9.96E-06
Cyp309a2	0.41	6.49E-14	1.52	0.00049
PPO1	0.13	1.23E-67	2.11	1.29E-10
TotA	0.09	1.36E-88	4.03	8.11E-36
Cyp6g1	0.52	8.23E-10	1.51	0.00012
CG2233	0.53	2.23E-09	1.60	1.23E-05
CG3246	0.51	3.56E-10	1.41	0.0015
Spn42Dd	0.49	6.51E-11	1.47	0.00045
Tep4	0.49	3.95E-11	1.45	0.00054
Hml	0.47	1.91E-12	1.53	6.22E-06
NimC1	0.87	0.00032	1.48	0.00057

Figure 4. Age and IF associated changes to stress and inflammation pathway components in adult neural tissues. RNA-Seq and DAVID analysis of *Drosophila* head transcriptomes identified multiple genes in stress and inflammation related pathways that demonstrated both an age and IF-dependent difference in expression levels. (**A**) The heatmap and corresponding table represents the clustering of scaled gene expression profiles and includes the respective fold change in expression (FC, *p* values) that occur between 4W/1W and 4W/4W-IF tissue cohorts. Scaled expression values (z-scores) were plotted using a red–blue color scale represented dynamic differences expression differences to individual genes. Red indicates relative elevated expression and blue indicates relative reduced expression levels. The mRNA expression levels of *Hsp22-1* in head (**H**) and thorax (**T**) tissues measured in RPKM values by (**B**) RNA-seq or (**C**) qRT-PCR analyses. The relative mRNA expression of 1W flies (w^{1118}/+) was set to 1.0 and subsequent RNA values were also normalized to the housekeeping *CXba* transcript.

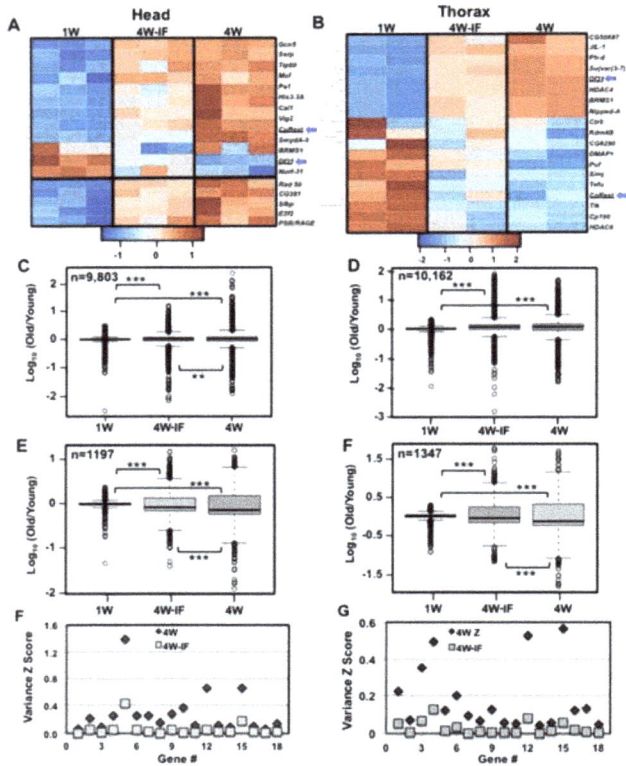

Figure 5. Age and IF-dependent changes to epigenetic pathway components and tissue specific transcriptional drift-variance profiles. Heatmaps of chromatin remodeling pathway components, showing dynamic FC to expression levels in aged (**A**) head or (**B**) thoracic tissue samples (n = number of genes). Arrows highlight the *Df31* and *CoRest* transcripts (blue arrows) that have significant but opposing tissue specific FC to expression profiles. Representative drift-plots showing the global variance changes or VC (~10,000 genes) to (**C**) head or (**D**) thorax tissue transcriptomes, isolated from 1-week 1W, 4W to 4W-IF adult cohorts. Drift plots representing (**E**) head and (**F**) thoracic transcripts (~1200 genes) that showed significant age-dependent FC to expression levels. *** $p \leq 0.001$. See Table 2 for additional details. DAVID analyses were used to determine the functional pathways of head and thoracic genes that demonstrated elevated variance Z scores (4W/4W-IF VC > 3.75). (**G**) The scaled variance Z scores (SD/Ave) for individual transcripts involved with neuronal function were used to generate scatter plots of 4W (◆) and 4W-IF (■) head cohorts. (**H**) The scaled variance Z scores (SD/Ave) for lipid metabolic genes were plotted for 4W (◆) and 4W-IF (■) thoracic transcriptome samples.

2.6. Age and IF-Dependent Changes to Transcriptional Drift

Neural PCA values (4-week, Figure 1A) and muscle *Hsp22* mRNA profiles (Figure 4B,C) suggested that middle-aged fly tissues showed an increased variability between matching replicate transcriptomes [23]. Therefore, the impact of aging and IF-treatment on transcriptional drift variance (TD) was assessed for both tissues. The global variance differences were examined for those genes with replicate RNA-seq values of >1.0 RPKM across all datasets for a particular tissues. The corrected log-fold change in expression variance was determined for each transcriptional cohort, using the 1W transcriptomes as 'young' reference values ($\log_{10}[\text{old/young reference}_{1W}]$) [23,26,40]. Global TD profiles for head (9803) and thorax (10,163) mRNA cohorts are outlined in Table 2 and illustrated as drift-plots (Figure 5C–F, www.r-project.org) [23,26]. When compared to young flies (1W), total transcriptome profiles from aged cohorts (4W, 4W-IF) demonstrated a significant increase in expression variance or TD (Figure 5C,D) [23]. However, when the two aged profiles were compared, global TD levels were suppressed in 4W-IF neural samples (Figure 5C, Table S4). When a subset of genes that had age-dependent FCs in expression was examined, IF-treatment reduced the variance levels for both tissues (Figure 5E,F, Table S4). This analysis indicated that expression variance significantly increased in middle-aged *Drosophila*, while IF-treatment promoted more youthful TD or variance change (VC) profiles for both tissues, suggestive of alteration in global transcriptional regulation [23,26].

Table 2. Variance Changes * with Age and IF.

Variance > 3.75	4W/1W	4W/4W-IF	1W/4W-IF
Thorax genes	1457	1592	1441
Head genes	1156	1322	520

* Based on replicate RPKM values and corrected Z scores (SD/Ave).

Further characterization of mRNA fold changes suggested that differences in expression patterns could be classified in terms of variability between replicate mRNA samples. Therefore, gene variance Z scores (**VZ**) from normalized replicate RNA-seq values were generated (VZ = SD/Ave RPKM). Comparisons were made between the different age (4W/1W) and treatment (4W/4W-IF) cohorts for both tissues, and individual genes with fold VCs were identified. Nearly 1300 neural and 1600 thoracic transcripts demonstrated a significant reduction in variance profiles following IF-exposure (4W/4W-IF VC > 3.75, Table 2). Often, gene specific FC to message levels did not coincide with VCs, which itself may reflect a heretofore-underappreciated feature of aging. DAVID analysis was performed on those transcripts in both tissues with high VC levels (4W), and that were suppressed in 4W-IF cohorts. In the aging CNS, genes with both FC and VC differences included odorant-binding and synaptic proteins and key components of the circadian pathway (*tim, per*) (Figure 5G, Table S5) [2,16]. Together with the age-dependent behavioral changes (Figure 3B,C), these finding suggest that examining both FC and VC differences could serve to clarify the genes and functional pathways that have malleable expression patterns that can be influenced by aging, diets or other modifying factors [23,26,29,41].

DAVID analysis was also performed on thoracic genes demonstrating age-related TD or expression variance patterns (4W/4W-IF). Several lipid pathway components showed substantial VCs that were significantly reduced following IF-exposure (Figure 5H), while the FC to expression levels remained largely unchanged (Table S6). As previously reported, basal and acute autophagy responses are influenced by age and IF-treatment, suggesting that whole animal fasting responses and rates of metabolic catabolism may also be similarly altered [1,2,32,42]. Therefore, the global starvation responses of 1W, 4W and 4W-IF treated flies were examined. Male flies were placed on fasting media (1% agar) and the number of dead flies counted every 8 h [43]. Middle-aged adults showed a heightened sensitivity to starvation that was partly suppressed in age matched IF-treated cohorts (Figure S4A,B). Since aged thoraces showed both substantial fold and variance changes to metabolic pathways, we examined the global catabolism rates of stored metabolites. Young (1W) and middle-aged (4W, 4W-IF) flies were fasted for 0 or 8 h, flash frozen and tissue homogenates used to determine whole body triglyceride (TG), glycogen and glucose levels (mg/mg protein, Figure S4C–H) [8,42]. Young flies had the highest basal stores and catabolism rates for all three metabolites (Figure S4C–H). Interestingly, older adults (4W) showed similar TG levels but lower consumption rates following and 8-h fast (14% decrease). In fasted 4W-IF cohorts, the TG catabolism profiles were more youthful (27.5% decrease, Figure S4C,D), while turnover of carbohydrate stores showed different profiles (Figure S4E–H) [42,44]. Together, this suggests that age and IF-dependent changes that are reflected as FC and VC expression differences in thoracic tissues have functional consequences in metabolic and survival responses in older individuals (Figure S4) [8,42].

2.7. Altered Proteolytic Pathways and Protein Aggregate Profiles in Aging Fly Tissues

Using a sequential detergent extraction and Western analysis, we have consistently shown that under normal culturing conditions that by middle-age, adult flies have a significant buildup of insoluble ubiquitinated protein (IUP) aggregates in neural tissues [1,16,17,19,45]. We have also demonstrated that the rate at which protein aggregates increase (IUP, Ref(2)P) closely correlates with the tissue specific decline in autophagic capacity, which can be manipulated by genetic, transgenic and dietary factors [1,16,17,19,45]. Close examinations of FCs found relatively few proteolytic pathway components with significant expression differences in either head or thoracic tissues (Figure S3B). In sharp contrast, DAVID analysis of neural genes with altered variance scores identified multiple elements in proteolytic pathways that became hyper-variable with age (4W), which were largely suppressed by IF-treatment (4W/4W-IF ratio > 3.5, Tables S7 and S8). Individually, this included proteasomal (18), lysosomal (21), autophagy (7) and ubiquitin (7) pathway components, which in neural samples had remained undetected due to minimal FCs to message levels (Figure 6A; Tables S7 and S8). Following IF-treatment (4W-IF) there was substantial suppression of variance profiles for the same set of genes (Figure 6A) [23,26,43,45,46]. In matching thoracic samples, these genes showed minimal FC or VC differences, indicating proteolytic pathway components were largely unaffected by aging or dietary conditions in muscle tissues (Figure 6B; Tables S7 and S8).

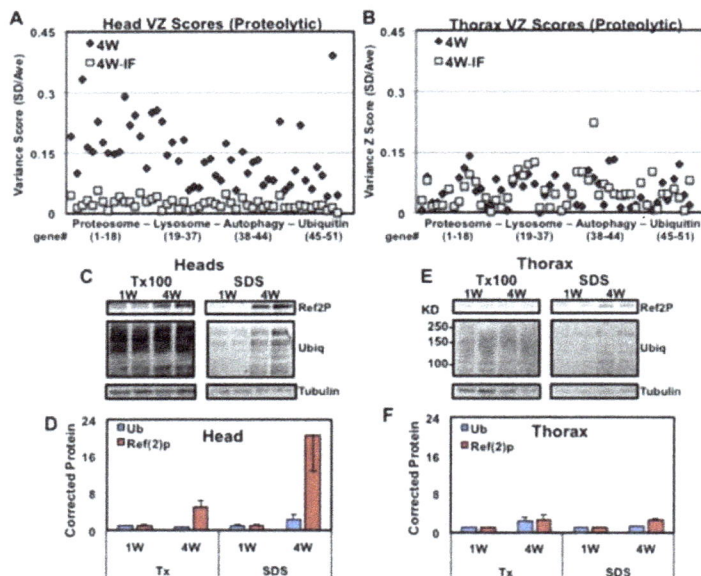

Figure 6. Age and IF-dependent tissue specific changes to proteolytic pathway variance and aggregate profiles. DAVID analysis was performed on neural transcripts that showed an age-related elevation and IF-reduction in TD variance scores (4W/4W-IF, >3.75). (**A**) Scatter plot of scaled variance Z scores (SD/Ave) of proteolytic pathway components from 4W (◆) to 4W-IF (■) head transcriptomes. (**B**) The scaled variance Z scores (SD/Ave) for the same cohort of genes from replicate 4W (◆) to 4W-IF (■) thoracic transcriptomes. See Tables S7 and S8 for individual gene information. Head and thoracic tissues from 1W to 4W flies underwent sequential detergent fractionation before Western analyses. Blots containing the Triton-X100 and SDS soluble protein fractions were probed with ant-UB, anti-Ref(2)P and anti-Tubulin antibodies and corrected protein levels quantified for both (**C,D**) head and (**E,F**) thoracic tissues. At 4 weeks of age, neural tissues demonstrated the buildup of protein aggregates, which is limited in age-matched thoracic samples.

Given the marked tissue-specific differences in proteolytic expression patterns, the age-dependent buildup of aggregate profiles between head and thoracic tissue samples was directly compared. Sequential Triton-X100 (1%) and SDS (2%) detergent protein extracts were isolated from replicate 1W to 4W old fly heads and thoraxes and Westerns probed to establish protein aggregate profiles (Figure 6C–F). In neural tissues, Western blots of the SDS soluble protein fractions showed the normal age-dependent buildup of neural protein aggregates (4W) (Figure 6C,D). Age-matched thoracic samples showed minimal IUP and Ref(2)P level differences (Figure 6E,F). Together this was consistent with our studies showing the timeline and tissue specific development of protein aggregates in the aging fly CNS, which also coincides with altered transcriptional regulation for proteolytic pathway components [1]. These results also underscored the concept that therapeutic changes, including modified diets, can have a positive influence on a diverse array of tissue-specific cellular processes including proteolysis [1,23,26,29]. The transcriptional changes observed with the intermittent fasting based diet also appears to reflect the long-term positive consequences for aged individuals in terms of global and tissue-specific expression patterns and physiological responses [38,39].

3. Discussion

The progressive changes to multiple interconnected cellular systems has been mechanistically linked to the aging [1,8,16,39]. It has consequently been challenging to separate the primary cause(s) leading to the progressive deterioration of cells and tissues from the downstream or secondary collateral effects of aging [3,8,9]. Both genetic and environmental studies have established that multiple cellular pathways are involved with and can influence the normal aging process as well as long-term maintenance and function of the nervous system [1,12,16,25,34]. The implications from other longevity studies are that net positive changes due to diet likely involved subtle alterations to signaling, epigenetic, mitochondrial and proteolytic based pathways [2,3,25,28,31,47]. In this report, we have used RNA-seq analysis to continue our investigation into the dynamic expression and phenotypic changes that occur in adult *Drosophila* as a consequence of aging and exposure to intermittent fasting [1]. Previously, we have demonstrated that IF-treatment had a positive impact on middle-aged flies by slowing the progression of several global and neurodegenerative phenotypes [1]. This included longer average lifespans, the preservation of adult behaviors (geotaxis), lower neural aggregate profiles (IUP, Ref(2)P) and enhance basal and acute autophagic responses in the aging fly nervous system [1]. This implied that even modest dietary changes could have a significant impact on the regulation of cellular pathways and that are essential for the long-term functional maintenance of adult tissues.

For this study, the selection of mRNA samples (time, treatment, tissue) was based on well-defined age-dependent alterations to the behavioral and physiological profiles of wild type adult flies [1,17,45]. The over primary goal of these studies was to identify early changes to transcriptome profiles that could serve as the nexus to assess early changes to molecular mechanisms involved with progressive age-related defects, and to highlight cellular pathways that are responsive to environmental treatments [1,2,8]. Our initial comparisons of PCA and expression clustering profiles for individual RNA-sequenced transcriptomes confirmed the similarities between replicate samples, and highlighted the dynamic tissue-specific changes occurring to basal expression profiles as consequence of aging or IF-treatment (Figure 1 and Figures S1 and S2; Tables 1 and 2) [29]. Further analysis confirmed that adult head and thoracic tissues have unique expression profiles that are selectively enriched for genes normally produced in neural, glial, or muscle cell types (Table S2). Global tissue-specific responses also included a relatively higher fold expression for thoracic transcripts, though neural tissues showed a larger number of genes responding to IF-treatment (4W/1W, Table 1). The relative number of genes that showed significant age-dependent FC and VC differences were similar between the two tissue types, although relatively few genes showed matching mRNA fluctuation patterns (Figure S2B).

IF-treatment may have a greater impact on the regulation of the adult head transcriptome profiles in part due to the level of tissue heterogeneity found within this *Drosophila* body segment. The majority of the adult thorax represents a relatively homogenous tissue, primarily comprised of flight and jump skeletal muscle cell type (see Table S2). Conversely, the adult fly head is comprised of diverse neural and glial cells, which have multiple functions and unique tissue specific transcriptional responses as a result of aging or IF exposure. The difference in the number of replicate transcriptome data sets used in our analysis of head ($n = 3$) and thorax ($n = 2$) tissues could have resulted in an under representation of expression differences between the two tissue types. However, the PCA and expression clustering analyses (Figure 1A,B and Figure S1) confirmed replicate sample similarities as well as the divergence expression trends that occurred as a result of age or dietary conditions for each tissue type. In addition, the relatively uniform number of genes showing age and IF-dependent changes between the two tissues (Tables 1 and 2) suggested that analysis of duplicate RNA-seq data sets, while not optimal could be used to identify significant fold and variance changes to individual transcripts.

From these initial studies, non-directed DAVID analysis identified multiple functional groups and pathways that demonstrated unique tissue-specific FC in expression that could be attributed to both aging and nutritional conditions (Figures 2–5 and Figures S2 and S3; Table 1 and Table S7) [23,29]. Each tissue showed unique age or IF-dependent changes to multiple pathway components, with only few genes demonstrating similar expression trends. One notable exception was the *Hsp22* gene, which in *Drosophila* serves as a chaperone protein involved with mitochondrial homeostasis, ROS detoxification and longevity-based functions [15,21]. Previously, fly tissues showed a basal age-dependent FC to *Hsp22* message and protein levels that closely correlates with enhanced longevity and stress resistance [15,21]. Consistent with these findings, RNA-seq and qRT-PCR analysis for both tissues showed an aged-dependent increase (4W) and IF-dependent reduction in the *Hsp22* message (Figure 4B,C). Together with elevated variance profiles in thoracic tissues (4W), this data suggests that IF-treatment reduces endogenous levels of cellular stress and may improve mitochondrial function in 4W-IF flies [15,21].

Recent studies examining the impact that aging and therapeutic compounds have on the regulation of adult *C. elegans* and murine transcriptome profiles, has highlighted the occurrence of transcriptional drift-variance in model organism [23,26]. This work has strengthened the concept that along with average FCs to mRNA levels, there exists an age-dependent increase in global expression drift-variance, likely reflecting the dysregulation of transcriptional stoichiometry [23,26]. Treatment with the drug mianserin (serotonergic pathway inhibitor) promoted longevity in adult *C. elegans* and resulted in a dose dependent reduction in TD and more youthful serotonergic pathway function [23,25]. Older mice exposed to the neural protective compound J147 also showed suppressed TD in neural tissues as well as the reduction in plasma metabolomics drift profiles [26]. Further, detecting TD variance in tissue transcriptome profiles from older human tissues suggests this is a conserved phenomenon, likely to involve progressive changes to epigenetic regulatory pathways [23,38,48,49]. In this study, we determined that tissue-specific changes to chromatin-remodeling pathway components (Figure 5A,B), closely correlated with the significant increase in TD profiles (4W) and the partial suppression of both following IF-treatment (Figure 5C–H). While both head and thoracic tissues showed a similar number of transcripts with an age-dependent increase in TD variance, the individual components were tissue-specific and largely did not overlap (Table 2). In addition, neural tissues appeared to have a greater response to IF-treatment and lower global expression variance profiles (4W-IF) than those found for matching thoracic transcriptome profiles (>3.75, Figure 5C–F; Table 2).

Since middle-aged *Drosophila* tissues showed unique VC transcriptional patterns DAVID analysis was used to characterized genes and functional pathways influenced by aging with IF-treatment and restored to more youthful pattern following IF exposure. In neural tissues genes showing lower VC have circadian (*tim, per*), olfactory (*Obp99a*), and neural (*neur, mnb, Fmr1, Sap47*) functions (Figure 5G; Table S5). The increase in TD profiles coincided with neuronal FC differences and the progressive decline and partial rescue of adult behaviors (Figure 3) [1,16]. For thoracic tissues, DAVID analyses revealed a very different set of genes showing dynamic FC and VC differences (Figure 5B,H and Figure 6B; Tables S6–S8). This included proteins involved with metabolism (Figures S2C and S4) as well as significant VCs to lipid homeostatic components (Figure 5H; Table S6) [32,42]. The functional potential impact of aging and IF-treatment on metabolic-based FC and VC in expression patterns closely correlated with improved starvation responses (Figure S4A,B) and more youthful catabolism rates for lipids (TG) in fasting 4W-IF flies (Figure S4C,D) [1,42]. Overall, the metabolic differences seen in IF-treated flies were consistent with previous dietary studies showing improved metabolic and physiological metrics in aged animals [2,32].

Along with age-related changes to key neuronal systems, the decline in CNS function is often associated with progressive proteostasis defects that include functional changes to the ubiquitin-proteasome and autophapy-lysosomal systems [1,16,17,25,34,50]. Indeed, the age-dependent formation of protein inclusions is often closely associated with the decline of motor behaviors, sensory perception (olfaction), sleep/circadian patterns and cognition [1,16,29,34]. Human studies and work using model systems have shown modified diets can slow the functional decline of the nervous system and may promote synaptic plasticity and neurogenesis in older adults [3,37,50]. In previous studies, we demonstrated that middle-aged flies exposed to IF-treatment demonstrated improved autophagy profiles including basal and acute fasting responses [1]. Consistent with this finding was the pronounced VC detected for multiple proteolytic components within the proteasome (*Rpn1,*), ubiquitin (*UBE4A*), lysosomal (*Cath-L*), and autophagy (*Atg16L2, PI3K59F*) pathways (Figure 6A; Tables S7 and S8). This appeared to be dynamic neural-specific VC, which was primarily not detected in matching thoracic samples (Figure 6B; Tables S7 and S8) [43,45,46,51]. Of particular note are key components involved with regulating proteostasis by the proteasome (subunits *Rpt, Rpn*) [46] and autophagy (*PI3K59F, Atg16L2*) [51], which showed significant VC and minimal FC in expression in contrasting transcriptome samples from aged fly cohorts (4W/4W-IF, Figure 6A; Tables S7 and S8). Taken together, the selective neuronal increase in proteolytic VC profiles was consistent with tissue-specific disparities observed in the timing and relative level of ubiquitinated protein aggregate accumulation (Figure 6C–F).

Overall this study indicates that conditions that alter in vivo aging phenotypes can be closely linked to expression difference to pathway components associated with individual functions. In addition, this analysis can also be used to identify upstream regulatory factors that coordinate the expression of key downstream functional components [39]. Therefore, observing the suppression of FC and TD profiles by select treatment conditions could not only highlight "sensitive" gene targets but also identify the cellular mechanisms that facilitate or promote their long-term regulation. This could include individual changes to signaling systems, transcription factors or epigenetic pathway components. In this study, IF-associated basal changes to expression patterns primarily highlighted differences to epigenetic components for both adult *Drosophila* tissues (Figure 5A,B). This implies that individual treatments or environmental conditions that alter aging phenotypes may be reflected as unique tissue-specific transcriptional "fingerprints" that are impacted by nucleosome and heterochromatin modifications [18,38,39,52]. In terms of healthy human aging, there is a growing appreciation that modest dietary interventions may be an effective method to preserve more youthful transcription patterns and facilitate the treatment of chronic progressive disorders [27,34,53,54]. The overarching goal for this type of treatment would be to maximize the benefit, while minimizing the side effects and compliance issues associated with harsher dietary regimens, including caloric restriction [34,54,55]. Given these attributes, applying targeted dietary modifications to aging population may have a lasting beneficial effect upon disorders linked to the functional decline of the nervous system [6,37,47,56–58].

4. Material and Methods

4.1. Drosophila Stocks, Culturing Conditions and Starvation Responses

Canton-S (CS) and w^{1118} fly stocks (Bloomington *Drosophila* Stock Center, Bloomington, IN, USA), crosses and F1 offspring have been previously described and were originally obtained from the Bloomington Stock center (Bloomington, IN, USA) [1]. Flies used for each study represented F1 offspring generated from crosses between CS virgin females and w^{1118} males (w^{1118}/+) [16]. Adult male flies were collected within four hours of eclosion and maintained on standard fly media (25 per vial) and culturing conditions (25 °C, 65% humidity, 12-h light:dark, LD cycle) [1,16]. For IF-treatment studies, flies were maintained using standard conditions and media until 1-week of age, then were exposed to either IF or *ad libitum* culturing conditions [1]. IF-treated flies were turned onto fasting vials (1% agar) three times per week from 9:00 a.m. to 5:00 p.m. (8-h) [1]. For all studies, IF-treated flies were placed on *ad libitum* conditions for two full days prior to being used for behavioral and starvation studies or being flash frozen in liquid nitrogen for subsequent tissue isolation [1]. Flies were stored at −80 °C for before tissues were isolated and processed for mRNA, protein or metabolic analyses [1].

4.2. RNA Isolation, Library Construction, RNA-Sequencing and Bioinformatics Analyses

F1 fly cohorts (w^{1118}/+) exposed to IF-treatment or *ad libitum* conditions for three weeks were collected and flash frozen at 1-week or at 4-week of age. *Drosophila* tissue processing, RNA isolation and cDNA library construction are outlined in Supplemental Information [16]. Libraries were sequenced using the Illumina HiSeq2000 technology (Illumina, Inc., San Diego, CA, USA) for single-end 100-bp format (7-base index). Briefly, staff at The Scripps Research Institute Sequencing Core Facility (La Jolla, CA, USA) processed sequencing data to generate FASTQ files demultiplexed based on index sequences [2,24]. Sequenced reads were mapped to a *Drosophila* reference genome (UCSC Genome and Browser (vmm9.gtf, UCSC, Santa Cruz, CA, USA) and alignments generated using TopHat (v2.0.9) and Bowtie 2 (v2.1.0, John Hopkins University, Baltimore, MD, USA) software., Cufflinks (http://cufflinks.cbcb.umd.edu/) was used to assemble sequence reads for individual transcripts [2,24,29]. For each sequenced library, the number of reads per gene was normalized and reported as RPKM values as an estimation of expression levels. To determine the fold changes, both the average RPKM and $\log_2$ RPKM scaled values for each gene was determined from replicate reads (FC = Ave log2 RPKM 4W/Ave log2 RPKM 1W). Significant expression differences due to aging were set at >1.4 (4W/1W) and IF-treatment at >1.3 (4W/4W-IF). In addition, normalized RPKM values used to generate heat maps (fold), variance fold changes and drift-variance plots (box plots) from different transcript cohorts and are further detailed in Supplemental Information. The AltAnalyse software (v2.1.0, Cincinnati Children's Hospital, Cincinnati, OH, USA) was used for PCA analysis and to generate expression-clustering profiles for each RNA-seq data set. The DAVID bioinformatics program (https://david.ndifcrf.gov) was used to identify functional gene clusters and associations for transcripts showing fold or variance differences due to age or IF. A custom designed Python program (Python 3.6.4, Copyright© 2001–2018 Python Software Foundation, https://www.python.org/psf/) was used to refine functional pathway gene lists that were used to generate heatmaps, expression tables and variance scatter plots. Methods used to normalize values for heatmaps, transcriptional drift and drift-variance profiles and normalized drift-variance values are further detailed in Supplemental Information.

4.3. 24-h Activity Profiles of Aged Group-Housed Male Flies

Outcrossed male flies were collected within 4 h of eclosion and aged in cohorts of 25 using standard husbandry and IF conditions. Flies were entrained using 12-h light and 12-h dark conditions (LD), with lights-on starting at 8:00 a.m. and lights-off at 8:00 p.m. [16,20]. The activity profiles of group-housed male flies were examined using the LAM25 systems (Trikinetics Inc., Waltham, MA, USA) and established protocols [16]. The LAM25 system detects fly movement (infrared beam breaks) and activity events were detected using the DAM System3 program (Trikinetics Inc., Waltham, MA, USA) [16]. Following an overnight recovery, the activity profiles of 1W, 4W and 4W-IF male fly groups (10 per vial) were monitored for 48 consecutive hours, using 12-h LD conditions [16]. Collected data sets were further analyzed using a custom-designed Python program (https://www.python.org/psf/, Python v3.6.4, Copyright © 2001-2018 Python Software Foundation) and Microsoft Excel software (Microsoft Office®, Mac 2011 v14.7.1) to generate activity graphs during the 12-light (ZT0–12), 12-dark (ZT12–24) and 6-h mid-dark (ZT15–21) time periods [16].

4.4. Quantitative RT-PCR

Flies from different ages and treatment conditions were collected, flash frozen and stored at −80 °C. Head and thoracic tissues were isolated, cDNA libraries generated quantitative Real-Time PCR studies performed using established technique [16,20]. The relative mRNA expression of 1W WT flies (w^{1118}/+) was set to 1.0 and subsequent RNA values were also normalized to the housekeeping *CXba* transcript [1,16,19]. Primer sequences will be provided upon request and additional methods details are included in Supplemental Information.

4.5. Metabolite Profiles

Adult male flies from 1W, 4W, or 4W-IF conditions were fasted for 0-h, 4-h or 8-h on 1% agar prior to final collection, flash freezing and storage at −80 °C. See Supplemental Information for additional detailed methods used to determine whole body triglyceride, glycogen and glucose levels and Figure S4 for results [1,42,43].

4.6. Western Blot Analysis

Outcrossed male flies were aged, collected and flash frozen at 1 and 3-week of age. Adult heads and thoraces were collected and individual tissues homogenized using standardized reagents and techniques. [1,16,20]. See Supplemental Information for detailed methods.

4.7. Statistical Analysis

Quantified Western blot, activity profiles, and metabolic graphs were generated using Microsoft Excel and figures assembled in PowerPoint (Microsoft Office®, Mac 2011 v14.7.1) [16]. Unless otherwise stated statistical analyses between groups were performed using the GraphPad software (https://www.graphpad.com/) and Student's *t*-test (two-tailed, unpaired). All values are reported as mean values ± SEM [16,20].

5. Conclusions

This and previous studies have demonstrated that even modest dietary modifications can have a significant beneficial impact on the cellular processes and physiological responses of aging worms, flies, mice and humans [1–3,5,47]. Along with promoting longevity and healthy neural aging, we have shown that IF-treatment can have a net positive impact on the progressive decline of several additional phenotypes that are reflected by the tissue-specific fluctuations to adult *Drosophila* transcriptome profiles [1,2]. This includes more youthful metabolic responses, the maintenance of olfactory-based courtship behavior, and the neural specific buildup of proteins aggregates [1,3,4,7,16,34]. The work detailed in this study indicates that even modest dietary changes can have a significant impact on progression of aging phenotypes, which in turn is reflected by significant differences in basal expression patterns. Our results indicate that protein homeostasis and clearance pathways in neural tissues are influenced and that changes to upstream epigenetic factors may be mediating these responses [59,60].

Supplementary Materials: The following are available online at http://www.mdpi.com/1422-0067/19/4/1140/s1.

Acknowledgments: We would like to thank Steven Head and other members of the Next Generation Sequencing Core facility at The Scripps Research Institute (TSRI). We would also like to thank Michael Petrascheck (TSRI) for helpful discussions regarding the analysis and modeling of transcriptional drift-variance in *Drosophila* tissues. This work was supported by grants from NIH/NIA including: R21AG030187, R01AG039628 and R44AG033427.

Author Contributions: Eric P. Ratliff, Roxanne W. Kotzebue, Sharon Zhang, Kim D. Finley conceived of these studies. Eric P. Ratliff, Sharon Zhang, Brandon Molina, Ruth E. Mauntz, Arysa Gonzalez, Madhulika Achal, Nadja El-Mecharrafie, Roxanne W. Kotzebue, Ayeh Barekat, Kim D. Finley carried out the analyses of expression data (RNA sequencing, qRT-PCR) as well as experiments involving Western analysis, behaviors and metabolic experiments. Sharon Zhang, Daniel Daugherty, Eric P. Ratliff, Ruth E. Mauntz, Madhulika Achal, Robert A. Edwards, Kim D. Finley conducted bioinformatics analysis and interpretation of data sets. Sharon Zhang,

Andrew M. Macias, Eric P. Ratliff, Greg L. Harris, Jessica Mastroianni, William A. Bray, Kim D. Finley prepared the manuscript.

Conflicts of Interest: The authors declare no conflict of interest.

References

1. Ratliff, E.P.; Kotzebue, R.W.; Molina, B.; Mauntz, R.E.; Gonzalez, A.; Barekat, A.; El-Mecharrafie, N.; Garza, S.; Gurney, M.A.; Achal, M.; et al. Assessing Basal and Acute Autophagic Responses in the Adult *Drosophila* Nervous System: The Impact of Gender, Genetics and Diet on Endogenous Pathway Profiles. *PLoS ONE* **2016**, *11*, e0164239. [CrossRef] [PubMed]
2. Gill, S.; Le, H.D.; Melkani, G.C.; Panda, S. Time-restricted feeding attenuates age-related cardiac decline in *Drosophila*. *Science* **2015**, *347*, 1265–1269. [CrossRef] [PubMed]
3. Brandhorst, S.; Choi, I.Y.; Wei, M.; Cheng, C.W.; Sedrakyan, S.; Navarrete, G.; Dubeau, L.; Yap, L.P.; Park, R.; Vinciguerra, M.; et al. A Periodic Diet that Mimics Fasting Promotes Multi-System Regeneration, Enhanced Cognitive Performance, and Healthspan. *Cell Metab.* **2015**, *22*, 86–99. [CrossRef] [PubMed]
4. Hansen, M.; Chandra, A.; Mitic, L.L.; Onken, B.; Driscoll, M.; Kenyon, C. A role for autophagy in the extension of lifespan by dietary restriction in *C. elegans*. *PLoS Genet.* **2008**, *4*, e24. [CrossRef] [PubMed]
5. Longo, V.D.; Antebi, A.; Bartke, A.; Barzilai, N.; Brown-Borg, H.M.; Caruso, C.; Curiel, T.J.; de Cabo, R.; Franceschi, C.; Gems, D.; et al. Interventions to Slow Aging in Humans: Are We Ready? *Aging Cell* **2015**, *14*, 497–510. [CrossRef] [PubMed]
6. Longo, V.D.; Panda, S. Fasting, Circadian Rhythms, and Time-Restricted Feeding in Healthy Lifespan. *Cell Metab.* **2016**, *23*, 1048–1059. [CrossRef] [PubMed]
7. Min, K.J.; Tatar, M. Restriction of amino acids extends lifespan in *Drosophila melanogaster*. *Mech. Ageing Dev.* **2006**, *127*, 643–646. [CrossRef] [PubMed]
8. Wang, L.; Karpac, J.; Jasper, H. Promoting longevity by maintaining metabolic and proliferative homeostasis. *J. Exp. Biol.* **2014**, *217*, 109–118. [CrossRef] [PubMed]
9. Hansen, M.; Kennedy, B.K. Does Longer Lifespan Mean Longer Healthspan? *Trends Cell Biol.* **2016**, *26*, 565–568. [CrossRef] [PubMed]
10. Sala, A.J.; Bott, L.C.; Morimoto, R.I. Shaping proteostasis at the cellular, tissue, and organismal level. *J. Cell Biol.* **2017**, *216*, 1231–1241. [CrossRef] [PubMed]
11. Grandison, R.C.; Wong, R.; Bass, T.M.; Partridge, L.; Piper, M.D. Effect of a standardised dietary restriction protocol on multiple laboratory strains of *Drosophila melanogaster*. *PLoS ONE* **2009**, *4*, e4067. [CrossRef] [PubMed]
12. Hwangbo, D.S.; Gershman, B.; Tu, M.P.; Palmer, M.; Tatar, M. *Drosophila* dFOXO controls lifespan and regulates insulin signalling in brain and fat body. *Nature* **2004**, *429*, 562–566. [CrossRef] [PubMed]
13. Yamamoto, R.; Tatar, M. Insulin receptor substrate chico acts with the transcription factor FOXO to extend *Drosophila* lifespan. *Aging Cell* **2011**, *10*, 729–732. [CrossRef] [PubMed]
14. Stegeman, R.; Weake, V.M. Transcriptional Signatures of Aging. *J. Mol. Biol.* **2017**. [CrossRef] [PubMed]
15. Tower, J.; Landis, G.; Gao, R.; Luan, A.; Lee, J.; Sun, Y. Variegated expression of Hsp22 transgenic reporters indicates cell-specific patterns of aging in *Drosophila* oenocytes. *J. Gerontol. A Biol. Sci. Med. Sci.* **2014**, *69*, 253–259. [CrossRef] [PubMed]
16. Ratliff, E.P.; Mauntz, R.E.; Kotzebue, R.W.; Gonzalez, A.; Achal, M.; Barekat, A.; Finley, K.A.; Sparhawk, J.M.; Robinson, J.E.; Herr, D.R.; et al. Aging and Autophagic Function Influences the Progressive Decline of Adult *Drosophila* Behaviors. *PLoS ONE* **2015**, *10*, e0132768. [CrossRef] [PubMed]
17. Bartlett, B.J.; Isakson, P.; Lewerenz, J.; Sanchez, H.; Kotzebue, R.W.; Cumming, R.C.; Harris, G.L.; Nezis, I.P.; Schubert, D.R.; Simonsen, A.; et al. p62, Ref(2)P and ubiquitinated proteins are conserved markers of neuronal aging, aggregate formation and progressive autophagic defects. *Autophagy* **2011**, *7*. [CrossRef]
18. Wood, J.G.; Hillenmeyer, S.; Lawrence, C.; Chang, C.; Hosier, S.; Lightfoot, W.; Mukherjee, E.; Jiang, N.; Schorl, C.; Brodsky, A.S.; et al. Chromatin remodeling in the aging genome of *Drosophila*. *Aging Cell* **2010**, *9*, 971–978. [CrossRef] [PubMed]
19. Simonsen, A.; Cumming, R.C.; Brech, A.; Isakson, P.; Schubert, D.R.; Finley, K.D. Promoting basal levels of autophagy in the nervous system enhances longevity and oxidant resistance in adult *Drosophila*. *Autophagy* **2008**, *4*, 176–184. [CrossRef] [PubMed]

20. Barekat, A.; Gonzalez, A.; Mauntz, R.E.; Kotzebue, R.W.; Molina, B.; El-Mecharrafie, N.; Conner, C.J.; Garza, S.; Melkani, G.C.; Joiner, W.J.; et al. Using *Drosophila* as an integrated model to study mild repetitive traumatic brain injury. *Sci. Rep.* **2016**, *6*, 25252. [CrossRef] [PubMed]
21. Morrow, G.; Tanguay, R.M. *Drosophila melanogaster* Hsp22: A mitochondrial small heat shock protein influencing the aging process. *Front. Genet.* **2015**, *6*, 103. [CrossRef] [PubMed]
22. Ivanisevic, J.; Stauch, K.L.; Petrascheck, M.; Benton, H.P.; Epstein, A.A.; Fang, M.; Gorantla, S.; Tran, M.; Hoang, L.; Kurczy, M.E.; et al. Metabolic drift in the aging brain. *Aging* **2016**, *8*, 1000–1020. [CrossRef] [PubMed]
23. Rangaraju, S.; Solis, G.M.; Thompson, R.C.; Gomez-Amaro, R.L.; Kurian, L.; Encalada, S.E.; Niculescu, A.B., 3rd; Salomon, D.R.; Petrascheck, M. Suppression of transcriptional drift extends *C. elegans* lifespan by postponing the onset of mortality. *Elife* **2015**, *4*, e08833. [CrossRef] [PubMed]
24. Landis, G.N.; Salomon, M.P.; Keroles, D.; Brookes, N.; Sekimura, T.; Tower, J. The progesterone antagonist mifepristone/RU486 blocks the negative effect on life span caused by mating in female *Drosophila*. *Aging* **2015**, *7*, 53–69. [CrossRef] [PubMed]
25. Rangaraju, S.; Solis, G.M.; Andersson, S.I.; Gomez-Amaro, R.L.; Kardakaris, R.; Broaddus, C.D.; Niculescu, A.B., 3rd; Petrascheck, M. Atypical antidepressants extend lifespan of Caenorhabditis elegans by activation of a non-cell-autonomous stress response. *Aging Cell* **2015**, *14*, 971–981. [CrossRef] [PubMed]
26. Goldberg, J.; Currais, A.; Prior, M.; Fischer, W.; Chiruta, C.; Ratliff, E.; Daugherty, D.; Dargusch, R.; Finley, K.; Esparza-Molto, P.B.; et al. The mitochondrial ATP synthase is a shared drug target for aging and dementia. *Aging Cell* **2018**. [CrossRef] [PubMed]
27. Schafer, M.J.; Dolgalev, I.; Alldred, M.J.; Heguy, A.; Ginsberg, S.D. Calorie Restriction Suppresses Age-Dependent Hippocampal Transcriptional Signatures. *PLoS ONE* **2015**, *10*, e0133923. [CrossRef] [PubMed]
28. Min, K.J.; Yamamoto, R.; Buch, S.; Pankratz, M.; Tatar, M. *Drosophila* lifespan control by dietary restriction independent of insulin-like signaling. *Aging Cell* **2008**, *7*, 199–206. [CrossRef] [PubMed]
29. Currais, A.; Goldberg, J.; Farrokhi, C.; Chang, M.; Prior, M.; Dargusch, R.; Daugherty, D.; Armando, A.; Quehenberger, O.; Maher, P. A comprehensive multiomics approach toward understanding the relationship between aging and dementia. *Aging* **2015**, *7*, 937–955. [CrossRef] [PubMed]
30. Tariq, M.A.; Kim, H.J.; Jejelowo, O.; Pourmand, N. Whole-transcriptome RNAseq analysis from minute amount of total RNA. *Nucleic Acids Res.* **2011**, *39*, e120. [CrossRef] [PubMed]
31. Buch, S.; Melcher, C.; Bauer, M.; Katzenberger, J.; Pankratz, M.J. Opposing effects of dietary protein and sugar regulate a transcriptional target of *Drosophila* insulin-like peptide signaling. *Cell Metab.* **2008**, *7*, 321–332. [CrossRef] [PubMed]
32. Chatterjee, D.; Katewa, S.D.; Qi, Y.; Jackson, S.A.; Kapahi, P.; Jasper, H. Control of metabolic adaptation to fasting by dILP6-induced insulin signaling in *Drosophila* oenocytes. *Proc. Natl. Acad. Sci. USA* **2014**, *111*, 17959–17964. [CrossRef] [PubMed]
33. Fann, D.Y.; Santro, T.; Manzanero, S.; Widiapradja, A.; Cheng, Y.L.; Lee, S.Y.; Chunduri, P.; Jo, D.G.; Stranahan, A.M.; Mattson, M.P.; et al. Intermittent fasting attenuates inflammasome activity in ischemic stroke. *Exp. Neurol.* **2014**, *257*, 114–119. [CrossRef] [PubMed]
34. Halagappa, V.K.; Guo, Z.; Pearson, M.; Matsuoka, Y.; Cutler, R.G.; Laferla, F.M.; Mattson, M.P. Intermittent fasting and caloric restriction ameliorate age-related behavioral deficits in the triple-transgenic mouse model of Alzheimer's disease. *Neurobiol. Dis.* **2007**, *26*, 212–220. [CrossRef] [PubMed]
35. Lee, S.; Notterpek, L. Dietary restriction supports peripheral nerve health by enhancing endogenous protein quality control mechanisms. *Exp. Gerontol.* **2013**, *48*, 1085–1090. [CrossRef] [PubMed]
36. Mattson, M.P. Lifelong brain health is a lifelong challenge: From evolutionary principles to empirical evidence. *Ageing Res. Rev.* **2015**, *20*, 37–45. [CrossRef] [PubMed]
37. Murphy, T.; Dias, G.P.; Thuret, S. Effects of diet on brain plasticity in animal and human studies: Mind the gap. *Neural Plast.* **2014**, *2014*, 563160. [CrossRef] [PubMed]
38. Xia, B.; Gerstin, E.; Schones, D.E.; Huang, W.; Steven de Belle, J. Transgenerational programming of longevity through E(z)-mediated histone H3K27 trimethylation in *Drosophila*. *Aging* **2016**, *8*, 2988–3008. [CrossRef] [PubMed]
39. Sierra, M.I.; Fernandez, A.F.; Fraga, M.F. Epigenetics of Aging. *Curr. Genom.* **2015**, *16*, 435–440. [CrossRef] [PubMed]

40. Anders, S.; McCarthy, D.J.; Chen, Y.; Okoniewski, M.; Smyth, G.K.; Huber, W.; Robinson, M.D. Count-based differential expression analysis of RNA sequencing data using R and Bioconductor. *Nat. Protoc.* **2013**, *8*, 1765–1786. [CrossRef] [PubMed]

41. Currais, A.; Fischer, W.; Maher, P.; Schubert, D. Intraneuronal protein aggregation as a trigger for inflammation and neurodegeneration in the aging brain. *FASEB J.* **2017**, *31*, 5–10. [CrossRef] [PubMed]

42. Walls, S.M., Jr.; Attle, S.J.; Brulte, G.B.; Walls, M.L.; Finley, K.D.; Chatfield, D.A.; Herr, D.R.; Harris, G.L. Identification of sphingolipid metabolites that induce obesity via misregulation of appetite, caloric intake and fat storage in *Drosophila*. *PLoS Genet.* **2013**, *9*, e1003970. [CrossRef] [PubMed]

43. Simonsen, A.; Cumming, R.C.; Finley, K.D. Linking lysosomal trafficking defects with changes in aging and stress response in *Drosophila*. *Autophagy* **2007**, *3*, 499–501. [CrossRef] [PubMed]

44. Zirin, J.; Nieuwenhuis, J.; Perrimon, N. Role of autophagy in glycogen breakdown and its relevance to chloroquine myopathy. *PLoS Biol.* **2013**, *11*, e1001708. [CrossRef] [PubMed]

45. Simonsen, A.; Cumming, R.C.; Lindmo, K.; Galaviz, V.; Cheng, S.; Rusten, T.E.; Finley, K.D. Genetic modifiers of the *Drosophila* blue cheese gene link defects in lysosomal transport with decreased life span and altered ubiquitinated-protein profiles. *Genetics* **2007**, *176*, 1283–1297. [CrossRef] [PubMed]

46. Tsakiri, E.N.; Sykiotis, G.P.; Papassideri, I.S.; Terpos, E.; Dimopoulos, M.A.; Gorgoulis, V.G.; Bohmann, D.; Trougakos, I.P. Proteasome dysfunction in *Drosophila* signals to an Nrf2-dependent regulatory circuit aiming to restore proteostasis and prevent premature aging. *Aging Cell* **2013**, *12*, 802–813. [CrossRef] [PubMed]

47. Longo, V.D.; Mattson, M.P. Fasting: Molecular mechanisms and clinical applications. *Cell Metab.* **2014**, *19*, 181–192. [CrossRef] [PubMed]

48. Tower, J. Mitochondrial maintenance failure in aging and role of sexual dimorphism. *Arch. Biochem. Biophys.* **2015**, *576*, 17–31. [CrossRef] [PubMed]

49. Lapierre, L.R.; Kumsta, C.; Sandri, M.; Ballabio, A.; Hansen, M. Transcriptional and epigenetic regulation of autophagy in aging. *Autophagy* **2015**, *11*, 867–880. [CrossRef] [PubMed]

50. Van Praag, H.; Fleshner, M.; Schwartz, M.W.; Mattson, M.P. Exercise, energy intake, glucose homeostasis, and the brain. *J. Neurosci.* **2014**, *34*, 15139–15149. [CrossRef] [PubMed]

51. Lipinski, M.M.; Zheng, B.; Lu, T.; Yan, Z.; Py, B.F.; Ng, A.; Xavier, R.J.; Li, C.; Yankner, B.A.; Scherzer, C.R.; et al. Genome-wide analysis reveals mechanisms modulating autophagy in normal brain aging and in Alzheimer's disease. *Proc. Natl. Acad. Sci. USA* **2010**, *107*, 14164–14169. [CrossRef] [PubMed]

52. Jiang, N.; Du, G.; Tobias, E.; Wood, J.G.; Whitaker, R.; Neretti, N.; Helfand, S.L. Dietary and genetic effects on age-related loss of gene silencing reveal epigenetic plasticity of chromatin repression during aging. *Aging* **2013**, *5*, 813–824. [CrossRef] [PubMed]

53. Hanzen, S.; Vielfort, K.; Yang, J.; Roger, F.; Andersson, V.; Zamarbide-Fores, S.; Andersson, R.; Malm, L.; Palais, G.; Biteau, B.; et al. Lifespan Control by Redox-Dependent Recruitment of Chaperones to Misfolded Proteins. *Cell* **2016**, *166*, 140–151. [CrossRef] [PubMed]

54. Huffman, D.M.; Schafer, M.J.; LeBrasseur, N.K. Energetic interventions for healthspan and resiliency with aging. *Exp. Gerontol.* **2016**, *86*, 73–83. [CrossRef] [PubMed]

55. Rogina, B.; Helfand, S.L. Sir2 mediates longevity in the fly through a pathway related to calorie restriction. *Proc. Natl. Acad. Sci. USA* **2004**, *101*, 15998–16003. [CrossRef] [PubMed]

56. Wang, H.B.; Loh, D.H.; Whittaker, D.S.; Cutler, T.; Howland, D.; Colwell, C.S. Time-Restricted Feeding Improves Circadian Dysfunction as well as Motor Symptoms in the Q175 Mouse Model of Huntington's Disease. *eNeuro* **2018**, *5*. [CrossRef] [PubMed]

57. Schumacher, B.; van der Pluijm, I.; Moorhouse, M.J.; Kosteas, T.; Robinson, A.R.; Suh, Y.; Breit, T.M.; van Steeg, H.; Niedernhofer, L.J.; van Ijcken, W.; et al. Delayed and accelerated aging share common longevity assurance mechanisms. *PLoS Genet.* **2008**, *4*, e1000161. [CrossRef] [PubMed]

58. Melkani, G.C.; Panda, S. Time-restricted feeding for prevention and treatment of cardiometabolic disorders. *J. Physiol.* **2017**, *595*, 3691–3700. [CrossRef] [PubMed]

59. Pal, S.; Tyler, J.K. Epigenetics and aging. *Sci. Adv.* **2016**, *2*, e1600584. [CrossRef] [PubMed]
60. Wood, J.G.; Jones, B.C.; Jiang, N.; Chang, C.; Hosier, S.; Wickremesinghe, P.; Garcia, M.; Hartnett, D.A.; Burhenn, L.; Neretti, N.; et al. Chromatin-modifying genetic interventions suppress age-associated transposable element activation and extend life span in *Drosophila. Proc. Natl. Acad. Sci. USA* **2016**, *113*, 11277–11282. [CrossRef] [PubMed]

International Journal of
Molecular Sciences

MDPI

Article

Electronegative Low-Density Lipoprotein L5 Impairs Viability and NGF-Induced Neuronal Differentiation of PC12 Cells via LOX-1

Jiz-Yuh Wang [1,2,*] ⓘ, Chiou-Lian Lai [3,4], Ching-Tien Lee [5] and Chen-Yen Lin [1]

[1]   Graduate Institute of Medicine, College of Medicine, Kaohsiung Medical University, Kaohsiung 80708,
     Taiwan; k00511882@gmail.com
[2]   Department of Medical Research, Kaohsiung Medical University Hospital, Kaohsiung 80708, Taiwan
[3]   Department of Neurology, Faculty of Medicine, College of Medicine, Kaohsiung Medical University,
     Kaohsiung 80708, Taiwan; cllai@kmu.edu.tw
[4]   Department of Neurology, Kaohsiung Medical University Hospital, Kaohsiung 80756, Taiwan
[5]   Department of Nursing, Hsin-Sheng College of Medical Care and Management, Taoyuan 32544, Taiwan;
     chingtien1213@gmail.com
*   Correspondence: jizyuhwang@cc.kmu.edu.tw; Tel.: +886-7-3121101 (ext. 5092#432)

Received: 20 July 2017; Accepted: 7 August 2017; Published: 11 August 2017

Abstract: There have been striking associations of cardiovascular diseases (e.g., atherosclerosis) and hypercholesterolemia with increased risk of neurodegeneration including Alzheimer's disease (AD). Low-density lipoprotein (LDL), a cardiovascular risk factor, plays a crucial role in AD pathogenesis; further, L5, a human plasma LDL fraction with high electronegativity, may be a factor contributing to AD-type dementia. Although L5 contributing to atherosclerosis progression has been studied, its role in inducing neurodegeneration remains unclear. Here, PC12 cell culture was used for treatments with human LDLs (L1, L5, or oxLDL), and subsequently cell viability and nerve growth factor (NGF)-induced neuronal differentiation were assessed. We identified L5 as a neurotoxic LDL, as demonstrated by decreased cell viability in a time- and concentration-dependent manner. Contrarily, L1 had no such effect. L5 caused cell damage by inducing ATM/H2AX-associated DNA breakage as well as by activating apoptosis via lectin-like oxidized LDL receptor-1 (LOX-1) signaling to p53 and ensuring cleavage of caspase-3. Additionally, sublethal L5 long-termly inhibited neurite outgrowth in NGF-treated PC12 cells, as evidenced by downregulation of early growth response factor-1 and neurofilament-M. This inhibitory effect was mediated via an interaction between L5 and LOX-1 to suppress NGF-induced activation of PI3k/Akt cascade, but not NGF receptor TrkA and downstream MAPK pathways. Together, our data suggest that L5 creates a neurotoxic stress via LOX-1 in PC12 cells, thereby leading to impairment of viability and NGF-induced differentiation. Atherogenic L5 likely contributes to neurodegenerative disorders.

Keywords: electronegative LDL; lectin-like oxidized LDL receptor-1 (LOX-1); nerve growth factor (NGF); neuronal differentiation; apoptosis

---

1. Introduction

Dementia is the most common neurodegenerative disorder afflicting the aged. Growing evidence suggests a strong and likely causal association of cardiovascular disease (CVD) with cognitive decline and Alzheimer's disease (AD) [1]. Actually, a high prevailing rate of often neglected cardiovascular problems is found in the AD population [1,2]. Epidemiological studies have established a strong relation between CVD and AD, but it remains unclear whether this relation is due to shared risk factors that independently influence disease progression, or if certain CVD risk factors contribute to AD pathogenesis by inducing neuronal damage or promoting the accumulation of plaques and tangles. Among CVD risk factors, elevated low-density lipoprotein (LDL) is reported to reduce cerebral perfusion, increase oxidative stress, and activate neuroinflammatory responses, all of which promote AD progression [3,4]. An association of high LDL with cognitive impairment and AD has been demonstrated; however, how plasma LDL induces dementia has yet to be clearly defined.

Hypercholesterolemia has been emphasized a risk factor for the development of AD [5,6]. Cholesterol is transported inside LDL particles in circulation, so high LDL-cholesterol (LDL-C) level may increase the individuals suffering from cognitive decline and dementia. Cholesterol-related studies suggest that there is a negative correlation between plasma LDL-C level and cognition [7,8], and that high LDL-C level is associated with faster cognitive degeneration in AD patients [9]. Moreover, LDL particles are known to cause mischief to both cardiovasculature and cerebrovasculature if they become oxidized and invade the endothelium. Oxidized LDL (oxLDL) can promote AD progression through inducing cerebrovascular dysfunction [10,11]. Thus, lowering LDL-C with statins has been an excellent strategy for reducing dementia risk [12].

One plausible mechanisms by which LDLs contribute to cognitive dysfunction and dementia progression is through the modification to electronegative LDL [LDL(−)], a class of naturally occurring atherogenic lipoproteins [13]. Similar to oxLDL, plasma LDL(−) is increased in patients with high CVD risk, including patients with hyperlipidemia, diabetes, severe renal disease, and nonalcoholic steatohepatitis [14]. Of note, L5 has the relatively highest electronegative charge among chromatographically resolved human LDL subfractions (L1–L5), and evidence has shown that L5 induces marked atherogenic changes and apoptosis in cultured vascular endothelial cells (ECs) [15]. Further, circulating L5 is moderately elevated in asymptomatic individuals with increased CVD risk, such as those with metabolic syndrome, hypercholesterolaemia and type 2 diabetes [16,17]. Nevertheless, to our knowledge, the role of L5 in neurodegenerative pathogenesis remains uncharacterized. Specifically, it remains uncertain whether L5 leakage into brain tissue under the situations of aberrant cerebrovasculature caused by cerebral amyloid angiopathy or increased blood-brain barrier (BBB) permeability induced by CVD-related vascular factors can directly induce neuronal dysfunction or death.

We hypothesized that L5 is neurotoxic, thereby leading to neurodegeneration. Thus, our purpose in this in vitro study was to examine whether L5 negatively impacts on neuron-like PC12 cell reactivities, such as cell viability and nerve growth factor (NGF)-induced differentiation. The related signaling cascades that L5 disturbs were assessed after the desired treatments. We found that L5 induces ataxia-telangiectasia mutated (ATM) protein/H2AX-associated DNA damage, causes cell apoptosis via lectin-like oxidized LDL receptor-1 (LOX-1)/p53/caspase-3, and inhibits neurite outgrowth by interacting with LOX-1 and reducing NGF-stimulated Akt activation. Here, we present novel findings showing that L5 induces neurotoxicity directly, thereby identifying elevated plasma L5 as a possible causal link between hypercholesterolemia-related CVD and neurodegeneration, such as AD.

2. Results

2.1. L5 Shows a Cytotoxic Property and Reduces the Viability of PC12 Cells

To characterize the action of L5 on neuronal cell reactivity, both cell viability and NGF-induced neuronal differentiation were assessed in cultured neuron-like PC12 cells. Two controls were included in most experiments, L1 as a negative control and oxLDL as a positive control. Both L5 and L1 are naturally occurring lipoproteins in human circulation; in contrast to L5, L1 possesses the lowest electronegative charge among fractions and is regarded as non-cytotoxic [15,18]. OxLDL is a known toxic lipoprotein that plays a pivotal role in the progression of atheromatous plaque via mechanisms, including the destruction of the arterial wall and induction of vascular cell apoptosis [19,20].

We first examined whether L5 decreases the survival of PC12 cells by treating cultures with 0 to 50 µg/mL L1, L5, or oxLDL for 24 h, followed by MTT reduction assays (Figure 1A). Compared to the untreated control, L5 ≥ 30 µg/mL and oxLDL ≥ 10 µg/mL significantly reduced cell viability, while no such effects were observed in L1-treated cultures. Thus, we defined 30 µg/mL L5 as the threshold lethal concentration for PC12 cells. Next, cultures were exposed to 30 µg/mL of each LDL for times ranging from 0 to 48 h. Both L5 and oxLDL induced measurable cell death at 12 h and progressive cytotoxicity thereafter. In contrast, L1 entirely exerted no damaged effect on cell viability even after a 48 h exposure (Figure 1B). Regarding cell morphological changes in response to LDLs, a large amount of fragmented cell debris was detectable in L5- or oxLDL-treated cultures, but not in L1-treated cultures (Figure 1C). These results indicate that L5 induces both concentration- and time-dependent PC12 cell death.

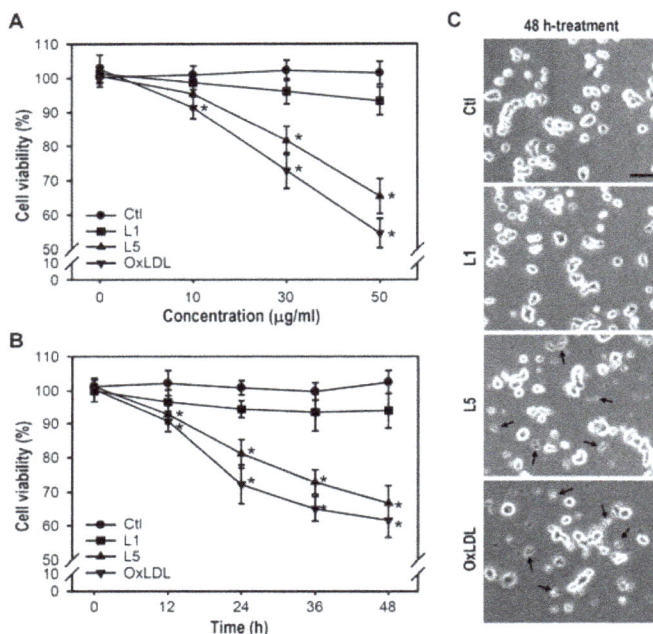

Figure 1. LDL fraction L5 shows a concentration- and time-dependent cytotoxicity to cultured PC12 cells. (**A**) Concentration-response relationships for L1 (negative control), L5, and oxLDL (positive control) were established. Indicated LDLs were applied for 24 h at various concentrations (0 to 50 µg/mL); (**B**) Time course was determined for L1, L5, and oxLDL cytotoxicity. Cultures were treated with the indicated LDL at 30 µg/mL for different times (0 to 48 h). In (**A**,**B**), the MTT reduction assay was applied to measure PC12 cell survival (viability) with untreated cells used as the control (Ctl). The number of live cells per well was calculated as a percentage of untreated control cell number. Each point is presented as the mean ± SEM from at least four independent experiments performed in triplicate. * $p < 0.05$ vs. Ctl at equal concentration (**A**) or at the same time point (**B**); and (**C**) Representative photos show an untreated culture and cultures treated with 30 µg/mL of the indicated LDL for 48 h. Fragmented cell debris (putative apoptotic bodies) are indicated by arrows. Scale bar = 50 µm.

2.2. L5 Induces Genotoxicity via ATM/H2AX Activation in PC12 Cells

Next, we assessed whether L5-induced cell death involves DNA damage, such as the formation of double-strand breaks (DSBs) in chromatin, using a neutral comet assay. Under the experimental conditions tested in our study, both L5 and oxLDL, but not L1, caused DSBs as evidenced by the appearance of migrating DNA fragments (reflected by the length of the comet tail) (Figure 2A). Further, compared to control cultures, clear comet tails were observed in cultures treated with L5 or oxLDL at 6 and 12 h (Figure 2B), indicating that L5 has a genotoxic property that leads to destruction of DNA structure.

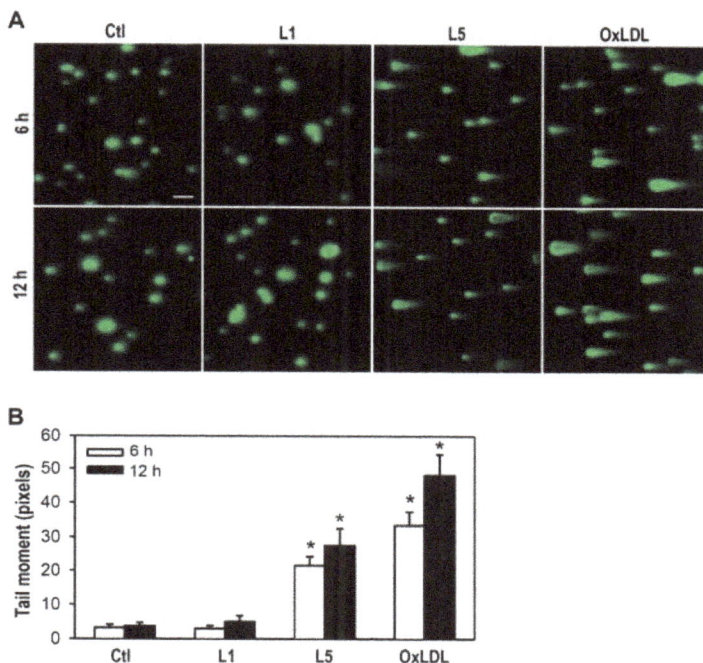

Figure 2. L5 induces DNA double-strand breaks (DSBs) in cultured PC12 cells. (**A**) Cultures treated with the indicated LDL (30 μg/mL) for 6 or 12 h were analyzed for DSBs using a neutral comet assay kit. Untreated cells were used as a control. Representative images from one of four independent experiments are presented. Scale bar = 50 μm; and (**B**) The comet tail moment used to determine DNA damage was quantified by COMETscore.v1.5 software. Data are presented as mean ± SEM from four independent experiments. * $p < 0.05$ vs. Ctl at the same time point. At least 200 cells were analyzed in each treatment group per experiment.

Studies have suggested that DSBs formation is closely correlated with activation of both ATM and H2AX [21]. Hence, we examined whether DSBs induction by L5 is associated with ATM/H2AX activation. Immunostaining revealed the formation of γH2AX (H2AX phosphorylation on serine 139), which positively corresponds to DNA DSBs. Both L5 and oxLDL induced γH2AX foci formation in the nucleus as evidenced by punctate immunofluorescence signals, while such γH2AX immunoreactivity was not observed in untreated control or L1-treated cultures (Figure 3A,B). To further verify L5-induced genotoxicity, ATM and H2AX activation were determined by western blotting. Consistent with immunostaining results, a rapid and marked increase in both pATM and γH2AX was observed after only 10 min of L5 treatment, but not L1 treatment (Figure 3C). Together with the data presented in Figure 2, these results suggest that L5 injures PC12 cells by inducing ATM/H2AX-associated DNA DSBs.

2.3. L5 Induces Apoptotic Death of PC12 Cells via the LOX-1/p53/Caspase-3 Pathway

Damage to genomic integrity is known to cause apoptosis [22,23], and indeed many of the debris fragments observed following L5 or oxLDL treatment resembled apoptotic bodies (Figure 1C). Accordingly, we examined whether L5 causes apoptotic death in PC12 cells and activates apoptosis-associated signal transduction pathways. Compared to control, L5 increased cleaved caspase-3 (a key apoptotic caspase) expression after 6 or 12 h of treatment while L1 had no such effect (Figure 4A); moreover, this apoptotic insult was evidently inhibited by the pan-caspase inhibitor Z-VAD-FMK (Figure 4B). Subsequently, p53 activity was examined due to its close association with apoptosis and activation upon cellular stresses such as DNA damage. Furthermore, p53 is a direct substrate of ATM [24]. ATM rapidly phosphorylates p53 at Ser15 during early phase of DNA damage response, thereby reducing the interaction of p53 with MDM2, a negative p53 regulator that inhibits accumulation of p53 by targeting it for ubiquitination and proteasomal degradation [25,26]. Compared to the control, L5-treated cultures (but not L1-treated cultures) exhibited phosphorylated p53 elevation after as little as 10 min of exposure (Figure 4C). Moreover, the p53 inhibitor pifithrin-α (PFT-α) electively reduced the cleaved caspase-3 level in L5-treated cultures without altering the L5-phosphorylated p53 level, indicating that p53-dependent apoptosis is a major mediator of L5-induced PC12 cell death (Figure 4D).

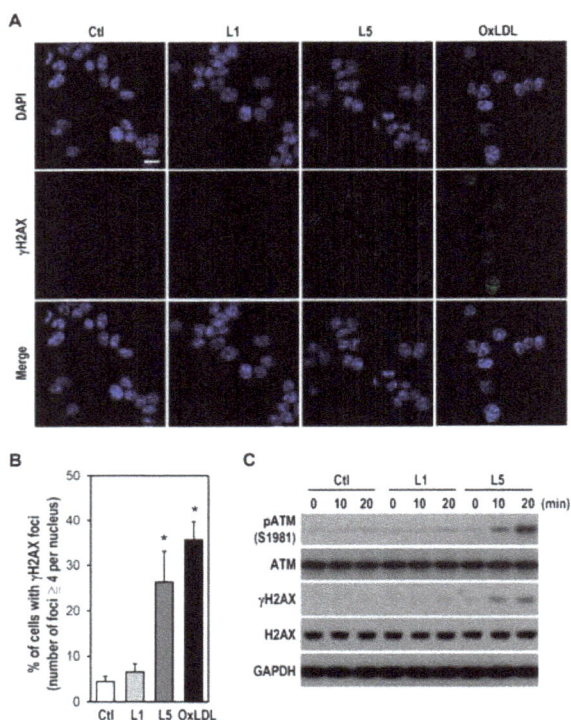

Figure 3. L5 induces ATM phosphorylation and formation of γH2AX foci in cultured PC12 cells. (**A**) Representative immunofluorescence images of γH2AX foci are shown. Cells grown on coverslips were treated with the indicated LDL (30 μg/mL) for 30 min. Untreated cells were used as a control. Immunostaining was performed using an antibody against γH2AX (green). Cells were counterstained with DAPI to recognize the nucleus (blue). Scale bar = 10 μm; (**B**) The histogram shows the quantification of (**A**). At least 20 randomly selected fields for each treatment group were examined. Cells with $\geq 4\gamma$H2AX foci per nucleus were regarded as positive and then scored. Value expressed as a percentage by calculating the number of positive cells (% of total). Data are presented as mean $\pm$ SEM from four independent experiments. * $p < 0.05$ vs. Ctl; and (**C**) After treating the cultures with L1 or L5 at 30 μg/mL for different times as indicated, western blotting was performed to detect pATM, ATM, γH2AX, and H2AX expression. GAPDH was a protein loading control. Representative blots selected from one of four independent experiments are presented.

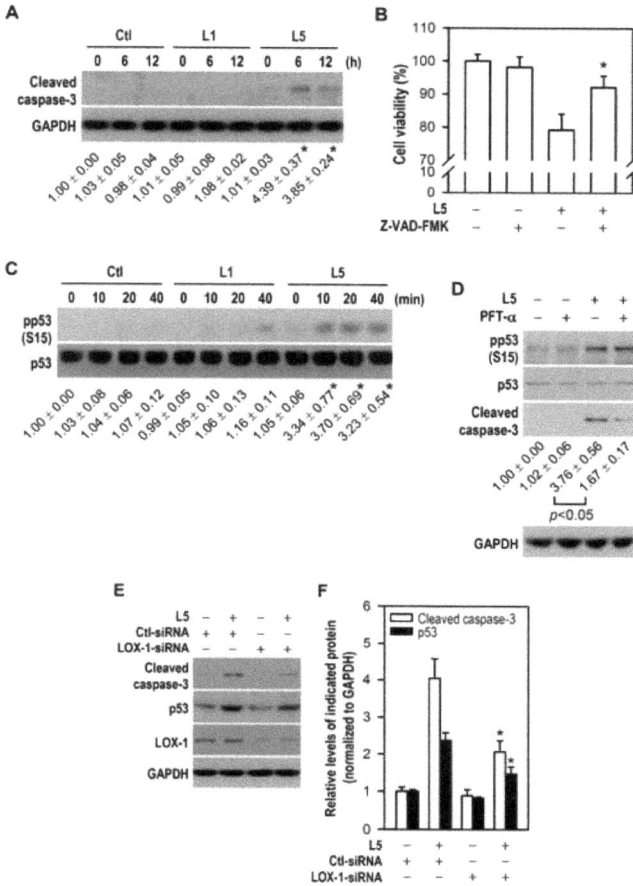

Figure 4. Reduced viability of L5-treated PC12 cells is associated with LOX-1 expression, p53 activation, and caspase-3 cleavage. (**A**) Cultures were treated with L1 or L5 (30 μg/mL) for the indicated times and then western blotting was used to detect cleaved caspase-3. Untreated cells were used as a control. GAPDH was a protein loading control. Quantitative values are expressed relative to untreated control at a time point of 0 h (assigned a value of 1); the means ± SEM from four independent experiments are shown below the representative blots. * $p < 0.05$ vs. Ctl at the same time point; (**B**) Cells were treated with or without L5 (30 μg/mL) for 12 h in the absence or presence of a pan-caspase inhibitor Z-VAD-FMK (20 μM). The histogram shows the cell viability of each treatment group relative to the untreated control (assigned a value of 100%). Data are presented as mean ± SEM from at least three independent experiments performed in triplicate. * $p < 0.05$ vs. L5 alone; (**C**) After the indicated treatments, western blotting was used to detect pp53 and p53 expression. Quantitative results of relative pp53 levels from four independent experiments are shown below the representative blots. * $p < 0.05$ vs. Ctl at the same time point; (**D**) Before sham or L5 (30 μg/mL) treatment for either 20 min or 12 h, cells were pretreated with or without PFT-α (10 μM) for 1 h. The expression of pp53 and p53 (at 20 min), and cleaved caspase-3 (at 12 h) was detected by western blotting. Relative quantitative results of cleaved caspase-3 from four independent experiments are shown below the representative blots; (**E**) One day after transfection with a scrambled siRNA (Ctl-siRNA) or a LOX-1-siRNA, cultures were treated with or without L5 for 12 h. Cleaved caspase-3, p53, and LOX-1 were detected by western blotting and representative blots are presented; and (**F**) Relative quantitative expression of cleaved caspase-3 and p53 in (**E**) is shown by a histogram. Data are from four independent experiments. * $p < 0.05$ vs. Ctl-siRNA + L5.

To identify the upstream factors leading to activation of the p53/caspase-3 pathway, we first focused on possible receptors involved. It has been shown that L5 signals and is internalized through LOX-1, and that the binding complex of L5/LOX-1 plays a critical role in L5-induced apoptosis of vascular ECs [15]. Accordingly, we examined whether LOX-1 participates in L5-induced induction of p53/caspase-3-dependent apoptosis in PC12 cells. Compared to L5-treated cultures transfected with control small interfering RNA (Ctl-siRNA), silencing of endogenous LOX-1 by RNA interference using a LOX-1-siRNA not only downregulated LOX-1 protein expression, as determined by western blotting, but also markedly reduced both cleaved caspase-3 level and p53 accumulation in response to L5 (Figure 4E,F). Noteworthily, this reduced p53 accumulation might be primarily because LOX-1-siRNA disturbs L5-induced p53 phosphorylation and then MDM2-mediated p53 degradation occurs. This finding suggests that LOX-1 is required for L5-induced downstream activation of p53/caspase-3. Taken together, these results strongly suggest that LOX-1/p53/caspase-3 signaling mediates L5-induced PC12 cell apoptosis.

2.4. L5 Inhibits NGF-Induced Differentiation of PC12 Cells

Neurite outgrowth in PC12 cells in response to NGF stimulation is a widely used model to examine neuronal differentiation and development. In addition to assessing L5 effects on PC12 cell viability, we further examined whether L5 disturbs NGF-induced neuronal differentiation of PC12 cells. According to above results indicating that L5 < 30 µg/mL did not cause detectable PC12 cell death (Figure 1A), we defined 10 µg/mL as the sublytic concentration to assess the impact of L5 on PC12 cell differentiation induced by NGF.

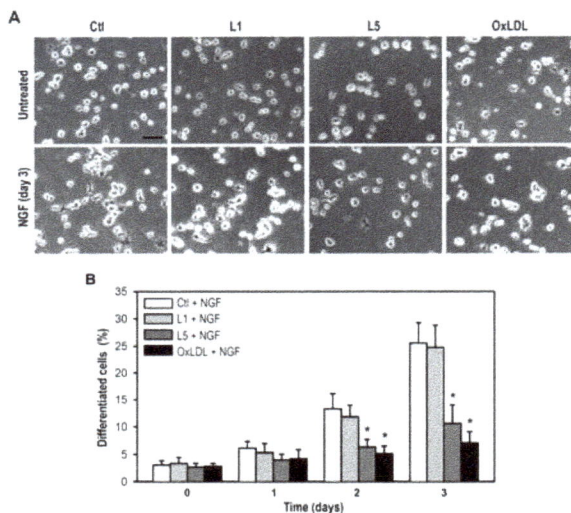

Figure 5. L5 inhibits NGF-induced neuronal differentiation of cultured PC12 cells. (**A**) Representative photos show PC12 cells treated with the indicated LDLs (10 µg/mL) in the absence or presence of NGF (50 ng/mL) for 3 days. Cells untreated with LDLs were used as a control. NGF-induced neuronal differentiation is indicated by the extent of neurite outgrowth. Scale bar = 50 µm; and (**B**) The percentage of neuronal differentiation in PC12 cells exposed to LDLs (10 µg/mL) plus NGF (50 ng/mL) for 1, 2, and 3 days was determined by calculating the number of cells (% of total) bearing neurites longer than one cell body in diameter. Ten randomly selected fields were examined and at least 300 cells were counted for each treatment group at each time point. Data are presented as mean ± SEM of one representative experiment; two other experiments revealed consistent results. * $p < 0.05$ vs. Ctl + NGF at the same time point.

As shown in Figure 5A, both L5 and oxLDL interfered with neurite outgrowth in PC12 cultures subjected to long-term NGF exposure, as evidenced by generally shorter neurites compared to control NGF-treated cultures on day 3 of treatment. As expected, cultured cells treated with L1 plus NGF displayed normal outgrowth of neurites. This finding implies that L5 at sublytic concentration is capable of disturbing NGF-induced neurite outgrowth. Indeed, both L5 and oxLDL but not L1 markedly lowered the proportion cells showing NGF-induced morphological differentiation (defined as cells with neurites longer than one soma in length) on days 2 and 3 of treatment compared to control cultures incubated with NGF (Figure 5B).

Early growth response factor-1 (Egr-1) is a NGF-responsive transcription factor and exerts long-term effects on neural cell growth, while neurofilament is critical for radial axon growth and determines axon caliber [27,28]. Therefore, we further determined the expression levels of both Egr-1 and neurofilament-medium (NF-M) under the conditions of NGF combined with distinct LDLs. The expression of Egr-1 was induced within 1 h after NGF addition, and this response was markedly reduced by both L5 and oxLDL but not by L1 (Figure 6A). Further, NGF induced an elevation in neurite growth-associated NF-M expression level, and again this response was evidently attenuated by L5 over a 4-day experimental period (Figure 6B). Thus, L5 appears to have a negative impact on neurodifferentiation-associated upregulation of Egr-1 and NF-M. Collectively, these data suggest that a sublytic concentration of L5 (10 µg/mL) remains still neurotoxic and impairs neuronal development by inhibiting NGF-induced differentiation.

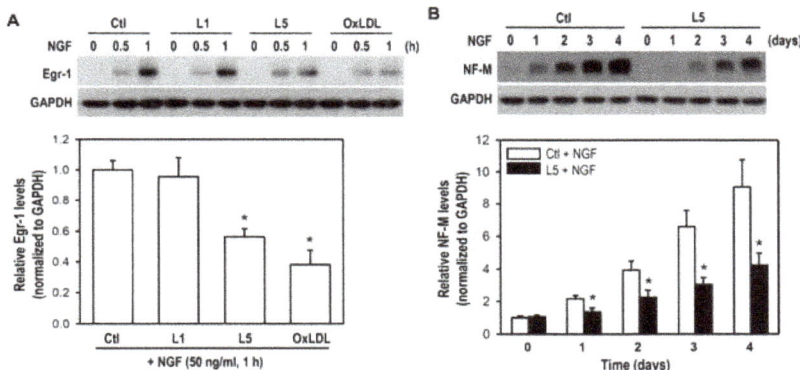

Figure 6. L5 inhibits NGF-induced upregulation of Egr-1 and NF-M. (**A**) Cultures were treated with or without the indicated LDLs (10 µg/mL) for 1 h, followed by NGF (50 ng/mL) addition for the indicated time periods. Cells untreated with LDLs were used as a control. Egr-1 expression was detected by western blotting (representative blot is shown). GAPDH was a protein loading control. The histogram shows relative Egr-1 expression values of each treatment group at a time point of 1 h. Data are presented as mean ± SEM from four independent experiments. * $p < 0.05$ vs. Ctl + NGF; and (**B**) Cultures were treated with L5 (10 µg/mL) for 1 h, followed by NGF (50 ng/mL) addition for another times (days) as indicated. Cells untreated with LDLs were used as a control. Western blotting was performed to detect NF-M expression (representative blot of NF-M is shown). The histogram shows relative NF-M levels between Ctl + NGF and L5 + NGF groups from day 0 to 4. Data are from four independent experiments. * $p < 0.05$ vs. Ctl + NGF on the same day.

2.5. L5 Weakens NGF-Induced Akt but Not TrkA and MAPKs Activation

We then examined the upstream signaling pathways responsible for L5 suppression of NGF-induced differentiation in PC12 cells. Given that many NGF effects are mediated by activation of its receptor TrkA and downstream signaling kinases such as MAPKs and PI3K/AKT, we examined whether L5 weakens NGF-induced TrkA phosphorylation and resultant activation of MAPKs

(ERK, JNK, and p38) and Akt in PC12 cultures. We first measured the phosphorylation status of several crucial residues within the tyrosine kinase domain of TrkA because autophosphorylation at these sites reflects TrkA activity or is required for activation of downstream kinase cascades [29]. In response to NGF stimulation, all tested tyrosine residues within TrkA including Tyr490, Tyr674/675, Tyr751, and Tyr785 were rapidly phosphorylated; however, no obvious difference in phosphorylation was found at any one of these tyrosine sites between cultures treated with NGF alone versus NGF plus L1, L5, or oxLDL (Figure 7A). This indicates that L5 has no influence on TrkA catalytic activity or on TrkA coupling to downstream Ras/MAPK, PI3K/Akt, and phospholipase Cγ. Further, no detectable difference was observed in the phosphorylation levels of ERK, JNK, and p38 between cultures treated with NGF alone and those treated with NGF plus L1, L5, or oxLDL. In contrast, NGF-induced Akt phosphorylation was markedly reduced in L5- and oxLDL-treated cultures at both 10 and 30 µg/mL compared to cultures treated with NGF alone, while such a repressive effect was not found in cultures incubated with L1 plus NGF (Figure 7B). These results suggest that L5 represses NGF-induced neuronal differentiation in PC12 cells by interfering with the TrkA/PI3K/Akt pathway.

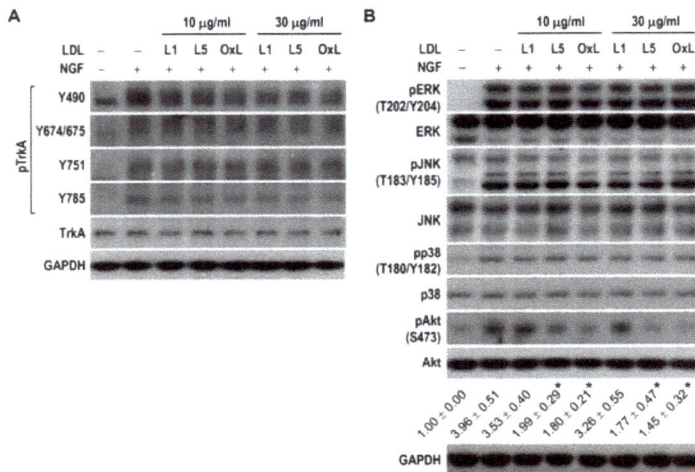

Figure 7. L5 decreases the level of phospho-Akt (pAkt), but not pTrkA and pMAPKs, in NGF-treated PC12 cells. (**A**) Cultures were treated with L1, L5, or oxLDL (OxL) at 10 or 30 µg/mL for 1 h, followed by NGF (50 ng/mL) addition for another 20 min. Untreated cells were used as a control. Western blotting was used to detect the expression of TrkA phosphorylated at various tyrosine sites and total TrkA protein. GAPDH was a protein loading control. Representative blots selected from one of four independent experiments are presented; and (**B**) Alternatively, western blotting was performed to detect pERK, ERK, pJNK, JNK, pp38, p38, pAkt, and Akt expression. Quantitative values of pAkt relative to the level of untreated control (assigned a value of 1) are expressed; the means ± SEM from four independent experiments are shown below the representative blots. * $p < 0.05$ vs. NGF alone.

2.6. LOX-1 Is Required for L5 Suppression of NGF-Induced Akt Phosphorylation

To verify abovementioned findings that NGF-induced Akt activation is reduced in L5-treated cultures, cultures were treated with NGF alone or NGF combined with L1 or L5 for brief periods and Akt phosphorylation/activation status was measured. As shown in Figure 8A, phosphorylated Akt became detectable within 10–40 min after the onset of NGF treatment, and this response was reduced in the presence of L5 but not L1. The suppressive effect of L5 on NGF-induced Akt phosphorylation reached significance at time points of 10, 20, and 40 min compared to control group treated with NGF (Ctl + NGF) (Figure 8B).

Results present in Figure 4 strongly suggest that LOX-1 is involved in PC12 cell apoptosis induced by L5. We also explored the possibility that LOX-1 participates in L5-induced suppression of neuronal differentiation in NGF-treated PC12 cells. Because NGF-stimulated Akt activation is impaired by L5, we examined if L5 weakens the TrkA/PI3K/Akt signal cascade via LOX-1. As shown in Figure 8C, NGF-induced stimulation of Akt phosphorylation was not reduced by L5 in cultures transfected with a LOX-1-siRNA, but was weakened by L5 in cultures transfected with a Ctl-siRNA. Therefore, these results suggest that L5 attenuates NGF-activated Akt via LOX-1, and that LOX-1 acts as a mediator coupling extrinsic L5 to intracellular PI3K/Akt pathway.

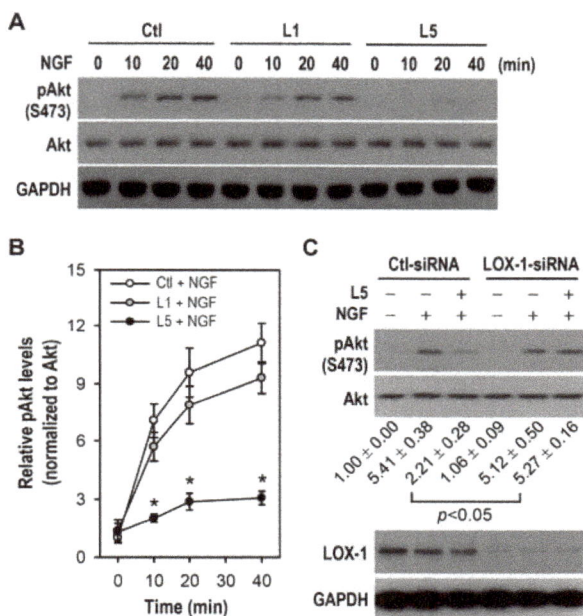

Figure 8. LOX-1 knockdown abolishes L5-induced Akt phosphorylation decline in NGF-treated PC12 cells. (**A**) Cultures were treated with L1 or L5 (10 µg/mL) for 1 h, followed by NGF (50 ng/mL) addition for the indicated time periods. Cells untreated with LDLs were used as a control. Western blotting was used to detect pAkt and Akt expression. GAPDH was a protein loading control. Representative blots selected from one of four independent experiments are presented; (**B**) According to the results in (**A**), quantitative analyses of relative pAkt levels were done in each treatment group. Data are presented as the mean ± SEM from four independent experiments. * $p < 0.05$ vs. Ctl + NGF at the same time point; and (**C**) One day after transfection with siRNA as indicated, cultures were treated with or without L5 (10 µg/mL) for 1 h and incubated in the absence or presence of NGF (50 ng/mL) for another 20 min. Western blotting was conducted to detect pAkt, Akt, and LOX-1 expression. Quantitative results of relative pAkt levels were obtained from four independent experiments and are shown below the representative blots.

3. Discussion

Frequent comorbidity of AD with CVD suggests that vascular problems contribute to cognitive decline and AD-associated dementia. High circulating LDL levels, a strong CVD risk factor, are associated with increased production of cerebrovascular amyloid β (Aβ) peptide and increased risk of AD [4]. Further, more studies also suggest that elevated blood cholesterol and triacylglycerol increase the likelihood of AD [6]. Cerebrovasculature damage induced by circulating LDL or other vascular mechanisms may enhance access of detrimental LDLs to brain parenchyma, resulting in neurotoxicity. Indeed, this study suggests that in a situation of BBB breach, L5 may directly impair neuronal viability and NGF effects, such as neurodifferentiation. Thus, it is a feasible strategy for prevention of AD and other forms of age-related cognitive impairment by aggressively targeting hypercholesterolemia.

Numerous studies have highlighted a negative role of L5 in the progression of CVDs, especially atherosclerosis, and thus L5 is regarded as a potent atherogenic LDL. It has been indicated that L5 is abundant in dyslipidemic but not normolipidemic human plasma, and hyperlipidemia is also a shared factor for AD [3]. Shen et al. found a functional link between pro-atherogenic lipoproteins and platelet-mediated thrombus formation in stroke, and pointed out that L5 is substantially elevated (40-fold higher than control subjects) in ischemic stroke patients [30]. Moreover, L5 can cause Aβ peptide secretion from platelets, which contributes to platelet hyper-reactivity and stroke complications [31]. These clinical results suggest the possibility of L5 directly impacting neurological functions, such as cognition, under the premise of impaired BBB. In the present work, we identify a negative role of L5 in basal viability and NGF-induced neuronal differentiation of PC12 cells, consistent with a neurotoxic effect in aging humans. Further investigations aimed at establishing the clinical relevance of plasma L5 to AD-related neurodegeneration are clearly warranted.

It is highly possible that plasma LDL penetrates the BBB due to a broken architecture of cerebral microvasculature during the progression of AD or CVD, which in turn leads to neuronal damage and neuroinflammation-related glial activation. In fact, apolipoprotein B (apoB) or LDL has been detected in cerebrospinal fluid (CSF) or/and central nervous system under pathological conditions. For example, apoB/LDL was found or obviously elevated in CSF collected from patients with subarachnoid hemorrhage, cerebrotendinous xanthomatosis, or tuberculous meningitis [32–34]. Plasma apoB was co-localized with cerebral Aβ in the plaques of transgenic AD mice as revealed by a 3-dimensional immunomicroscopy; further, plaque abundance in these mice positively correlated with apoB [35,36]. Moreover, BBB breakdown has been described in chronic CVDs, and peripheral inflammatory response is a cause of enhanced BBB permeability [37]. CVDs like arteriosclerosis and hypertension are characterized by increased production of pro-inflammatory cytokines in blood [38], and administration of pro-inflammatory cytokines such as TNF-α, IL-1β, and IL-6 to monolayers of cerebral ECs or pericyte/endothelial cell co-cultures leads to BBB disruption [39,40]. Clinical studies show that diseases compromising the vascular system, such as hypertension, hypercholesterolemia, and diabetes, are able to disrupt BBB integrity and increase the risk of AD [41]. Autopsy studies have also shown that vascular abnormalities, including endothelial damage-associated microangiopathy, were universally observed in AD brain tissue [42]. Together, these studies revealing the occurrence of BBB dysfunction in CVDs or encephalopathies set a stage for this study and suggest that L5 and other neurotoxic agents in plasma may directly access neurons via a damaged BBB.

Actually, L5 does not exist in healthy individuals with clinically normolipidemia; certainly, L5 does not appear in the brain under normally physiological conditions. However, plasma L5 likely extravasates directly into brain milieu in certain pathological cases, such as BBB dysfunction. Moreover, the L5 concentration (10 or 30 μg/mL) used in this study should not be present in humans with normal or pathological CNS states, but is approximately equal to or less than other in vitro studies examining the impact of L5 or LDL(−) on the pathogenic process of atherosclerosis. For example, Chu et al. used 50 μg/mL of L5 for 24 h to examine C-reactive protein expression in vascular ECs [43]. Estruch et al. assessed cytokine release in monocytes treated with LDL(−) at 150 μg apoB/mL [44]. Because in vitro L5 toxicity has a tendency towards a concentration- and time-dependency, we can reasonably consider that neurotoxicity is attainable in vivo if using lower L5 concentrations (<10 μg/mL) for longer treatment periods (a few weeks or months) under the premise of impaired BBB. To elucidate this thought, a long-term exposure to low plasma L5 levels will be applied to future animal studies using transgenic AD mice, and hypertensive or hypercholesterolemic rodents.

It is worthily noticed that the differential effects of L1 and L5 on PC12 cell viability and NGF-induced neuronal differentiation are mainly associated with activation of different cell surface receptors and intracellular signaling pathways. Several properties ascribed to L5, such as highly electronegative charge, higher aggregation potential, conformationally distinct apoB, and greater content in inflammatory lipids, suggest that L5 interacts with different receptors than L1. Due to this higher electronegativity than L1, it was thought that L5 binds to the LDL receptor (LDLR) with less affinity. Avogaro et al. observed that the more electronegative LDL subfraction exhibited a lower binding capacity for LDLR [45]. This finding is similar to Benitez et al. who found that LDLR affinity was three-fold lower for LDL(−) than for native LDL [46]. The low binding affinity of L5 to LDLR may be partly explained by high nonesterified fatty acid content [46], increased degree of aggregation [47], and abnormal conformation of apoB [48]. The consequence of this loss of LDLR affinity is a diminished L5 clearance from human circulation, allowing for further deleterious effects on vascular cells and neurons.

Regarding cell surface receptor for L5 binding, Chen and coworkers reported that L5, a highly pure form of LDL(−), is a ligand for LOX-1, but not normal LDLR. L5 recognizes, signals through, and is internalized by LOX-1, which has high affinity for negatively charged ligands, whereas the less electronegative LDL subfractions L1–L4 rely on the LDLR for biological effects [15,18]. In addition to L5, LOX-1 is also activated by oxLDL, reactive oxygen species (ROS), endothelin-1, angiotensin II, advanced glycation end products, and shear stress. Previous studies have demonstrated activation of diverse downstream signaling pathways upon binding of L5 to LOX-1. For example, L5 suppresses Akt phosphorylation via LOX-1 to impair Akt-mediated growth and survival signals in bovine aortic ECs [15]. The L5/LOX-1 complex also triggers ROS production to induce oxidative stress in human aortic ECs [43], and stimulates overexpression of various adhesion molecules and inflammatory chemokines, thus promoting monocyte adherence to vascular endothelium, an early pathogenic event for atherosclerosis [49]. L5 binding to LOX-1 induces the apoptosis of human aortic ECs via p38 MAPK/caspase-3 signaling [50]. L5/LOX-1-activated platelets switch on a signaling pathway including IκB kinase 2 and nuclear factor-κB activation, which is critical for stroke pathobiology [31]. Further, we found that L5 signals to ATM/H2AX and p53/caspase-3 cascades through LOX-1, resulting in PC12 cell DNA breaks and apoptosis. Based on these findings, we conclude that LOX-1 transmits damage signals from L5 but not L1, thereby accounting for the unique cytotoxic effects of L5 in PC12 cells.

Once DNA damage such as DSBs is elicited by genotoxic insults like radiation and free radicals, apoptosis may be activated if this damage is not repaired [22]. We identified L5 as a genotoxic agent that injures PC12 cells through DNA DSBs and apoptotic death, and further found out the intermediary signaling pathways. Several lines of evidence suggest that ATM is activated in response to DNA damage, which then promotes H2AX phosphorylation at DSB sites (manifested as nuclear foci) for initiation of DSB repair [21,22]. As expected, γH2AX foci formation was observed by immunofluorescence, and increased phosphorylation levels of both ATM and H2AX were revealed by western blotting after L5 stimulation. In addition, activated p53 exerts pro-apoptotic effects by transcriptionally activating genes involved in cell cycle arrest, inhibition of proliferation, and induction of senescence, all of which are conducive to the maintenance of genomic integrity following genotoxic insults [51]. Indeed, we found that L5 rapidly triggers p53 phosphorylation accompanied by caspase-3 activation and PC12 cell apoptosis. This finding was further verified by the p53 inhibitor PFT-α, which effectively lowered the cleaved caspase-3 level and inhibited apoptosis.

In this study, L5-induced PC12 cell apoptosis also relied on LOX-1 expression, as evidenced by the reduction of L5-elevated caspase-3 cleavage and p53 accumulation levels in cells transfected with LOX-1 siRNA. These results concur with those of Li et al. who reported that LOX-1 is mainly found in cortical neurons, and its upregulation is involved in neuronal apoptosis [52]. To mediate L5 cytotoxicity, we speculate that unknown intracellular transmitters triggered by L5/LOX-1 complex should signal to ATM/H2AX and p53/caspase-3 cascades, followed by DNA breaks and apoptotic death in PC12 cells. Of note, L5 but not L1 has been indicated to rapidly elicit ROS production via LOX-1 and functions similarly to oxLDL in inducing oxidative stress [43]. Moreover, ROS is able to directly evoke p53-dependent apoptosis in PC12 cells [53], and the accumulation of ROS caused by polyQ proteins activates ATM/ATR-dependent DNA damage response [54]. ATM acts upstream of p53 in a signal transduction pathway initiated by ionizing radiation and can directly activate/phosphorylate p53 in vivo [24]. Thus, it is possible that ROS serves as a stress signal transmitter that connects the L5/LOX-1 complex to ATM/H2AX and p53/caspase-3 pathways in our PC12 cell model. Further work is required to verify this inference.

Morphological differentiation of neurons is essential for cell-cell communication and information processing. Differentiation begins with neurite sprouting, followed by progressive elongation of axons and elaboration of dendrites. Regarding this, we found that L5 impaired NGF-induced neurite outgrowth through weakening Akt phosphorylation. Actually, PI3K/Akt cascade has been suggested to be critical for NGF-induced PC12 cell differentiation [55,56]. Moreover, LOX-1 knockdown abolished L5-mediated inhibition of Akt activation by NGF, indicating that L5 requires interaction with LOX-1 to inhibit NGF-elevated Akt activity. Given that NGF exerts trophic effects via TrkA binding, it was expected that L5/LOX-1 would interfere with NGF-induced neuronal differentiation by disrupting TrkA. TrkA phosphorylation at specific sites is required for activation of various kinase pathways (Tyr490 for Shc/Ras/MAPK, Tyr751 for PI3K/Akt, Tyr785 for PLCγ/PKC/MAPK) and elevated TrkA kinase activity (Tyr674/675) [29]. However, L5 had no effect on overall phosphotyrosine status after NGF stimulation or on NGF-induced ERK, JNK, and p38 activation. This illustrates that L5/LOX-1 selectively weakens intracellular PI3K/Akt signaling, but not TrkA receptor activity or downstream MAPK pathways. Our findings agree with previous evidence indicating that L5 executes its biological effects via PI3K/Akt. For example, L5 activates CXCR2/PI3K/NFκB signaling, indirectly inducing cardiomyocyte apoptosis [57]. In ECs and endothelial progenitor cells as well, the PI3k/Akt pathway is inhibited by L5/LOX-1 [15,58]. Although our findings indicate that L5/LOX-1 suppresses TrkA/PI3K/Akt-associated neuronal differentiation in PC12 cells, more detailed mechanisms are still unclear at the present stage. Profound studies are needed in the future.

In summary, our findings reveal, for the first time, the direct cytotoxicity of LDL(−), especially fraction L5, to neuron-like PC12 cells. We demonstrate that L5 concentration- and time-dependently reduces cell viability by inducing ATM/H2AX-associated DNA damage and LOX-1/p53/caspase-3-dependent apoptosis; moreover, sublytic concentration of L5 inhibits NGF-induced neuronal differentiation by LOX-1-mediated suppression of the PI3K/Akt cascade (Figure 9). High circulating LDL is strongly implicated in the pathophysiology of neurodegenerative disorders such as AD-type dementia. Therefore, lowering plasma LDL(−) and inhibiting electronegative modification of LDL should be considered primary prevention measures for cognitive impairment and AD.

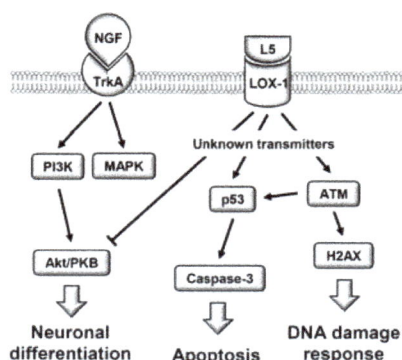

Figure 9. A schematic illustration shows the L5-induced cytotoxicity in PC12 cells. Through LOX-1 and unknown transmitters, L5 particle triggers ATM/H2AX-associated DNA damage response, activates p53/caspase 3-dependent apoptosis, and impairs NGF-induced neuronal differentiation (i.e., the suppression of NGF-activated PI3K/Akt cascade).

4. Materials and Methods

4.1. Chemicals, Antibodies, Kits and Reagents

All chemicals were from Sigma-Aldrich (Saint Louis, MO, USA) unless stated otherwise. Comet assay kits, reagents and slides were purchased from Trevigen (Gaithersburg, MD, USA). Antibodies, including phospho-ERK (pERK), pJNK, pp38, pAkt, pp53, pTrkA, ERK, JNK, p38, p53, ATM, H2AX, Egr-1, NF-M, TrkA, and cleaved caspase-3 were purchased from Cell Signaling Technology (Beverly, MA, USA). The pATM and γH2AX antibodies were obtained from Millipore (Billerica, MA, USA). Akt, LOX-1 and GAPDH antibodies, and siRNA were obtained from Santa Cruz Biotechnology (Santa Cruz, CA, USA). Alexa Fluor 488-conjugated antibody and Lipofectamine 2000 reagent were purchased from ThermoFisher Scientific (Waltham, MA, USA). Recombinant human β-NGF and caspase inhibitor Z-VAD-FMK were purchased from R & D System (Minneapolis, MN, USA). Horseradish peroxidase-conjugated anti-mouse and anti-rabbit IgG antibodies were purchased from Jackson ImmunoResearch Laboratories (West Grove, PA, USA). All cell culture reagents were obtained from Gibco-BRL/Invitrogen (Carlsbad, CA, USA).

4.2. Obtainment of Human Plasma LDL

Plasma LDL was obtained from blood samples of hypercholesterolemic patients (diagnosed by physicians) in Kaohsiung Medical University (KMU) Hospital. All participants provided written consent, and ethics committees (Institutional Review Board of KMU Hospital) reviewed and approved the consent procedure. The preparation of LDL particles was at Lipid Science and Aging Research Center (LSARC) of KMU, Kaohsiung, Taiwan. Initially, LDL isolation was done by sequential potassium bromide density centrifugation. The yielding LDL at a density of 1.019 to 1.063 g/mL was treated with 5 mM EDTA and nitrogen to avoid ex vivo oxidation. After removing salts by dialysis (20 mM Tris-HCl [pH 8.0], 0.5 mM EDTA and 0.01% NaN_3), LDL was further separated into subfractions of L1 to L5 according to electrical charge by using a fast protein liquid chromatography with a UnoQ12 column (Bio-Rad Laboratories, Inc., Hercules, CA, USA), as described previously [18,59]. OxLDL was prepared by incubating L5-free LDL (L1) with 5 μM $CuSO_4$ at 37 °C for 24 h. All LDL particles were sterilized by passing through 0.22 μm filters before using for cell experiments.

4.3. Cell Culture

Rat PC12 cells (American Type Culture Collection, Rockville, MD, USA) were grown as monolayer cultures and maintained in Dulbecco's modified Eagle's medium (DMEM) containing 10% horse serum, 5% heat-inactivated fetal bovine serum, and 1% penicillin/streptomycin. In order to induce morphological differentiation, cells (5×10^4 cells/mL) were first subcultured in multiwell culture plates previously coated with poly-L-lysine (0.1 mg/mL). One day later, cells were rinsed once with fresh DMEM, followed by NGF (50 ng/mL) treatment for the indicated time periods in low serum medium composed of DMEM, 1% horse serum, and 1% penicillin/streptomycin. The medium containing NGF was replaced every 2 days when long-term treatment was executed.

4.4. MTT Reduction Assay

For the determination of cell viability, a biochemical method based on 3-(4, 5-dimethylthianol-2-yl)-2, 5-diphenyltetrazolium bromide (MTT) reduction was used to assay the metabolic activity of cultured cells. Stock MTT solution (5 mg/mL) was added to all cell-containing wells and diluted to a final concentration of 0.5 mg/mL. After incubation at 37 °C for 2 h, the MTT-containing cell medium was removed. The purple formazan crystals yielded by the action of mitochondrial dehydrogenase was dissolved with dimethyl sulfoxide and quantified spectrophotometrically. The light absorbance values at 570 nm against a 630 nm reference wavelength (ΔOD; OD: optical density) were recorded. The cell survival percentage is calculated by the formula: (viable cells) % = (ΔOD of treated sample/ΔOD of untreated control) × 100. The untreated control well was assigned a percentage value of 100% cell survival.

4.5. siRNA Transfection

Cells were cultured in growth medium without antibiotics for 24 h prior to transfection. According to the manufacturer's protocol, transfection of siRNA duplex (100 pmol) in 6-well culture plates was performed using Lipofectamine 2000 reagent and Opti-MEM. After maintaining transfected cells for 6 h, the culture medium was replaced with fresh growth medium. Western blotting was used for determining the knockdown effect of siRNA on target protein. All siRNA oligos were designed and synthesized by Santa Cruz Biotechnology. Rat LOX-1 siRNA (sc-156076) is a pool of 3 target-specific 19–25 nt siRNAs. The specific sequences are as follows: (i) LOX-1 siRNA-1 sense, 5′-CUGUAAGCUACUACAUAGATT-3′ and antisense, 5′-UCUAUGUAGUAGCUUACAGTT-3′; (ii) LOX-1 siRNA-2 sense, 5′-CCAUUAUGCUAGAGGUAAUTT-3′ and antisense, 5′-AUUACCUCUAGCAUAAUGGTT-3′; (iii) LOX-1 siRNA-3 sense, 5′-CUAGAACAAACACCAAUCUTT-3′ and antisense, 5′-AGAUUGGUGUUUGUUCUAGTT-3′. Control siRNA (sc-37007) is a non-targeting 20–25 nt siRNA designed as a negative control and leads to no degradation of any known cellular mRNA. The manufacturer does not offer the sequence of scrambled control siRNA.

4.6. Comet Assay

A neutral comet assay (single cell gel electrophoresis assay) was used to detect DSBs, as per the manufacturer's instructions. After finishing electrophoresis, at least 50 randomly selected images were analyzed from each treatment group under a fluorescence microscope (Nikon ECLIPSE Ts2 microscope, Nikon, Inc., Melville, NY, USA) equipped with a fluorescein isothiocyanate filter set. Images were captured using a cooled CCD camera (Spot, Diagnostic Instruments, Sterling Heights, MI, USA). As an informative and reliable DNA damage parameter, the Olive tail moment, which is defined as the product of the amount of DNA in the tail (fraction of total DNA in the comet tail) and the mean distance of migration in the tail (comet tail length), was calculated. The quantifications of the comet tail length and olive tail moment were conducted using the COMETscore.v1.5 image processing software.

4.7. Immunofluorescence Detection of DNA DSBs

Cultures grown on coverslips were subjected to immunostaining. To detect DNA DSBs, cells were fixed in 4% paraformaldehyde, permeabilized with 0.3% Triton X-100 in phosphate buffered saline, and then incubated with 5% normal goat serum to block nonspecific binding. Subsequently, an antibody against phosphorylated H2AX (γH2AX, 1:200 dilution) was used overnight at 4 °C, followed by the corresponding secondary antibody conjugated to Alexa Fluor 488. The nuclei were observable by labelling with 4′, 6-diamidino-2-phenylindole (DAPI). After rinsing the labelled cells with phosphate buffered saline and mounting slides with an anti-fading aqueous medium, observation was carried out under a fluorescent microscope equipped with appropriate filters and a digital camera.

4.8. Western Blotting Analysis

Cells were collected and homogenized in ice-cold RIPA buffer as described previously [59]. After a 20-min incubation at 4 °C, cell extracts were centrifuged at 14,000× *g* for 10 min and protein concentration in the supernatant was determined with a Bio-Rad protein assay (Bio-Rad, Hercules, CA, USA). Cell lysates (40 μg/lane) were separated by electrophoresis through 8–12% SDS-polyacrylamide gels and then electroblotted to 0.45 μm PVDF membranes (Millipore, Bedford, MA, USA) using a semi-dry transfer apparatus (Hoefer Scientific Instruments, San Francisco, CA, USA). Membranes were blocked in 5% non-fat dry skim milk for 1 h at room temperature, followed by immunoblotting for desired proteins with the assist of specific primary a ntibodies and appropriate HRP-conjugated secondary antibodies. Protein bands developed on the X-ray film were visualized by an enhanced chemiluminescence kit (Amersham Biosciences, Piscataway, NJ, USA). Analyses of protein band densities were performed using an Image J software (NIH, Bethesda, MD, USA).

4.9. Statistical Analysis

Treatment group means were compared by ANOVA, followed by Dunnett's test or Bonferroni's *t*-test for pair-wise comparisons. Paired groups were compared by Student's *t*-test. All results are expressed as the mean ± standard error of the mean (SEM). Differences were considered significant when the probability value was less than 0.05 ($p < 0.05$). SigmaStat version 4.0 software (Jandel Scientific, San Diego, CA, USA) was used for all statistical analyses.

Acknowledgments: We are grateful to KMU-LSARC for providing us with LDL particles, including L1, L5, and oxLDL, and excellent technical assistances. This work was supported by grants (MOST 104-2314-B-037-069 owned by Jiz-Yuh Wang and MOST 104-2320-B-570-002 owned by Ching-Tien Lee from the Ministry of Science and Technology, Taipei, Taiwan; and grants (KMU-M106005 owned by Jiz-Yuh Wang and KMU-TP103D02 owned by Chiou-Lian Lai from KMU, Kaohsiung, Taiwan.

Author Contributions: Jiz-Yuh Wang designed and supervised experiments and wrote the paper. Ching-Tien Lee and Chen-Yen Lin performed experiments and data analysis. Chiou-Lian Lai contributed valuable suggestions and critically revised the manuscript for important intellectual content. Submission of the final manuscript was endorsed by all authors.

Conflicts of Interest: The authors declare no conflict of interest.

Abbreviations

AD	Alzheimer's disease
ApoB	Apolipoprotein B
ATM	Ataxia-telangiectasia mutated
BBB	Blood-brain barrier
CSF	Cerebrospinal fluid
CVD	Cardiovascular disease
DAPI	4′,6-Diamidino-2-phenylindole
DMEM	Dulbecco's modified Eagle's medium
DSBs	Double strand breaks
EC	Endothelial cell
Egr-1	Early growth response-1
ERK	Extracellular signal-regulated kinase
JNK	c-JUN NH2-terminal protein kinase
LDL	Low-density lipoprotein
LDL(−)	Electronegative low-density lipoprotein
LDL-C	LDL-cholesterol
LDLR	LDL receptor
LOX-1	Lectin-like oxidized low-density lipoprotein receptor-1
MAPKs	Mitogen-activated protein kinases
MTT	3-(4,5-Dimethylthianol-2-yl)-2,5 diphenyltetrazolium bromide
NGF	Nerve growth factor
NF-M	Neurofilament-medium
OxLDL	Oxidized LDL
PFT-α	Pifithrin-α

References

1. Stampfer, M.J. Cardiovascular disease and Alzheimer's disease: Common links. *J. Intern. Med.* **2006**, *260*, 211–223. [CrossRef] [PubMed]
2. Kalaria, R.N. Arteriosclerosis, apolipoprotein E, and Alzheimer's disease. *Lancet* **1997**, *349*, 1174. [CrossRef]

3. Luchsinger, J.A.; Mayeux, R. Cardiovascular risk factors and Alzheimer's disease. *Curr. Atheroscler. Rep.* **2004**, *6*, 261–266. [CrossRef] [PubMed]

4. Kuo, Y.M.; Emmerling, M.R.; Bisgaier, C.L.; Essenburg, A.D.; Lampert, H.C.; Drumm, D.; Roher, A.E. Elevated low-density lipoprotein in Alzheimer's disease correlates with brain abeta 1–42 levels. *Biochem. Biophys. Res. Commun.* **1998**, *252*, 711–715. [CrossRef] [PubMed]

5. Refolo, L.M.; Malester, B.; LaFrancois, J.; Bryant-Thomas, T.; Wang, R.; Tint, G.S.; Sambamurti, K.; Duff, K.; Pappolla, M.A. Hypercholesterolemia accelerates the Alzheimer's amyloid pathology in a transgenic mouse model. *Neurobiol. Dis.* **2000**, *7*, 321–331. [CrossRef] [PubMed]

6. Kivipelto, M.; Laakso, M.P.; Tuomilehto, J.; Nissinen, A.; Soininen, H. Hypertension and hypercholesterolaemia as risk factors for Alzheimer's disease: Potential for pharmacological intervention. *CNS Drugs* **2002**, *16*, 435–444. [CrossRef] [PubMed]

7. Lai, C.L.; Hsu, C.Y.; Liou, L.M.; Hsieh, H.Y.; Hsieh, Y.H.; Liu, C.K. Effect of cholesterol and CYP46 polymorphism on cognitive event-related potentials. *Psychophysiology* **2011**, *48*, 1572–1577. [CrossRef] [PubMed]

8. Taylor, V.H.; MacQueen, G.M. Cognitive dysfunction associated with metabolic syndrome. *Obes. Rev.* **2007**, *8*, 409–418. [CrossRef] [PubMed]

9. Helzner, E.P.; Luchsinger, J.A.; Scarmeas, N.; Cosentino, S.; Brickman, A.M.; Glymour, M.M.; Stern, Y. Contribution of vascular risk factors to the progression in Alzheimer disease. *Arch. Neurol.* **2009**, *66*, 343–348. [CrossRef] [PubMed]

10. Smith, C.C.; Stanyer, L.; Betteridge, D.J.; Cooper, M.B. Native and oxidized low-density lipoproteins modulate the vasoactive actions of soluble beta-amyloid peptides in rat aorta. *Clin. Sci.* **2007**, *113*, 427–434. [CrossRef] [PubMed]

11. Griffiths, H.; Irundika, D.; Lip, G.; Spickett, C.; Polidori, C. Oxidised LDL lipids, statins and a blood-brain barrier. *Free Radic. Biol. Med.* **2014**, *75*, S15–S16. [CrossRef] [PubMed]

12. Kuller, L.H. Statins and dementia. *Curr. Atheroscler. Rep.* **2007**, *9*, 154–161. [CrossRef] [PubMed]

13. Sanchez-Quesada, J.L.; Benitez, S.; Ordonez-Llanos, J. Electronegative low-density lipoprotein. *Curr. Opin. Lipidol.* **2004**, *15*, 329–335. [CrossRef] [PubMed]

14. Sanchez-Quesada, J.L.; Estruch, M.; Benitez, S.; Ordonez-Llanos, J. Electronegative LDL: A useful biomarker of cardiovascular risk? *Clin. Lipidol.* **2012**, *7*, 345–359. [CrossRef]

15. Lu, J.; Yang, J.H.; Burns, A.R.; Chen, H.H.; Tang, D.; Walterscheid, J.P.; Suzuki, S.; Yang, C.Y.; Sawamura, T.; Chen, C.H. Mediation of electronegative low-density lipoprotein signaling by LOX-1: A possible mechanism of endothelial apoptosis. *Circ. Res.* **2009**, *104*, 619–627. [CrossRef] [PubMed]

16. Chen, H.H.; Hosken, B.D.; Huang, M.; Gaubatz, J.W.; Myers, C.L.; Macfarlane, R.D.; Pownall, H.J.; Yang, C.Y. Electronegative LDLs from familial hypercholesterolemic patients are physicochemically heterogeneous but uniformly proapoptotic. *J. Lipid Res.* **2007**, *48*, 177–184. [CrossRef] [PubMed]

17. Chen, C.H.; Jiang, T.; Yang, J.H.; Jiang, W.; Lu, J.; Marathe, G.K.; Pownall, H.J.; Ballantyne, C.M.; McIntyre, T.M.; Henry, P.D.; et al. Low-density lipoprotein in hypercholesterolemic human plasma induces vascular endothelial cell apoptosis by inhibiting fibroblast growth factor 2 transcription. *Circulation* **2003**, *107*, 2102–2108. [CrossRef] [PubMed]

18. Ke, L.Y.; Engler, D.A.; Lu, J.; Matsunami, R.K.; Chan, H.C.; Wang, G.J.; Yang, C.Y.; Chang, J.G.; Chen, C.H. Chemical composition-oriented receptor selectivity of L5, a naturally occurring atherogenic low-density lipoprotein. *Pure Appl. Chem.* **2011**, *83*, 1731–1740. [CrossRef] [PubMed]

19. Lee, T.; Chau, L. Fas/Fas ligand-mediated death pathway is involved in oxLDL-induced apoptosis in vascular smooth muscle cells. *Am. J. Physiol. Cell Physiol.* **2001**, *280*, C709–C718. [PubMed]

20. Maziere, C.; Meignotte, A.; Dantin, F.; Conte, M.A.; Maziere, J.C. Oxidized LDL induces an oxidative stress and activates the tumor suppressor p53 in MRC5 human fibroblasts. *Biochem. Biophys. Res. Commun.* **2000**, *276*, 718–723. [CrossRef] [PubMed]

21. Burma, S.; Chen, B.P.; Murphy, M.; Kurimasa, A.; Chen, D.J. ATM phosphorylates histone H2AX in response to DNA double-strand breaks. *J. Biol. Chem.* **2001**, *276*, 42462–42467. [CrossRef] [PubMed]

22. Roos, W.P.; Kaina, B. DNA damage-induced cell death by apoptosis. *Trends Mol. Med.* **2006**, *12*, 440–450. [CrossRef] [PubMed]

23. Kaina, B. DNA damage-triggered apoptosis: Critical role of DNA repair, double-strand breaks, cell proliferation and signaling. *Biochem. Pharmacol.* **2003**, *66*, 1547–1554. [CrossRef]

24. Canman, C.E.; Lim, D.S.; Cimprich, K.A.; Taya, Y.; Tamai, K.; Sakaguchi, K.; Appella, E.; Kastan, M.B.; Siliciano, J.D. Activation of the ATM kinase by ionizing radiation and phosphorylation of p53. *Science* **1998**, *281*, 1677–1679. [CrossRef] [PubMed]

25. Shieh, S.Y.; Ikeda, M.; Taya, Y.; Prives, C. DNA damage-induced phosphorylation of p53 alleviates inhibition by MDM2. *Cell* **1997**, *91*, 325–334. [CrossRef]

26. Honda, R.; Tanaka, H.; Yasuda, H. Oncoprotein MDM2 is a ubiquitin ligase E3 for tumor suppressor p53. *FEBS Lett.* **1997**, *420*, 25–27. [CrossRef]

27. Beckmann, A.M.; Wilce, P.A. Egr transcription factors in the nervous system. *Neurochem. Int.* **1997**, *31*, 477–510. [CrossRef]

28. Al-Chalabi, A.; Miller, C.C. Neurofilaments and neurological disease. *Bioessays* **2003**, *25*, 346–355. [CrossRef] [PubMed]

29. Longo, F.M.; Massa, S.M. Small-molecule modulation of neurotrophin receptors: A strategy for the treatment of neurological disease. *Nat. Rev. Drug Discov.* **2013**, *12*, 507–525. [CrossRef] [PubMed]

30. Shen, M.Y.; Chen, F.Y.; Hsu, J.F.; Fu, R.H.; Chang, C.M.; Chang, C.T.; Liu, C.H.; Wu, J.R.; Lee, A.S.; Chan, H.C.; et al. Plasma L5 levels are elevated in ischemic stroke patients and enhance platelet aggregation. *Blood* **2016**, *127*, 1336–1345. [CrossRef] [PubMed]

31. Podrez, E.A.; Byzova, T.V. Prothrombotic lipoprotein patterns in stroke. *Blood* **2016**, *127*, 1221–1222. [CrossRef] [PubMed]

32. Kay, A.D.; Day, S.P.; Nicoll, J.A.; Packard, C.J.; Caslake, M.J. Remodelling of cerebrospinal fluid lipoproteins after subarachnoid hemorrhage. *Atherosclerosis* **2003**, *170*, 141–146. [CrossRef]

33. Salen, G.; Berginer, V.; Shore, V.; Horak, I.; Horak, E.; Tint, G.S.; Shefer, S. Increased concentrations of cholestanol and apolipoprotein B in the cerebrospinal fluid of patients with cerebrotendinous xanthomatosis. Effect of chenodeoxycholic acid. *N. Engl. J. Med.* **1987**, *316*, 1233–1238. [CrossRef] [PubMed]

34. Mu, J.; Yang, Y.; Chen, J.; Cheng, K.; Li, Q.; Wei, Y.; Zhu, D.; Shao, W.; Zheng, P.; Xie, P. Elevated host lipid metabolism revealed by iTRAQ-based quantitative proteomic analysis of cerebrospinal fluid of tuberculous meningitis patients. *Biochem. Biophys. Res. Commun.* **2015**, *466*, 689–695. [CrossRef] [PubMed]

35. Takechi, R.; Galloway, S.; Pallebage-Gamarallage, M.; Wellington, C.; Johnsen, R.; Mamo, J.C. Three-dimensional colocalization analysis of plasma-derived apolipoprotein B with amyloid plaques in APP/PS1 transgenic mice. *Histochem. Cell Biol.* **2009**, *131*, 661–666. [CrossRef] [PubMed]

36. Takechi, R.; Galloway, S.; Pallebage-Gamarallage, M.M.; Wellington, C.L.; Johnsen, R.D.; Dhaliwal, S.S.; Mamo, J.C. Differential effects of dietary fatty acids on the cerebral distribution of plasma-derived apo B lipoproteins with amyloid-beta. *Br. J. Nutr.* **2010**, *103*, 652–662. [CrossRef] [PubMed]

37. Yang, Y.; Rosenberg, G.A. Blood-brain barrier breakdown in acute and chronic cerebrovascular disease. *Stroke* **2011**, *42*, 3323–3328. [CrossRef] [PubMed]

38. Szczepanska-Sadowska, E.; Cudnoch-Jedrzejewska, A.; Ufnal, M.; Zera, T. Brain and cardiovascular diseases: Common neurogenic background of cardiovascular, metabolic and inflammatory diseases. *J. Physiol. Pharmacol.* **2010**, *61*, 509–521. [PubMed]

39. Kerkar, S.; Williams, M.; Blocksom, J.M.; Wilson, R.F.; Tyburski, J.G.; Steffes, C.P. TNF-alpha and IL-1beta increase pericyte/endothelial cell co-culture permeability. *J. Surg. Res.* **2006**, *132*, 40–45. [CrossRef] [PubMed]

40. de Vries, H.E.; Blom-Roosemalen, M.C.; van Oosten, M.; de Boer, A.G.; van Berkel, T.J.; Breimer, D.D.; Kuiper, J. The influence of cytokines on the integrity of the blood-brain barrier in vitro. *J. Neuroimmunol.* **1996**, *64*, 37–43. [CrossRef]

41. Zenaro, E.; Piacentino, G.; Constantin, G. The blood-brain barrier in Alzheimer's disease. *Neurobiol. Dis.* **2016**. [CrossRef] [PubMed]

42. Kalaria, R.N.; Hedera, P. beta-Amyloid vasoactivity in Alzheimer's disease. *Lancet* **1996**, *347*, 1492–1493. [PubMed]

43. Chu, C.S.; Wang, Y.C.; Lu, L.S.; Walton, B.; Yilmaz, H.R.; Huang, R.Y.; Sawamura, T.; Dixon, R.A.; Lai, W.T.; Chen, C.H.; et al. Electronegative low-density lipoprotein increases C-reactive protein expression in vascular endothelial cells through the LOX-1 receptor. *PLoS ONE* **2013**, *8*, e70533. [CrossRef] [PubMed]

44. Estruch, M.; Bancells, C.; Beloki, L.; Sanchez-Quesada, J.L.; Ordonez-Llanos, J.; Benitez, S. CD14 and TLR4 mediate cytokine release promoted by electronegative LDL in monocytes. *Atherosclerosis* **2013**, *229*, 356–362. [CrossRef] [PubMed]

45. Avogaro, P.; Bon, G.B.; Cazzolato, G. Presence of a modified low density lipoprotein in humans. *Arteriosclerosis* **1988**, *8*, 79–87. [CrossRef] [PubMed]

46. Benitez, S.; Villegas, V.; Bancells, C.; Jorba, O.; Gonzalez-Sastre, F.; Ordonez-Llanos, J.; Sanchez-Quesada, J.L. Impaired binding affinity of electronegative low-density lipoprotein (LDL) to the LDL receptor is related to nonesterified fatty acids and lysophosphatidylcholine content. *Biochemistry* **2004**, *43*, 15863–15872. [CrossRef] [PubMed]

47. Bancells, C.; Villegas, S.; Blanco, F.J.; Benitez, S.; Gallego, I.; Beloki, L.; Perez-Cuellar, M.; Ordonez-Llanos, J.; Sanchez-Quesada, J.L. Aggregated electronegative low density lipoprotein in human plasma shows a high tendency toward phospholipolysis and particle fusion. *J. Biol. Chem.* **2010**, *285*, 32425–32435. [CrossRef] [PubMed]

48. Bancells, C.; Benitez, S.; Ordonez-Llanos, J.; Oorni, K.; Kovanen, P.T.; Milne, R.W.; Sanchez-Quesada, J.L. Immunochemical analysis of the electronegative LDL subfraction shows that abnormal N-terminal apolipoprotein B conformation is involved in increased binding to proteoglycans. *J. Biol. Chem.* **2011**, *286*, 1125–1133. [CrossRef] [PubMed]

49. Abe, Y.; Fornage, M.; Yang, C.Y.; Bui-Thanh, N.A.; Wise, V.; Chen, H.H.; Rangaraj, G.; Ballantyne, C.M. L5, the most electronegative subfraction of plasma LDL, induces endothelial vascular cell adhesion molecule 1 and CXC chemokines, which mediate mononuclear leukocyte adhesion. *Atherosclerosis* **2007**, *192*, 56–66. [CrossRef] [PubMed]

50. Chen, W.Y.; Chen, F.Y.; Lee, A.S.; Ting, K.H.; Chang, C.M.; Hsu, J.F.; Lee, W.S.; Sheu, J.R.; Chen, C.H.; Shen, M.Y. Sesamol reduces the atherogenicity of electronegative L5 LDL in vivo and in vitro. *J. Nat. Prod.* **2015**, *78*, 225–233. [CrossRef] [PubMed]

51. Symonds, H.; Krall, L.; Remington, L.; Saenz-Robles, M.; Lowe, S.; Jacks, T.; van Dyke, T. p53-dependent apoptosis suppresses tumor growth and progression in vivo. *Cell* **1994**, *78*, 703–711. [CrossRef]

52. Li, Y.; Duan, Z.; Gao, D.; Huang, S.; Yuan, H.; Niu, X. The new role of LOX-1 in hypertension induced neuronal apoptosis. *Biochem. Biophys. Res. Commun.* **2012**, *425*, 735–740. [CrossRef] [PubMed]

53. Zhou, L.J.; Zhu, X.Z. Reactive oxygen species-induced apoptosis in PC12 cells and protective effect of bilobalide. *J. Pharmacol. Exp. Ther.* **2000**, *293*, 982–988. [PubMed]

54. Giuliano, P.; De Cristofaro, T.; Affaitati, A.; Pizzulo, G.M.; Feliciello, A.; Criscuolo, C.; De Michele, G.; Filla, A.; Avvedimento, E.V.; Varrone, S. DNA damage induced by polyglutamine-expanded proteins. *Human Mol. Genet.* **2003**, *12*, 2301–2309. [CrossRef] [PubMed]

55. Kim, Y.; Seger, R.; Suresh Babu, C.V.; Hwang, S.Y.; Yoo, Y.S. A positive role of the PI3-K/Akt signaling pathway in PC12 cell differentiation. *Mol. Cells* **2004**, *18*, 353–359. [PubMed]

56. Read, D.E.; Gorman, A.M. Involvement of Akt in neurite outgrowth. *Cell Mol. Life Sci.* **2009**, *66*, 2975–2984. [CrossRef] [PubMed]

57. Lee, A.S.; Wang, G.J.; Chan, H.C.; Chen, F.Y.; Chang, C.M.; Yang, C.Y.; Lee, Y.T.; Chang, K.C.; Chen, C.H. Electronegative low-density lipoprotein induces cardiomyocyte apoptosis indirectly through endothelial cell-released chemokines. *Apoptosis* **2012**, *17*, 1009–1018. [CrossRef] [PubMed]

58. Tang, D.; Lu, J.; Walterscheid, J.P.; Chen, H.H.; Engler, D.A.; Sawamura, T.; Chang, P.Y.; Safi, H.J.; Yang, C.Y.; Chen, C.H. Electronegative LDL circulating in smokers impairs endothelial progenitor cell differentiation by inhibiting Akt phosphorylation via LOX-1. *J. Lipid Res.* **2008**, *49*, 33–47. [CrossRef] [PubMed]

59. Yu, L.E.; Lai, C.L.; Lee, C.T.; Wang, J.Y. Highly electronegative low-density lipoprotein L5 evokes microglial activation and creates a neuroinflammatory stress via Toll-like receptor 4 signaling. *J. Neurochem.* **2017**, *142*, 231–245. [CrossRef] [PubMed]

International Journal of
Molecular Sciences

MDPI

Review

Activating the Anaphase Promoting Complex to Enhance Genomic Stability and Prolong Lifespan

Troy A. A. Harkness

Department of Anatomy and Cell Biology, University of Saskatchewan, Saskatoon, SK S7N 5E5, Canada;
troy.harkness@usask.ca; Tel.: +1-(306)-966-1995

Received: 15 May 2018; Accepted: 20 June 2018; Published: 27 June 2018

Abstract: In aging cells, genomic instability is now recognized as a hallmark event. Throughout life, cells encounter multiple endogenous and exogenous DNA damaging events that are mostly repaired, but inevitably DNA mutations, chromosome rearrangements, and epigenetic deregulation begins to mount. Now that people are living longer, more and more late life time is spent suffering from age-related disease, in which genomic instability plays a critical role. However, several major questions remain heavily debated, such as the following: When does aging start? How long can we live? In order to minimize the impact of genomic instability on longevity, it is important to understand when aging starts, and to ensure repair mechanisms remain optimal from the very start to the very end. In this review, the interplay between the stress and nutrient response networks, and the regulation of homeostasis and genomic stability, is discussed. Mechanisms that link these two networks are predicted to be key lifespan determinants. The Anaphase Promoting Complex (APC), a large evolutionarily conserved ubiquitin ligase, can potentially serve this need. Recent work demonstrates that the APC maintains genomic stability, mounts a stress response, and increases longevity in yeast. Furthermore, inhibition of APC activity by glucose and nutrient response factors indicates a tight link between the APC and the stress/nutrient response networks.

Keywords: Anaphase Promoting Complex; lifespan; cancer; yeast; human cell culture

1. Introduction

When does the aging process begin? How long can we live? Why do we age? These questions are highly debated with no distinct, definitive answers. Does aging begin when our skin starts to wrinkle, or when our hair commences to turn grey? Or perhaps aging begins after the completion of growth [1]. Aging has also been defined as a shift in an organism's aging reality. The aging reality has been described as a mutually enslaved system of DNA and its environment in which signaling failures within this DNA environment occur over time [2]. Much of the debate reflects the fact that we have not perceived children as aging; however, acquired somatic mutations are recognized in infants and children, leading to the development of childhood cancers [3–5]. More compelling support for childhood aging comes from premature aging syndromes, such as progeria, in which children age in a very similar manner to normal aging individuals but at an 8-fold accelerated rate [6]. Even Hayflick considered when aging begins, weighing the possibilities that aging starts before or at conception or, alternatively, when maximum strength and stamina is achieved [7]. Hayflick eventually concluded that aging is a stochastic program that begins after reproductive maturity in animals, resulting in the loss of molecular fidelity. This loss of fidelity ultimately surpasses repair capacity, leaving individuals prone to age-related diseases [8,9]. The idea that aging is a random stochastic program is supported by many researchers in the field [10,11]. The stochastic idea of aging gained traction when the free radical theory of aging was proposed. This theory states that aging occurs due to the natural wear and tear of cellular machinery and biological substances due to exposure to free radicals generated within the cell [12]. Biological systems are constantly fighting a battle with its environment, both internally and externally, to ward off damage. The simple generation of mitochondrial-dependent energy and DNA replication expose cells to damage that must be repaired. Evidence for a stochastic program of aging also comes from inorganic compounds that age over time; for example, rusting of metal and peeling of paint (discussed in [13]), implying that something beyond genetics controls aging. From this work comes the idea that entropy drives aging, while genes drive longevity.

This review will summarize the current ideas describing our thoughts on the aging process. Molecular mechanisms will then be described that facilitate cellular aging. The notion that genomic instability is the driving force leading to aging and age-related disease will be discussed. Finally, the novel concept that activation of a cell cycle regulator, the Anaphase Promoting Complex, which is required for maintenance of cell health, inhibition of cancer progression, and enhanced longevity, will be explored.

2. Genetic Control of Longevity

Longevity depends on how long our cells remain functional, which is countered by the many insults faced by cells. Pathways that maintain cellular homeostasis are genetically controlled; thus, it clearly follows that a genetic program would be in place to control longevity. A recent review links proteostasis (a housekeeping phenomenon that controls the integrity of protein structure and function) with lifespan determination, and suggests the failure of the proteostatic network occurs early in life and marks the beginning of aging [14]. A global network encompassing maintenance of genomic stability, as well as enhancing proteostasis, would involve, for example, genes and proteins that repair DNA, scavenge free radicals, and the proteins that run these programs. Thus, the genetic and stochastic models appear to oppose one another. On the one hand, the stochastic model dictates that over time, our cells randomly accumulate damage, such as the accumulation of DNA mutations and genomic instability, and eventually succumb to the damage. On the other hand, the genetic program is in place to provide cells with options to survive the intrinsic and extrinsic environmental assaults that chronically bombard the cell and the biological system as a whole. Evolutionary biologists have argued that selection of lifespan-extending genes is unlikely, since the effects of these genes would only be seen long after reproduction has ceased, with the force of natural selection that declines with age leaving no valid reason to remain alive [2,15]. However, when these genes are viewed as controlling cell repair in the face of a damaging environment in order to survive, then it becomes clear that enhanced longevity may only be an indirect benefit accompanying the ability to survive unfavorable life events. Thus, evolution may not be selecting for longevity genes at all, but simply looking for genes that increase survival under trying times, with increased longevity simply a lucky side effect. This idea forms the foundation of the theory describing the response to nutrients versus stress as a driving force defining one's lifespan, as suggested earlier [16]. The concept of Hormesis, in which a potentially life-threatening stress, when given in a low dose, increases health or extends lifespan, is an example of a stress response providing a favorable and beneficial reaction [17,18].

One genetic theory of aging that is appealing to evolutionary biologists is the idea of pleiotropic antagonism. Antagonistic pleiotropy suggests that genes that are beneficial in the early years become harmful in later life [19–21]. A similar idea is described in the Disposable Soma theory [15]. It was proposed that because of high environmental mortality, resources are primarily spent on growth and reproduction, rather than on the soma, which would leave the soma exposed to environmental and intrinsic stresses. Nonetheless, the idea has been used to describe a Darwinian-evolutionary concept of aging in which the deleterious effects of previously beneficial genes in later life leads to the development of age-related disease [22]. Examples of antagonistic pleiotropy and how it could contribute to a Darwinian-evolutionary concept include the calcification of bones, which enables strength in early life, but eventually leads to deleterious calcification of arteries in late life. The erosion of telomeres, often considered a clear mechanism of aging, is also considered as support of a Darwinian-evolutionary model. While expression of telomerase extends telomere length in stem cell populations, it also contributes to tumor formation [23]. Thus, in early life, longer telomeres maintain the health and vitality of the cell. However, in later life, the inactivation of telomerase is proposed to ward off cancer, but at the cost of continued telomere erosion, and likely cellular senescence [24]. Darwinian selection of stress sensing and Darwinian selection of antagonistic pleiotropy genes are both used as examples of alternative mechanisms of lifespan determination, as both provide benefits in early life. However, the selection of stress sensing genes, but not antagonistic pleiotropy genes, provides an indirect longevity benefit in later life.

3. Genetic Control of Aging

The genetic model of lifespan determination is not at odds with the stochastic model, and involves a genetic program that determines the length of time that our cells and bodies can function. In the end, the more capable the cell is at damage repair, occurring through stochastic intrinsic and extrinsic events, the better the odds of surviving harsh environments and living to pass on genes to descendants. Longevity of the organism starts with the health of the cells. If cell health cannot be maintained, then health of the tissues and the animal itself will suffer. Cell type, in particular, is determined by programmed genetic and epigenetic networks. In the human body, for example, each cell harbors the same sequence of DNA, yet different cells carry out very different functions. Alterations to any of these networks can signal the end to that cell. Maintenance of the cellular equilibrium supporting tissue renewal is critical to the longevity of the organism. Over time, equilibrium and cell renewal begins to fail, leading to reduced replacement of cells lost due to attrition or senescence. Thus, the failing of the genetic system, contributing to the loss of cell equilibrium and renewal through accumulating mutations, is considered a hallmark of aging. Hallmarks of aging consist of the following attributes: genomic instability, telomere attrition, epigenetic alterations, and loss of proteostasis, leading to deregulated nutrient sensing, altered mitochondrial function, and cellular senescence [25,26].

Specific genes, many of which were first described and characterized in the simple lower eukaryotic yeast system (i.e., *SIR2* (yeast *SIRT* gene), *SNF1* (yeast *AMPK*), *FKH1/2* (yeast *FOXO*), *SCH9* (yeast *AKT/S6K*), *TOR1*, and *RAS2*, have been shown to be evolutionarily conserved genes that respond to stress or nutrients, influencing longevity [27–32]. Thus, genes (such as *SIR2*) clearly play a significant role in promoting lifespan from single cells to humans, but do genes also drive the loss of homeostasis and the aging process? While the stress response genes drive cell health and longevity in the presence of low-level stress, an example of Hormesis [17,18], the nutrient response genes do the opposite, and act as pro-aging genes. Thus, genes provide the impetus for both longevity and aging. Unlike the stochastic model, which relies on random factors to drive the aging process, the genetic program promoting aging relies on the activation of a web of nutrient response genes that inhibit the stress response network in the presence of usable resources [33]. Lifespan extension through caloric restriction is a classic example of Hormesis, and can be mimicked in yeast by mutating genes, such as *SCH9*, *TOR1*, or *RAS2*, which encode nutrient response proteins [29]. Ultimately, the longevity of an organism depends on the effectiveness of the counterbalanced stress and nutrient sensing pathways.

4. A Ceiling on a Maximum Lifespan?

Support for a predetermined program delineating our lifespan has been around for decades and is derived from the knowledge that human mean lifespan has increased dramatically over the past century, but the maximum lifespan has not [34]. The longest verified living human, Jeanne Calment, died at age 122 in 1997 [35], which is consistent with the idea that the maximum attainable human lifespan is not increasing and is likely capped at 125 years [36,37]. Indeed, few humans have ever lived past 115 years [38]. A maximum of 125 years is nonetheless controversial, as some feel there is no limit to our lifespan [8,39–43]. Further evidence supporting the idea that a predetermined genetic program dictates our maximum lifespan is provided by observations that maximum achievable lifespans are observed across evolutionary boundaries, as each specific organism seems to have a built in maximum possible lifespan [44]. The ever-increasing mean lifespan observed over the past century has also been used as evidence that human maximum lifespan will also continue unimpeded. Of course, this rise in expected lifespan is largely attributable to new developments in medical care, improved diet, less exposure to toxins, and regular exercise, which may only increase healthspan, and not lifespan. Regardless of how people feel about the dramatic rise in global expected lifespans, time is needed to fully realize the effect of improved human well-being, and thought should be put into policy development to deal with the likelihood that people will be living longer, healthier lives. Thus, if maximum lifespans have reached a ceiling, with mean life expectancy continuing to rise, lifespan curves may soon be considered lifespan cliffs, with increased human productivity a likely benefit.

5. So When Does Aging Begin?

It now seems quite clear that cellular aging is largely dependent on the degree to which genomic instability has affected DNA-dependent processes. Many studies, from yeast to humans, have repeatedly shown that during aging, senescent cells that exit the cell cycle or cease to function harbor large accumulations of DNA mutation, rearrangements, and epigenetic alterations. There are numerous sources of DNA damage, both endogenous and exogenous, that the cell must deal with. It is thought that a somatic cell may receive as many as 100,000 lesions daily [45,46]. It is not a coincidence that most age-dependent diseases, such as cancer, type II diabetes, and cardiopulmonary and neurodegenerative diseases are associated with increasingly elevated levels of genomic instability that occur over time [47–51]. When a cell is born, it is presumably at its functional apex, performing at its highest level. In yeast, the mother cell sequesters damage so that the daughter does not receive it, having a much better chance to begin life in a pristine state [52,53]. However, eventually the damage is too much for the yeast mother cell to fully sequester, with the daughter born with accumulating damage. If similar mechanisms that occur in yeast are occurring in higher eukaryotic systems, then it is easier to understand how a newly born cell would be at its best to repair damage and maintain proteostasis. With this in mind, the answer for when aging begins might be when the cells that form the zygote are first born; thus, aging of an individual may begin much earlier than conception, such as at the very moment when the mother develops oocytes in utero [54].

6. Connecting Stress Sensing with Nutrient Sensing

Genomic instability appears to be the gateway to aging and age-related disease. Genomic stability is threatened as soon as a cell is born due to the intrinsic damage caused by energy generation and the errors inflicted by DNA replication. The damage repair processes are presumably functioning at their best in these new cells, so genomic instability likely does not become an obstacle until much later in life. As discussed above, multiple antagonistic molecular networks are vying for available resources to respond to either stress and/or nutrients. It should be clear that the opposition of these pathways should not be all or none, as aspects of nutrient availability may be present even in an unfavorable environment. Thus, the question becomes how are nutrient and stress sensing networks regulated? What mediates the end of stress signaling when the stress is gone, or the stalling of the nutrient sensing pathways when the food source is used up?

7. The Anaphase Promoting Complex, Using Chromatin Assembly during Mitosis to Maintain Genome Stability

To answer these questions, it is important to identify components that connect stress and nutrient-sensing pathways. The Anaphase Promoting Complex (APC) has come to light as a potential link between the stress and nutrient sensing networks. The APC is an evolutionarily conserved large ubiquitin-protein ligase (E3) that targets proteins that inhibit mitotic entrance and exit, as well as proteins that inhibit G1 maintenance, for ubiquitin and proteasome-dependent degradation [55]. The APC is controlled by 2 co-activators, CDC20 and CDH1, which control mitotic progression, and G1 maintenance (Figures 1 and 2). CDC20 binds with the APC to initiate mitosis, and is then targeted for degradation by the APC^{CDH1} complex at the M/G1 transition [56,57]. CDH1 is then targeted for degradation at the G1/S transition by a second large E3 complex called the SCF (Skp-Cullin-F-box complex) [58]. The APC is largely known for its role in cell cycle progression, but we and others have identified it as a central player in stress sensing and lifespan determination using the simple brewing yeast eukaryotic model system (Figure 3) [31,59–65]. Mitosis is a time during the cell cycle when DNA damage can become permanent and lead to further chromosome erosion and genomic instability [66]. The APC is also required for replication-independent chromatin assembly and histone modifications [60,67–70]. Considering that replication-independent chromatin assembly is required for DNA repair [71,72], we speculate that the APC may be involved in repair of DNA damage incurred during chromosome segregation (Figure 3). The chromatin assembly factors Asf1, and the CAF-1 complex, have been shown in yeast and human cells to be involved in assembly of histones onto repaired DNA duplexes [73–76]. The link between repair of DNA during mitosis and the APC may be the CAF-1 and Asf1 chaperones, as the APC genetically interacts with both Asf1 and CAF-1 mutants in yeast (mutant combinations have worse phenotypes), and increased expression of any one of the CAF-1 subunits, or Asf1, rescues APC defects [67]. Consistent with a role in maintaining genomic stability, APC defects result in elevated sensitivity to UV radiation, increased loss of centromere based plasmids, and increased rDNA instability [60,64].

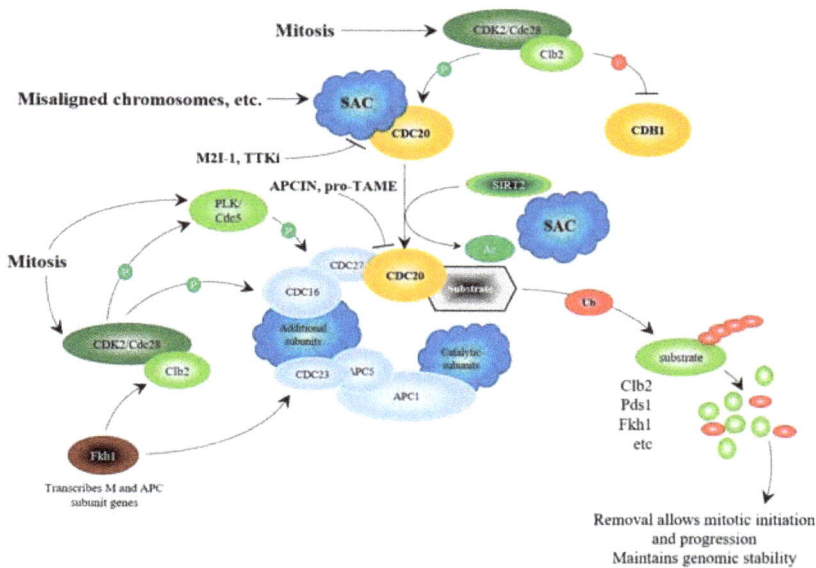

Figure 1. Regulation of the APC at the G2/M transition. The schematic considers results from mammalian and yeast studies. Yeast proteins are written as Cdc20, whereas mammalian proteins are written as CDC20. Genomic stability and segregation of replicated and repaired chromosomes is established via the Spindle Assembly Checkpoint (SAC) that sequesters Cdc20/CDC20 away from the APC, thus inhibiting APC function as cells enter mitosis [77,78]. When the SAC is satisfied, the cyclin Clb2 (Cyclin B), synthesized during G2 by Fkh1 [79], interacts with cyclin-dependent kinase Cdc28 (CDK2) to phosphorylate a series of proteins needed for mitotic progression: Cdc5 (PLK), Cdc16, Cdc23, and Cdc27 [80,81]. Once PLK is active, it further activates the APC by phosphorylating Apc9 (or APC1 in mammalian cells), Cdc16 and Cdc27 [80,81]. Cdc28-Clb2 also phosphorylates the co-activators Cdc20 for activation [80], and Cdh1 for inhibition [82]. A further activating stimulus is provided by SIRT2, which deacetylates CDC20 [83]. APC^{Cdc20} then targets proteins for degradation, such as Pds1 (PTTG1/Securin), to allow chromosome segregation, and Clb2 and Fkh1 to complete a negative feedback loop that prepares the cell for mitotic exit and G1 maintenance [65,84,85]. Degradation of Clb2

stops inhibition of Cdh1, allowing replacement of the APC^{Cdc20} complex with APC^{Cdh1}. APC chemical inhibitors, APCIN and pro-TAME, disrupt the CDC20-APC interaction [86,87], whereas the small molecule APC activators (M2I-1, TTKi) disrupt the CDC20-SAC interaction [88–90]. Protein degradation is shown by Ub, shaded with a red oval, attached to the target protein to build poly-Ub chains, followed by break down of the protein, shown in smaller circles. Inhibitory phosphorylation is shown with a red shaded "P", and activating phosphorylation is shown with a green shaded "P".

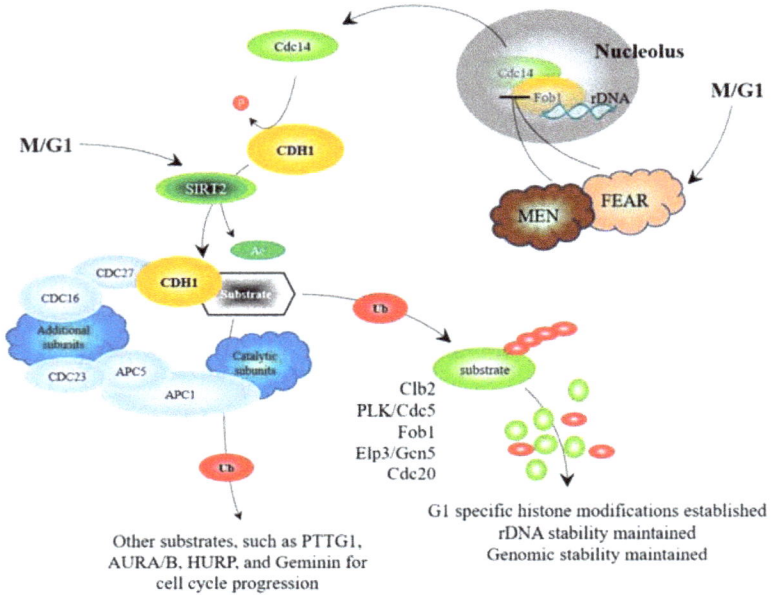

Figure 2. Regulation of the APC at the M/G1 transition. As mitosis comes to an end, the phosphatase Cdc14 is activated and released from sequestration within the nucleolus by Fob1, through a biphasic interaction involving the FEAR and MEN pathways [91,92]. Cdc14 dephosphorylates Cdh1, thus facilitating the interaction between Cdh1 and the APC [93]. Further activation is accomplished by deacetylation of CDH1 by SIRT2 [83]. APC^{Cdh1} function then leads to wholesale changes required for mitotic exit and transition into G1. Residual Pds1 and Clb2 are targeted for degradation by APC^{Cdh1}, as are Cdc20. Cdc5. and other targets, which puts an end to the pattern of proteins required for mitotic progression [56,57,94–100]. Degradation of Fob1, a negative regulator of FEAR, is required for G1 progression, as Fob1 [64] is required for rDNA condensation during mitosis. Gcn5 (and likely Elp3) is also required for G1 progression [69], as it presumably acetylates histones during mitosis to establish an epigenetic pattern required for G1 progression. Once this pattern is established during mitosis, Gcn5 (and likely Elp3) must be degraded. Ubiquitinated and degraded proteins are depicted as described above.

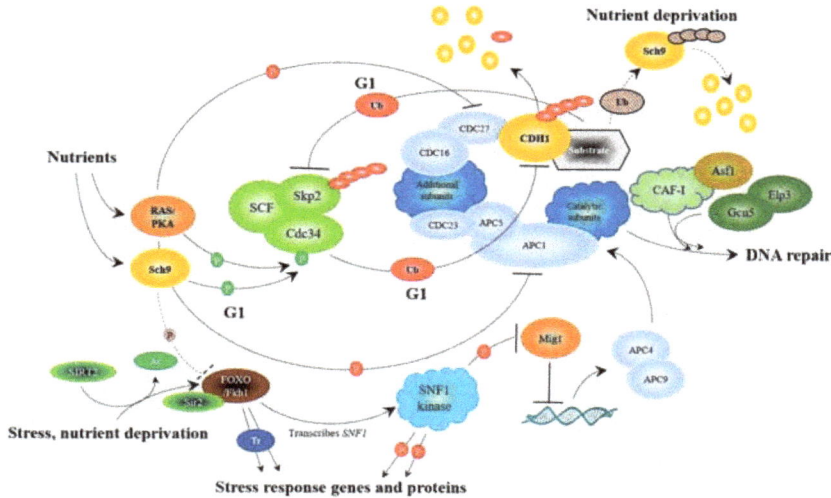

Figure 3. APC activity is decreased under nutrient conditions, and increased when nutrients are limiting. Inhibition of APC activity by glucose is mediated by the Ras/PKA pathway [81,101–103]. Recent work suggests that this is accomplished by driving the activity of the SCF ubiquitin ligase by the phosphorylation and activation of the SCF E2 component Cdc34 by the nutrient response kinases PKA and Sch9 (AKT/S6K) [104]. This could mediate APC inhibition, as it has been shown that the SCF targets the degradation of CDH1 during mitosis [58,105]. Our unpublished data also reveals that Sch9 likely inhibits Fkh1 function, and the subsequent induction of the stress pathways. Upon encountering stress, SIRT2 deacetylates and activates FOXO proteins [106], and in yeast, Sir2 physically associates with Fkh1 to facilitate inhibition of *CLB2* transcription in late M/G1 [107]. Fkh1 transcribes stress response genes (depicted by a a blue shaded "Tr"), including *SNF1*, which encodes the catalytic component of the SNF1 kinase, the yeast AMPK [79,108]. FOXO and AMPK interact across evolutionary boundaries to deal with stress [108–110]. The SNF1 kinase then enters the nucleus and inhibits the glucose responsive repressor Mig1, which represses the expression of the APC subunits *APC4* and *APC9* under nutrient conditions [61]. DNA repair is likely mediated, at least in part, by the APC, which controls the deposition and modification of histones during mitosis, which plays a pivotal role in DNA repair [67,69,71–74]. Inhibition of SCF-Cdc34 following APC activation is accomplished in two ways: first, the APC targets the SCF F-box protein Skp2 for degradation in G1 [111,112], and second, our unpublished data shows that the APC targets Sch9 for degradation once nutrients are depleted. Preliminary unpublished data is shown using dashed lines.

8. Maintaining Genomic Stability via APC-Mediated Histone Modifications

Histone post-translational modifications are involved in cell cycle progression, particularly mitosis [113], and in DNA repair. In yeast, DNA repair requires Asf1, CAF-1, and acetylation of H3 Lys56 (H3K56^{Ac}), mediated by the Asf1/Rtt109 complex [71,74]. Cells with impaired APC function have reduced H3K9^{Ac}, H3K79^{Me}, and H3K56^{Ac} [69]. H3K79^{Me} accumulates during mitosis [114], while H3K56^{Ac} and H3K9^{Ac} are reduced during mitosis but increase as cells enter G1 [115,116]. H3K9^{Ac} is important for transcriptional activation [117,118], H3K56^{Ac} is involved in histone deposition and DNA repair [74,119], while H3K79^{Me} is required for a variety of activities including transcriptional elongation, DNA repair, and cell cycle checkpoints [120,121]. Thus, the loss of these modifications due to impaired APC has a dramatic impact on chromatin and chromosome structure, transcription, and DNA repair. Furthermore, the histone acetyltransferase (HAT) that mediates H3K9^{Ac}, Gcn5, interacts genetically and functionally with the APC [69,70]. Increased expression of *GCN5* rescued APC defects and deletion of *GCN5* in APC mutants exacerbated growth defects. Furthermore, Gcn5 is targeted by the APC for degradation at the M/G1 transition [69]. Acetylation of histones during mitosis may be important to reset the epigenome as cells re-enter G1, leading to the appropriate activation of specific genes. The correlation of Gcn5 degradation at G1, just after the accumulation of H3K9^{Ac} as cells exit mitosis, with APC mitotic function, is at the crux of establishing an active transcriptome for continued cell cycle progression. Furthermore, if targeted degradation of Gcn5 by the APC is conserved from yeast to humans, then this may be critical for tumor suppression and maintenance of genomic stability, as increased H3K9^{Ac} is associated with DNA damage, genomic instability, and progression of multiple myeloma [122]. Consistent with this, APC defects lead to elevated genomic instability in yeast [60,64,65] and in human cells [123,124]. Thus, although the APC is required for mitotic progression, it is also required to guard against damage that can occur during chromosome segregation, and to ensure that histones are acetylated to enable proper transcription as cells enter G1. These activities are all critical to ensure that cells remain healthy, leading to enhanced lifespan. On the other hand, the inability to maintain cellular homeostasis is linked with genomic instability associated with cancer development and progression

9. Targeting APC Inhibition for Anticancer Therapy

Because of the role the APC plays in cell cycle progression, initial work focused on the inhibition of the APC as a means to block tumor growth [125–127]. The evolutionarily conserved Spindle Assembly Checkpoint (SAC) complex, consisting of the proteins MAD1, MAD2, BUB1, BUBR1, BUB3 and MPS1, binds and sequesters the APC co-activator CDC20 prior to mitosis [77,78], inhibiting APC activation until all chromosomes are ready for segregation (Figure 1). It was suggested that activation of the SAC, and inhibition of the APC, would protect the cell from inappropriate chromosome segregation and mitotic catastrophe in the presence of damaged chromosomes, which is often observed in cancer cells. Furthermore, *CDC20* mRNA expression is observed to be elevated in cancer cells, which is associated with a poor prognosis; *CDC20* knockdown is required for mitotic arrest and inhibition of cell growth [94,128,129]. Specific (APCIN and pro-TAME [86,87]) and non-specific (Velcade [126]) APC inhibitors have been developed recently and inhibit tumor growth in vitro [86,130]. Both APCIN and pro-TAME act by inhibiting the interaction of CDC20 with the APC (Figure 1). Thus, inhibition of the APC was believed to be a viable anti-tumor strategy.

10. Targeting APC Activation for Anticancer Therapy

Recent work in mammalian cancer cells provides evidence that APC activation, rather than inhibition, may be a potent anticancer therapy that antagonizes genomic instability. As discussed above, CDC20 is an APC coactivator, and high APC^{CDC20} may be inappropriately driving cells through mitosis to promote genomic instability and cancer progression, inferring that APC inhibition will be beneficial. Regulation of CDC20 is highly coordinated (Figure 1). As discussed above, CDC20 is sequestered and inhibited by the SAC until all chromosomes are aligned along the metaphase plate and ready for segregation [77,78]. Cdc20 in yeast is activated by Cdc28-Clb2-dependent phosphorylation [80]. Cdc28-Clb2 also phosphorylates the APC subunits Cdc16, Cdc23, and Cdc27 [81], the yeast Polo-like kinase, Cdc5 [131], and Cdh1 to maintain its inactivity [82]. Once Cdc5 is activated, it then potentially targets Cdc16, Cdc27, and Apc9 for phosphorylation to further activate the APC [80,81]. Cdc5 is later targeted by APC^{Cdh1} to exit mitosis [132]. CDC20 is also deacetylated by SIRT2, adding another level of activation [83]. Additional activation signals in yeast come from the Forkhead transcription factors Fkh1 and Fkh2. The *FKH1* and *FKH2* genes are transcribed during G2 by Hcm1 [133], and are required for the transcription of the "CLB2 cluster" of genes, which contains genes required for APC activity, such as *CLB2*, *CDC5*, *CDC20*, and *APC1* [79].

The APC is essential, and this is conserved from yeast to humans, as yeast deletion mutants are lethal and mouse models lacking APC subunits, or CDC20, die in embryogenesis [84,134–136]. The APC is also essential for the prevention of aneuploidy, which contributes to tumorigenesis [84].

Thus, the systemic in vivo use of APC inhibitors may be highly toxic, limiting this approach to cancer therapy. However, an alternative interpretation is possible to explain why CDC20 accumulates in cancer cells. CDC20 itself is targeted by APC^{CDH1} for degradation once mitosis is complete [56,57]. Therefore, elevated *CDC20* expression could reflect APC^{CDH1} impairment in cancer cells, inferring that APC activation will be beneficial to cell health. Our in vitro and in vivo work (Davies, Arnason and Harkness, unpublished), and findings from others, have noted that many APC^{CDH1} mitotic substrate genes and proteins are elevated in cancer cells, including CDC20 [94], PLK1 [95], AURA/B [96,97], HURP (*DLGAP* gene [98]), Securin (*PTTG1* gene [99]), and Geminin [100], hinting that impaired APC activity as a whole is involved, rather than isolated CDC20 elevation. Moreover, using the Cancer Genome Atlas database [137], we observed that the expression of the APC substrate genes *PTTG1* and *DLGAP5* in cancer patients is differentially regulated between normal tissues and tumor tissues, across 24 different types of cancer (Figure 4).

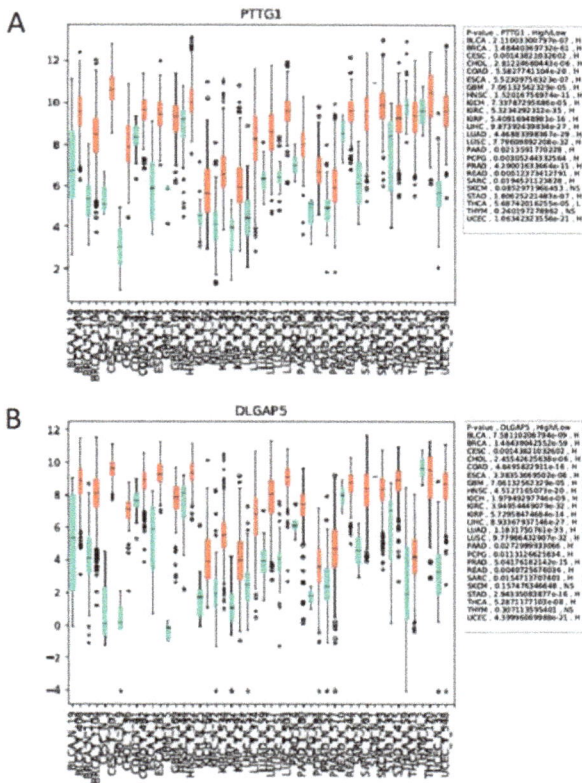

Figure 4. The APC substrate mRNAs PTTG1 and DLGAP5 are overexpressed in multiple cancer types. Expression scores for (**A**) PTTG1 and (**B**) DLGAP5 within 24 different types of cancer and normal tissue from TCGA [134]. The numbers in x-axis labels denote the number of patient samples in each cancer type. Statistical significance of the difference in expression between the normal and tumor samples is depicted for each cancer type. N.S. not significant. The abbreviation of each cancer in the axis label is represented as described in the TCGA portal [137].

While CDC20 has been linked to cancer progression, the second APC co-activator, CDH1, has been linked to tumor suppression, with earlier work demonstrating that cells lacking CDH1 have a shortened G1 phase, accumulate DNA damage, and undergo apoptosis [85]. CDH1 is also regulated through a complex web of interactions (Figure 2). As discussed above, the yeast Cdh1 is maintained in an inactive form by Cdc28-Clb2 phosphorylation until the end of mitosis, when Clb2 is targeted for degradation and the phosphatase Cdc14 is released from the nucleolus to undo the work of Cdc28-Clb2 [82,91,93]. The mammalian CDH1 is further activated by deacetylation by SIRT2 [83]. Recent work has demonstrated that cells with low levels of CDH1 accumulated in G1 with elevated mitotic APC substrates, causing genome instability [123,124]. Furthermore, entire loss of CDH1 increased DNA damage accumulation, driving progression of murine and human B-cell acute leukemia [138]. It was also revealed that many cancer cell lines lack the ability to activate APC^{Cdh1} when under replication stress [139,140], and that CDH1-depleted cells undergo senescence in G2, suggesting that APC^{Cdh1} may normally act as a barrier to genome instability [123]. Support for this idea comes from studies using SIRT2, an antitumor and lifespan-extending protein, which activates the APC by deacetylating CDC20 and CDH1; SIRT2-deficient mice exhibited higher levels of cancer and elevated levels of APC substrates [83]. Thus, impaired APC function appears linked with genomic instability and cancer development, providing strong therapeutic potential through targeted activation in cancer cells.

APC dysfunction and cancer development could occur in several ways. Loss of either CDC20 or CDH1 is deleterious; CDC20 deletion is lethal, while loss of CDH1 leads to genomic instability [123,124]. In addition, mutations have been observed in several APC subunit genes (APC3, APC6/CDC16, and APC8/CDC23) in cancer cells [141]. Inappropriate expression of the CDC23ΔTPR mutant disrupted cell cycle progression and led to elevated levels of APC substrates. Loss of the APC7 subunit has also been implicated in various tumors [142,143]. Furthermore, silencing of a variety of APC subunits causes cells to survive treatment with compounds that inhibit the SAC, providing a mechanism for the development of drug resistance [88,89]. Thus, evidence is accumulating to support the idea that APC activity is required for cell health, while loss of normal APC function leads to genomic instability and cancer.

11. APC Activation Reduces Substrate Levels and Inhibits Cancer Cell Growth

Recently, focus has shifted towards the creation of compounds that activate the APC. To do so, SAC inhibition has been targeted. Prolonged SAC, or impaired APC activity, can lead to inappropriate mitotic progression in a process called mitotic slippage [144,145]. This potentially provides time for cells to respond to increased toxic levels of genomic instability common in cancer cells. Furthermore, because of the aneuploid nature of cancer cells, cancer cells are heavily reliant on the SAC for proper segregation of chromosomes; inhibition of the SAC in cancer cells produces intolerable levels of genomic instability, killing these cells [146,147]. One compound, called Mad2-inhibitor-1, or M2I-1, blocks the MAD2/CDC20 interaction (Figure 1) and weakens the SAC, leading to early activation of the APC [90]. We have subsequently used M2I-1 in vitro and in vivo, and have found that, in vitro, M2I-1 synergizes with Doxorubicin to reduce the growth of drug resistant MCF7 breast cancer cells, while growth of patient-derived triple negative breast cancer cells in mice was stalled by M2I-1 (Davies, Arnason, and Harkness, unpublished). Both in vitro and in vivo, APC substrate mRNA and protein levels were reduced, showing that M2I-1 does indeed activate the APC. Additional SAC inhibitors have been developed that inhibit the kinase MPS1/TTK (TTKi's), a SAC component [88,89]. Kaplan-Meier plots revealed that overexpression of MPS1/TTK is correlated with poor overall and relapse-free survival in breast cancer patients [148]. Interestingly, as mentioned above, silencing of APC subunits generates resistance to the MPS1/TTK inhibitors (TTKi's) reversine and CFI-402257 [88,89]. This suggests that the lethal mitotic segregation errors induced by TTK inhibition can be overcome by prolonging the onset of anaphase.

12. APC Activity, via the Fkh/SNF Kinase/Sir2 Pathway, is Required for Prolonged Longevity

We have reported that the yeast APC prolongs longevity (increased expression of only *APC10* increased replicative lifespan [61]), responds to stress, and interacts with multiple conserved stress response pathways highlighted by the Forkhead (FOXO) and Snf1 (AMPK) pathways [31,32,61,63–65,108] (Figure 3). It is already clear that the FOXO and AMPK pathways intersect under stress in mammalian cells and drive the activity of several other stress response networks [109,110]. In yeast, *snf1Δ* mutants were also shown to interact genetically with the *apc5^{CA}* mutant; deletion of *SNF1* worsened the *apc5^{CA}* defect, whereas overexpression rescued it [61]. Furthermore, Mig1, a glucose responsive transcriptional repressor inhibited by Snf1 phosphorylation, repressed the expression of the APC subunits *APC4* and *APC9* [61]. Subsequent work showed that Fkh1 transcribed *SNF1*, and that increased longevity observed in the Snf1^{UBA} mutant depended on Fkh1 or Fkh2 [32]. This stress response network is further bolstered by the anti-aging protein deacetylase SIRT2, which deacetylates FOXO3a to increase its DNA binding ability in mammalian cells [106]. SIRT2 also binds to the APC^{CDC20} and APC^{CDH1} complexes and deacetylates both CDC20 and CDH1 to turn on the APC [83]. The SIRT2-FOXO interaction is also conserved in yeast, as the yeast Forkhead proteins, Fkh1 and Fkh2, physically associate with Sir2 during late M and G1 to repress the expression of the Fkh target gene Clb2 [107]. In addition, under stress conditions, Sir2 assists in APC function by inhibiting *CLB2* transcription; overexpression of *CLB2* under stress conditions is toxic [107]. However, it was not shown whether Sir2 deacetylates the Fkh proteins in this study. In yeast, the Fkh1 and Fkh2 transcription factors, like in mammalian cells, are involved in cell cycle progression, stress response, and longevity [63]. *FKH1* and *FKH2* are expressed during G2 to drive the expression of mitotic specific genes [79,149]. The *FKH* genes are activated by a third Forkhead protein called Hcm1, which is expressed at the G1/S boundary [133]. Interestingly, Hcm1 nuclear translocation is facilitated by the SNF1 kinase [150], defining a positive feedforward loop involving Snf1, Hcm1, and the Fkh proteins. Furthermore, the ubiquitin conjugating enzyme, Ubc1, interacts with the APC [151] and is required for SNF1 kinase function [108]. It was revealed that in yeast *ubc1Δ* mutants, Hcm1 remains cytosolic, *FKH1* and *FKH2* transcription is reduced, and SNF1 kinase activity is decreased [108]. Fkh1 action is then reduced at the onset of mitosis, as the bulk of Fkh1 is targeted for degradation by the APC^{Cdc20} complex [65] (Figure 1). Interestingly, Fkh1 and the APC subunit Apc5 physically interacted throughout the cell cycle [65]. Deletion of both *FKH1* and *FKH2* in APC defective cells worsened the already short replicative and chronological lifespans [31], and mutation of a single, conserved lysine in Fkh1 (K_{373}) mimicked the null *FKH1* allele, reduced chronological lifespan, and increased genomic instability [65]. Thus, it appears that ubiquitination of Fkh1 at K_{373}, mediated by APC^{Cdc20} at the onset of mitosis, is required to maintain normal lifespan and genomic stability.

In addition to Fkh1, the APC also targets a second lifespan determinant, Fob1, for degradation [64]. Fob1 in yeast is an rDNA replication fork blocking protein [152,153]. Fob1 condenses rDNA and stalls replication fork progression during mitosis, creating free DNA ends that produce extra chromosomal circles [92,154]. Fob1 also sequesters the Cdc14 phosphatase within the nucleolus at the rDNA locus during early mitosis [91,92]. Cdc14 is released from Fob1 by the combined activity of the FEAR (Cdc14 early anaphase release) and MEN (mitotic exit network) complexes during late mitosis, enabling activation of Cdh1 via Cdc14 dephosphorylation of Cdc28-Clb2 [93]. Deletion of *FOB1* enhances yeast replicative lifespan [64,154], while increased *FOB1* expression reduces replicative lifespan [64]. We identified Fob1 as a binding partner for Apc5 in a yeast 2-hybrid screen. Mutation of an amino acid required for Fob1-Apc5 interactions (E_{420}V) stabilized Fob1, increased rDNA instability, and abolished the accumulation of modified Fob1 species. We observed that Fob1 was specifically unstable during G1 and targeted for degradation by APC^{Cdh1} [64]. Deletion of *FOB1*, like that of *FKH1*, rescued the lifespan defect observed in APC mutants [64,65]. Taken together, the APC target substrates we have identified (Fkh1, Fob1, and Gcn5) function during mitosis and G1 to elicit wide-ranging effects on genomic stability and longevity (Figure 2).

13. The APC Triggers the End of Nutrient Signaling in the Presence of Stress

In order to fully maximize longevity, from the beginning to the end, coupling the stress and nutrient sensing pathways may be critical. The APC may be in a position to recognize both stress and nutrients. The APC is activated by phosphorylation to promote cell cycle progression. Using mouse fibroblast NIH/3T3 cells, it was shown that the Polo-like kinase, Plk, activates the APC by phosphorylating CDC16, CDC27, and APC1 [81] (Figure 1). Plk in yeast (Cdc5) also phosphorylates the APC, as does the cyclin-dependent kinase Cdc28 on Cdc16, Cdc23, and Cdc27 to activate APC^{Cdc20} function [80]. Conversely, mammalian protein kinase A (PKA) phosphorylates CDC27 and APC1 to inhibit APC function [81] (Figure 3). It is known in yeast that nutrients, such as glucose, and nutrient signaling networks involving Ras/PKA inhibit the APC [101–103,155]. The cell cycle proceeds in the presence of nutrients, so it remains unresolved how the positive and negative phosphorylation events on APC subunits using the nutrient response and cell cycle promoting kinases are coordinated. It remains possible that the APC's role in cell cycle progression and stress response are controlled via different mechanisms. If this were the case, PKA inhibition of the APC may be specific to its stress response activity, whereas activation by the cyclin-dependent and Polo-like kinases may be more geared towards the APC's cell cycle role. These observations suggest that the nutrient-sensing pathway plays a pivotal role in shutting down the APC and its stress-sensing functions.

The yeast nutrient-sensing kinases Sch9 (similar to the AKT/S6K homologues in humans [156]) and PKA also control APC activity in the presence of nutrients by phosphorylating the ubiquitin conjugating enzyme, Cdc34, the E2 component of the ubiquitin-ligase (E3) SCF [104]. Work in mammalian cells shows that the two E3 enzymes, the APC and the SCF, work to counterbalance one another during G1, with the SCF targeting the CDH1 for degradation [58,105], and the APC targeting the SCF F-box subunit SKP2 for degradation [111,112]. Thus, the nutrient response kinases inhibit APC activity in the presence of nutrients. Furthermore, our preliminary results indicate that the long life observed in *sch9Δ* and *tor1Δ* mutants requires functional Fkh1 or Fkh2, suggesting that Sch9 and/or Tor1 inhibit Fkh function (Postnikoff and Harkness, unpublished; Figure 3), leading to further inactivation of the APC.

However, how does the nutrient sensing pathway shut down when nutrients are limited? A recent report described the turnover of the nutrient sensing kinase Sch9 in yeast [157]. Deletion of *SCH9* in yeast increases yeast replicative and chronological lifespan [28,158], and, as mentioned above, deletion of both *FKH1* and *FKH2* in either the *sch9Δ* or *tor1Δ* background eliminates the observed long life (Postnikoff and Harkness, unpublished). As cells entered stationary phase, it was observed that total ubiquitinated protein decreased, as did total Sch9 protein levels [157]. In the presence of the proteasome poison MG132, it was observed that Sch9 protein levels increased [157], supporting the idea that Sch9 is ubiquitinated and degraded as nutrient levels decrease. We therefore asked whether Sch9 is targeted for ubiquitination by the APC, as a means to inactivate this arm of the nutrient response network when nutrient levels decline. Our preliminary experiments show that deletion of *SCH9* in APC mutants suppressed the chronological lifespan and oxidative stress sensitive defects in APC mutants (Postnikoff and Harkness, unpublished). We also confirmed that Sch9 turnover occurs as cells enter stationary phase, and that this is blocked in APC mutants (Malo and Harkness, unpublished). Taken as a whole, the published and unpublished literature supports the idea that the APC sits at the apex of the stress and nutrient-sensing pathways, controlling cell cycle progression, DNA repair, and chromosome maintenance (Figure 3).

14. Conclusions

The positioning of the APC at the intersection point of the stress and nutrient sensing pathways confers importance upon this complex, as it may have the potential to protect the cells that come together to form the zygote from the aging process. The potential for aging likely begins for an individual as soon as the germ cells responsible for them are born. For the oocyte, that means during the mother's in utero development. It will be many years before that oocyte is fertilized; therefore,

Int. J. Mol. Sci. **2018**, *19*, 1888

plenty of time exists for damaging side effects of cell metabolism to rear their ugly heads. It is critical that the repair mechanisms within these cells are functioning optimally. As long as the APC is at its peak function, protection against cellular damage should be high. With continued proper function of the APC through the life of the germ cells and the subsequent offspring, increased healthspan may be possible.

Funding: Troy A. A. Harkness is supported by grants from NSERC, CFI, CCS, and CIHR for the purpose of this work.

Acknowledgments: Franco Vizeacoumar and Frederick Vizeacoumar generated the data displayed in Figure 4. Christopher Eskiw edited drafts of this manuscript, and Terra Arnason provided fruitful discussions during the writing of this manuscript.

Conflicts of Interest: The author declares no conflicts of interest.

References

1. Sobel, H. When does human aging start? *Gerontologist* **1966**, *6*, 17–22. [CrossRef] [PubMed]
2. Lakatta, E.G. So! What's aging? Is cardiovascular aging a disease? *J. Mol. Cell. Cardiol.* **2015**, *83*, 1–13. [CrossRef] [PubMed]
3. Paashuis-Lew, Y.R.; Heddle, J.A. Spontaneous mutation during fetal development and post-natal growth. *Mutagenesis* **1998**, *13*, 613–617. [CrossRef] [PubMed]
4. Relton, C.L.; Daniel, C.P.; Hammal, D.M.; Parker, L.; Janet Tawn, E.; Burn, J. DNA repair gene polymorphisms, pre-natal factors and the frequency of somatic mutations in the glycophorin-A gene among healthy newborns. *Mutat. Res.* **2004**, *545*, 49–57. [CrossRef] [PubMed]
5. Milne, E.M. When does human ageing begin? *Mech. Ageing Dev.* **2006**, *127*, 290–297. [CrossRef] [PubMed]
6. Kubben, N.; Misteli, T. Shared molecular and cellular mechanisms of premature ageing and ageing-associated diseases. *Nat. Rev. Mol. Cell Biol.* **2017**, *18*, 595–609. [CrossRef] [PubMed]
7. Hayflick, L. When does aging begin? *Res. Aging* **1984**, *6*, 99–103. [CrossRef] [PubMed]
8. Hayflick, L. The future of ageing. *Nature* **2000**, *408*, 267–269. [CrossRef] [PubMed]
9. Hayflick, L. The not-so-close relationship between biological aging and age-associated pathologies in humans. *J. Gerontol. A Biol. Sci. Med. Sci.* **2004**, *59*, B547–B550. [CrossRef] [PubMed]
10. Kirkwood, T.B.; Austad, S.N. Why do we age? *Nature* **2000**, *408*, 233–238. [CrossRef] [PubMed]
11. Nemoto, S.; Finkel, T. Ageing and the mystery at Arles. *Nature* **2004**, *429*, 149–152. [CrossRef] [PubMed]
12. Harman, D. Aging: A theory based on free radical and radiation chemistry. *J. Gerontol.* **1956**, *11*, 298–300. [CrossRef] [PubMed]
13. Hayflick, L. Entropy explains aging, genetic determination explains longevity, and undefined terminology explains misunderstanding both. *PLoS Genet.* **2007**, *3*, e220. [CrossRef] [PubMed]
14. Labbadia, J.; Morimoto, R.I. Proteostasis and longevity: When does aging really begin? *F1000Prime Rep.* **2014**, *6*, 7. [CrossRef] [PubMed]
15. Kirkwood, T.B.; Holliday, R. The evolution of ageing and longevity. *Proc. R. Soc. Lond. B Biol. Sci.* **1979**, *205*, 531–546. [CrossRef] [PubMed]
16. Kenyon, C.J. The genetics of ageing. *Nature* **2010**, *464*, 504–512. [CrossRef] [PubMed]
17. Salminen, A.; Kaarniranta, K. ER stress and hormetic regulation of the aging process. *Ageing Res. Rev.* **2010**, *9*, 211–217. [CrossRef] [PubMed]
18. Martins, I.; Galluzzi, L.; Kroemer, G. Hormesis, cell death and aging. *Aging* **2011**, *3*, 821–828. [CrossRef] [PubMed]
19. Williams, G.C. Pleiotrophy, natural selection and the evolution of senescence. *Evolution* **1957**, *22*, 406–421.
20. Kirkwood, T.B.; Rose, M.R. Evolution of senescence: Late survival sacrificed for reproduction. *Philos. Trans. R Soc. Lond. B Biol. Sci.* **1991**, *332*, 15–24. [CrossRef] [PubMed]
21. Nesse, R.M.; Williams, G.C. Evolution and the origins of disease. *Sci. Am.* **1998**, *279*, 86–93. [CrossRef] [PubMed]
22. Wick, G.; Berger, P.; Jansen-Dürr, P.; Grubeck-Loebenstein, B.A. Darwinian-evolutionary concept of age-related diseases. *Exp. Gerontol.* **2003**, *38*, 13–25. [CrossRef]

23. Artandi, S.E.; DePinho, R.A. A critical role for telomeres in suppressing and facilitating carcinogenesis. *Curr. Opin. Genet. Dev.* **2000**, *10*, 39–46. [CrossRef]
24. Shay, J.W.; Wright, W.E. Ageing and cancer: The telomere and telomerase connection. *Novartis Found. Symp.* **2001**, *235*, 116–125. [PubMed]
25. López-Otín, C.; Blasco, M.A.; Partridge, L.; Serrano, M.; Kroemer, G. The hallmarks of aging. *Cell* **2013**, *153*, 1194–1217. [CrossRef] [PubMed]
26. Aunan, J.R.; Watson, M.M.; Hagland, H.R.; Søreide, K. Molecular and biological hallmarks of ageing. *Br. J. Surg.* **2016**, *103*, e29–e46. [CrossRef] [PubMed]
27. Kaeberlein, M.; McVey, M.; Guarente, L. The SIR2/3/4 complex and SIR2 alone promote longevity in *Saccharomyces cerevisiae* by two different mechanisms. *Genes Dev.* **1999**, *13*, 2570–2580. [CrossRef] [PubMed]
28. Fabrizio, P.; Pozza, F.; Pletcher, S.D.; Gendron, C.M.; Longo, V.D. Regulation of longevity and stress resistance by Sch9 in yeast. *Science* **2001**, *292*, 288–290. [CrossRef] [PubMed]
29. Wei, M.; Fabrizio, P.; Hu, J.; Ge, H.; Cheng, C.; Li, L.; Longo, V.D. Life span extension by calorie restriction depends on Rim15 and transcription factors downstream of Ras/PKA, Tor, and Sch9. *PLoS Genet.* **2008**, *4*, e13. [CrossRef] [PubMed]
30. Lu, J.Y.; Lin, Y.Y.; Sheu, J.C.; Wu, J.T.; Lee, F.J.; Chen, Y.; Lin, M.I.; Chiang, F.T.; Tai, T.Y.; Berger, S.L.; et al. Acetylation of yeast AMPK controls intrinsic aging independently of caloric restriction. *Cell* **2011**, *146*, 969–979. [CrossRef] [PubMed]
31. Postnikoff, S.D.; Malo, M.M.; Wong, B.; Harkness, T.A. The yeast forkhead transcription factors fkh1 and fkh2 regulate lifespan and stress response together with the anaphase-promoting complex. *PLoS Genet.* **2012**, *8*, e1002583. [CrossRef] [PubMed]
32. Jiao, R.; Postnikoff, S.; Harkness, T.A.; Arnason, T.G. The SNF1 Kinase Ubiquitin-associated Domain Restrains Its Activation, Activity, and the Yeast Life Span. *J. Biol. Chem.* **2015**, *290*, 15393–15404. [CrossRef] [PubMed]
33. Mirisola, M.G.; Taormina, G.; Fabrizio, P.; Wei, M.; Hu, J.; Longo, V.D. Serine- and threonine/valine-dependent activation of PDK and Tor orthologs converge on Sch9 to promote aging. *PLoS Genet.* **2014**, *10*, e1004113. [CrossRef] [PubMed]
34. Dong, X.; Milholland, B.; Vijg, J. Evidence for a limit to human lifespan. *Nature* **2016**, *538*, 257–259. [CrossRef] [PubMed]
35. Robine, J.M.; Allard, M. The oldest human. *Science* **1998**, *279*, 1834–1835. [CrossRef] [PubMed]
36. Weon, B.M.; Je, J.H. Theoretical estimation of maximum human lifespan. *Biogerontology* **2009**, *10*, 65–71. [CrossRef] [PubMed]
37. Modig, K.; Andersson, T.; Vaupel, J.; Rau, R.; Ahlbom, A. How long do centenarians survive? Life expectancy and maximum lifespan. *J. Intern. Med.* **2017**, *282*, 156–163. [CrossRef] [PubMed]
38. Gerontology Research Group. Available online: http://supercentenarian-research-foundation.org/TableE.aspx (accessed on 30 April 2018).
39. De Grey, A. A strategy for postponing aging indefinitely. *Stud. Health Technol. Inform.* **2005**, *118*, 209–219. [PubMed]
40. Lenart, A.; Vaupel, J.W. Questionable evidence for a limit to human lifespan. *Nature* **2017**, *546*, E13–E14. [CrossRef] [PubMed]
41. Hughes, B.G.; Hekimi, S. Many possible maximum lifespan trajectories. *Nature* **2017**, *546*, E8–E9. [CrossRef] [PubMed]
42. Takahashi, Y.; Kuro-o, M.; Ishikawa, F. Aging mechanisms. *Proc. Natl. Acad. Sci. USA* **2000**, *97*, 12407–12408. [CrossRef] [PubMed]
43. Lucke, J.C.; Hall, W. Who wants to live forever? *EMBO Rep.* **2005**, *6*, 98–102. [CrossRef] [PubMed]
44. Tacutu, R.; Thornton, D.; Johnson, E.; Budovsky, A.; Barardo, D.; Craig, T.; Diana, E.; Lehmann, G.; Toren, D.; Wang, J.; et al. Human Ageing Genomic Resources: New and updated databases. *Nucleic Acids Res.* **2018**, *46*, D1083–D1090. [CrossRef] [PubMed]
45. Lindahl, T. Instability and decay of the primary structure of DNA. *Nature* **1993**, *362*, 709–715. [CrossRef] [PubMed]
46. Hoeijmakers, J.H. DNA damage, aging, and cancer. *N. Engl. J. Med.* **2009**, *361*, 1475–1485. [CrossRef] [PubMed]
47. Cervelli, T.; Borghini, A.; Galli, A.; Andreassi, M.G. DNA damage and repair in atherosclerosis: Current insights and future perspectives. *Int. J. Mol. Sci.* **2012**, *13*, 16929–16944. [CrossRef] [PubMed]

48. Vijg, J.; Suh, Y. Genome instability and aging. *Annu. Rev. Physiol.* **2013**, *75*, 645–668. [CrossRef] [PubMed]
49. Grindel, A.; Brath, H.; Nersesyan, A.; Knasmueller, S.; Wagner, K.H. Association of genomic instability with HbA1c levels and medication in diabetic patients. *Sci. Rep.* **2017**, *7*, 41985. [CrossRef] [PubMed]
50. Yuza, K.; Nagahashi, M.; Watanabe, S.; Takabe, K.; Wakai, T. Hypermutation and microsatellite instability in gastrointestinal cancers. *Oncotarget* **2017**, *8*, 112103–112115. [CrossRef] [PubMed]
51. Barzilai, A.; Schumacher, B.; Shiloh, Y. Genome instability: Linking ageing and brain degeneration. *Mech. Ageing Dev.* **2017**, *161*, 4–18. [CrossRef] [PubMed]
52. Henderson, K.A.; Gottschling, D.E. A mother's sacrifice: what is she keeping for herself? *Curr. Opin. Cell Biol.* **2008**, *20*, 723–728. [CrossRef] [PubMed]
53. Gottschling, D.E.; Nyström, T. The Upsides and Downsides of Organelle Interconnectivity. *Cell* **2017**, *169*, 24–34. [CrossRef] [PubMed]
54. Sun, Y.C.; Sun, X.F.; Dyce, P.W.; Shen, W.; Chen, H. The role of germ cell loss during primordial follicle assembly: A review of current advances. *Int. J. Biol. Sci.* **2017**, *13*, 449–457. [CrossRef] [PubMed]
55. Sivakumar, S.; Gorbsky, G.J. Spatiotemporal regulation of the anaphase-promoting complex in mitosis. *Nat. Rev. Mol. Cell Biol.* **2015**, *16*, 82–94. [CrossRef] [PubMed]
56. Zhang, S.; Chang, L.; Alfieri, C.; Zhang, Z.; Yang, J.; Maslen, S.; Skehel, M.; Barford, D. Molecular mechanism of APC/C activation by mitotic phosphorylation. *Nature* **2016**, *533*, 260–264. [CrossRef] [PubMed]
57. Zhou, Z.; He, M.; Shah, A.A.; Wan, Y. Insights into APC/C: From cellular function to diseases and therapeutics. *Cell Div.* **2016**, *11*, 9. [CrossRef] [PubMed]
58. Fukushima, H.; Ogura, K.; Wan, L.; Lu, Y.; Li, V.; Gao, D.; Liu, P.; Lau, A.W.; Wu, T.; Kirschner, M.W.; et al. SCF-mediated Cdh1 degradation defines a negative feedback system that coordinates cell-cycle progression. *Cell Rep.* **2013**, *4*, 803–816. [CrossRef] [PubMed]
59. Simpson-Lavy, K.J.; Sajman, J.; Zenvirth, D.; Brandeis, M. APC/CCdh1 specific degradation of Hsl1 and Clb2 is required for proper stress responses of *S. cerevisiae. Cell Cycle* **2009**, *8*, 3003–3009. [CrossRef] [PubMed]
60. Harkness, T.A.; Davies, G.F.; Ramaswamy, V.; Arnason, T.G. The ubiquitin-dependent targeting pathway in Saccharomyces cerevisiae plays a critical role in multiple chromatin assembly regulatory steps. *Genetics* **2002**, *162*, 615–632. [PubMed]
61. Harkness, T.A.; Shea, K.A.; Legrand, C.; Brahmania, M.; Davies, G.F. A functional analysis reveals dependence on the Anaphase Promoting Complex for prolonged life span in yeast. *Genetics* **2004**, *168*, 759–774. [CrossRef] [PubMed]
62. Lindsay, D.L.; Bonham-Smith, P.C.; Postnikoff, S.; Gray, G.R.; Harkness, T.A. A role for the anaphase promoting complex in hormone regulation. *Planta* **2011**, *233*, 1223–1235. [CrossRef] [PubMed]
63. Postnikoff, S.D.; Harkness, T.A. Mechanistic insights into aging, cell-cycle progression, and stress response. *Front. Physiol.* **2012**, *3*, 183. [CrossRef] [PubMed]
64. Menzel, J.; Malo, M.E.; Chan, C.; Prusinkiewicz, M.; Arnason, T.G.; Harkness, T.A. The anaphase promoting complex regulates yeast lifespan and rDNA stability by targeting Fob1 for degradation. *Genetics* **2014**, *196*, 693–709. [CrossRef] [PubMed]
65. Malo, M.E.; Postnikoff, S.D.; Arnason, T.G.; Harkness, T.A. Mitotic degradation of yeast Fkh1 by the Anaphase Promoting Complex is required for normal longevity, genomic stability and stress resistance. *Aging* **2016**, *8*, 810–830. [CrossRef] [PubMed]
66. Ganem, N.J.; Pellman, D. Linking abnormal mitosis to the acquisition of DNA damage. *J. Cell Biol.* **2012**, *199*, 871–881. [CrossRef] [PubMed]
67. Harkness, T.A.; Arnason, T.G.; Legrand, C.; Pisclevich, M.G.; Davies, G.F.; Turner, E.L. Contribution of CAF-I to anaphase-promoting-complex-mediated mitotic chromatin assembly in *Saccharomyces cerevisiae. Eukaryot Cell* **2005**, *4*, 673–684. [CrossRef] [PubMed]
68. Arnason, T.G.; Pisclevich, M.G.; Dash, M.D.; Davies, G.F.; Harkness, T.A. Novel interaction between Apc5p and Rsp5p in an intracellular signaling pathway in *Saccharomyces cerevisiae. Eukaryot Cell* **2005**, *4*, 134–146. [CrossRef] [PubMed]
69. Turner, E.L.; Malo, M.E.; Pisclevich, M.G.; Dash, M.D.; Davies, G.F.; Arnason, T.G.; Harkness, T.A. The *Saccharomyces cerevisiae* Anaphase Promoting Complex interacts with multiple histone-modifying enzymes to regulate cell cycle progression. *Eukaryot Cell* **2010**, *9*, 1418–1431. [CrossRef] [PubMed]
70. Islam, A.; Turner, E.L.; Menzel, J.; Malo, M.E.; Harkness, T.A. Antagonistic Gcn5-Hda1 interactions revealed by mutations to the Anaphase Promoting Complex in yeast. *Cell Div.* **2011**, *6*, 13. [CrossRef] [PubMed]

71. Linger, J.G.; Tyler, J.K. Chromatin Disassembly and Reassembly during DNA Repair. *Mutat. Res.* **2007**, *618*, 52–64. [CrossRef] [PubMed]

72. Li, X.; Tyler, J.K. Nucleosome disassembly during human non-homologous end joining followed by concerted HIRA- and CAF-1-dependent reassembly. *eLife* **2016**, *5*, e15129. [CrossRef] [PubMed]

73. Mello, J.A.; Silljé, H.H.W.; Roche, D.M.J.; Kirschner, D.B.; Nigg, E.A.; Almouzni, G. Human Asf1 and CAF-1 interact and synergize in a repair-coupled nucleosome assembly pathway. *EMBO Rep.* **2002**, *3*, 329–334. [CrossRef] [PubMed]

74. Chen, C.C.; Tyler, J. Chromatin reassembly signals the end of DNA repair. *Cell Cycle* **2008**, *7*, 3792–3797. [CrossRef] [PubMed]

75. Kim, J.A.; Haber, J.E. Chromatin assembly factors Asf1 and CAF-1 have overlapping roles in deactivating the DNA damage checkpoint when DNA repair is complete. *Proc. Natl. Acad. Sci. USA* **2009**, *106*, 1151–1156. [CrossRef] [PubMed]

76. Rodriges Blanko, E.; Kadyrova, L.Y.; Kadyrov, F.A. DNA Mismatch Repair Interacts with CAF-1- and ASF1A-H3-H4-dependent Histone (H3-H4)$_2$ Tetramer Deposition. *J. Biol. Chem.* **2016**, *291*, 9203–9217. [CrossRef] [PubMed]

77. Izawa, D.; Pines, J. The mitotic checkpoint complex binds a second CDC20 to inhibit active APC/C. *Nature* **2015**, *517*, 631–634. [CrossRef] [PubMed]

78. Alfieri, C.; Chang, L.; Zhang, Z.; Yang, J.; Maslen, S.; Skehel, M.; Barford, D. Molecular basis of APC/C regulation by the spindle assembly checkpoint. *Nature* **2016**, *536*, 431–436. [CrossRef] [PubMed]

79. Zhu, G.; Spellman, P.T.; Volpe, T.; Brown, P.O.; Botstein, D.; Davis, T.N.; Futcher, B. Two yeast forkhead genes regulate the cell cycle and pseudohyphal growth. *Nature* **2000**, *406*, 90–94. [CrossRef] [PubMed]

80. Rudner, A.D.; Murray, A.W. Phosphorylation by Cdc28 activates the Cdc20-dependent activity of the anaphase-promoting complex. *J. Cell Biol.* **2000**, *149*, 1377–1390. [CrossRef] [PubMed]

81. Kotani, S.; Tugendreich, S.; Fujii, M.; Jorgensen, P.M.; Watanabe, N.; Hoog, C.; Hieter, P.; Todokoro, K. PKA and MPF-activated polo-like kinase regulate anaphase-promoting complex activity and mitosis progression. *Mol. Cell* **1998**, *1*, 371–380. [CrossRef]

82. Crasta, K.; Lim, H.H.; Giddings, T.H., Jr.; Winey, M.; Surana, U. Inactivation of Cdh1 by synergistic action of Cdk1 and polo kinase is necessary for proper assembly of the mitotic spindle. *Nat. Cell Biol.* **2008**, *10*, 665–675. [CrossRef] [PubMed]

83. Kim, H.S.; Vassilopoulos, A.; Wang, R.H.; Lahusen, T.; Xiao, Z. SIRT2 maintains genome integrity and suppresses tumorigenesis through regulating APC/C activity. *Cancer Cell* **2011**, *20*, 487–499. [CrossRef] [PubMed]

84. Zhang, J.; Wan, L.; Dai, X.; Sun, Y.; Wei, W. Functional characterization of Anaphase Promoting Complex/Cyclosome (APC/C) E3 ubiquitin ligases in tumorigenesis. *Biochim. Biophys. Acta* **2014**, *1845*, 277–293. [PubMed]

85. Peters, J.M. The anaphase promoting complex/cyclosome: A machine designed to destroy. *Nat. Rev. Mol. Cell Biol.* **2006**, *7*, 644–656. [CrossRef] [PubMed]

86. Sackton, K.L.; Dimova, N.; Zeng, X.; Tian, W.; Zhang, M.; Sackton, T.B.; Meaders, J.; Pfaff, K.L.; Sigoillot, F.; Yu, H.; et al. Synergistic blockade of mitotic exit by two chemical inhibitors of the APC/C. *Nature* **2014**, *514*, 646–649. [CrossRef] [PubMed]

87. Wang, L.; Zhang, J.; Wan, L.; Zhou, X.; Wang, Z.; Wei, W. Targeting Cdc20 as a novel cancer therapeutic strategy. *Pharmacol. Ther.* **2015**, *151*, 141–151. [CrossRef] [PubMed]

88. Sansregret, L.; Patterson, J.O.; Dewhurst, S.; López-García, C.; Koch, A.; McGranahan, N.; Chao, W.C.H.; Barry, D.J.; Rowan, A.; Instrell, R.; et al. APC/C Dysfunction Limits Excessive Cancer Chromosomal Instability. *Cancer Discov.* **2017**, *7*, 218–233. [CrossRef] [PubMed]

89. Thu, K.L.; Silvester, J.; Elliott, M.J.; Ba-Alawi, W.; Duncan, M.H.; Elia, A.C.; Mer, A.S.; Smirnov, P.; Safikhani, Z.; Haibe-Kains, B.; et al. Disruption of the anaphase-promoting complex confers resistance to TTK inhibitors in triple-negative breast cancer. *Proc. Natl. Acad. Sci. USA* **2018**, *115*, E1570–E1577. [CrossRef] [PubMed]

90. Kastl, J.; Braun, J.; Prestel, A.; Möller, H.M.; Huhn, T.; Mayer, T.U. Mad2 Inhibitor-1 (M2I-1): A Small Molecule Protein-Protein Interaction Inhibitor Targeting the Mitotic Spindle Assembly Checkpoint. *ACS Chem. Biol.* **2015**, *10*, 1661–1666. [CrossRef] [PubMed]

91. Stegmeier, F.; Huang, J.; Rahal, R.; Zmolik, J.; Moazed, D.; Amon, A. The replication fork block protein Fob1 functions as a negative regulator of the FEAR network. *Curr. Biol.* **2004**, *14*, 467–480. [CrossRef] [PubMed]

92. Waples, W.G.; Chahwan, C.; Ciechonska, M.; Lavoie, B.D. Putting the brake on FEAR: Tof2 promotes the biphasic release of Cdc14 phosphatase during mitotic exit. *Mol. Biol. Cell* **2009**, *20*, 245–255. [CrossRef] [PubMed]

93. Visintin, R.; Craig, K.; Hwang, E.S.; Prinz, S.; Tyers, M.; Amon, A. The phosphatase Cdc14 triggers mitotic exit by reversal of Cdk-dependent phosphorylation. *Mol. Cell* **1998**, *2*, 709–718. [CrossRef]

94. Karra, H.; Repo, H.; Ahonen, I.; Löyttyniemi, E.; Pitkänen, R.; Lintunen, M.; Kuopio, T.; Söderström, M.; Kronqvist, P. Cdc20 and securin overexpression predict short-term breast cancer survival. *Br. J. Cancer* **2014**, *110*, 2905–2913. [CrossRef] [PubMed]

95. Schmit, T.L.; Ledesma, M.C.; Ahmad, N. Modulating polo-like kinase 1 as a means for cancer chemoprevention. *Pharm. Res.* **2010**, *27*, 989–998. [CrossRef] [PubMed]

96. Staff, S.; Isola, J.; Jumppanen, M.; Tanner, M. Aurora-A gene is frequently amplified in basal-like breast cancer. *Oncol. Rep.* **2010**, *23*, 307–312. [CrossRef] [PubMed]

97. Heredia, F.F.; de Sousa, J.C.; Ribeiro Junior, H.L.; Carvalho, A.F.; Magalhaes, S.M.; Pinheiro, R.F. Proteins related to the spindle and checkpoint mitotic emphasize the different pathogenesis of hypoplastic MDS. *Leuk. Res.* **2014**, *38*, 218–224. [CrossRef] [PubMed]

98. Kuo, T.C.; Lu, H.P.; Chao, C.C. The tyrosine kinase inhibitor sorafenib sensitizes hepatocellular carcinoma cells to taxol by suppressing the HURP protein. *Biochem. Pharmacol.* **2011**, *82*, 184–194. [CrossRef] [PubMed]

99. Xiang, W.; Wu, X.; Huang, C.; Wang, M.; Zhao, X.; Luo, G.; Li, Y.; Jiang, G.; Xiao, X.; Zeng, F. PTTG1 regulated by miR-146a-3p promotes bladder cancer migration, invasion, metastasis and growth. *Oncotarget* **2017**, *8*, 664–678. [CrossRef] [PubMed]

100. Zhang, L.; Cai, M.; Gong, Z.; Zhang, B.; Li, Y.; Guan, L.; Hou, X.; Li, Q.; Liu, G.; Xue, Z.; et al. Geminin facilitates FoxO3 deacetylation to promote breast cancer cell metastasis. *J. Clin. Investig.* **2017**, *127*, 2159–2175. [CrossRef] [PubMed]

101. Irniger, S.; Bäumer, M.; Braus, G.H. Glucose and ras activity influence the ubiquitin ligases APC/C and SCF in *Saccharomyces cerevisiae*. *Genetics* **2000**, *154*, 1509–1521. [PubMed]

102. Bolte, M.; Dieckhoff, P.; Krause, C.; Braus, G.H.; Irniger, S. Synergistic inhibition of APC/C by glucose and activated Ras proteins can be mediated by each of the Tpk1-3 proteins in *Saccharomyces cerevisiae*. *Microbiology* **2003**, *149*, 1205–1216. [CrossRef] [PubMed]

103. Searle, J.S.; Schollaert, K.L.; Wilkins, B.J.; Sanchez, Y. The DNA damage checkpoint and PKA pathways converge on APC substrates and Cdc20 to regulate mitotic progression. *Nat. Cell Biol.* **2004**, *6*, 138–145. [CrossRef] [PubMed]

104. Cocklin, R.; Goebl, M. Nutrient sensing kinases PKA and Sch9 phosphorylate the catalytic domain of the ubiquitin-conjugating enzyme Cdc34. *PLoS ONE* **2011**, *6*, e27099. [CrossRef] [PubMed]

105. Choudhury, R.; Bonacci, T.; Arceci, A.; Lahiri, D.; Mills, C.A.; Kernan, J.L.; Branigan, T.B.; DeCaprio, J.A.; Burke, D.J.; Emanuele, M.J. APC/C and SCF(cyclin F) constitute a reciprocal feedback circuit controlling S-phase entry. *Cell Rep.* **2016**, *16*, 3359–3372. [CrossRef] [PubMed]

106. Wang, F.; Nguyen, M.; Qin, F.X.; Tong, Q. SIRT2 deacetylates FOXO3a in response to oxidative stress and caloric restriction. *Aging Cell* **2007**, *6*, 505–514. [CrossRef] [PubMed]

107. Linke, C.; Klipp, E.; Lehrach, H.; Barberis, M.; Krobitsch, S. Fkh1 and Fkh2 associate with Sir2 to control CLB2 transcription under normal and oxidative stress conditions. *Front. Physiol.* **2013**, *4*, 173. [CrossRef] [PubMed]

108. Jiao, R.; Lobanova, L.; Waldner, A.; Fu, A.; Xiao, L.; Harkness, T.A.; Arnason, T.G. The ubiquitin-conjugating enzyme, Ubc1, indirectly regulates SNF1 kinase activity via Forkhead-dependent transcription. *Microb. Cell* **2016**, *3*, 540–553. [CrossRef] [PubMed]

109. Chiacchiera, F.; Simone, C. The AMPK-FoxO3A axis as a target for cancer treatment. *Cell Cycle* **2010**, *9*, 1091–1096. [CrossRef] [PubMed]

110. Salminen, A.; Kaarniranta, K. AMP-activated protein kinase (AMPK) controls the aging process via an integrated signaling network. *Ageing Res. Rev.* **2012**, *11*, 230–241. [CrossRef] [PubMed]

111. Wei, W.; Ayad, N.G.; Wan, Y.; Zhang, G.J.; Kirschner, M.W.; Kaelin, W.G., Jr. Degradation of the SCF component Skp2 in cell-cycle phase G1 by the anaphase-promoting complex. *Nature* **2004**, *428*, 194–198. [CrossRef] [PubMed]

112. Bashir, T.; Dorrello, N.V.; Amador, V.; Guardavaccaro, D.; Pagano, M. Control of the SCF(Skp2-Cks1) ubiquitin ligase by the APC/C(Cdh1) ubiquitin ligase. *Nature* **2004**, *428*, 190–193. [CrossRef] [PubMed]

113. Wang, F.; Higgins, J.M. Histone modifications and mitosis: countermarks, landmarks, and bookmarks. *Trends Cell Biol.* **2013**, *23*, 175–184. [CrossRef] [PubMed]

114. Guppy, B.J.; McManus, K.J. Mitotic accumulation of dimethylated lysine 79 of histone H3 is important for maintaining genome integrity during mitosis in human cells. *Genetics* **2015**, *199*, 423–433. [CrossRef] [PubMed]

115. Recht, J.; Tsubota, T.; Tanny, J.C.; Diaz, R.L.; Berger, J.M.; Zhang, X.; Garcia, B.A.; Shabanowitz, J.; Burlingame, A.L.; Hunt, D.F.; et al. Histone chaperone Asf1 is required for histone H3 lysine 56 acetylation, a modification associated with S phase in mitosis and meiosis. *Proc. Natl. Acad. Sci. USA* **2006**, *103*, 6988–6993. [CrossRef] [PubMed]

116. Jeong, Y.S.; Cho, S.; Park, J.S.; Ko, Y.; Kang, Y.K. Phosphorylation of serine-10 of histone H3 shields modified lysine-9 selectively during mitosis. *Genes Cells* **2010**, *15*, 181–192. [CrossRef] [PubMed]

117. Karmodiya, K.; Krebs, A.R.; Oulad-Abdelghani, M.; Kimura, H.; Tora, L.H. 3K9 and H3K14 acetylation co-occur at many gene regulatory elements, while H3K14ac marks a subset of inactive inducible promoters in mouse embryonic stem cells. *BMC Genom.* **2012**, *13*, 424. [CrossRef] [PubMed]

118. Gates, L.A.; Shi, J.; Rohira, A.D.; Feng, Q.; Zhu, B.; Bedford, M.T.; Sagum, C.A.; Jung, S.Y.; Qin, J.; Tsai, M.J.; et al. Acetylation on histone H3 lysine 9 mediates a switch from transcription initiation to elongation. *J. Biol. Chem.* **2017**, *292*, 14456–14472. [CrossRef] [PubMed]

119. Downs, J.A. Histone H3 K56 acetylation, chromatin assembly, and the DNA damage checkpoint. *DNA Repair* **2008**, *7*, 2020–2024. [CrossRef] [PubMed]

120. Farooq, Z.; Banday, S.; Pandita, T.K.; Altaf, M. The many faces of histone H3K79 methylation. *Mutat. Res. Rev. Mutat. Res.* **2016**, *768*, 46–52. [CrossRef] [PubMed]

121. Wood, K.; Tellier, M.; Murphy, S. DOT1L and H3K79 Methylation in Transcription and Genomic Stability. *Biomolecules* **2018**, *8*, E11. [CrossRef] [PubMed]

122. Cea, M.; Cagnetta, A.; Adamia, S.; Acharya, C.; Tai, Y.T.; Fulciniti, M.; Ohguchi, H.; Munshi, A.; Acharya, P.; Bhasin, M.K.; et al. Evidence for a role of the histone deacetylase SIRT6 in DNA damage response of multiple myeloma cells. *Blood* **2016**, *127*, 1138–1150. [CrossRef] [PubMed]

123. Ercilla, A.; Llopis, A.; Feu, S.; Aranda, S.; Ernfors, P. New origin firing is inhibited by APC/CCdh1 activation in S-phase after severe replication stress. *Nucleic Acids Res.* **2016**, *44*, 4745–4762. [CrossRef] [PubMed]

124. Garzón, J.; Rodríguez, R.; Kong, Z.; Chabes, A.; Rodríguez-Acebes, S.; Méndez, J.; Moreno, S.; García-Higuera, I. Shortage of dNTPs underlies altered replication dynamics and DNA breakage in the absence of the APC/C cofactor Cdh1. *Oncogene* **2017**, *36*, 5808–5818. [CrossRef] [PubMed]

125. Wäsch, R.; Engelbert, D. Anaphase-promoting complex-dependent proteolysis of cell cycle regulators and genomic instability of cancer cells. *Oncogene* **2005**, *24*, 1–10. [CrossRef] [PubMed]

126. Cardozo, T.; Pagano, M. Wrenches in the works: Drug discovery targeting the SCF ubiquitin ligase and APC/C complexes. *BMC Biochem.* **2007**, *8* (Suppl. 1), S9. [CrossRef] [PubMed]

127. Bolanos-Garcia, V.M. Assessment of the mitotic spindle assembly checkpoint (SAC) as the target of anticancer therapies. *Curr. Cancer Drug Targets* **2009**, *9*, 131–141. [CrossRef] [PubMed]

128. Taniguchi, K.; Momiyama, N.; Ueda, M.; Matsuyama, R.; Mori, R.; Fujii, Y.; Ichikawa, Y.; Endo, I.; Togo, S.; Shimada, H. Targeting of CDC20 via Small Interfering RNA Causes Enhancement of the Cytotoxicity of Chemoradiation. *Anticancer Res.* **2008**, *28*, 1559–1563. [PubMed]

129. Chang, D.Z.; Ma, Y.; Ji, B.; Liu, Y.; Hwu, P.; Abbruzzese, J.L.; Logsdon, C.; Wang, H. Increased CDC20 expression is associated with pancreatic ductal adenocarcinoma differentiation and progression. *J. Hematol. Oncol.* **2012**, *5*, 15. [CrossRef] [PubMed]

130. Lub, S.; Maes, A.; Maes, K.; De Veirman, K.; De Bruyne, E.; Menu, E.; Fostier, K.; Kassambara, A.; Moreaux, J.; Hose, D.; et al. Inhibiting the anaphase promoting complex/cyclosome induces a metaphase arrest and cell death in multiple myeloma cells. *Oncotarget* **2016**, *7*, 4062–4076. [CrossRef] [PubMed]

131. Mortensen, E.M.; Haas, W.; Gygi, M.; Gygi, S.P.; Kellogg, D.R. Cdc28-dependent regulation of the Cdc5/Polo kinase. *Curr. Biol.* **2005**, *15*, 2033–2037. [CrossRef] [PubMed]

132. Visintin, C.; Tomson, B.N.; Rahal, R.; Paulson, J.; Cohen, M.; Taunton, J.; Amon, A.; Visintin, R. APC/C-Cdh1-mediated degradation of the Polo kinase Cdc5 promotes the return of Cdc14 into the nucleolus. *Genes Dev.* **2008**, *22*, 79–90. [CrossRef] [PubMed]

133. Pramila, T.; Wu, W.; Miles, S.; Noble, W.S.; Breeden, L.L. The Forkhead transcription factor Hcm1 regulates chromosome segregation genes and fills the S-phase gap in the transcriptional circuitry of the cell cycle. *Genes Dev.* **2006**, *20*, 2266–2278. [CrossRef] [PubMed]

134. Lamb, J.R.; Michaud, W.A.; Sikorski, R.S.; Hieter, P.A. Cdc16p, Cdc23p and Cdc27p form a complex essential for mitosis. *EMBO J.* **1994**, *13*, 4321–4328. [PubMed]

135. Irniger, S.; Nasmyth, K. The anaphase-promoting complex is required in G1 arrested yeast cells to inhibit B-type cyclin accumulation and to prevent uncontrolled entry into S-phase. *J. Cell Sci.* **1997**, *110*, 1523–1531. [PubMed]

136. Li, M.; York, J.P.; Zhang, P. Loss of Cdc20 causes a securin-dependent metaphase arrest in two-cell mouse embryos. *Mol. Cell. Biol.* **2007**, *27*, 3481–3488. [CrossRef] [PubMed]

137. TCGA. Available online: https://portal.gdc.cancer.gov (accessed on 1 January 2018).

138. Ishizawa, J.; Sugihara, E.; Kuninaka, S.; Mogushi, K.; Kojima, K.; Benton, C.B.; Zhao, R.; Chachad, D.; Hashimoto, N.; Jacamo, R.O.; et al. FZR1 loss increases sensitivity to DNA damage and consequently promotes murine and human B-cell acute leukemia. *Blood* **2017**, *129*, 1958–1968. [CrossRef] [PubMed]

139. Villa-Hernández, S.; Bueno, A.; Bermejo, R. The Multiple Roles of Ubiquitylation in Regulating Challenged DNA Replication. *Adv. Exp. Med. Biol.* **2017**, *1042*, 395–419. [PubMed]

140. Kitao, H.; Iimori, M.; Kataoka, Y.; Wakasa, T.; Tokunaga, E. DNA replication stress and cancer chemotherapy. *Cancer Sci.* **2018**, *109*, 264–271. [CrossRef] [PubMed]

141. Wang, Q.; Moyret-Lalle, C.; Couzon, F.; Surbiguet-Clippe, C.; Saurin, J.C.; Lorca, T.; Navarro, C.; Puisieux, A. Alterations of anaphase-promoting complex genes in human colon cancer cells. *Oncogene* **2003**, *22*, 1486–1490. [CrossRef] [PubMed]

142. Park, K.H.; Choi, S.E.; Eom, M.; Kang, Y. Downregulation of the anaphase-promoting complex (APC)7 in invasive ductal carcinomas of the breast and its clinicopathologic relationships. *Breast Cancer Res.* **2005**, *7*, R238–R247. [CrossRef] [PubMed]

143. Kim, I.Y.; Kwon, H.Y.; Park, K.H.; Kim, D.S. Anaphase-Promoting Complex 7 is a Prognostic Factor in Human Colorectal Cancer. *Ann. Coloproctol.* **2017**, *33*, 139–145. [CrossRef] [PubMed]

144. Tsuda, Y.; Iimori, M.; Nakashima, Y.; Nakanishi, R.; Ando, K.; Ohgaki, K.; Kitao, H.; Saeki, H.; Oki, E.; Maehara, Y. Mitotic slippage and the subsequent cell fates after inhibition of Aurora B during tubulin-binding agent-induced mitotic arrest. *Sci. Rep.* **2017**, *7*, 16762. [CrossRef] [PubMed]

145. Haschka, M.; Karbon, G.; Fava, L.L.; Villunger, A. Perturbing mitosis for anti-cancer therapy: Is cell death the only answer? *EMBO Rep.* **2018**, *19*, e45440. [CrossRef] [PubMed]

146. Kops, G.J.; Foltz, D.R.; Cleveland, D.W. Lethality to human cancer cells through massive chromosome loss by inhibition of the mitotic checkpoint. *Proc. Natl. Acad. Sci. USA* **2004**, *101*, 8699–8704. [CrossRef] [PubMed]

147. Bharadwaj, R.; Yu, H. The spindle checkpoint, aneuploidy, and cancer. *Oncogene* **2004**, *23*, 2016–2027. [CrossRef] [PubMed]

148. Lee, M.; Rivera-Rivera, Y.; Moreno, C.S.; Saavedra, H.I. The E2F activators control multiple mitotic regulators and maintain genomic integrity through Sgo1 and BubR1. *Oncotarget* **2017**, *8*, 77649–77672. [CrossRef] [PubMed]

149. Hollenhorst, P.C.; Bose, M.E.; Mielke, M.R.; Müller, U.; Fox, C.A. Forkhead genes in transcriptional silencing, cell morphology and the cell cycle. Overlapping and distinct functions for FKH1 and FKH2 in *Saccharomyces cerevisiae*. *Genetics* **2000**, *154*, 1533–1548. [PubMed]

150. Rodríguez-Colman, M.J.; Sorolla, M.A.; Vall-Llaura, N.; Tamarit, J.; Ros, J.; Cabiscol, E. The FOX transcription factor Hcm1 regulates oxidative metabolism in response to early nutrient limitation in yeast. Role of Snf1 and Tor1/Sch9 kinases. *Biochim. Biophys. Acta* **2013**, *1833*, 2004–2015. [CrossRef] [PubMed]

151. Rodrigo-Brenni, M.C.; Morgan, D.O. Sequential E2s drive polyubiquitin chain assembly on APC targets. *Cell* **2007**, *130*, 127–139. [CrossRef] [PubMed]

152. Kobayashi, T.; Horiuchi, T. A yeast gene product, Fob1 protein, required for both replication fork blocking and recombinational hotspot activities. *Genes Cells* **1996**, *1*, 465–474. [CrossRef] [PubMed]

153. Kobayashi, T.; Heck, D.J.; Nomura, M.; Horiuchi, T. Expansion and contraction of ribosomal DNA repeats in *Saccharomyces cerevisiae*: Requirement of replication fork blocking (Fob1) protein and the role of RNA polymerase I. *Genes Dev.* **1998**, *12*, 3821–3830. [CrossRef] [PubMed]

154. Defossez, P.A.; Prusty, R.; Kaeberlein, M.; Lin, S.J.; Ferrigno, P.; Silver, P.A.; Keil, R.L.; Guarente, L. Elimination of replication block protein Fob1 extends the life span of yeast mother cells. *Mol. Cell* **1999**, *3*, 447–455. [CrossRef]

155. Yamashita, Y.M.; Nakaseko, Y.; Samejima, I.; Kumada, K.; Yamada, H.; Michaelson, D.; Yanagida, M. 20S cyclosome complex formation and proteolytic activity inhibited by the cAMP/PKA pathway. *Nature* **1996**, *384*, 276–279. [CrossRef] [PubMed]

156. Madia, F.; Gattazzo, C.; Wei, M.; Fabrizio, P.; Burhans, W.C.; Weinberger, M.; Galbani, A.; Smith, J.R.; Nguyen, C.; Huey, S.; et al. Longevity mutation in SCH9 prevents recombination errors and premature genomic instability in a Werner/Bloom model system. *J. Cell Biol.* **2008**, *180*, 67–81. [CrossRef] [PubMed]

157. Qie, B.; Lyu, Z.; Lyu, L.; Liu, J.; Gao, X.; Liu, Y.; Duan, W.; Zhang, N.; Du, L.; Liu, K. Sch9 regulates intracellular protein ubiquitination by controlling stress responses. *Redox Biol.* **2015**, *5*, 290–300. [CrossRef] [PubMed]

158. Kaeberlein, M.; Powers, R.W., 3rd; Steffen, K.K.; Westman, E.A.; Hu, D.; Dang, N.; Kerr, E.O.; Kirkland, K.T.; Fields, S.; Kennedy, B.K. Regulation of yeast replicative life span by TOR and Sch9 in response to nutrients. *Science* **2005**, *310*, 1193–1196. [CrossRef] [PubMed]

International Journal of
Molecular Sciences

MDPI

Article

Network-Driven Proteogenomics Unveils an Aging-Related Imbalance in the Olfactory IκBα-NFκB p65 Complex Functionality in Tg2576 Alzheimer's Disease Mouse Model

Maialen Palomino-Alonso [1], Mercedes Lachén-Montes [1,2,3], Andrea González-Morales [1,2,3], Karina Ausín [2,3], Alberto Pérez-Mediavilla [3,4], Joaquín Fernández-Irigoyen [1,2,3,†] and Enrique Santamaría [1,2,3,*,†] (ORCID)

[1]   Clinical Neuroproteomics Group, Navarrabiomed, Departamento de Salud,
      Universidad Pública de Navarra, 31008 Pamplona, Spain; mpalomino@alumni.unav.es (M.P.-A);
      mercedes.lachen.montes@navarra.es (M.L.-M.); andrea.gonzalez.morales@navarra.es (A.G.-M.);
      jokfer@gmail.com (J.F.-I.)
[2]   Proteored-ISCIII, Proteomics Unit, Navarrabiomed, Departamento de Salud,
      Universidad Pública de Navarra, 31008 Pamplona, Spain; karina.ausin.perez@navarra.es
[3]   Instituto de Investigación Sanitaria de Navarra (IdiSNA), Navarra Institute for Health Research,
      31008 Pamplona, Spain; lamediav@unav.es
[4]   Neurobiology of Alzheimer's Disease, Neurosciences Division, Center for Applied Medical
      Research (CIMA), Department of Biochemistry, University of Navarra, 31008 Pamplona, Spain
*   Correspondence: esantamma@navarra.es; Tel.: +34-848-425-740; Fax: +34-848-422-200
†   These authors contributed equally to this work.

Received: 27 September 2017; Accepted: 25 October 2017; Published: 27 October 2017

Abstract: Olfaction is often deregulated in Alzheimer's disease (AD) patients, and is also impaired in transgenic Tg2576 AD mice, which overexpress the Swedish mutated form of human amyloid precursor protein (APP). However, little is known about the molecular mechanisms that accompany the neurodegeneration of olfactory structures in aged Tg2576 mice. For that, we have applied proteome- and transcriptome-wide approaches to probe molecular disturbances in the olfactory bulb (OB) dissected from aged Tg2576 mice (18 months of age) as compared to those of age matched wild-type (WT) littermates. Some over-represented biological functions were directly relevant to neuronal homeostasis and processes of learning, cognition, and behavior. In addition to the modulation of CAMP responsive element binding protein 1 (CREB1) and APP interactomes, an imbalance in the functionality of the IκBα-NFκB p65 complex was observed during the aging process in the OB of Tg2576 mice. At two months of age, the phosphorylated isoforms of olfactory IκBα and NFκB p65 were inversely regulated in transgenic mice. However, both phosphorylated proteins were increased at 6 months of age, while a specific drop in IκBα levels was detected in 18-month-old Tg2576 mice, suggesting a transient activation of NFκB in the OB of Tg2576 mice. Taken together, our data provide a metabolic map of olfactory alterations in aged Tg2576 mice, reflecting the progressive effect of APP overproduction and β-amyloid (Aβ) accumulation on the OB homeostasis in aged stages.

Keywords: Tg2576 mice; olfactory bulb; proteogenomics; mass-spectrometry

1. Introduction

Olfactory dysfunction has been related to aging and Alzheimer's disease (AD) [1,2]. The smell impairment is considered an early event of AD, preceding the appearance of dementia symptoms. The Tg2576 transgenic mice express the hAPPSw via the hamster prion promoter, an isoform of the human amyloid precursor protein (APP) with double mutation K670N, M671L [3]. These mice displayed an increase of APP production with consequent overproduction of β-amyloid (Aβ) 40 and Aβ42 and plaques formation in the frontal, temporal, and entorhinal cortices, hippocampus, presubiculum, and cerebellum at about 11–13 months of age [4]. There is strong evidence that accumulation of Aβ peptide is responsible for age-related memory decline in these mice [5–7]. Other than the increase in Aβ production, these mice can also display hyperphosphorylated tau at old age. Synaptic deficits, and mitochondrial imbalance have been reported for this late-plaque model [8,9]. Metabolic and posttranslational modification alterations occur long before the onset of behavioral impairment [10–12]. However, Tg2576 mice did not present a profound cognitive impairment, even at old ages [13].

The olfactory bulb (OB) is the first brain structure of the olfactory pathway [14]. APP processing products have been observed in the OB of 1-month-old Tg2576 mice, as has Aβ deposition at 13.5 months of age [15]. Moreover, Tg2576 mice (between ages of 6.5 and 8 months) present a reduced rate of OB neurogenesis, a reduction in the volume of the granular cell layers of the OB [16], and some olfactory memory deficits [16,17]. A detailed analysis of olfaction in Tg2576 mice also revealed behavioral deficits in odor habituation and discrimination [18,19]. Interestingly, the appearance of these behavioral impairments corresponds with a progressive Aβ deposition in specific olfactory structures [18]. In view of these data, an in depth biochemical characterization of the OB is necessary to reveal the missing links in the biochemical understanding of smell impairments in Tg2576 AD mice.

In this study, we used a discovery platform, applying mass-spectrometry based quantitative proteomics and transcriptome-wide analyses, to decipher the pathophysiological mechanisms that are disturbed in the OB from aged Tg2576 mice (18 months of age) as compared to those of age matched background strain control mice. 107 differential genes and 25 differentially expressed proteins were detected, pinpointing specific molecular pathways, protein interactomes, and potential olfactory therapeutic targets.

2. Results and Discussion

OB perturbations are responsible for olfactory dysfunction in neurological syndromes [2], however, few studies have examined this structure using high throughput molecular approaches [20–23]. Focusing on the transgenic Tg2576 mouse AD model, different proteomic and transcriptomic studies were performed to characterize novel molecular mediators associated with AD pathophysiology in brain structures affected during the disease progression [9,24–30]. To our knowledge, this is the first study that characterizes olfactory-associated molecular changes in this late-plaque model using omics technologies.

2.1. Molecular Alterations Detected in the Olfactory Bulb (OB) of 18-Month-Old Tg2576 Mice

Tg2576 transgenic mice suffer from memory deficits accompanied by β-amyloid plaques that increase with disease progression [5–7]. We have applied a dual-omic approach to analyze the molecular imbalance induced by the hAPPSw isoform at the olfactory level, with the final goal to reveal novel information about the OB site-specific molecular signature at late AD stages in 18-month-old Tg2576 mice. To analyze the potential differences in olfactory molecular expression profiles, OB specimens for each experimental group (Tg2576 and WT mice) were subjected to chemical tags (isobaric tag for relative and absolute quantitation, iTRAQ) coupled to tandem mass spectrometry (3 mice/condition) and into RNA microarray platform (3 mice/condition) (Figure 1).

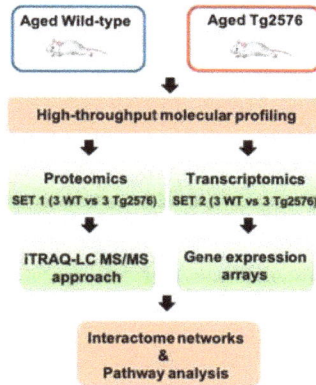

Figure 1. An overview of the workflow used for the molecular characterization of the olfactory bulbs (OBs) derived from aged Tg2576 mice.

Among 2466 quantified proteins (Table S1), differential analysis revealed 25 de-regulated proteins in Tg2576 OBs with respect to wild-type (WT) OBs (11 down- and 14 up-regulated in aged Tg2576 mice) (Figure 2A,B and Table S2). The up-regulation of our intrinsic positive control (APP) was verified by Western-blotting (Figure 2A). According to the STRING Database [31], this subproteome is mainly involved in membrane organization (False discovery rate (FDR): 1.21×10^{-5}; e.g., *SYNE2*, *NOL3*, *AP1G1*), protein transport (FDR: 0.047; e.g., *RAB5B*, *RPL28*, *SRPR*, *RPL18*, *CHMP3*, *COPE*), and negative regulation of neuron differentiation (FDR: 0.041; e.g., *APP*, *APOE*, *GFAP*, *MT3*). In the transcriptomic phase, 107 protein-coding genes were differentially regulated in the OB of aged Tg2576 mice (16 down- and 91 up-regulated genes with respect to WTs) (Table S3). Gene interactome networks suggested an alteration in the response to cyclic adenosine monophosphate (cAMP) (down regulation of *EGR1*, *EGR2*, *NR4A1*, *JUNB*, and *FOSB* genes) and in the olfactory transduction signaling due to the up-regulation of *ADCY3*, and *GNAL* genes, together with the overexpression of some olfactory receptors (OR) like *OLFR553*, *OLFR1312*, and *OLFR597* genes (dashed circles in Figure 2C).

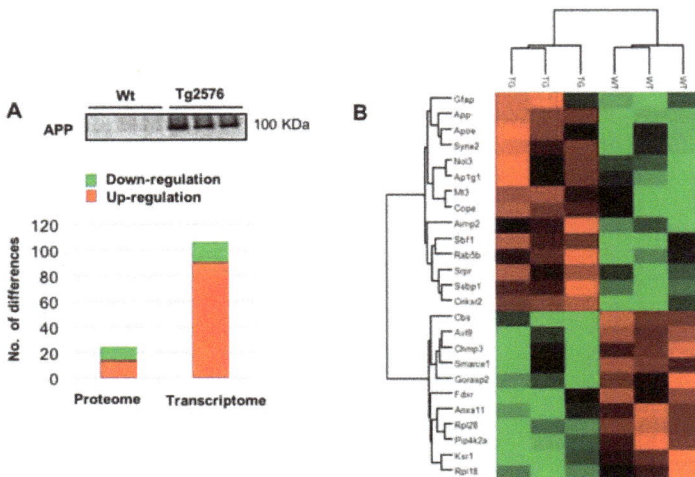

Figure 2. *Cont.*

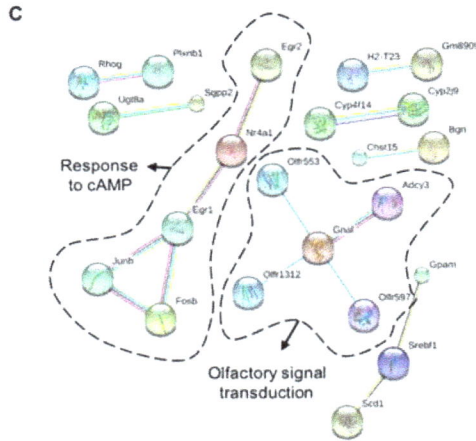

Figure 2. Multi-omic approach to decipher the OB site-specific molecular signature in aged Tg2576 mice. (**A**) Differential molecular profiling detected by the dual-omic approach in Tg2576 OBs. The olfactory protein expression levels of amyloid precursor protein (APP) at late Alzheimer's disease (AD) stages in 18-month-old Tg2576 mice is shown. Equal loading control and quantitation values have been included in Table S5; (**B**) Heat map representing the degree of change for the differentially expressed proteins (Table S2) between 18-month-old wild-type (WT) and Tg2576 mice. Red and green, up- and down-regulated proteins, respectively; (**C**) Gene interactome networks for the differentially expressed genes detected in aged Tg2576 mice. Network analysis was performed submitting the corresponding gene IDs to the STRING software (v. 10.5) (Available online: https://string-db.org/). Only interactions tagged as "high confidence" (>0.7) in STRING database were considered. Dashed circles highlight the potential alteration in the response to cyclic adenosine monophosphate (cAMP) and in the olfactory transduction signaling.

Moreover, *RTP2* (Receptor-transporter-protein 2) was also up-regulated in the OB of Tg2576 mice. RTP2 promotes OR cell-surface expression and activation in response to odorant stimulation [32]. Based on transcriptomic information from the prefrontal cortex, it has been suggested as a cause of the alteration in the smell perception of AD subjects [33]. According to our data, *OR* gene dysregulation has been also demonstrated in the OB, entorhinal, and frontal cortex in human AD subjects [34,35]. On the other hand, there was not an evident RNA-protein correlation derived from the differential datasets obtained from Tg2576 OBs. This may be due to the use of different sets of animals for each technology platform. Moreover, other reasons may also explain the observed discrepancy, such as the spatial and temporal delayed synthesis between mRNA and protein [36], post-transcriptional events, and the different hydrophobicity and solubility of the "missing proteome" during the proteomic phase, hampering its characterization and quantitation by mass-spectrometry (e.g., ORs) [37].

2.2. Biological Functions and Neuronal-Specific Processes Altered in the OB of Aged Tg2576 Mice

To obtain a more detailed description of the proteogenomic modulation in the Tg2576 OBs, differential datasets were analyzed for higher-level organization of genes and proteins into common biological pathways. For that, differential proteomic and transcriptomic datasets were merged and functionally analyzed across specific biological functions using the Ingenuity Pathway Analysis (IPA) software (V. 36601845, Release Date: 22 June 2017, Ingenuity Systems®, Redwood City, CA, USA). Some statistically over-represented processes were directly relevant to cell movement (*p*-value: 0.0003), cell survival (*p*-value: 0.00015), and cell death (*p*-value: 1.08×10^{-6}) (Figure 3A and Table S4).

Moreover, molecular clusters involved in learning (*p*-value: 0.004), cognition (*p*-value: 0.0007), behavior (*p*-value: 0.003), and dementia (*p*-value: 0.003) were also significantly represented (Figure 3B and Table S4).

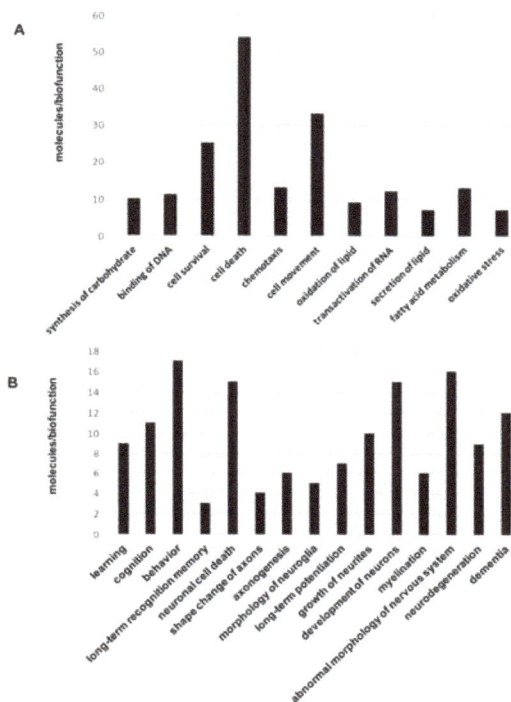

Figure 3. Significantly represented biofunctions in the OB of aged Tg2576 mice. Canonical (**A**) and neuronal-specific (**B**) over-represented biofunctions in omic datasets derived from 18-month-old Tg2576 mice. In (**A**), 45 out of 54 molecules present in cell death category are related to apoptosis (See Table S4).

Although the role of β-amyloid in olfactory deficits detected in Tg2576 mice has been extensively studied [18,19], there is no information about the survival potential of OB neurons in aged Tg2576 mice. To complement our proteogenomic workflow, survival and apoptotic pathways were monitored to analyze the effect of the β-amyloid burden on the viability of the olfactory neurons in aged Tg2576 mice. For that, steady-state levels of survival proteins like Bcl-xL and activated forms of caspase-3, -9, and -12 were measured in protein extracts from Tg2576 OBs at 18 months. The characterization of pro- and anti-apoptotic factors revealed no activation of mitochondrial or endoplasmic reticulum apoptotic routes in aged Tg2576 mice at the level of OB. Subsequent experiments were performed to monitor specific survival pathways at the level of OB. Total and residue-specific phosphorylation of focal adhesion kinase (FAK), protein kinase B (Akt), ERK activator kinase 1 (MEK)/extracellular signal-regulated kinase (ERK), Phosphoinositide-dependent protein kinase 1 (PDK1), protein kinase C (PKC), p38 mitogen-activated protein kinase (p38 MAPK), and mitogen-activated protein kinase kinase 4 (SEK1/MKK4)were measured in Tg2576 and WT OBs. As shown in Figure 4, no changes in the activation state of this survival panel were observed between transgenic and WT mice at 18 months of age, except a slight increase (non-significant) in the activation state of SEK1/MKK4 kinase (Figure 4).

However, deregulation of MAP kinases (MEK, ERK) and the PDK1/PKC axis has been observed in human OB at advanced AD stages [22,23]. These differences may be partially explained by stage, and species-dependent responses [23] and differences in molecular mechanisms associated to β-amyloidogenesis.

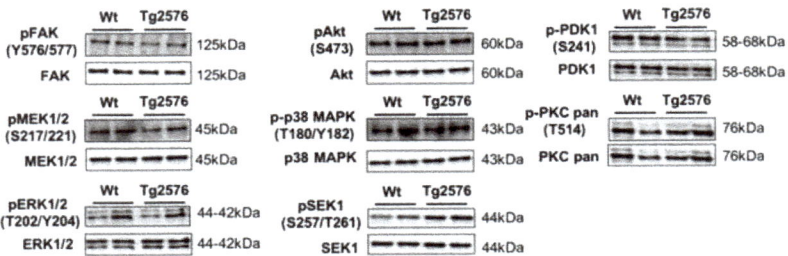

Figure 4. Activation state of specific survival pathways in the OB of aged Tg2576 mice. Levels and residue-specific phosphorylation of focal adhesion kinase (FAK), ERK Activator Kinase 1 (MEK)/extracellular signal-regulated kinase (ERK), protein kinase B (Akt), phosphoinositide-dependent protein kinase 1 (PDK1), protein kinase C (PKC), p38 mitogen-activated protein kinase (p38 MAPK), and mitogen-activated protein kinase kinase 4 (SEK1/MKK4), in the OB of aged Tg2576 mice. Equal loading of the gels was assessed by stain free digitalization. Equal loading control and quantitation values have been included in Table S5.

2.3. Functional Interactome of the hAPPSw Isoform at the Olfactory Level: Characterization of Potential Hubs by Network-Driven Proteogenomics

To explore the cooperative action among differentially expressed OB genes/proteins in aged Tg2576 mice, we performed molecular interaction networks, merging the olfactory targets that tend to be de-regulated in this model. We consider the discovery of unexpected relationships between apparently unrelated proteins and AD-causing neuropathological substrates as a powerful strategy for the characterization of novel AD causative/susceptibility targets with a central role during olfactory neurodegeneration. Functional interactome maps were generated using IPA software (Figures 5 and 6). We explored whether the hAPPSw isoform, highly expressed in Tg2576 mice, may potentially be interconnected with differential molecular targets detected by our dual-omic approach. As shown in Figure 5, differential functional interactors for APP protein were identified in the OB from aged transgenic mice. The functional APP interactome was composed by targets potentially distributed in different cellular compartments: (i) *APOE*, *SERPINH1*, and *BGN* in the extracellular space; (ii) *VAPA* at the plasma membrane level; (iii) *Irgm1*, *GFAP*, *MT3*, *ARC*, *GAB1*, *ZDHHC23*, and *GORASP2* in the cytoplasm; and (iv) *FOSB*, *JUNB*, *ZFP36L1*, *NR4A2*, *HEY2*, *EGR1*, *TP63*, and *Mmp* in the nuclear compartment (Figure 5).

Moreover, integrative network analysis also allowed us to establish a framework to map interactions between differentially expressed targets and network hubs. According to IPA analysis, CREB1, NFKBIA (IκBα), and NFκB were postulated as potential upstream regulators of part of the differential targets detected in our study (Table 1).

Even though changes in their expression were not detected in our system-wide approaches, the alteration of some of their targets may correspond to a dysregulation in their functionality in the OB of aged Tg2576 mice. For that, subsequent experiments were performed to monitor the protein expression of CREB1 as well as the activation state of IκBα-NFκB p65 complex in the OB of aged Tg2576 mice. CREB1 is a pivotal molecule during synaptic strengthening and memory formation processes [38–40]. Protein expression levels of transcription factor CREB1 were significantly reduced in the OB of Tg2576 mice (Figure 6) in parallel with the down-regulation of *FOSB*, *NR4A2*, and *EGR1*, well-known CREB target genes [41] (Figure 6 and Table S3). Our data obtained in aged Tg2576 mice reinforce the direct relationship between a reduction of CREB1 activation and AD pathology [41,42].

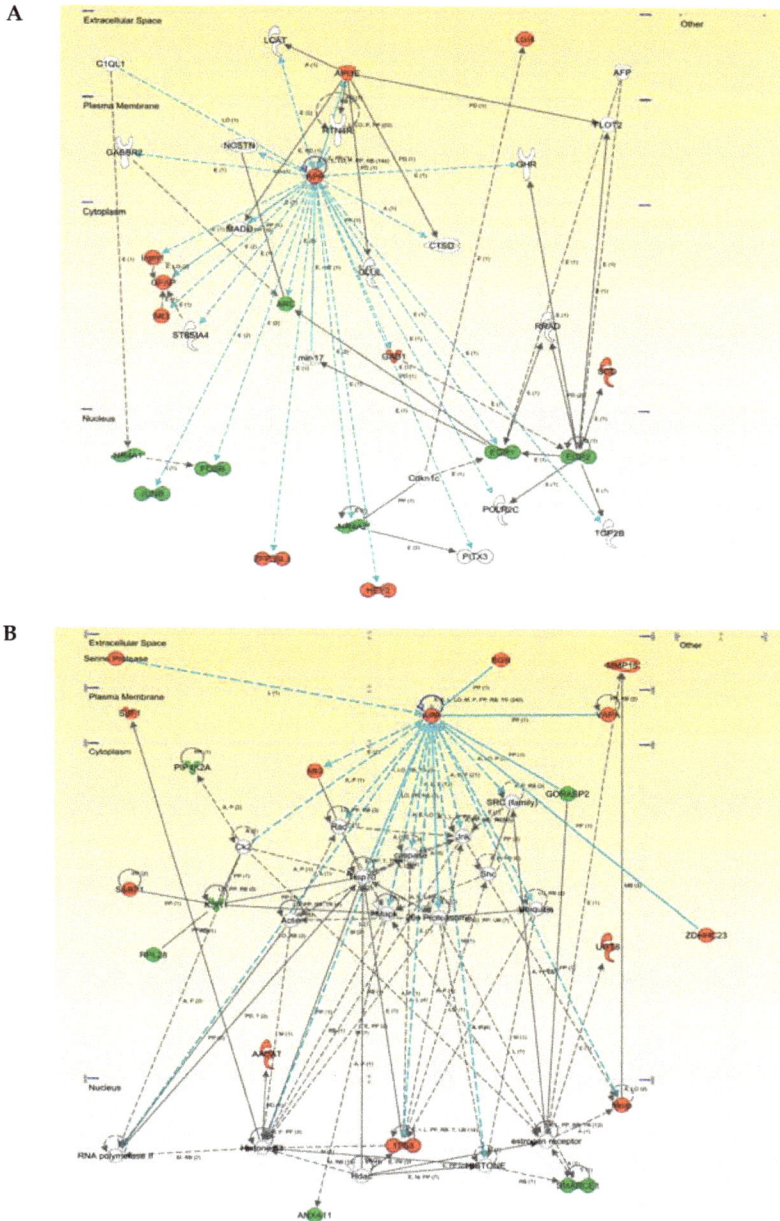

Figure 5. Modulation of APP molecular interaction network in the OB of aged Tg2576 mice. Visual representation of the relationships between differentially expressed genes/proteins and APP functional interactors (blue lines) are shown in both potentially deregulated networks in Tg2576 OBs (**A** and **B**). Continuous lines represent direct interactions, while discontinuous lines correspond to indirect functional interactions. Up-regulated molecules are shown in red, down-regulated molecules in green, and proteins proposed by the software in white. The complete legend of this type of interactomes may be found at http://ingenuity.force.com/ipa/articles/Feature_Description/Legend (accessed on 10 October 2017).

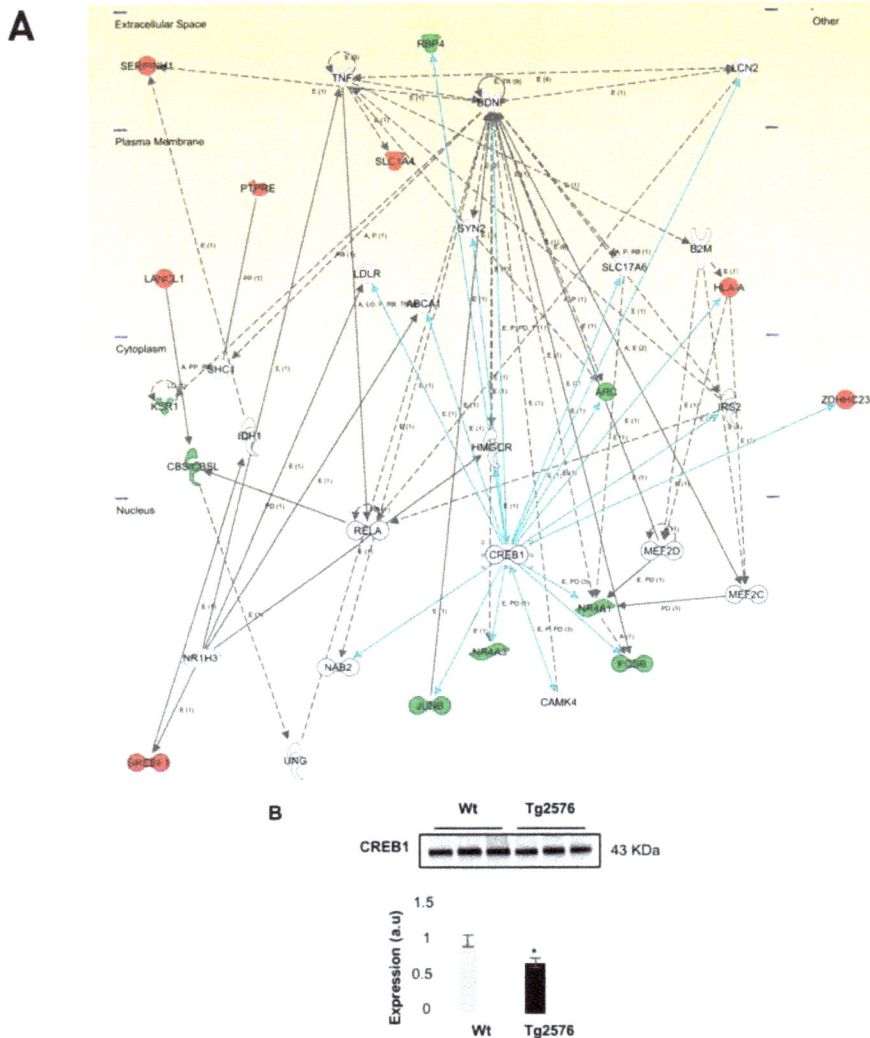

Figure 6. Modulation of CREB1 molecular interactome in the OB of aged Tg2576 mice. Visual representation of the relationships between differential expressed genes/proteins and CREB1 functional interactors (blue lines) are shown in the deregulated network in Tg2576 OBs (**A**); Continuous lines represent direct interactions, while discontinuous lines correspond to indirect functional interactions. Up-regulated molecules are shown in red, down-regulated molecules in green, and molecules proposed by the software in white. The complete legend of this type of interactomes may be found at http://ingenuity.force.com/ipa/articles/Feature_Description/Legend (accessed on 17 October 2017). CREB1 protein levels were down-regulated in the OB of aged Tg2576 mice. * $p < 0.05$ vs. WT group (**B**). Equal loading control and quantitation values have been included in Table S5. a.u., arbitrary units.

Table 1. Potential upstream regulators of differential targets detected in our study.

Upstream Regulator	Molecule Type	*p*-Value	Target Molecules in OB Omics Dataset
CREB1	transcription regulator	0.00000209	*APOE, ARC, EGR1, EGR2, FOSB, GNAL, HLA-A, JUNB, NR4A1, NR4A2, NR4A3, RBP4, SCD, ZDHHC23*
NFKBIA	transcription regulator	0.00916	*BCL2A1, CD82, HLA-A, JUNB, MMP15, NID1, NR4A1*
NFKB1	transcription regulator	0.0000748	*APOE, APP, BCL2A1, CD82, EGR1, HLA-DMB, NR4A1, TBX21*
NFκB (complex)	complex	0.00174	*APOE, APP, BCL2A1, CD82, EGR1, GFAP, HLA-A, HLA-DMB, JUNB, KLF3*

2.4. An Impairment in the Olfactory IκBα-NFκB p65 Complex Functionality during the Progression of Alzheimer's Disease (AD) in Tg2576 Mice

The nuclear factor NFκB controls the transcription of a wide variety of genes including pro-apoptotic and pro-survival genes, proinflammatory cytokines, antioxidant enzymes, pro-oxidant enzymes, and many others [43]. In general, the formation of a complex of NFκB dimer with a typical member of IkB family prevents the nuclear translocation and gene activation function of NFκB [44]. The phosphorylation of serine 32 of IκBα leads to ubiquitination and proteasomal degradation of IκBα, allowing for the phosphorylation of the NFκB p65 subunit and the enhancement of the p65 transactivation potential [44]. In our study, we monitored the functionality of the IκBα-NFκB p65 complex at three stages of AD: long before (2 months of age), immediately before (6 months), and after (18 months) the appearance of Aβ plaques [9]. At two months of age, the phosphorylated isoforms of IκBα and NFκB p65 were inversely regulated in Tg2576 transgenic mice (Figure 7A,B). However, both phosphorylated proteins were increased at 6 months of age, whereas a specific drop in IκBα levels was detected in 18-month-old Tg2576 mice (Figure 7A,B). Based on these data, it may be hypothesized that APP overproduction induces a transient activation of NFκB in the OB of Tg2576 mice compared to that of WT mice. It has been previously observed that there is an increased NFκB activity in different hippocampal and cortical structures in post-mortem AD brains [45–47], probably due to increased oxidative stress, inflammatory reactions, and toxicity of accumulated Aβ peptides [43]. To deepen our understanding of the functional dynamics of the olfactory IκBα-NFκB p65 complexes during the aging process, steady-state levels and phosphorylated isoforms were independently evaluated in WT and Tg2576 mice during aging. For that, protein profiles were quantified in a time-dependent manner. With respect to data obtained at 2 months of age, a specific drop in NFκB p65 protein levels and a progressive decrease in the phosphorylated NFκB p65 subunit were observed in WT mice, whereas an increase in NFκB p65 activity at 6 months was exclusively observed in Tg2576 mice (Figure 7C). In addition, the increase in total and phosphorylated levels of IkB α observed in 18-month-old WT mice, was blocked in aged Tg2576 mice (Figure 7C). Previous reports have shown that NFκB inhibitors prevent Aβ-induced toxicity in vivo and in vitro AD experimental models [48]. To our knowledge, the early impairment in NFκB functionality observed in Tg2576 mice at the level of OB, might open new avenues of targeting the NFκB signaling cascade at the olfactory level, gaining new insight into disease pathogenesis and identifying potential disease modifying agents. However, it is important to note that the phosphoproteome of IκBα-NFκB complexes is highly complicated [44], and many post-translational modifications (PTMs) have been characterized (see http://www.uniprot.org/uniprot/Q04206 (accessed on 10 October 2017), and http://www.uniprot.org/uniprot/P25963 (accessed on 10 October 2017)), but it is still unclear how the tangled crosstalk between all PTMs regulates the ability of NFκB proteins to induce or to repress defined target genes. Due to the early olfactory imbalance detected in the IκBα-NFκB p65 complex in Tg2576 mice, additional studies are necessary to decipher the effects of APP-dependent NFκB dysregulation on the OB molecular landscape at early stages of AD pathology to understand the smell impairment that appears during the AD progression in Tg2576 mice.

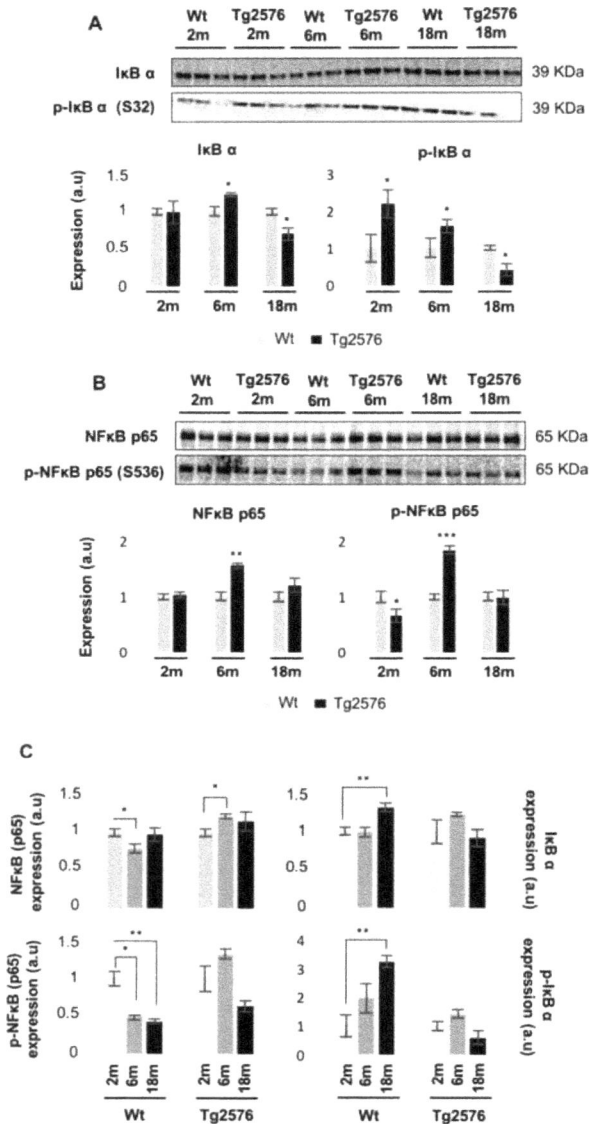

Figure 7. Time-dependent functionality of the OB IκBα-NFκB p65 complex in Tg2576 mice. Levels and residue-specific phosphorylation of the IκBα (**A**); levels and residue-specific phosphorylation of NFκB p65 subunit (**B**); Equal loading of the gels was assessed by stain-free digitalization. Panels show histograms of band densities. Data are presented as mean ± SEM from three independent OB samples per group. * $p < 0.05$ vs. control group; ** $p < 0.01$ vs. control group; *** $p < 0.001$ vs. control group. The expression of the IκBα-NFκB p65 complex was also evaluated during the aging process in WT and Tg2576 mice. * $p < 0.05$ vs. 2-month-old mice; ** $p < 0.01$ vs. 2-month-old mice (**C**). Equal loading control and quantitation values have been included in Table S5. m, month; a.u., arbitrary units.

Although the accumulation of Aβ oligomers in specific OB regions results in impaired neural integrity in specific OB cell layers in Tg2576 mice during AD progression [49], our results indicate that: (i) 1% of the 2.466 quantified proteins are differentially expressed; and (ii) 0.5% of the 20.900 protein-coding genes are differentially modulated in the OB of 18-month-old Tg2576 mice. These data pointed out that AD-related effects on the OB transcriptome and proteome composition at the bulk level are not as massive at RNA and protein levels at they are at old stages. It is important to note that due to technical reasons, only the most abundant OB proteins corresponding to approximately 10% of the mouse proteome were explored. Consequently, alterations other than those reported in this study might also participate in the AD neurodegeneration at the level of the OB in aged Tg2576 mice. Despite the analysis of the OB proteogenome providing a unique window into their biochemistry and dysfunction in aged Tg2576 mice, there are limitations of our study that warrant discussion. The OB is composed by multiple cell types with a tangled connectivity and architecture [50]. In our case, the information about specific-cell types from where proteins and mRNAs originated is lost due to the processing of the bulk OB in our omic strategy. The implementation of novel approaches that allow the exploration of olfactory cell-type specific molecular profiling would complement the output of our nonbiased exploration of the OB transcriptome/proteome, reducing the effect of multiple neuronal populations in the aged Tg2576 OBs.

3. Materials and Methods

3.1. Materials

The following reagents and materials were used. From Cell Signaling technology (Danvers, MA, USA): anti-APP (ref. 2450), anti-FAK (ref. 3285), anti-phospho-FAK (Y576/577) (ref. 3281), anti-Akt (ref. 4685), anti-phospho-Akt (S473) (ref. 4060), anti-PDK1 (ref. 3062), anti-phospho-PDK1 (S241) (ref. 3061), anti-phospho-PKC pan (T514) (ref. 9379), anti-p38 MAPK (ref. 9212), anti-phospho-p38 MAPK (T180/Y182) (ref. 9211), anti-MEK1/2 (ref. 9126), anti-phospho-MEK1/2 (S217/221) (ref. 9154), anti-ERK1/2 (ref. 9102), anti-phospho-ERK1/2 (T202/y204) (ref. 4370), anti-CREB (ref. 9104), anti-IκBα (ref. 4814S), anti-phospho-IκBα (S32) (ref. 2859), anti-NFκB p65 (ref. 8242), anti-phospho-NFκB p65 (ref. 3033S), anti-SEK1 (ref. 9152), and anti-phospho-SEK1 (S257/T261) (Ref. 9156). Anti-PKC-pan was from SigmaAldrich (St Louis, MO, USA) (ref. SAB4502356). Electrophoresis reagents were purchased from Bio-Rad (Hercules, CA, USA) and trypsin from Promega (Madison, WI, USA).

3.2. Animals

Female Tg2576 transgenic mice were used [3]. Animals were housed 4–5 per cage with free access to water and food, and they were maintained in a temperature-controlled environment on a 12 h light–dark cycle. The progressive development of AD signs in our colony has been previously described [51]. Animal care procedures were conducted in accordance with the European Community Council Directive (2010/63/EU, 22 September 2010) and approved by the local ethics committee. Twelve aged animals (6 WT and 6 Tg2576 mice of 18 months of age), divided into two sets, were used for proteomics and transcriptomics analysis (3 mice/group/platform).

3.3. Olfactory Proteomics

OB protein extraction, protein digestion, and peptide iTRAQ labelling was performed as previously described [20–23]. Tryptic digests were labelled according to the manufacturer's instructions with one isobaric amine-reactive tag as follows: Tag113, WT-1; Tag114, WT-2; Tag115, WT-3; Tag116, Tg2576-1; Tag117, Tg2576-2; Tag118, Tg2576-3. After 2 h incubation, the set of labelled samples were pooled and evaporated in a vacuum centrifuge. To increase the proteome coverage, the peptide pool was submitted to cation exchange chromatography using spin Columns (Pierce, Rockford, IL, USA). Twelve fractions were collected (from 10 mM to 150 mM KCl), evaporated under vacuum, and reconstituted into 10 μL of 2% acetonitrile, 0.1% formic acid, and 98% MilliQ-H_2O prior to mass spectrometric analysis. Peptide mixtures were separated by reverse phase chromatography and analyzed by mass-spectrometry as previously described [20–23]. The raw MS/MS spectra search were processed using the MaxQuant software (v. 1.5.8.3) [52]. The parameters used were as follows: initial maximum precursor (25 ppm), fragment mass deviations (40 ppm); variable modification (methionine oxidation and N-terminal acetylation) and fixed modification (MMTS); enzyme (trypsin) with a maximum of one missed cleavage; minimum peptide length (7 amino acids); false discovery rate (FDR) for peptide spectrum match (PSM), and protein identification (1%). The frequently observed laboratory contaminants were removed. Protein identification was considered valid with at least one unique or "razor" peptide. The protein quantification was calculated using at least two razor + unique peptides, and statistical significance was calculated by a two-way Student-*t* test ($p < 0.05$). A 1.3-fold change cut-off was used. Proteins with iTRAQ ratios below the low range (0.77) were considered to be down-regulated, whereas those above the high range (1.3) were considered to be upregulated. The Perseus software (version 1.5.6.0) [53] was used for statistical analysis and data visualization. Search results files and MS raw data were deposited to the ProteomeXchange Consortium (Available online: http://proteomecentral.proteomexchange.org; accessed on 20 September 2017) via the PRIDE partner repository [54] with the identifier PXD007795.

3.4. OB Transcriptomics

Maxwell® 16 simplyRNA Kit (Promega) was used to extract the OB mitochondrial RNAs (mRNAs) from aged Tg2576 mice and WT littermates. The sense complementary DNA (cDNA) was fragmented and biotinylated using the Affymetrix Clarion S Pico assay (902932; ThermoFisher Scientific, Waltham, MA, USA). Affymetrix mouse Clarion S chips (ThermoFisher Scientific) were used according to the manufacturer protocols. Hybridization, washing, staining, scanning, and data analysis [55] were performed as previously described [34,35]. As in other transcriptomic studies performed in AD brains [56,57], we worked with a *p*-value < 0.01 (without using any method for multiple testing correction). Microarray data files were submitted to the GEO (Gene Expression Omnibus) database and are available under accession number GSE103835.

3.5. Bioinformatics

The identification of specifically dysregulated regulatory/metabolic networks was analyzed using QIAGEN's Ingenuity Pathway Analysis (IPA) (Available online: www.qiagen.com/ingenuity; accessed on 10 October 2017) and STRING software [31]. The IPA software considers signaling pathway/biofunctions according to the calculated *p*-value and reports it hierarchically.

3.6. Immunoblotting Analysis

Equal amounts of OB protein (5 µg) were resolved in 4–15% TGX stain-free gels (Bio-Rad). OB proteins derived from murine samples were electrophoretically transferred onto nitrocellulose membranes using a Trans-blot Turbo transfer system (up to 25 V, 7 min) (Bio-Rad). Equal loading of the gels was assessed by stain free digitalization and by Ponceau staining. Western-blotting was performed as previously described [20–23]. After densitometric analyses (Image Lab Software Version 5.2; Bio-Rad), optical density values were expressed as arbitrary units and normalized to total stain in each gel lane.

4. Conclusions

The largely similar OB proteogenome of aged WT and Tg2576 mice suggests that abnormal protein-protein interactions or post-translational modifications, defective intracellular trafficking, or misfolding of proteins could play a pivotal part in driving the neurodegeneration that occurs at the olfactory level in aged Tg2576 mice. Moreover, omics sciences have partially revealed the potential interactome of the hAPPSw isoform at the olfactory level as well as the disruption of the IκBα-NFκB p65 complex during the neurodegenerative process, providing molecular features that may be used as novel olfactory drug target candidates to treat AD.

Supplementary Materials: Supplementary materials can be found at www.mdpi.com/1422-0067/18/11/2260/s1.

Acknowledgments: This work was funded by grants from the Spanish Ministry of Economy and Competitiveness (MINECO) (Ref. SAF2014-59340-R), Department of Economic Development from Government of Navarra (Ref. PC023-24) and Obra Social la Caixa to Enrique Santamaría. Andrea González-Morales and Karina Ausín were supported by PEJ-2014-A-61949 and PEJ-2014-A-72151 (MINECO). Mercedes Lachén-Montes is supported by a predoctoral fellowship from the Public University of Navarra (UPNA). Authors thank all of the proteomics identifications (PRIDE) Team for helping with the mass spectrometric data deposit in ProteomeXChange/PRIDE. The Proteomics Unit of Navarrabiomed is a member of Proteored, PRB2-ISCIII, and is supported by grant PT13/0001, of the PE I + D + I 2013–2016 funded by ISCIII and FEDER. This project is part of the HUPO Brain Proteome Project and these results are lined up with the Spanish Initiative on the Human Proteome Project (SpHPP).

Author Contributions: Joaquín Fernández-Irigoyen and Enrique Santamaría designed and supervised the complete study. Maialen Palomino-Alonso, Mercedes Lachen-Montes, Andrea González-Morales, Karina Ausín, and Alberto Pérez-Mediavilla performed sample selection and sample preparation for proteomic and transcriptomic phases. Maialen Palomino-Alonso performed proteomic analysis. Maialen Palomino-Alonso, Mercedes Lachen-Montes, and Andrea González-Morales performed functional assays. Joaquín Fernández-Irigoyen and Enrique Santamaría performed liquid chromatography-tandem mass spectrometry analysis and data interpretation. Enrique Santamaría wrote the paper.

Conflicts of Interest: The authors declare no conflict of interest.

Abbreviations

AD	Alzheimer's disease
APP	Amyloid precursor protein
Aβ	β-amyloid
ERK	Extracellular signal-regulated kinase
FAK	Focal adhesion kinase
OB	Olfactory bulb
MS/MS	Tandem mass-spectrometry
PDK1	Phosphoinositide-dependent protein kinase 1
PKC	Protein kinase C
p38 MAPK	p38 Mitogen-activated protein kinase
WT	Wild-type

References

1. Bahar-Fuchs, A.; Chetelat, G.; Villemagne, V.L.; Moss, S.; Pike, K.; Masters, C.L.; Rowe, C.; Savage, G. Olfactory deficits and amyloid-β burden in Alzheimer's disease, mild cognitive impairment, and healthy aging: A PIB PET study. *J. Alzheimers Dis.* **2010**, *22*, 1081–1087. [CrossRef] [PubMed]

2. Attems, J.; Walker, L.; Jellinger, K.A. Olfactory bulb involvement in neurodegenerative diseases. *Acta Neuropathol.* **2014**, *127*, 459–475. [CrossRef] [PubMed]

3. Hsiao, K.; Chapman, P.; Nilsen, S.; Eckman, C.; Harigaya, Y.; Younkin, S.; Yang, F.; Cole, G. Correlative memory deficits, Aβ elevation, and amyloid plaques in transgenic mice. *Science* **1996**, *274*, 99–102. [CrossRef] [PubMed]

4. Puzzo, D.; Gulisano, W.; Palmeri, A.; Arancio, O. Rodent models for Alzheimer's disease drug discovery. *Expert Opin. Drug Discov.* **2015**, *10*, 703–711. [CrossRef] [PubMed]

5. Westerman, M.A.; Cooper-Blacketer, D.; Mariash, A.; Kotilinek, L.; Kawarabayashi, T.; Younkin, L.H.; Carlson, G.A.; Younkin, S.G.; Ashe, K.H. The relationship between Aβ and memory in the Tg2576 mouse model of Alzheimer's disease. *J. Neurosci.* **2002**, *22*, 1858–1867. [PubMed]

6. Janus, C.; Pearson, J.; McLaurin, J.; Mathews, P.M.; Jiang, Y.; Schmidt, S.D.; Chishti, M.A.; Horne, P.; Heslin, D.; French, J.; et al. Aβ peptide immunization reduces behavioural impairment and plaques in a model of Alzheimer's disease. *Nature* **2000**, *408*, 979–982. [CrossRef] [PubMed]

7.   Chen, G.; Chen, K.S.; Knox, J.; Inglis, J.; Bernard, A.; Martin, S.J.; Justice, A.; McConlogue, L.; Games, D.; Freedman, S.B.; et al. A learning deficit related to age and β-amyloid plaques in a mouse model of Alzheimer's disease. *Nature* **2000**, *408*, 975–979. [CrossRef] [PubMed]

8.   Jacobsen, J.S.; Wu, C.C.; Redwine, J.M.; Comery, T.A.; Arias, R.; Bowlby, M.; Martone, R.; Morrison, J.H.; Pangalos, M.N.; Reinhart, P.H.; et al. Early-onset behavioral and synaptic deficits in a mouse model of Alzheimer's disease. *Proc. Natl. Acad. Sci. USA* **2006**, *103*, 5161–5166. [CrossRef] [PubMed]

9.   Reddy, P.H.; McWeeney, S.; Park, B.S.; Manczak, M.; Gutala, R.V.; Partovi, D.; Jung, Y.; Yau, V.; Searles, R.; Mori, M.; et al. Gene expression profiles of transcripts in amyloid precursor protein transgenic mice: Up-regulation of mitochondrial metabolism and apoptotic genes is an early cellular change in Alzheimer's disease. *Hum. Mol. Genet.* **2004**, *13*, 1225–1240. [CrossRef] [PubMed]

10.   Lalande, J.; Halley, H.; Balayssac, S.; Gilard, V.; Dejean, S.; Martino, R.; Frances, B.; Lassalle, J.M.; Malet-Martino, M. 1H NMR metabolomic signatures in five brain regions of the AβPPswe Tg2576 mouse model of Alzheimer's disease at four ages. *J. Alzheimers Dis.* **2014**, *39*, 121–143. [PubMed]

11.   Nistico, R.; Ferraina, C.; Marconi, V.; Blandini, F.; Negri, L.; Egebjerg, J.; Feligioni, M. Age-related changes of protein sumoylation balance in the AβPP Tg2576 mouse model of Alzheimer's disease. *Front. Pharmacol.* **2014**, *5*, 63. [PubMed]

12.   Fodero, L.R.; Saez-Valero, J.; McLean, C.A.; Martins, R.N.; Beyreuther, K.; Masters, C.L.; Robertson, T.A.; Small, D.H. Altered glycosylation of acetylcholinesterase in APP (SW) Tg2576 transgenic mice occurs prior to amyloid plaque deposition. *J. Neurochem.* **2002**, *81*, 441–448. [CrossRef] [PubMed]

13.   King, D.L.; Arendash, G.W. Behavioral characterization of the Tg2576 transgenic model of Alzheimer's disease through 19 months. *Physiol. Behav.* **2002**, *75*, 627–642. [CrossRef]

14.   Doty, R.L. The olfactory vector hypothesis of neurodegenerative disease: Is it viable? *Ann. Neurol.* **2008**, *63*, 7–15. [CrossRef] [PubMed]

15.   Lehman, E.J.; Kulnane, L.S.; Lamb, B.T. Alterations in β-amyloid production and deposition in brain regions of two transgenic models. *Neurobiol. Aging* **2003**, *24*, 645–653. [CrossRef]

16.   Guerin, D.; Sacquet, J.; Mandairon, N.; Jourdan, F.; Didier, A. Early locus coeruleus degeneration and olfactory dysfunctions in Tg2576 mice. *Neurobiol. Aging* **2009**, *30*, 272–283. [CrossRef] [PubMed]

17.   Young, J.W.; Sharkey, J.; Finlayson, K. Progressive impairment in olfactory working memory in a mouse model of mild cognitive impairment. *Neurobiol. Aging* **2009**, *30*, 1430–1443. [CrossRef] [PubMed]

18.   Wesson, D.W.; Levy, E.; Nixon, R.A.; Wilson, D.A. Olfactory dysfunction correlates with amyloid-β burden in an Alzheimer's disease mouse model. *J. Neurosci.* **2010**, *30*, 505–514. [CrossRef] [PubMed]

19.   Wesson, D.W.; Borkowski, A.H.; Landreth, G.E.; Nixon, R.A.; Levy, E.; Wilson, D.A. Sensory network dysfunction, behavioral impairments, and their reversibility in an Alzheimer's β-amyloidosis mouse model. *J. Neurosci.* **2011**, *31*, 15962–15971. [CrossRef] [PubMed]

20.   Fernandez-Irigoyen, J.; Corrales, F.J.; Santamaria, E. Proteomic atlas of the human olfactory bulb. *J. Proteom.* **2012**, *75*, 4005–4016. [CrossRef] [PubMed]

21.   Zelaya, M.V.; Perez-Valderrama, E.; de Morentin, X.M.; Tunon, T.; Ferrer, I.; Luquin, M.R.; Fernandez-Irigoyen, J.; Santamaria, E. Olfactory bulb proteome dynamics during the progression of sporadic Alzheimer's disease: Identification of common and distinct olfactory targets across alzheimer-related co-pathologies. *Oncotarget* **2015**, *6*, 39437–39456. [CrossRef] [PubMed]

22.   Lachen-Montes, M.; Gonzalez-Morales, A.; Zelaya, M.V.; Perez-Valderrama, E.; Ausin, K.; Ferrer, I.; Fernandez-Irigoyen, J.; Santamaria, E. Olfactory bulb neuroproteomics reveals a chronological perturbation of survival routes and a disruption of prohibitin complex during Alzheimer's disease progression. *Sci. Rep.* **2017**, *7*, 9115. [CrossRef] [PubMed]

23.   Lachen-Montes, M.; Gonzalez-Morales, A.; de Morentin, X.M.; Perez-Valderrama, E.; Ausin, K.; Zelaya, M.V.; Serna, A.; Aso, E.; Ferrer, I.; Fernandez-Irigoyen, J.; et al. An early dysregulation of FAK and MEK/ERK signaling pathways precedes the β-amyloid deposition in the olfactory bulb of APP/PS1 mouse model of Alzheimer's disease. *J. Proteom.* **2016**, *148*, 149–158. [CrossRef] [PubMed]

24.   Stein, T.D.; Johnson, J.A. Lack of neurodegeneration in transgenic mice overexpressing mutant amyloid precursor protein is associated with increased levels of transthyretin and the activation of cell survival pathways. *J. Neurosci.* **2002**, *22*, 7380–7388. [PubMed]

25. George, A.J.; Gordon, L.; Beissbarth, T.; Koukoulas, I.; Holsinger, R.M.; Perreau, V.; Cappai, R.; Tan, S.S.; Masters, C.L.; Scott, H.S.; et al. A serial analysis of gene expression profile of the Alzheimer's disease Tg2576 mouse model. *Neurotox. Res.* **2010**, *17*, 360–379. [CrossRef] [PubMed]

26. Tan, L.; Wang, X.; Ni, Z.F.; Zhu, X.; Wu, W.; Zhu, L.Q.; Liu, D. A systematic analysis of genomic changes in Tg2576 mice. *Mol. Neurobiol.* **2013**, *47*, 883–891. [CrossRef] [PubMed]

27. Shin, S.J.; Lee, S.E.; Boo, J.H.; Kim, M.; Yoon, Y.D.; Kim, S.I.; Mook-Jung, I. Profiling proteins related to amyloid deposited brain of Tg2576 mice. *Proteomics* **2004**, *4*, 3359–3368. [CrossRef] [PubMed]

28. Gillardon, F.; Rist, W.; Kussmaul, L.; Vogel, J.; Berg, M.; Danzer, K.; Kraut, N.; Hengerer, B. Proteomic and functional alterations in brain mitochondria from Tg2576 mice occur before amyloid plaque deposition. *Proteomics* **2007**, *7*, 605–616. [CrossRef] [PubMed]

29. Shevchenko, G.; Wetterhall, M.; Bergquist, J.; Hoglund, K.; Andersson, L.I.; Kultima, K. Longitudinal characterization of the brain proteomes for the Tg2576 amyloid mouse model using shotgun based mass spectrometry. *J. Proteome Res.* **2012**, *11*, 6159–6174. [CrossRef] [PubMed]

30. Cuadrado-Tejedor, M.; Cabodevilla, J.F.; Zamarbide, M.; Gomez-Isla, T.; Franco, R.; Perez-Mediavilla, A. Age-related mitochondrial alterations without neuronal loss in the hippocampus of a transgenic model of Alzheimer's disease. *Curr. Alzheimer Res.* **2013**, *10*, 390–405. [CrossRef] [PubMed]

31. Szklarczyk, D.; Morris, J.H.; Cook, H.; Kuhn, M.; Wyder, S.; Simonovic, M.; Santos, A.; Doncheva, N.T.; Roth, A.; Bork, P.; et al. The string database in 2017: Quality-controlled protein-protein association networks, made broadly accessible. *Nucleic Acids Res.* **2017**, *45*, D362–D368. [CrossRef] [PubMed]

32. Saito, H.; Kubota, M.; Roberts, R.W.; Chi, Q.; Matsunami, H. RTP family members induce functional expression of mammalian odorant receptors. *Cell* **2004**, *119*, 679–691. [CrossRef] [PubMed]

33. Zhang, B.; Gaiteri, C.; Bodea, L.G.; Wang, Z.; McElwee, J.; Podtelezhnikov, A.A.; Zhang, C.; Xie, T.; Tran, L.; Dobrin, R.; et al. Integrated systems approach identifies genetic nodes and networks in late-onset Alzheimer's disease. *Cell* **2013**, *153*, 707–720. [CrossRef] [PubMed]

34. Ansoleaga, B.; Garcia-Esparcia, P.; Llorens, F.; Moreno, J.; Aso, E.; Ferrer, I. Dysregulation of brain olfactory and taste receptors in AD, PSP and CJD, and AD-related model. *Neuroscience* **2013**, *248*, 369–382. [CrossRef] [PubMed]

35. Lachen-Montes, M.; Zelaya, M.; Segura, V.; Fernández-Irigoyen, J.; Santamaría, E. Progressive modulation of the human olfactory bulb transcriptome during Alzheimer's disease evolution: Novel insights into the olfactory signaling across proteinopathies. *Oncotarget* **2017**. [CrossRef] [PubMed]

36. Liu, Y.; Beyer, A.; Aebersold, R. On the dependency of cellular protein levels on mRNA abundance. *Cell* **2016**, *165*, 535–550. [CrossRef] [PubMed]

37. Baker, M.S.; Ahn, S.B.; Mohamedali, A.; Islam, M.T.; Cantor, D.; Verhaert, P.D.; Fanayan, S.; Sharma, S.; Nice, E.C.; Connor, M.; et al. Accelerating the search for the missing proteins in the human proteome. *Nat. Commun.* **2017**, *8*, 14271. [CrossRef] [PubMed]

38. Sakamoto, K.; Karelina, K.; Obrietan, K. CREB: A multifaceted regulator of neuronal plasticity and protection. *J. Neurochem.* **2011**, *116*, 1–9. [CrossRef] [PubMed]

39. Tully, T.; Bourtchouladze, R.; Scott, R.; Tallman, J. Targeting the CREB pathway for memory enhancers. *Nat. Rev. Drug Discov.* **2003**, *2*, 267–277. [CrossRef] [PubMed]

40. Teich, A.F.; Nicholls, R.E.; Puzzo, D.; Fiorito, J.; Purgatorio, R.; Fa, M.; Arancio, O. Synaptic therapy in Alzheimer's disease: A CREB-centric approach. *Neurotherapeutics* **2015**, *12*, 29–41. [CrossRef] [PubMed]

41. Saura, C.A. CREB-regulated transcription coactivator 1-dependent transcription in Alzheimer's disease mice. *Neurodegener. Dis.* **2012**, *10*, 250–252. [CrossRef] [PubMed]

42. Saura, C.A.; Valero, J. The role of CREB signaling in Alzheimer's disease and other cognitive disorders. *Rev. Neurosci.* **2011**, *22*, 153–169. [CrossRef] [PubMed]

43. Kaur, U.; Banerjee, P.; Bir, A.; Sinha, M.; Biswas, A.; Chakrabarti, S. Reactive oxygen species, redox signaling and neuroinflammation in Alzheimer's disease: The NF-κB connection. *Curr. Top. Med. Chem.* **2015**, *15*, 446–457. [CrossRef] [PubMed]

44. Viatour, P.; Merville, M.P.; Bours, V.; Chariot, A. Phosphorylation of NF-κB and IκB proteins: Implications in cancer and inflammation. *Trends Biochem. Sci.* **2005**, *30*, 43–52. [CrossRef] [PubMed]

45. Kaltschmidt, B.; Uherek, M.; Volk, B.; Baeuerle, P.A.; Kaltschmidt, C. Transcription factor NF-κB is activated in primary neurons by amyloid β peptides and in neurons surrounding early plaques from patients with alzheimer disease. *Proc. Natl. Acad. Sci. USA* **1997**, *94*, 2642–2647. [CrossRef] [PubMed]

46. Terai, K.; Matsuo, A.; McGeer, P.L. Enhancement of immunoreactivity for NF-kappa B in the hippocampal formation and cerebral cortex of Alzheimer's disease. *Brain Res.* **1996**, *735*, 159–168. [CrossRef]

47. Ferrer, I.; Marti, E.; Lopez, E.; Tortosa, A. NF-κB immunoreactivity is observed in association with β A4 diffuse plaques in patients with Alzheimer's disease. *Neuropathol. Appl. Neurobiol.* **1998**, *24*, 271–277. [CrossRef] [PubMed]

48. Paris, D.; Patel, N.; Quadros, A.; Linan, M.; Bakshi, P.; Ait-Ghezala, G.; Mullan, M. Inhibition of Aβ production by NF-κB inhibitors. *Neurosci. Lett.* **2007**, *415*, 11–16. [CrossRef] [PubMed]

49. Yoo, S.J.; Lee, J.H.; Kim, S.Y.; Son, G.; Kim, J.Y.; Cho, B.; Yu, S.W.; Chang, K.A.; Suh, Y.H.; Moon, C. Differential spatial expression of peripheral olfactory neuron-derived BACE1 induces olfactory impairment by region-specific accumulation of β-amyloid oligomer. *Cell Death Dis.* **2017**, *8*, e2977. [CrossRef] [PubMed]

50. Nagayama, S.; Homma, R.; Imamura, F. Neuronal organization of olfactory bulb circuits. *Front. Neural Circuits* **2014**, *8*, 98. [CrossRef] [PubMed]

51. Cuadrado-Tejedor, M.; Garcia-Osta, A. Current animal models of Alzheimer's disease: Challenges in translational research. *Front. Neurol.* **2014**, *5*, 182. [CrossRef] [PubMed]

52. Tyanova, S.; Temu, T.; Cox, J. The maxquant computational platform for mass spectrometry-based shotgun proteomics. *Nat. Protoc.* **2016**, *11*, 2301–2319. [CrossRef] [PubMed]

53. Tyanova, S.; Temu, T.; Sinitcyn, P.; Carlson, A.; Hein, M.Y.; Geiger, T.; Mann, M.; Cox, J. The perseus computational platform for comprehensive analysis of (prote)omics data. *Nat. Methods* **2016**, *13*, 731–740. [CrossRef] [PubMed]

54. Vizcaino, J.A.; Deutsch, E.W.; Wang, R.; Csordas, A.; Reisinger, F.; Rios, D.; Dianes, J.A.; Sun, Z.; Farrah, T.; Bandeira, N.; et al. Proteomexchange provides globally coordinated proteomics data submission and dissemination. *Nat. Biotechnol.* **2014**, *32*, 223–226. [CrossRef] [PubMed]

55. Irizarry, R.A.; Bolstad, B.M.; Collin, F.; Cope, L.M.; Hobbs, B.; Speed, T.P. Summaries of Affymetrix GeneChip probe level data. *Nucleic Acids Res.* **2003**, *31*, e15. [CrossRef] [PubMed]

56. Silva, A.R.; Grinberg, L.T.; Farfel, J.M.; Diniz, B.S.; Lima, L.A.; Silva, P.J.; Ferretti, R.E.; Rocha, R.M.; Filho, W.J.; Carraro, D.M.; et al. Transcriptional alterations related to neuropathology and clinical manifestation of Alzheimer's disease. *PLoS ONE* **2012**, *7*, e48751. [CrossRef] [PubMed]

57. Cuadrado-Tejedor, M.; Garcia-Barroso, C.; Sanchez-Arias, J.A.; Rabal, O.; Perez-Gonzalez, M.; Mederos, S.; Ugarte, A.; Franco, R.; Segura, V.; Perea, G.; et al. A first-in-class small-molecule that acts as a dual inhibitor of HDAC and PDE5 and that rescues hippocampal synaptic impairment in Alzheimer's disease mice. *Neuropsychopharmacology* **2016**, *42*, 524–539. [CrossRef] [PubMed]

International Journal of
Molecular Sciences

MDPI

Article

Ultrastructural and Molecular Analysis of Ribose-Induced Glycated Reconstructed Human Skin

Roberta Balansin Rigon [1,†,‡], Sabine Kaessmeyer [2,†], Christopher Wolff [1], Christian Hausmann [1], Nan Zhang [1], Michaela Sochorová [3], Andrej Kováčik [3], Rainer Haag [4], Kateřina Vávrová [3], Martina Ulrich [5], Monika Schäfer-Korting [1] and Christian Zoschke [1,*]

1   Institute of Pharmacy (Pharmacology & Toxicology), Freie Universität Berlin, Königin-Luise-Str. 2+4, 14195 Berlin, Germany; roberta_rigon@yahoo.com.br (R.B.R.); christopher.wolff@fu-berlin.de (C.W.); christian.hausmann@fu-berlin.de (C.H.); nan.zhang@fu-berlin.de (N.Z.); Monika.Schaefer-Korting@fu-berlin.de (M.S.-K.)
2   Institute of Veterinary Anatomy, Department of Veterinary Medicine, Freie Universität Berlin, Koserstr. 20, 14195 Berlin, Germany; sabine.kaessmeyer@fu-berlin.de
3   Faculty of Pharmacy, Charles University, Akademika Heyrovského 1203, 50005 Hradec Králové, Czech Republic; sochormi@faf.cuni.cz (M.S.); kovacika@faf.cuni.cz (A.K.); katerina.vavrova@faf.cuni.cz (K.V.)
4   Institute of Chemistry and Biochemistry, Freie Universität Berlin, Takustr. 3, 14195 Berlin, Germany; Germany; haag@chemie.fu-berlin.de
5   Collegium Medicum Berlin, Luisenstr. 54, 10117 Berlin, Germany; ulrich@dermatologie-am-regierungsviertel.de
*   Correspondence: christian.zoschke@fu-berlin.de; Tel.: +49-30-838-56189
†   These authors contributed equally to this work.
‡   Present address: UNESP—Universidade Estadual Paulista "Júlio de Mesquita Filho", Faculdade de Ciências Farmacêuticas, Rodovia Araraquara Jaú, Km 01-s/n, 14800-903 Araraquara, Brazil.

Received: 1 October 2018; Accepted: 4 November 2018; Published: 8 November 2018

Abstract: Aging depicts one of the major challenges in pharmacology owing to its complexity and heterogeneity. Thereby, advanced glycated end-products modify extracellular matrix proteins, but the consequences on the skin barrier function remain heavily understudied. Herein, we utilized transmission electron microscopy for the ultrastructural analysis of ribose-induced glycated reconstructed human skin (RHS). Molecular and functional insights substantiated the ultrastructural characterization and proved the relevance of glycated RHS beyond skin aging. In particular, electron microscopy mapped the accumulation and altered spatial orientation of fibrils and filaments in the dermal compartment of glycated RHS. Moreover, the epidermal basement membrane appeared thicker in glycated than in non-glycated RHS, but electron microscopy identified longitudinal clusters of the finest collagen fibrils instead of real thickening. The *stratum granulosum* contained more cell layers, the morphology of keratohyalin granules decidedly differed, and the *stratum corneum* lipid order increased in ribose-induced glycated RHS, while the skin barrier function was almost not affected. In conclusion, dermal advanced glycated end-products markedly changed the epidermal morphology, underlining the importance of matrix–cell interactions. The phenotype of ribose-induced glycated RHS emulated aged skin in the dermis, while the two to three times increased thickness of the *stratum granulosum* resembled poorer cornification.

Keywords: advanced glycated end products; aging; diabetes; electron microscopy; nanomedicine; ribose; reconstructed human skin; skin absorption

1. Introduction

Age-related changes in human physiology become increasingly relevant, because life expectancy is rising as fast as the number of geriatric and often multimorbid patients. Aging comprises the functional impairment of cells, tissues, and organs, which promotes diseases like diabetes or cancer [1]. However, drug treatment of elderly patients remains challenging, as drug development mainly focuses on tests in juvenile cells and humans. While drug metabolism and elimination are known to change over lifetime, alterations in the pharmacodynamics of drugs depict a significant knowledge gap—not only in dementia [2].

One hallmark of aging is the accumulation of advanced glycated end-products (AGEs), although increased amounts of AGEs are also found in diabetes and psoriasis [3,4]. Being exclusively removed by protein turn over, AGEs accumulate in proteins with long half-lives. Thus, the collagen half-life of 15 years in skin and 117 years in cartilage fosters the accumulation of AGEs [5]. As products of the non-enzymatic reaction of sugars or aldehydes with proteins, AGEs comprise a heterogeneous group of molecules including pentosidine and *N*-carboxymethyl-lysine (CML).

Several nonclinical models of skin aging tried to emulate the long process of aging in vitro, by exposing juvenile reconstructed human skin (RHS) either to ribose, glyoxal, glyceraldehyde, or specific AGEs [6–10]. As the consequences of extracellular matrix (ECM) glycation depend on the stressor in vitro and strongly deviate among the protocols for RHS glycation, we were intrigued in the unknown effects of AGEs in RHS on epithelial proliferation and differentiation.

Herein, we studied the effects of ECM glycation on the ultrastructure of cellular and extracellular parts of RHS by transmission electron microscopy (TEM) and compared the morphology to juvenile, non-glycated RHS. TEM proved already useful to study subcellular interactions during vascular morphogenesis in 3D cocultures between endothelial cells and fibroblasts [11]. Moreover, we investigated the morphology of ribose-induced glycated RHS by reflectance confocal microscopy, being clinically used for the evaluation of skin cancer and aging [12,13]. Next, we analyzed the protein expression as well as the *stratum corneum* lipid prolife and their spatial orientation. Finally, we studied the effect of collagen glycation on the absorption of reference compounds and the efficacy of an investigative new drug delivery system.

2. Results

2.1. Selection of Glycation Agent

We exposed normal human dermal fibroblasts (NHDF) and keratinocytes (NHK) either to ribose or to glyoxal for 48 h. Next, we tested the efficiency of both glycation agents in producing AGEs in collagen following a three-week exposure to either ribose or glyoxal. The viability of both fibroblasts and keratinocytes markedly declined by up to 87% even in trace concentrations of the reagents (Figure 1a). As ribose and glyoxal induced comparable amounts of AGEs in collagen (Figure 1b), we selected the collagen, glycated with ribose, to build glycated RHS. After three weeks of RHS culture, we still detected 1.4-fold increased levels of *N*-carboxymethyl-lysine (CML), a hallmark AGE, in the dermal compartment of glycated RHS. The accumulation of CML was even higher (1.9-fold increase) in the epidermal compartment, although ribose has been removed from the collagen by dialysis prior to RHS building. Accordingly, the expression of the receptor of AGEs (RAGE) was elevated in the entire glycated RHS (Section 2.5).

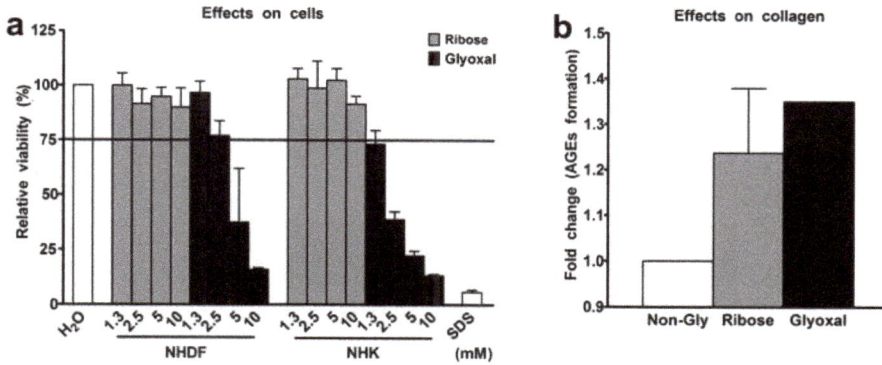

Figure 1. Selection of glycation agent. (**a**) Viability of normal human dermal fibroblasts (NHDF) and normal human keratinocytes (NHK) following exposure to ribose or glyoxal for 48 h. (**b**) Advanced glycation end-products (AGEs) in collagen following the exposure to 10 mM ribose or 10 mM glyoxal. *n* = 3, mean + SD.

2.2. RHS Morphology

Both glycated and non-glycated RHS showed a stratified epidermis, but the epidermal compartment of glycated RHS was 1.3 ± 0.18-fold thicker than the epidermis of non-glycated RHS. Moreover, glycated RHS appeared yellowish during the culture period (Figure 2).

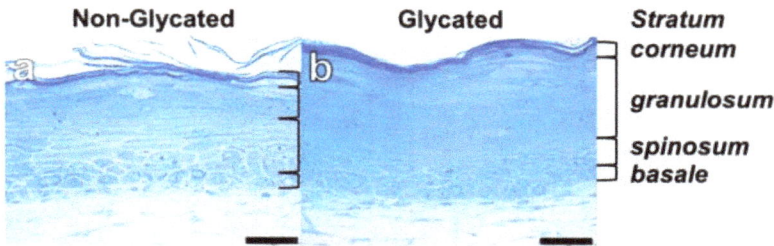

Figure 2. Reconstructed human skin (RHS) morphology. Semi-thin sections of (**a**) non-glycated and (**b**) glycated RHS. Light microscopy, bar = 50 μm.

2.3. Dermal Ultrastructure

Next, we analyzed the dermal morphology of glycated RHS and compared it with non-glycated RHS. The dermal compartment was divided into a *stratum papillare*, connected to the *stratum basale* of the epidermal compartment and the deeper *stratum reticulare*. Reflectance confocal microscopy (RCM) proved a homogenous dermal compartment in non-glycated RHS (Figure 3a) and collagen aggregates (white structures in Figure 3b) in the *stratum papillare* of glycated RHS. These findings were corroborated by TEM, where aggregates appeared as an accumulation of finest fibrils in the *stratum papillare* of glycated RHS. Those fibrils either orientated themselves longitudinally to the basal layer (mimicking a punctually developed basement membrane) or were cross-linked into densely organized ellipsoid structures, comprising an area of approximately 8.4 μm × 5.0 μm (Figure 3d). The fine fibrils within these structures were virtually uniformly parallel and appeared in a wave-like pattern (Figure 3f). Furthermore, areas with a locally higher density of collagen fibrils were identified in both *stratum papillare* and *stratum reticulare* of the glycated RHS. Fibrils of about 38 nm diameter appeared densely-packed while fibrils of about 160 nm were loosely packed. It is noteworthy that the collagen fibrils are aligned parallel to the *stratum basale* in the glycated RHS (Figure 3h,k). In contrast, filaments and collagen fibrils aligned randomly in both parallel and perpendicular directions to the *stratum basale* within *stratum reticulare* of non-glycated RHS (Figure 3g,i). Overall, the fibrils and filaments accumulated locally in the glycated RHS, while they were disseminated in the non-glycated RHS. Interestingly, fibroblasts within both RHS exhibited a spread shape and contained a large amount of rough endoplasmic reticulum. Moreover, on average, fibroblast numbers in the *stratum papillare* exceeded those in the *stratum reticulare* about two-fold in both RHS.

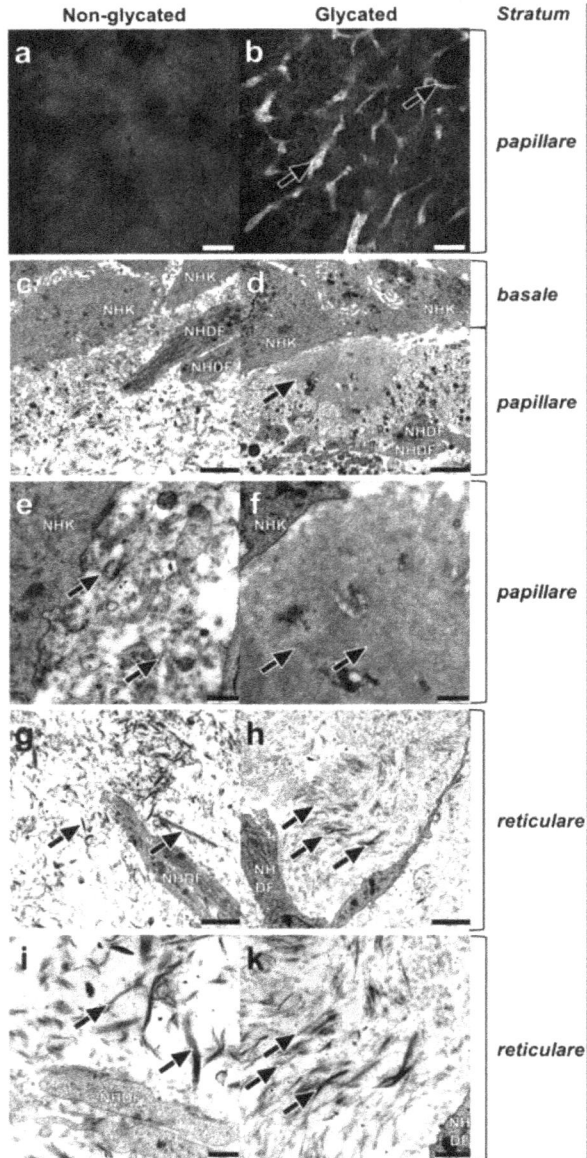

Figure 3. Dermal ultrastructure of (a,c,e,g,i) non-glycated and (b,d,f,h,k) glycated RHS. (a) Homogenous and (b) aggregated collagen in the *stratum papillare*. (c–f) Ultrastructure of the *stratum papillare*.

Fibroblasts (NHDF), cell fragments, and vesicles, as well as filaments and fibrils of various sizes bordering the keratinocytes (NHK) of the *stratum basale*. (**c,e**) Lower concentration of the structural elements in non-glycated dermal compartment filaments with ((**e**) magnification of (**c**)) fibrils of various sizes (arrows). (**d,f**) Local accumulations of thin, densely organized structures in glycated RHS, nearby the *stratum basale* (arrow) ((**f**) magnification of (**d**)), finest cross-linked filaments with a wave-like pattern constitute densely organized structures (arrows). (**g–k**) Ultrastructure of *stratum reticulare*. (**g,i**) Collagen fibers (arrows) show a disorientated pattern in all directions ((**i**) magnification of (**g**)). Collagen fibers with different sizes in non-glycated RHS (arrows). (**h,k**) Collagen fibrils axially orientated to the *stratum basale* ((**k**) magnification of (**h**)) the thinner the fibrils, the more densely they are packed. Densely packed fine fibrils and loosely packed thicker fibrils (arrows). (**a,b**) Reflectance confocal microscopy, bar = 100 µm; (**c,d,g,h**) transmission electron microscopy (TEM), bar = 2.5 µm; (**e,f,i,k**) TEM, bar = 1.0 µm.

2.4. Epidermal Ultrastructure

TEM analysis of the epidermal compartments revealed striking differences between glycated and non-glycated RHS, especially in the thickness and number of cell layers of the *stratum granulosum* (Figure 4c,d).

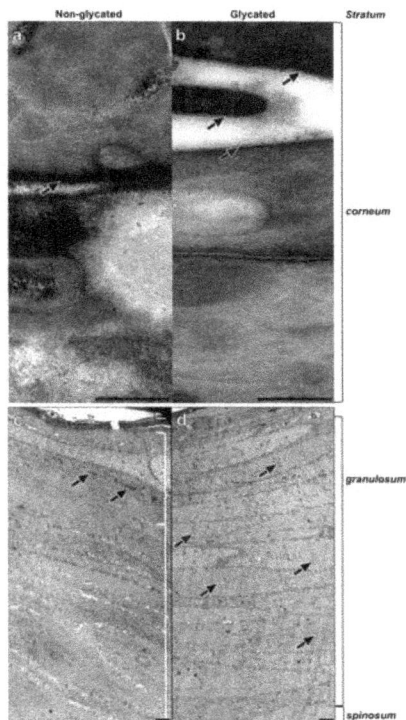

Figure 4. Epidermal ultrastructure of (**a,c**) non-glycated and (**b,d**) glycated RHS. (**a**) Lipid lamellae within the intercellular space of neighboring corneocytes. (**b**) Fine lines of the intercellular lipids are clearly visible (arrows). (**c**) Packed keratohyalin granule-like structures (arrows) in the upper layers of the viable epidermis, indicating the *stratum granulosum*. (**d**) The *stratum granulosum* dominates the epidermal compartment. Fine electron dense material evenly disseminated within the cytoplasm of the keratinocytes (arrows). Pictures are representative for two batches. (**a,b**). TEM, bar = 0.25 µm; (**c,d**) TEM, bar = 2.5 µm.

145

The *stratum spinosum* of non-glycated RHS contained seven layers on average, comprising a decidedly greater proportion of the viable epidermis than in glycated RHS. The *stratum granulosum* of non-glycated RHS constituted on average three layers of tightly packed cells with mitochondria and large, irregularly shaped keratohyalin granules (Figure 4c).

In the glycated RHS, the keratinocytes in the basal layer were columnar and cuboidal shaped. The *stratum spinosum* contained four layers on average, while more than 20 cell layers were designated as *stratum granulosum*, comprising the majority of keratinocytes within the epidermis of glycated RHS. These keratinocytes were increasingly flattened and densely packed. Atypical keratohyalin granules were present as fine, disseminated electron dense material. Although this material slightly differed from the keratohyalin granules in non-glycated RHS, we regarded this material as small keratohyalin granules. Moreover, the number of mitochondria and other cell organelles decreased, especially in the apical parts of the *stratum granulosum* (Figure 4d).

In both constructs, the keratinocytes transformed into flat, tightly packed corneocytes, attaching themselves to each other via corneodesmosomes. *Stratum corneum* of non-glycated RHS consisted of seven layers of corneocytes, which increased to twelve layers in glycated RHS. While a cornified envelope was developed in both types of RHS, the lipids in the glycated models were slightly clearer to visualize and distinguish from the plasma membrane than in non-glycated RHS (Figure 4a,b). This deviating ultrastructure points to different lipid organization in non-glycated and glycated RHS.

2.5. Molecular Analysis

The expressions of the adhesion molecule β1-integrin and laminin-5 slightly increased within the basal layer of the glycated RHS and faded in the suprabasal epidermal layers (Figure 5). In contrast, filaggrin expression markedly decreased and loricrin expression strongly increased in glycated compared with non-glycated RHS.

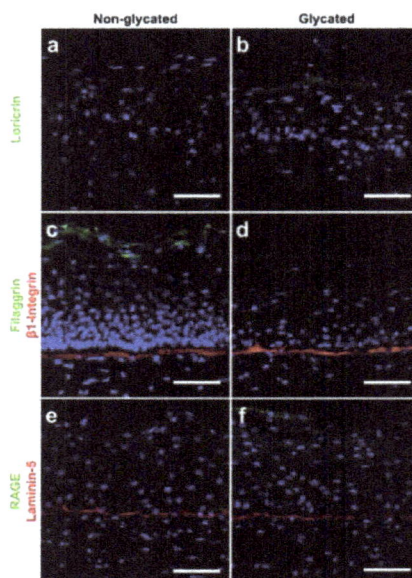

Figure 5. Immunolocalization of proteins in (**a,c,e**) non-glycated and (**b,d,f**) glycated RHS. Glycated RHS showed (**a,b**) loricrin upregulation, (**c,d**) filaggrin downregulation and slight increase in β1-integrin expression, and (**e,f**) slight upregulation of both receptor of AGEs (RAGE) and laminin-5 expression compared with non-glycated RHS. Proteins in green or red, cell nuclei (DAPI staining) in blue. Fluorescence microscopy, bar = 100 μm. Pictures are representative for two to four batches.

The arrangement of the intercellular lipids in the *stratum corneum* was probed by Fourier-transform infrared (FT-IR) spectroscopy. The lower wavenumbers of the methylene symmetric stretching vibrations (2851.8 ± 0.7 cm^{-1} in glycated compared with 2852.8 ± 1.0 cm^{-1} in non-glycated RHS, $p > 0.05$) indicated a higher proportion of all-trans chain conformers in the chain [14], and thus an improved lipid chain order in glycated RHS (Figure 6a). High performance thin layer chromatography of *stratum corneum* lipids revealed slight compositional differences. The *stratum corneum* lipids in the non-glycated RHS were composed of 7:5:1 FFA/Chol/Cer, while glycated RHS revealed a molar ratio of 5:3:1 (FFA/Chol/Cer). The absolute amounts of free fatty acids (FFA), cholesterol (Chol), and cholesterol sulphate (CholS) apparently decreased in glycated RHS ($p \le 0.05$), while the levels of the levels of ceramide precursors remained unchanged ($p > 0.05$). The respective values read as follows: FFA, 28.5 ± 21.0 and 20.9 ± 14.8 µg/mg *stratum corneum* in non-glycated and glycated RHS; Chol, 19.9 ± 8.6 and 16.9 ± 5.5 µg/mg; and CholS, 3.1 ± 1.4 and 1.8 ± 0.9 µg/mg ($p = 0.051$). The respective amounts of Cer precursors read sphingomyelin (SM) 8.3 ± 3.7 and 8.5 ± 3.7 µg/mg, glucosylceramides (GCer) 5.6 ± 0.8 and 7.2 ± 5.4 µg/mg, and phospholipids (PL) 9.9 ± 6.1 and 11.3 ± 5.1 µg/mg (Figure 6b). The amounts of ceramides (Cer) subclasses, Cer EO(d)S (0.39 ± 0.37 and 0.31 ± 0.15 µg/mg, $p > 0.05$), Cer NP (2.62 ± 0.71 and 2.69 ± 0.91 µg/mg, $p > 0.05$), Cer AS+NH (0.90 ± 0.46 and 0.98 ± 1.34 µg/mg, $p > 0.05$), and Cer AP+AH (0.68 ± 0.56 and 0.78 ± 1.05 µg/mg, $p > 0.05$) did not change. In contrast, the total amounts of Cer (8.7 ± 4.7 and 12.7 ± 7.9 µg/mg, $p > 0.05$) and, in particular, Cer N(d)S (4.06 ± 3.53 and 7.98 ± 7.18 µg/mg, $p > 0.05$) apparently increased in glycated RHS (Figure 6b,c).

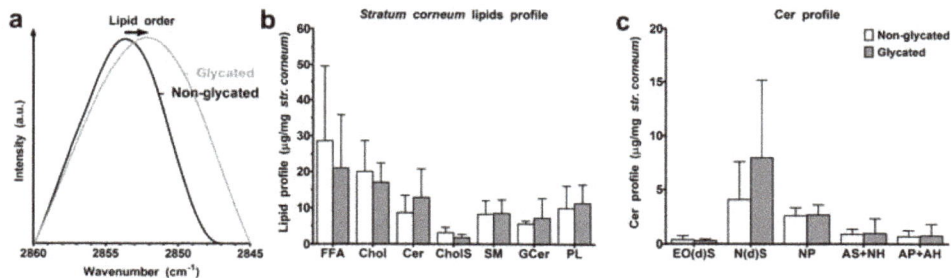

Figure 6. *Stratum corneum* lipids. (**a**) Lipid order. (**b**) *Stratum corneum* lipid profile of non-glycated and glycated RHS for free fatty acids (FFA), cholesterol (Chol), ceramides (Cer), cholesterol sulphate (CholS), sphingomyelin (SM), glucosylceramides (GCer), and phospholipids (PL). (**c**) Ceramide profile, for ceramide nomenclature, see Figure 8; $n = 3$–4, mean + SD.

2.6. Skin Barrier Function

Next, we studied the barrier function of glycated RHS. We found comparable caffeine permeation over 6 h (Figure 7a), with apparent permeability (P_{app}) values of 7.6 and 6.6×10^{-6} cm/s ($p > 0.05$) in non-glycated and glycated RHS, respectively. The lag-time increased from 26 min in non-glycated to 44 min in ribose-induced glycated RHS. However, this trend was not statistical significant.

Finally, we evaluated the delivery of the fluorescent drug surrogate Bodipy, either in base cream or loaded to dendritic core-multishell (CMS) nanotransporters into RHS (Figure 7b–g). Bodipy and CMS nanotransporters penetrated the *stratum corneum* of both glycated and non-glycated RHS. While Bodipy readily penetrated viable epidermal layers in the non-glycated RHS, this uptake appeared slightly reduced in glycated RHS (Figure 7c,f). However, Bodipy uptake was enhanced by CMS nanotransporters compared with cream in both non-glycated and glycated RHS (Figure 7b,c,e,f).

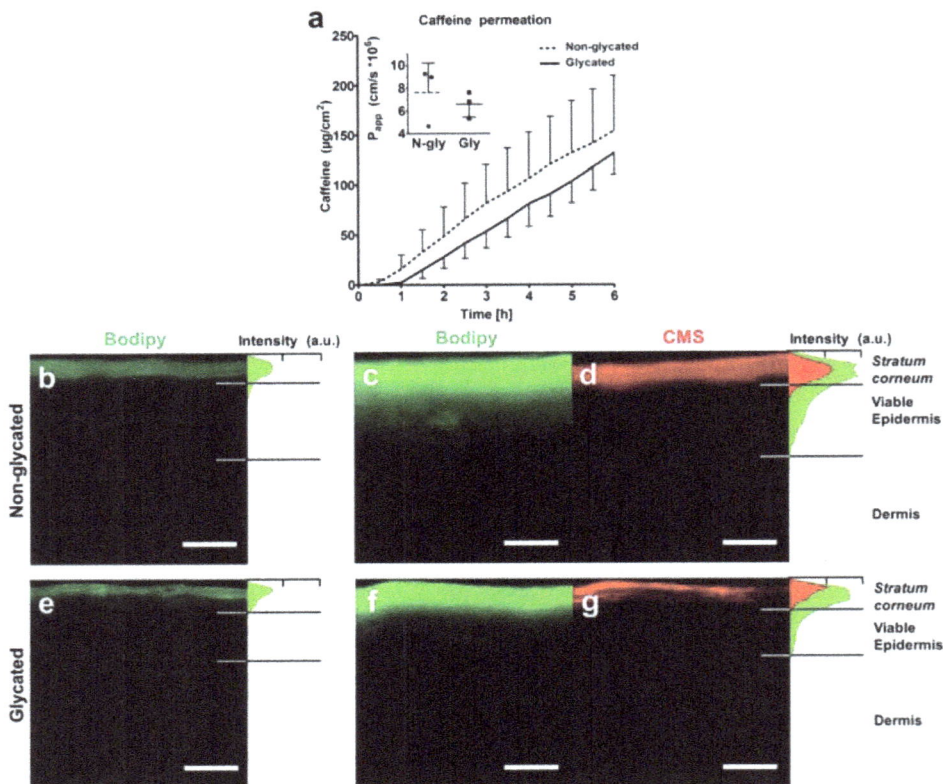

Figure 7. Skin barrier function. (**a**) Caffeine permeation through non-glycated and glycated RHS. The inset shows the respective P_{app} values. (**b**) Penetration of Bodipy cream and (**c,d**) Bodipy-core-multishell (CMS) into non-glycated RHS. (**e**) Penetration of Bodipy cream and (**f,g**) Bodipy-CMS into glycated RHS. (**c,d,f,g**) CMS labelled with red fluorescent indocarbocyanine. Fluorescence microscopy, bar = 100 μm, $n = 3$, mean ± SD.

3. Discussion

The complex and heterogeneous nature of skin aging challenges the relevant and reliable emulation of ageing hallmarks in vitro. However, RHS proved to be a powerful tool to mimic at least some important features of skin diseases like atopic dermatitis, psoriasis, and non-melanoma skin cancer [15–17]. In order to mimic aged skin, the ECM of juvenile RHS was exposed to either ribose glyoxal or specific AGEs. These artificial glycation stimuli are necessary to speed up the slow amino carbonylation by glucose. We selected ribose because of its biocompatibility and its glycation efficiency (Figure 1). In contrast to glyoxal, ribose did not affect the viability of skin cells. Moreover, ribose not only produces structurally similar AGEs [18], but also reacts faster than glucose [19]. Nevertheless, the effects of glycation appear to depend on the glycation agent; with ribose and glyoxal being the most frequently used compounds [8,9,20]. Granted that aged skin is sufficiently perfused in vivo, the oxidative environment should favor glycation by sugars [21].

The increased stiffness of the dermal ECM by collagen cross-linking depicts one hallmark of aged skin [22]. Ribose glycation reliably mimicked this morphology as reflectance confocal microscopy showed collagen aggregates and the ultrastructural analysis revealed the altered alignment and higher amounts of collagen fibers and fibrils in glycated RHS (Figure 3). Moreover, we proved the interdependency between fibril diameter and packing, as well as a hierarchy of fiber accumulation. While densely packed fibrils are smaller in diameter, loosely packed collagen fibrils showed larger diameters (Figure 3). The overall architecture of the dermal compartment in glycated RHS showed more or less axially orientated collagen fibrils. Cross-linking of the fibrils resulted in their dissociation to finer structures. Accumulations of increasingly finer cross-linked fibrils in densely packed structures resembled stiffly plaques within the skin. These changes in ECM also explain the lost dermal flexibility in diabetic patients [3] and are not limited to skin, but apply also to glycated vascular walls or tendons. In accordance, glycation in vivo increases fibril packing density, the range of fibril diameters, and the fusion of fibrils in tendons of diabetes patients [4].

The effects of dermal collagen glycation go beyond the mechanical alterations of dermal ECM. This is particular intriguing, as epidermal changes are induced either by traces of ribose (despite of the extensive dialysis to eliminate the glycation agent) or by AGEs or by altered paracrine signaling between fibroblasts and keratinocytes in glycated RHS. We argue for the latter, because the mechanical separation of glycated ECM and the epidermal compartments did not impair the effects of AGEs on keratinocytes [23]. Moreover, the constant paracrine signaling likely transduces stress signals from fibroblasts to keratinocytes [24].

In particular, the basement membrane appeared thicker in glycated RHS, although RHS in general lacks the sub-basal laminae of the basement membrane [25]. This alteration is well in accordance with the increased collagen-IV expression in glycated RHS [23], but the ultrastructural analysis identified longitudinal clusters of finest collagen fibrils to mimic a thick *lamina fibroreticularis* (fibroblast derived) of the basement membrane (Figure 3). Increased laminin-5 expression (Figure 4) nicely supports this hypothesis. As one of the principal structural elements of the basement membrane, laminin establishes networks with other proteins like collagens and determines the basement membranes common structure [26]. It is tempting to speculate that these structures emulate the age-specific reduplication of the lamina densa and its associated anchoring fibril complex as observed in vivo [27].

Moreover, we observed enhanced keratinocyte proliferation in glycated RHS, indicated by the increase in total epidermal thickness (Figures 2 and 4). In our study, more than 20 cell layers corresponded to the *stratum granulosum*, containing disseminated electron dense material, which we interpreted as atypical keratohyalin granules. Absent keratohyalin and increased *stratum granulosum* height were described in particular types of psoriatic epidermis [28,29]. The increased thickness of the *stratum granulosum* indicates defective differentiation of the transitional located keratinocytes of the *stratum granulosum* and thus resembles parakeratotic psoriatic skin. The increased expression of loricrin was well in accordance with the abundant fine granula in the epidermal compartment of glycated RHS (Figure 5) [30]. In fact, keratinocytes with atypical keratohyalin granules dominated the epidermal compartment as "*stratum intermedium*" of glycated RHS. Furthermore, we observed changes in the intercellular space of the *stratum corneum* (Figure 4), which intrigued us to investigate the *stratum corneum* lipid order and profile.

We found an apparent, although not statistically significant, trend to increased Cer amounts in glycated RHS, being accompanied by the significantly improved lipid order (Figure 6). The apparent increase in Cer levels without a change in the SM and GCer concentrations suggests enhanced Cer synthesis de novo, with similar activities of sphingomyelinase and β-glucocerebrosidase. The improved chain order of barrier lipids would explain the trend to reduced permeability of glycated RHS as the membranes with fully stretched all-trans lipid chains would provide a stronger barrier to permeating substances than that with more fluid mobile lipids. This was applied to small molecules like caffeine, a recommended test compound for percutaneous absorption experiments by the Organisation for Economic Co-operation and Development, as well as to Bodipy 493/503, which shares the physicochemical properties of glucocorticoids. As with non-glycated RHS, macromolecules like CMS nanotransporters do not surmount the *stratum corneum* (Figure 7). This is well in accordance with previous models with intact skin barrier and suggests a penetration enhancing effect of CMS nanotransporters beyond simply carrying their cargo [31].

Our results are well in accordance with previous ribose-based approaches, but are not in line with glyoxal induced effects [6,32]. Moreover, we found a deviating *stratum corneum* lipid profile from glycated reconstructed human epidermis [7]. These changes might be attributed to the glycation method or the fact that reconstructed human epidermis lack the cross-talk between fibroblasts and keratinocytes. Even glyoxal impaired the skin barrier function in excised murine skin only after 72 h of direct exposure [7]. Culturing reconstructed human epidermis on a basement membrane surrogate, glycated by glyceraldehyde, resulted in decreased filaggrin expression and skin barrier function [10]. Although we found similar changes in the protein expressions, we could not confirm an impaired barrier function due to collagen glycation. In fact, neither caffeine (M_r 194; $\log P$ 0.08), Bodipy (M_r 262.1; $\log P$ 3.5), nor CMS nanotransporter (M_r 74,000) were absorbed better into glycated RHS compared with non-glycated RHS. As fluorescence readings provide only a first insight, further studies on the trans-epidermal resistance of glycated human skin ex vivo might explain these deviating results. Taking into account the cytotoxic effects of aldehydes on keratinocytes, the constructs in both aldehyde-based studies might overestimate the impaired barrier function due to glycation.

Glycation induced aged or diabetic phenotypes in the dermis, while the glycated dermal ECM did not cause these phenotypes in the epidermal compartment of RHS. The increased epidermal thickness with incomplete differentiation resembles a psoriasis-like epidermal morphology. Correlating to increased granulocyte-macrophage colony-stimulating factor and interleukin-6 expression, the reduced filaggrin expression hinted towards an inflammatory RHS phenotype [33,34]. Moreover, keratinocytes in the level of *stratum granulosum* also lacked normal keratohyalin granules and pointed to a disturbed keratinization process with defective tonofibrillar differentiation—similar to psoriatic skin [35]. Nevertheless, glycation alone would not be suitable to emulate psoriatic skin, because neither immune cells nor Th-17 cytokines were added to the culture. Moreover, glycated RHS did not show the impaired barrier function of psoriatic skin [36]. Nevertheless, a clinical study proved increased amounts of AGEs in patients with severe psoriasis [3], and AGEs appeared to fuel cutaneous inflammation [37].

In conclusion, our study adds several missing links to the effects of AGEs on the cutaneous morphology and skin barrier function. Based on our results, glycation of the ECM markedly affects epithelial differentiation, and thus should be considered when emulating (skin) diseases like diabetes, psoriasis, and skin cancer. The better understanding of the interactions between matrix and cells will help to improve the pharmacological treatment of these diseases. Finally, the ultrastructural analysis proved indispensable, especially for matrix effects, and should be used complementarily with molecular analyses in the characterization of human cell-based models.

4. Materials and Methods

4.1. Collagen Glycation

Collagen was glycated as described previously [9]. In brief, collagen G (4 mg bovine collagen per mL; Biochrom, Berlin, Germany) was incubated with 10 mM D-ribose (Sigma Aldrich, München, Germany) in phosphate-buffered saline (PBS) at room temperature for three weeks and protected against light. Moreover, collagen G was incubated with PBS only as negative control and with 10 mM glyoxal (Sigma Aldrich, München, Germany) for comparison. Subsequently, glycated collagen was extensively dialyzed in sterile water at 4 °C for 48 h, using Spectra/Por® 4 Dry Standard RC Dialysis Tubing, 12–14 kD molecular weight cut-off, 32 mm flat width, 100 ft length (132703; Spectrum, Henderson, NV, USA). The sterile water was changed every 12 h. Following the dialysis, we subjected the samples to fluorescence analysis (Hitachi F-4500; Hitachi, Tokyo, Japan) to detect AGEs. Therefore, 1.5 mL glycated collagen was digested by pepsin (1 mg pepsin in 5 mL of 0.5 M acetic acid) at 37 °C overnight. The solution was centrifuged at 10,000× *g* for 5 min, the supernatant, containing the digested collagen, was measured at λ_{ex} 370 nm/λ_{em} 440 nm for total AGEs [9].

4.2. Cell Culture and Viability

Cell viability was assessed by 3-(4,5-dimethylthiazol-2-yl)-2,5 diphenyltetrazolium bromide (MTT; Sigma Aldrich, München, Germany) reduction as previously described [38]. In brief, NHK (passage 3) and NHDF (passage 3) were isolated from therapeutically indicated circumcision with permission (ethical approval EA1/081/13). Cell culture was performed according to standard operating procedures, and referring to good cell culture practice. We used 0.005% [w/v] sodium dodecyl sulphate (SDS) as positive control and 10% [v/v] double distilled, sterile water as solvent control.

4.3. Reconstructed Human Skin (RHS)

RHS was built in three steps [15]. First, 1 mL acellular collagen G or a 1:1 mixture of collagen G and glycated collagen G was poured into a cell-culture insert (Corning, Corning, NY, USA) and incubated for 2 h to form a gel. Second, fibroblasts (0.8×10^6 cells per construct) were added to 3 mL collagen G or a 1:1 mixture of collagen G and glycated collagen G, poured onto the acellular collagen, and incubated for 24 h. The dermal equivalents were submerse cultured for seven days in construct growth medium (CGM: Dulbecco's modified eagle's medium /F12 + GlutaMax supplemented with 10% fetal calf serum, 1% Pen/Strep, adenine HCl monohydrate 40 µM, amphotericin B 30 µg/L, choleratoxin 0.1 nM, epidermal growth factor 10 µg/L, hydrocortisone 3.5 mg/L, insulin 4.4 mg/L, non-essential amino acids 0.5%, transferrin 4.4 mg/L, triiodothyronine 2 nM). Third, NHK (3.0×10^6 per construct) were seeded onto the dermal compartment at day 8. RHS were raised to the air–liquid interface at day 10 and cultured in construct differentiation medium (CDM: CGM supplemented with ascorbic acid 0.25 mM, calcium chloride 2 mM). The culture medium was changed three times a week. At day 21, the constructs were subjected to further analyses as described below.

4.4. Reflectance Confocal Microscopy (RCM) and Immunolocalization of Proteins

RCM was performed with VivaScope® 3000 (MAVIG; München, Germany). Subsequently, the constructs were snap frozen, sectioned into 7 µm slices (Leica CM 1510S; Wetzlar, Germany), and analyzed by immunofluorescence staining [15]. Antibodies against the following antigens were purchased from Abcam (Cambridge, UK): filaggrin (1:1000; ab81468), carboxymethyl lysine (1:100; ab27684), human integrin beta 1 (1:100; ab3167); and from Thermo Fisher Scientific (Darmstadt, Germany): loricrin (1:500; PA5-34945). Immunofluorescence staining was performed according to standard protocols. Pictures were taken with a fluorescence microscope (BZ-8000, Keyence; Keyence, Neu-Isenburg, Germany) and analyzed with BZAnalyzer and ImageJ software. Five individual measurements per construct were performed observer-blinded (RBR, CH, CW, CZ).

4.5. Transmission Electron Microscopy (TEM)

RHS was fixed in Karnovsky solution (7.5% glutaraldehyde and 3% paraformaldehyde in phosphate buffered saline) for 4 h, then washed in 0.1 M cacodylate buffer (cacodylic acid sodium salt trihydrate; Roth, Karlsruhe, Germany) and incubated in 1% osmium tetroxide (Chempur; Karlsruhe, Germany) for 120 min. Dehydration was performed in an ascending series of ethanol. Finally, the models were washed in the intermedium propylene oxide (1,2 Epoxypropan; VWR, Darmstadt, Germany).

Embedding of RHS was performed in a mixture of agar 100 (epoxy resin), DDSA (softener), Epoxy embedding medium (hardener), and DMP 30 (catalyst) (all: Agar Scientific, Stansted, UK), followed by polymerization at 45 °C and 55 °C, each for 24 h. Semi- and ultrathin sections were cut using an ultra-microtome Reichert Ultracut S (Leica, Wetzlar, Germany) and the semi-thin sections (0.5 µm) were stained with modified Richardson solution [39] for 45 s at 80 °C.

Semi-thin sections were examined under light microscopy Olympus CX21 (Olympus, Stuttgart, Germany). Next, ultrathin (80 nm) sections were mounted on nickel-grids (Agar Scientific). The RHS morphology was assessed with an electron microscope (ZeissEM109, Oberkochen, Germany). Images were taken and processed using Adobe® Photoshop® (Adobe® Systems, San José, CA, USA).

4.6. Carboxymethyl Lysine Quantification

RHS was separated into epidermal and dermal compartments and dissected. Thereafter, constructs were transferred to tubes containing 300 µL (epidermis) or 700 µL (dermis) PBS and a 5 mm stainless steel bead and immediately disintegrated using Tissue Lyser II (Qiagen, Hilden, Germany) at 25 Hz for 5 min. The lysed suspensions were analyzed for *N*-carboxymethyl-lysine content by OxiSelect™ ELISA (Cell Biolabs; San Diego, CA, USA) according to the manufacturers' instructions.

4.7. Stratum Corneum Lipid Analysis

The *stratum corneum* of RHS was isolated, and subjected to infrared spectroscopy using a Nicolet 6700 spectrometer (Thermo Fisher Scientific, Waltham, MA, USA) equipped with a single-reflection MIRacle attenuated total reflectance ZnSe crystal (PIKE technologies, Madison, WI, USA) at 23 °C. The spectra were generated by co-addition of 256 scans collected at 4 cm^{-1} resolution and analyzed with the Bruker OPUS software. The exact peak positions were determined from second derivative spectra. The lipids were then extracted and analyzed by high performance thin layer chromatography (HPTLC; [40]). Briefly, the analysis was performed on silica gel 60 HPTLC plates (20 × 10 cm, Merck, Darmstadt, Germany). Cer subclasses, FFA, and Chol were separated using $CHCl_3$/MeOH/acetic acid 190:9:1.5 ($v/v/v$) mobile phase twice to the top of the plate, whereas GCer, SM, PL, and CholS were separated using $CHCl_3$/MeOH/acetic acid/H_2O 65:25:6:3. The lipids were visualized by dipping in a derivatization reagent (7.5% $CuSO_4$, 8% H_3PO_4, 10% MeOH in water) for 10 s, heating at 160 °C for 30 min, and quantitated by densitometry using TLC scanner 3 and WinCats software (Camag, Muttenz, Switzerland). For the used standard lipids and their calibration curve ranges, see Table 1; for ceramide nomenclature, see Figure 8. Cer and GCer were purchased from Avanti Polar Lipids (Alabaster, AL, USA). All other chemicals, including lignoceric acid, Chol, CholS, SM, and PL, were purchased from Sigma Aldrich (München, Germany). Non-commercially available Cer EOS and NH standards were synthesized according to previously published methods [41,42].

Table 1. Calibration curve ranges of lipid standards used for high performance thin layer chromatography (HPTLC) analysis. FFA—free fatty acid.

Lipid Standard		Calibration Curve Range [µg]
FFA	Lignoceric acid	0.5–12.5
Chol	Cholesterol	0.5–12.5
Cer	Ceramide EOS	0.05–1.25
	Ceramide NS	0.2–5
	Ceramide NP	0.4–9.5
	Ceramide AS	0.1–2.5
	Ceramide NH	0.15–4
	Ceramide AP	0.1–2.5
CholS	Cholesterol sulphate	0.105–2.474
SM	Sphingomyelin	0.211–4.947
GCer	Glucosylceramide	0.263–6.184
PL	Phospholipid	0.421–9.895

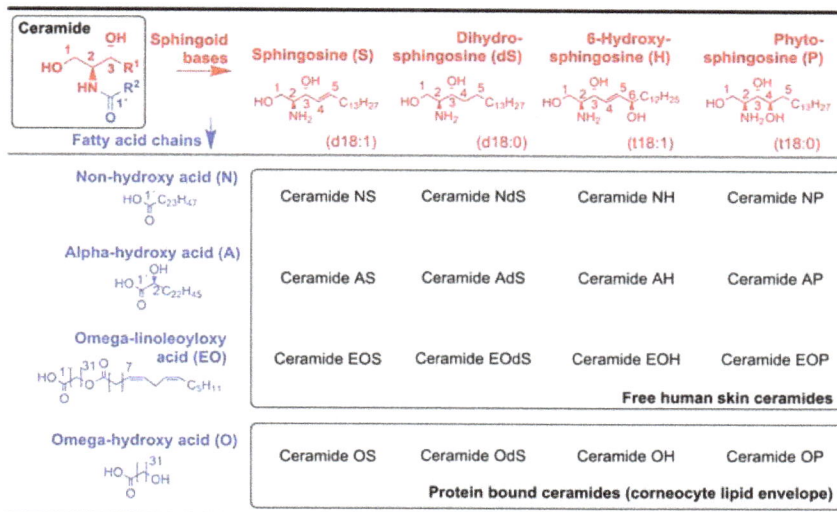

Figure 8. Ceramide nomenclature. Reprinted from the literature [15], Copyright 2016, with permission from Elsevier.

4.8. Skin Permeation

The permeation of radiolabeled [1-methyl^{14}C]caffeine (M_r 194; log$P$ 0.08; PerkinElmer, Waltham, MA, USA) was studied according to the validation study on the use of skin models for skin absorption testing [39]. In brief, tissue integrity of RHS was evaluated prior to mounting RHS in static Franz cells ($\varnothing$ = 15 mm, with receptor chamber volume = 12 cm^3; PermeGear, Bethlehem, PA, USA). The stratum corneum was placed facing the air and the dermal compartment of RHS was in contact with the supporting membrane and the receptor medium PBS. The receptor medium was kept at a constant temperature of 33.5 $\pm$ 0.5 °C using a temperature-controlled water jacket, and stirred with a magnetic bar at 500 rpm. The absence of air bubbles was monitored throughout the experiment. Then, 281.4 µg/cm^2 caffeine was applied in aqueous solution onto the RHS surface. The opening of the Franz cell was covered by Parafilm® and the caffeine permeation was quantified from the receptor medium by radiochemical detection (Hidex 300 SL liquid scintillation counter; HIDEX, Turku, Finland).

4.9. Penetration of Bodipy Loaded Dendritic Core-Multishell Nanotransporters (Bodipy-CMS)

Bodipy 493/503 (λ_{ex} 493 nm; λ_{em} 503, M_r 262.1, log$P$ 3.5, Thermo Fisher Scientific; Darmstadt, Germany) was dispersed in base cream (13.6 mg/L Bodipy; 40.0% purified water, 25.5% white vaseline; 10.0% propylenglycol; 7.5% medium chain triglycerides; 7.0% macrogol-1000-glycerol monostearate; 6.0% cetyl alcohol; and 4.0% glycerol monostearate 60). Additionally, CMS nanotransporters [43] were labelled with indocarbocyanine and loaded with Bodipy (13.6 mg/L). Then, 30 µL/cm^2 of test formulations was applied onto the RHS surface and remained there for 6 h according to a previously published protocol [31]. Pictures were taken with a fluorescence microscope (BZ-8000, Keyence; Keyence, Neu-Isenburg, Germany) and analyzed with ImageJ software. Five individual measurements per construct were performed observer-blinded (RBR, CW, CZ).

4.10. Statistical Analysis

Data are presented as the mean $\pm$ SD from two to four independent experiments. The Mann–Whitney test or Kruskal–Wallis test with Dunn's Post hoc test was used to indicate statistically significant differences ($p \leq 0.05$).

Author Contributions: Conceptualization, R.B.R. and C.Z.; Methodology, R.B.R. S.K., and C.Z.; Formal Analysis, R.B.R., S.K., K.V., M.U., and C.Z.; Investigation, R.B.R., S.K., C.W., C.H., N.Z., M.S., A.K., and M.U.; Resources, R.H. and M.S.-K.; Writing—Original Draft Preparation, R.B.R., S.K., and C.Z.; Writing—Review & Editing, R.B.R., S.K., M.S.-K., and C.Z.; Visualization, R.B.R., S.K., C.W., and C.Z.; Supervision, C.Z.; Project Administration, C.Z.; Funding Acquisition, R.B.R., C.H., K.V., M.S.-K., and C.Z.

Funding: This research was funded by the Bundesministerium für Bildung und Forschung (Berlin-Brandenburg research platform BB3R, 031A262A). R.B.R. thanks the Fundação de Amparo à Pesquisa do Estado de São Paulo (FAPESP, 2016/02195-1) for the international fellowship. C.H. gratefully acknowledges SFB 1112 and Elsa-Neumann doctoral scholarships. K.V., M.S., and A.K. were financially supported by Czech Science Foundation (16-25687J) and EFSA-CDN (No. CZ.02.1.01/0.0/0.0/16_019/0000841 co-funded by ERDF). Open Access funding provided by the Freie Universität Berlin.

Acknowledgments: We thank Verena Eckert-Funke, Franziska Ermisch, and Carola Kapfer for excellent technical assistance.

Conflicts of Interest: The authors declare no conflict of interest.

Abbreviations

AGEs	Advanced Glycated End-products
CML	N-CarboxyMethyl-Lysine
CMS	Dendritic Core-Multishell
DAPI	4′,6-DiAmidino-2-PhenylIndole
ECM	ExtraCellular Matrix
HPTLC	High-Performance Thin-Layer Chromatography
NHDF	Normal Human Dermal Fibroblast
NHK	Normal Human Keratinocytes
PBS	Phosphate-Buffered Saline
RAGE	Receptor of AGEs
RHS	Reconstructed Human Skin
SDS	Sodium Dodecyl Sulphate
TEM	Transmission Electron Microscopy

References

1. López-Otín, C.; Blasco, M.A.; Partridge, L.; Serrano, M.; Kroemer, G. The hallmarks of aging. *Cell* **2013**, *153*, 1194–1217. [CrossRef] [PubMed]
2. Reeve, E.; Trenaman, S.C.; Rockwood, K.; Hilmer, S.N. Pharmacokinetic and pharmacodynamic alterations in older people with dementia. *Expert Opin. Drug Metab. Toxicol.* **2017**, *13*, 651–668. [CrossRef] [PubMed]
3. Papagrigoraki, A.; Del Giglio, M.; Cosma, C.; Maurelli, M.; Girolomoni, G.; Lapolla, A. Advanced Glycation End Products are Increased in the Skin and Blood of Patients with Severe Psoriasis. *Acta Derm.-Venereol.* **2017**, *97*, 782–787. [CrossRef] [PubMed]
4. Lima, A.L.; Illing, T.; Schliemann, S.; Elsner, P. Cutaneous Manifestations of Diabetes Mellitus: A Review. *Am. J. Clin. Dermatol.* **2017**, *18*, 541–553. [CrossRef] [PubMed]
5. Verzijl, N.; DeGroot, J.; Thorpe, S.R.; Bank, R.A.; Shaw, J.N.; Lyons, T.J.; Bijlsma, J.W.; Lafeber, F.P.; Baynes, J.W.; TeKoppele, J.M. Effect of collagen turnover on the accumulation of advanced glycation end products. *J. Biol. Chem.* **2000**, *275*, 39027–39031. [CrossRef] [PubMed]
6. Pageon, H.; Zucchi, H.; Dai, Z.; Sell, D.R.; Strauch, C.M.; Monnier, V.M.; Asselineau, D. Biological Effects Induced by Specific Advanced Glycation End Products in the Reconstructed Skin Model of Aging. *BioRes. Open Access* **2015**, *4*, 54–64. [CrossRef] [PubMed]
7. Yokota, M.; Tokudome, Y. The Effect of Glycation on Epidermal Lipid Content, Its Metabolism and Change in Barrier Function. *Skin Pharmacol. Physiol.* **2016**, *29*, 231–242. [CrossRef] [PubMed]
8. Cadau, S.; Leoty-Okombi, S.; Pain, S.; Bechetoille, N.; Andre-Frei, V.; Berthod, F. In vitro glycation of an endothelialized and innervated tissue-engineered skin to screen anti-AGE molecules. *Biomaterials* **2015**, *51*, 216–225. [CrossRef] [PubMed]

9. Pennacchi, P.C.; de Almeida, M.E.; Gomes, O.L.; Faiao-Flores, F.; de Araujo Crepaldi, M.C.; Dos Santos, M.F.; de Moraes Barros, S.B.; Maria-Engler, S.S. Glycated Reconstructed Human Skin as a Platform to Study the Pathogenesis of Skin Aging. *Tissue Eng. Part A* **2015**, *21*, 2417–2425. [CrossRef] [PubMed]
10. Morimoto, H.; Gu, L.; Zeng, H.; Maeda, K. Amino Carbonylation of Epidermal Basement Membrane Inhibits Epidermal Cell Function and Is Suppressed by Methylparaben. *Cosmetics* **2017**, *4*, 38. [CrossRef]
11. Käßmeyer, S.; Sehl, J.; Khiao In, M.; Merle, R.; Richardson, K.; Plendl, J. Subcellular interactions during vascular morphogenesis in 3D cocultures between endothelial cells and fibroblasts. *Int. J. Mol. Sci.* **2017**, *18*, 2590. [CrossRef] [PubMed]
12. Longo, C.; Casari, A.; De Pace, B.; Simonazzi, S.; Mazzaglia, G.; Pellacani, G. Proposal for an in vivo histopathologic scoring system for skin aging by means of confocal microscopy. *Skin Res. Technol.* **2013**, *19*, e167–e173. [CrossRef] [PubMed]
13. Ulrich, M.; Lange-Asschenfeldt, S.; Gonzalez, S. Clinical applicability of in vivo reflectance confocal microscopy in dermatology. *G. Ital. Dermatol. Venereol.* **2012**, *147*, 171–178. [PubMed]
14. Lewis, R.N.; McElhaney, R.N. Fourier transform infrared spectroscopy in the study of lipid phase transitions in model and biological membranes: Practical considerations. *Methods Mol. Biol.* **2007**, *400*, 207–226. [CrossRef] [PubMed]
15. Zoschke, C.; Ulrich, M.; Sochorova, M.; Wolff, C.; Vavrova, K.; Ma, N.; Ulrich, C.; Brandner, J.M.; Schafer-Korting, M. The barrier function of organotypic non-melanoma skin cancer models. *J. Control. Release* **2016**, *233*, 10–18. [CrossRef] [PubMed]
16. van den Bogaard, E.H.; Tjabringa, G.S.; Joosten, I.; Vonk-Bergers, M.; van Rijssen, E.; Tijssen, H.J.; Erkens, M.; Schalkwijk, J.; Koenen, H.J. Crosstalk between keratinocytes and T cells in a 3D microenvironment: A model to study inflammatory skin diseases. *J. Investig. Dermatol.* **2014**, *134*, 719–727. [CrossRef] [PubMed]
17. Danso, M.O.; van Drongelen, V.; Mulder, A.; van Esch, J.; Scott, H.; van Smeden, J.; El Ghalbzouri, A.; Bouwstra, J.A. TNF-alpha and Th2 cytokines induce atopic dermatitis-like features on epidermal differentiation proteins and stratum corneum lipids in human skin equivalents. *J. Investig. Dermatol.* **2014**, *134*, 1941–1950. [CrossRef] [PubMed]
18. Khalifah, R.G.; Todd, P.; Booth, A.A.; Yang, S.X.; Mott, J.D.; Hudson, B.G. Kinetics of nonenzymatic glycation of ribonuclease A leading to advanced glycation end products. Paradoxical inhibition by ribose leads to facile isolation of protein intermediate for rapid post-Amadori studies. *Biochemistry* **1996**, *35*, 4645–4654. [CrossRef] [PubMed]
19. Luers, L.; Rysiewski, K.; Dumpitak, C.; Birkmann, E. Kinetics of advanced glycation end products formation on bovine serum albumin with various reducing sugars and dicarbonyl compounds in equimolar ratios. *Rejuv. Res.* **2012**, *15*, 201–205. [CrossRef] [PubMed]
20. Pageon, H.; Techer, M.P.; Asselineau, D. Reconstructed skin modified by glycation of the dermal equivalent as a model for skin aging and its potential use to evaluate anti-glycation molecules. *Exp. Gerontol.* **2008**, *43*, 584–588. [CrossRef] [PubMed]
21. Brings, S.; Fleming, T.; Freichel, M.; Muckenthaler, M.; Herzig, S.; Nawroth, P. Dicarbonyls and Advanced Glycation End-Products in the Development of Diabetic Complications and Targets for Intervention. *Int. J. Mol. Sci.* **2017**, *18*, 984. [CrossRef] [PubMed]
22. Rinnerthaler, M.; Streubel, M.K.; Bischof, J.; Richter, K. Skin aging, gene expression and calcium. *Exp. Gerontol.* **2015**, *68*, 59–65. [CrossRef] [PubMed]
23. Pageon, H.; Bakala, H.; Monnier, V.M.; Asselineau, D. Collagen glycation triggers the formation of aged skin in vitro. *Eur. J. Dermatol.* **2007**, *17*, 12–20. [CrossRef] [PubMed]
24. Maas-Szabowski, N.; Szabowski, A.; Stark, H.J.; Andrecht, S.; Kolbus, A.; Schorpp-Kistner, M.; Angel, P.; Fusenig, N.E. Organotypic cocultures with genetically modified mouse fibroblasts as a tool to dissect molecular mechanisms regulating keratinocyte growth and differentiation. *J. Investig. Dermatol.* **2001**, *116*, 816–820. [CrossRef] [PubMed]
25. Löwa, A.; Vogt, A.; Kaessmeyer, S.; Hedtrich, S. Generation of full-thickness skin equivalents using hair follicle-derived primary human keratinocytes and fibroblasts. *J. Tissue Eng. Regener. Med.* **2018**, *12*, e2134–e2146. [CrossRef] [PubMed]
26. Breitkreutz, D.; Koxholt, I.; Thiemann, K.; Nischt, R. Skin basement membrane: The foundation of epidermal integrity—BM functions and diverse roles of bridging molecules nidogen and perlecan. *BioMed Res. Int.* **2013**, *2013*, 179784. [CrossRef] [PubMed]

27. Candiello, J.; Cole, G.J.; Halfter, W. Age-dependent changes in the structure, composition and biophysical properties of a human basement membrane. *Matrix Biol.* **2010**, *29*, 402–410. [CrossRef] [PubMed]

28. Wolberink, E.A.; van Erp, P.E.; Teussink, M.M.; van de Kerkhof, P.C.; Gerritsen, M.J. Cellular features of psoriatic skin: Imaging and quantification using in vivo reflectance confocal microscopy. *Cytom. Part B Clin. Cytom.* **2011**, *80*, 141–149. [CrossRef] [PubMed]

29. Brody, I. The ultrastructure of the epidermis in psoriasis vulgaris as revealed by electron microscopy. 1. The dermo-epidermal junction and the stratum basale in parakeratosis without keratohyalin. *J. Ultrastruct. Res.* **1962**, *6*, 304–323. [CrossRef]

30. Steven, A.C.; Bisher, M.E.; Roop, D.R.; Steinert, P.M. Biosynthetic pathways of filaggrin and loricrin—Two major proteins expressed by terminally differentiated epidermal keratinocytes. *J. Struct. Biol.* **1990**, *104*, 150–162. [CrossRef]

31. Alnasif, N.; Zoschke, C.; Fleige, E.; Brodwolf, R.; Boreham, A.; Rühl, E.; Eckl, K.M.; Merk, H.F.; Hennies, H.C.; Alexiev, U.; et al. Penetration of normal, damaged and diseased skin—An in vitro study on dendritic core-multishell nanotransporters. *J. Control. Release* **2014**, *185c*, 45–50. [CrossRef] [PubMed]

32. Pageon, H.; Asselineau, D. An in vitro approach to the chronological aging of skin by glycation of the collagen: The biological effect of glycation on the reconstructed skin model. *Ann. N. Y. Acad. Sci.* **2005**, *1043*, 529–532. [CrossRef] [PubMed]

33. Turksen, K.; Kupper, T.; Degenstein, L.; Williams, I.; Fuchs, E. Interleukin 6: Insights to its function in skin by overexpression in transgenic mice. *Proc. Natl. Acad. Sci. USA* **1992**, *89*, 5068–5072. [CrossRef] [PubMed]

34. Sakai, T.; Hatano, Y.; Zhang, W.; Fujiwara, S.; Nishiyori, R. Knockdown of either filaggrin or loricrin increases the productions of interleukin (IL)-1alpha, IL-8, IL-18 and granulocyte macrophage colony-stimulating factor in stratified human keratinocytes. *J. Dermatol. Sci.* **2015**, *80*, 158–160. [CrossRef] [PubMed]

35. Brody, I. The ultrastructure of the epidermis in psoriasis vulgaris as revealed by electron microscopy: 4. Stratum corneum in parakeratosis without keretohyalin. *J. Ultrastruct. Res.* **1962**, *6*, 354–367. [CrossRef]

36. Gould, A.R.; Sharp, P.J.; Smith, D.R.; Steginik, A.J.; Chase, C.J.; Kovacs, J.C.; Penglis, S.; Chatterton, B.E.; Bunn, C.L. Increased permeability of psoriatic skin to the protein, plasminogen activator inhibitor 2. *Arch. Dermatol. Res.* **2003**, *295*, 249–254. [CrossRef] [PubMed]

37. Papagrigoraki, A.; Maurelli, M.; del Giglio, M.; Gisondi, P.; Girolomoni, G. Advanced Glycation End Products in the Pathogenesis of Psoriasis. *Int. J. Mol. Sci.* **2017**, *18*, 2471. [CrossRef] [PubMed]

38. Gerecke, C.; Edlich, A.; Giulbudagian, M.; Schumacher, F.; Zhang, N.; Said, A.; Yealland, G.; Lohan, S.B.; Neumann, F.; Meinke, M.C.; et al. Biocompatibility and characterization of polyglycerol-based thermoresponsive nanogels designed as novel drug-delivery systems and their intracellular localization in keratinocytes. *Nanotoxicology* **2017**, *11*, 267–277. [CrossRef] [PubMed]

39. Richardson, K.C.; Jarett, L.; Finke, E.H. Embedding in epoxy resins for ultrathin sectioning in electron microscopy. *Stain Technol.* **1960**, *35*, 313–323. [CrossRef] [PubMed]

40. Vavrova, K.; Henkes, D.; Struver, K.; Sochorova, M.; Skolova, B.; Witting, M.Y.; Friess, W.; Schreml, S.; Meier, R.J.; Schafer-Korting, M.; et al. Filaggrin deficiency leads to impaired lipid profile and altered acidification pathways in a 3D skin construct. *J. Investig. Dermatol.* **2014**, *134*, 746–753. [CrossRef] [PubMed]

41. Opálka, L.; Kováčik, A.; Sochorová, M.; Roh, J.; Kuneš, J.; Lenčo, J.; Vávrová, K. Scalable synthesis of human ultralong chain ceramides. *Org. Lett.* **2015**, *17*, 5456–5459. [CrossRef] [PubMed]

42. Kováčik, A.; Opálka, L.; Šilarová, M.; Roh, J.; Vávrová, K. Synthesis of 6-hydroxyceramide using ruthenium-catalyzed hydrosilylation-protodesilylation. Unexpected formation of a long periodicity lamellar phase in skin lipid membranes. *RSC Adv.* **2016**, *6*, 73343–73350. [CrossRef]

43. Radowski, M.R.; Shukla, A.; von Berlepsch, H.; Bottcher, C.; Pickaert, G.; Rehage, H.; Haag, R. Supramolecular aggregates of dendritic multishell architectures as universal nanocarriers. *Angew. Chem. Int. Ed. Engl.* **2007**, *46*, 1265–1269. [CrossRef] [PubMed]

International Journal of
Molecular Sciences

MDPI

Review

Aging, Melatonin, and the Pro- and Anti-Inflammatory Networks

Rüdiger Hardeland[ID]

Johann Friedrich Blumenbach Institute of Zoology and Anthropology, University of Göttingen, 37073 Göttingen, Germany; rhardel@gwdg.de; Tel.: +49-551-395414

Received: 11 February 2019; Accepted: 7 March 2019; Published: 11 March 2019

Abstract: Aging and various age-related diseases are associated with reductions in melatonin secretion, proinflammatory changes in the immune system, a deteriorating circadian system, and reductions in sirtuin-1 (SIRT1) activity. In non-tumor cells, several effects of melatonin are abolished by inhibiting SIRT1, indicating mediation by SIRT1. Melatonin is, in addition to its circadian and antioxidant roles, an immune stimulatory agent. However, it can act as either a pro- or anti-inflammatory regulator in a context-dependent way. Melatonin can stimulate the release of proinflammatory cytokines and other mediators, but also, under different conditions, it can suppress inflammation-promoting processes such as NO release, activation of cyclooxygenase-2, inflammasome NLRP3, gasdermin D, toll-like receptor-4 and mTOR signaling, and cytokine release by SASP (senescence-associated secretory phenotype), and amyloid-β toxicity. It also activates processes in an anti-inflammatory network, in which SIRT1 activation, upregulation of Nrf2 and downregulation of NF-κB, and release of the anti-inflammatory cytokines IL-4 and IL-10 are involved. A perhaps crucial action may be the promotion of macrophage or microglia polarization in favor of the anti-inflammatory phenotype M2. In addition, many factors of the pro- and anti-inflammatory networks are subject to regulation by microRNAs that either target mRNAs of the respective factors or upregulate them by targeting mRNAs of their inhibitor proteins.

Keywords: circadian; immunosenescence; inflammaging; melatonin; microRNAs; sirtuin-1

---

1. Introduction

Aging is associated with manifold changes. These comprise declined secretion of hormones such as melatonin [1,2], reduced activities of aging-related factors such as sirtuin-1 (SIRT1) [3], deterioration of the circadian oscillator system [4,5], multiple alterations in the immune system that is frequently shifted toward the proinflammatory side [6–10], and many more deviations of cell biological relevance. Importantly, the changes specifically mentioned are interrelated in multiple ways [3,5,9,10]. This is largely based on the pleiotropy of both melatonin [11] and the circadian system [12–14]. However, these relationships are highly complex, include actions in opposite directions, and cannot be interpreted in reductionist ways.

For example, the effects of melatonin in the immune system can be either pro- or anti-inflammatory [15–17]. Generally, melatonin acts as an immune stimulatory agent and the direction into which the balance is shifted has turned out to be highly conditional. The influence of melatonin on SIRT1 expression has also revealed effects of either down- or upregulation, in this case, with a remarkable difference between tumor and non-tumor cells [5]. While being strongly suppressive in cancer, melatonin mainly stimulated SIRT1 in nontransformed cells, especially in the context of aging. However, a considerable problem exists concerning the, unfortunately, prevailing determination of SIRT1 expression rather than activity. Assuming a correlation between expression and activity, which is questionable, is a profound misconception in the case of sirtuins [18]. Sirtuin activities are not primarily determined by their protein levels but rather by NAD$^+$ concentration, which depends on the activity of nicotinamide phosphoribosyltransferase (NAMPT) [19–22]. The contrast between SIRT1 expression and activity has become evident in a study on the effect of BRCA1 (breast cancer 1, early onset) in ovarian cancer [23]. Suppression of BRCA1 reduced SIRT1 expression, but it increased NAD$^+$ concentration and, therefore, SIRT1 activity. Moreover, BRCA1 overexpression upregulated SIRT1 expression and decreased NAD$^+$ levels and SIRT1 activity. This divergence also seems to be relevant to aging. SIRT1 expression was not generally shown to be decreased in the course of aging, but rather, it was shown to often increase. Nevertheless, SIRT1 activity was found to be reduced because of lowered NAD$^+$ levels [24,25]. However, this does not yet mean that positive correlations between SIRT1 expression and activity are generally excluded in the context of aging. In senescence-accelerated SAMP8 mice, SIRT1 expression was found to be reduced relative to the widely isogenic control strain SAMR1 [26]. Moreover, a number of studies have shown that effects of melatonin that increased SIRT1 expression were suppressed by sirtuin inhibitors such as sirtinol or EX527 [27–39]. Therefore, mandatory requirements are to either determine SIRT1 activity and NAD$^+$ concentration (recommended) or to at least test the effects of sirtuin inhibitors [18].

The changes in melatonin secretion and SIRT1 activity also have a circadian dimension. Both of them are under circadian control and exhibit cycles of high amplitudes. Moreover, either of them can influence circadian oscillators. Apart from its known chronobiotic actions via the suprachiasmatic nucleus (SCN), melatonin also influences peripheral oscillators [40]. SIRT1 has been identified as an accessory oscillator component [19–21] that increases circadian amplitudes of both central and peripheral clocks [21,41]. At least one of the mechanisms described is of relevance for antagonizing age-related decreases in the amplitudes of the SCN output [41]. An additional aspect concerns the conclusion that SIRT1 acts as a partial mediator of melatonin effects [18,42]. Finally, it seems important to remain aware of the different phases of increases and decreases observed within a circadian cycle. It is basic knowledge of chronobiology that a specific treatment applied in different phases leads to different, often opposite, effects [5]. In the context of aging, additional difficulty results from the fact of non-identical changes observed between the populations of oscillators within the body. In a senescent mammal, some oscillators exhibit phase changes, others reduced amplitudes, in the extreme, down to arrhythmicity, whereas others remain widely unchanged [4]. Collectively, all these variabilities summarized here oppose any expectation of finding exclusively unidirectional relationships by applying melatonin, sirtuin overexpression, or other modulators of circadian rhythms.

These reservations also have to be kept in mind when considering the effects of melatonin and SIRT1 in the regulation of pro- and anti-inflammatory processes. These are of particular relevance to aging and age-related diseases, especially as proinflammatory mechanisms gain increasing importance in the course of senescence. The overlapping effects of melatonin and SIRT1 in the field of inflammation regulation will be discussed in this article.

2. Melatonin and the Proinflammatory Network

In the course of studies that revealed numerous effects of melatonin beyond the control of circadian oscillators, immunological actions of melatonin were also discovered [9,15,43–46]. However, the statement that melatonin possesses properties that exceed the control of oscillators does not mean that the resulting effects are independent of circadian rhythms. Both melatonin secretion and signaling are subjected to circadian control. Parameters affected, including immunological ones, are also influenced by the circadian system [11,47]. With regard to the immune system, the situation is somewhat more complicated as melatonin is also synthesized by several types of leukocytes [11,44,48,49].

Many of the earlier investigations on melatonin in the immune system revealed proinflammatory effects, as repeatedly summarized [11,15,17,44]. Notably, these actions were predominantly discovered in studies using isolated leukocytes or transformed leukocyte-derived cell lines [16,17,44]. Although such an approach appears to be reasonable in the beginning, it may not sufficiently reflect the complexity of the immune system and the additional participation of, in classical terms, non-immune cells in both proinflammatory responses and the regulation of the pro-/anti-inflammatory balance. Data obtained in humans also indicated proinflammatory actions of melatonin in rheumatoid arthritis [50–52]. These findings were regarded as a general caveat concerning the use of the immune stimulator melatonin in autoimmune diseases. This view has received partial support by a study in a multiple sclerosis model [53]. Most of the earlier findings on proinflammatory effects in cell cultures concerned the upregulation of proinflammatory cytokines (IL-1β, IL-2, IL-6, IL-8, IL-12, IFNγ, and TNFα) and downregulation of their anti-inflammatory counterpart (IL-10), as summarized elsewhere [15,16,44]. Additionally, melatonin was shown to counteract the inhibition of IL-2 production by prostaglandin E_2 in human lymphocytes [54].

However, this short overview of the inflammation-promoting effects does not yet sufficiently describe the actions of melatonin within the proinflammatory network. In fact, these actions also comprise numerous suppressive changes in this network and, thereby, can turn out to result in an anti-inflammatory balance. To appropriately judge the complex influences of melatonin, it is necessary to view it under these different, additional perspectives: (1) changes in tissues and their main cellular constituents; (2) cell specificity; (3) effects on clonal expansion, differentiation, and polarization in the immune system; (4) contributing proinflammatory effects of non-immune cells; and (5) conditionality, with regard to high-grade or low-grade inflammatory challenges that may occur in the progression of aging or by experimental procedures.

As recently summarized, melatonin-induced elevation of proinflammatory cytokines, such as IL-1β, IL-2, IL-6, IL-12, TNFα, and IFNγ, has been repeatedly observed in monocytes, monocyte-derived cell lines, and type 1 T-helper cells [17]. One can assume that similar responses occur in M1 macrophages, M1 microglia and, especially concerning IL-6, IL-8, and TNFα, also in epithelial cells and various other cell types. Notably, these findings can be seen in the context of pro-oxidant and cytotoxic effects exerted by melatonin in monocytes, as observed above an activation threshold as low as 50 pM [55]. Moreover, melatonin was found to increase another proinflammatory cytokine, IL-17A, in Th17 cells [56], an effect that spreads proinflammatory responses by inducing the release of other mediators, such as IL-1β, IL-6, TNFα, the neutrophil-attracting Il-8, and by upregulating cyclooxygenase-2 (COX-2) and iNOS (inducible NO synthase) [57–65]. This host of secondary effects by IL-17A has been shown to be involved in autoimmune diseases and in neuroinflammation and is, therefore, of relevance to age-related health problems and to adverse effects of melatonin, especially in autoimmunity. Additionally, melatonin was shown to upregulate IL-1β, TNFα, and IFNγ in splenocytes [66]. Several other effects of melatonin concern the upregulation of cytokines that are primarily involved in differentiation and clonal expansion, such as M-CSF (macrophage colony-stimulating factor) and SCF (stem cell factor) in macrophages and splenocytes, TGFβ (transforming growth factor) in macrophages and dendritic cells, and thymosin-α and thymulin in thymocytes, as summarized elsewhere [16]. Melatonin was shown to promote differentiation of progenitor cells to Th lymphocytes, NK lymphocytes, granulocytes, and macrophages [16,44, 45]. All these effects may have secondary consequences to the pro-/anti-inflammatory balance. In bone marrow, an upregulation of GM-CSF (granulocyte/monocyte colony-stimulating factor) may be interpreted in a similar way, but, on the other hand, melatonin-induced release of the immune-opioids MIO15 (melatonin-induced opioid) and MIO67 [16,67–69] can be expected to favor the anti-inflammatory side of the immune system.

The proinflammatory network comprises numerous components, which are regulated differently by melatonin. Moreover, the responses are often highly contextual. Findings obtained by applying pro-oxidant or proinflammatory challenges, such as administration of oxidotoxins, mitotoxins, bacterial lipopolysaccharides (LPS) or experimental sepsis, may not lead to results comparable to those found in aging. Therefore, it seems important to focus on changes that really occur in aging or age-related diseases and can be mostly classified as low-grade inflammation [10,17,70]. Several sources of aging-related inflammation shall be mentioned as being of foremost importance: (1) enhanced release of proinflammatory cytokines as a consequence of immunosenescence; (2) neuronal overexcitation in an •NO-mediated interplay with microglia and astrocytes; (3) amyloid-β (Aβ) toxicity; (4) brain insulin resistance; (5) diabetes and metabolic syndrome; (6) senescence-associated secretory phenotype (SASP) of non-immune cells with DNA damage response (DDR); (7) garb-aging; (8) reduced antioxidant protection by decreased melatonin; and, presumably, also (9) metabolic malfunction because of poorly coordinated and weakened circadian rhythms. Notably, all these pathophysiological alterations are associated with oxidative and nitrosative/nitrative stress and mitochondrial malfunction. They have multiple interconnections and, thereby, display the potential of forming vicious cycles [10,17,70].

Immunosenescence, which is primarily associated with thymic involution and, additionally, exhaustion of leukocyte subpopulations upon lifelong exposure to foreign antigens, typically leads to a shift toward a proinflammatory phenotype, which is evident by increased levels of proinflammatory cytokines that may result in a so-called immune risk profile [7,10,70–73]. Neuronal overexcitation causes increased formation of •NO, which can activate microglia, astrocytes, and other neurons that also release •NO at elevated rates. These responses contribute to a spreading of excitation and to the release of oxidants and proinflammatory mediators, especially by microglia. High levels of •NO and superoxide anions ($O_2^{\bullet-}$) generate the formation of their adduct, peroxynitrite ($ONOO^-$), a highly reactive intermediate that causes damage to mitochondria and additionally generates free radicals ($ONOO^- + H^+ \rightarrow ONOOH \rightarrow {}^\bullet NO_2 + {}^\bullet OH$; $ONOO^- + CO_2 \rightarrow ONOOCO_2^- \rightarrow {}^\bullet NO_2 + CO_3^{\bullet-}$) [74,75]. These radicals, their precursors, and several other reactive nitrogen species that can be formed from them interfere with the mitochondrial electron transport chain (ETC) [74,76,77]. As recently summarized, melatonin protects the ETC against these reactive intermediates [77], in addition to various other mitochondria-protecting actions [78–83]. Aβ peptides have, besides other effects, pro-oxidant and proinflammatory properties. The main toxicity is caused by Aβ monomers and oligomers, with an additional contribution by amyloid plaques. Peptides and oligomers induce microglia activation [84,85] and, additionally, responses by astrocytes [86] and neurons [86,87], which all upregulate NADPH oxidase, thereby elevating $O_2^{\bullet-}$ formation. Notably, the release of proinflammatory mediators in response to Aβ is not restricted to microglia. Even neurons have been shown to respond to Aβ peptides by upregulating TNFα, IL-1β, COX-2, and the T-cell and monocyte attractant chemokine CX3CL1 [87]. Aβ toxicity should not only be seen in the context of established Alzheimer's disease (AD), since these peptides also appear in the CSF (cerebrospinal fluid) of healthy subjects, but are normally widely removed by clearance of CSF and ISF (interstitial fluid), especially during sleep [88–91]. Sleep disturbance impairs Aβ clearance, which has been discussed as a contribution to the development of AD pathology [92]. Circadian disruption, a major source of sleep disturbances, indicates that changes in melatonin secretion may be involved. In fact, melatonin has been recently shown to increase Aβ clearance [93,94]. Numerous other anti-amyloidogenic effects of melatonin have been also described and repeatedly reviewed [70,95–98]. A recent overview [17] has particularly focused on the anti-inflammatory aspect and the suppression of Aβ secretion by melatonin. This includes, in cellular test systems, the reduction of βAPP (β-amyloid precursor protein) mRNA expression [99], inhibition of β- and γ-secretases [100], and the upregulation of α-secretase, an enzyme that competes with β- and γ-secretases and produces the nonamyloidogenic and neuroprotective fragment sAPPα [101]. The shift from β- and γ-secretases to α-secretase was recently also confirmed in the hippocampus of senescent mice [102,103]. In transgenic AD mouse models, melatonin was found to substantially delay the accumulation of Aβ and to extend lifespan [104,105]. However, this was only observed after an early onset of treatment in the first months of life, but not at later age [106], findings that leave a pessimistic perspective for the treatment of humans. Another, surprising nexus to AD pathology and inflammation emerged, when brain insulin resistance was shown to be an early sign of neuroinflammation at the onset of AD [107,108]. For further details and references, see [17,109]. This aspect remains to be studied in the context of melatonin, but would require a strict discrimination between nocturnal rodents and humans, since the disregard of this difference has led to substantial misinterpretations concerning the beneficial or detrimental role of melatonin in type 2 diabetes [5,16]. Also, beyond the CNS, type 2 diabetes and, in a broader sense, metabolic syndrome, are associated with low-grade inflammation, a phenomenon known under the term of metaflammation [110–112], which contributes to inflammaging [111,112]. Numerous beneficial effects of melatonin in metabolic syndrome have been recently summarized [113], which have been observed in both experimental animals and clinical studies and indicate a reduction of metaflammation.

Another aspect of proinflammatory effects concerns the role of non-immune cells in inflammaging. This comprises mainly three processes, SASP, garb-aging, and release of macromolecules from dying cells. In SASP, DNA-damaged cells are arrested in terms of proliferation, but continue to participate in the metabolism of the tissue. However, they release several signal molecules, including proinflammatory cytokines and chemokines, thereby inducing local, low-grade inflammation [114–118]. SASP has been observed in many cells of peripheral tissues, but also in astrocytes and is, therefore, relevant to the CNS [70,119,120]. Notably, SASP has also been shown to be induced by sleep disruption [121]. As this was associated with increased oxidative stress, the observed changes may comprise a chronodisruption-related nocturnal loss of protection by melatonin. Studies on direct effects of melatonin against SASP are still in their infancy, but recent initial results have described this [122–124]. In mechanistic terms, melatonin was shown to suppress the PARP-1 [poly(ADP-ribose) polymerase-1]-induced expression of SASP genes [122]. Whether or not this role of PARP-1 can be generalized remains to be clarified, since this regulator, which is an indicator of DNA damage, has also been interpreted as an epigenomic safeguard that interferes with SASP-associated microRNAs [125]. In another context, melatonin was reported to suppress SASP by downregulating NF-κB and upregulating Nrf2 (nuclear factor erythroid 2-related factor 2) [124], i.e., two known effects of melatonin that are typically observed in mechanisms of antioxidative and anti-inflammatory protection [17,126]. Other sources of low-grade inflammation by non-immune cells, such as garb-aging [127] and release of components of dying cells such as histone H1 [128] and molecules of mitochondrial origin (nucleic acids, *N*-formylated peptides and proteins) to the cytosol [129–132], are presumably also of substantial relevance to inflammaging, but have not been sufficiently studied in relation to melatonin. Nevertheless, this connection exists with high likelihood with regard to mitochondrial DNA (mtDNA), especially if it is damaged or oxidized. Activation of the NLRP3 inflammasome by mtDNA has been repeatedly described, whereas counteractions by melatonin have been multiply documented and reviewed, as summarized elsewhere [133]. Melatonin administration was shown to be as effective as NLRP3 deficiency [134]. Unfortunately, the important proinflammatory pathway of cGAS/STING (cyclic GMP-AMP synthase/stimulator of IFN genes) signaling has, to date, been poorly considered in melatonin research. This would be of particular interest, as cGAS is a cytosolic DNA detector. Moreover, the depletion of cGAS and STING counteracted cell senescence and prevented SASP in human and murine fibroblasts [135]. cGAS appears as a mediator of cell stress reactions and has been assumed to be involved in aging-related diseases [131]. This interpretation has received support by recent findings on a human *Sting* gene polymorphism that is associated with low-risk of aging-related diseases, presumably by reducing inflammaging [136].

Activation of the NLRP3 inflammasome in various systems, under different conditions and counteractions by melatonin, have been recently reviewed [17]. These findings were widely related to the suppression of NF-κB signaling by melatonin, which is likewise important in the attenuation of oxidative damage [126]. NF-κB was also reported to induce pyroptosis via gasdermin D (GSDMD) in adipose tissue, which was likewise inhibited by melatonin [137]. Other inflammation-related and melatonin-sensitive effects of NF-κB concern the upregulation of iNOS and COX-2 [138–141]. Moreover, in the context of presenilin-1 upregulation and pathogenic βAPP processing, a pathway involving PIN1 (peptidyl-prolyl cis-trans isomerase NIMA-interacting 1) and GSK3β (glycogen synthase kinase 3β) was shown to activate NF-κB, which was, in accordance with many other findings on NF-κB suppression, inhibited by melatonin [142].

Another proinflammatory route is based on TLR4 (toll-like receptor 4) activation, e.g., via the IFNγ adaptor protein, TRIF (toll-receptor-associated activator of interferon). In the macrophage-like cell line RAW264.7, melatonin has been shown to suppress the release of proinflammatory cytokines, such as TNFα, IL-1β, IL-6, and IL-8, by TRIF and TLR4 inhibition [143]. As TLR4 also mediates pro-oxidant actions via NF-κB, more general effects by melatonin on this pathway may be assumed. This conclusion is supported by several pertinent findings describing protection by melatonin [17]. Similar anti-inflammatory effects were also obtained in an in vivo model of ovarian cancer [144]. Information on melatonin effects concerning other TLR subforms is still scarce. No effects were found in a single study on TLR2 [144], whereas inhibition of TLR3 was reported [145,146].

A further possible proinflammatory pathway that is inhibited by melatonin concerns mTOR (mechanistic target of rapamycin) activation. However, most respective information is not directly related to inflammation, but rather to mitophagy or apoptosis. Interestingly, an mTOR inhibiting action by melatonin was also shown to be suppressed by inhibition of PIN1 [123]. Moreover, the attenuation of microglial activation and neuroinflammation after traumatic brain injury by melatonin was also interpreted on the basis of interference with mTOR [147]. This route will be of further interest in the specific context of melatonin's anti-inflammatory actions.

3. Melatonin, SIRT1, and the Anti-Inflammatory Network

While melatonin is partially acting by either stimulating or inhibiting components of the proinflammatory network, it also upregulates molecules of an anti-inflammatory network. Some of them are negatively correlated with proinflammatory agents. For instance, NF-κB, a transcription factor involved in prooxidant and, thereby, proinflammatory responses, is inversely coupled to antioxidant and anti-inflammatory regulators, in particular, Nrf2 [17,126,139,148–151]. A similar correlation seems to exist in the case of PARK7 (parkinsonism associated deglycase; also known as DJ-1) [149,150], a protein that acts, beside other effects, as a redox-sensitive chaperone and stress sensor. In Parkinson's disease (PD), it has been shown to be neuroprotective [152].

An especially important anti-inflammatory regulator under control by melatonin is SIRT1. It has been classified as a secondary signaling molecule that mediates several effects of melatonin [18,42]. In non-tumor cells, it has been shown to be upregulated by melatonin and effects by melatonin have been repeatedly reported to be suppressed by sirtuin inhibitors or *Sirt1* siRNA [5], notably also in an anti-inflammatory context [17]. The relationship between melatonin and SIRT1 may be regarded as a mutual one, since SIRT1 can enhance circadian amplitudes in the SCN [41] and may, thereby, influence the melatonin rhythm [3]. With this background, the functional overlap of described melatonin and SIRT1 actions seems worthwhile to be recalled.

This overlap becomes obvious from two lines of evidence, (1) the interference of sirtuin-related agents with melatonin effects, and (2) similar actions of melatonin and SIRT1. In the former context, reductions of NLRP3 inflammasome activation and IL-1β levels by melatonin were blocked by the sirtuin inhibitor EX527 in a rat COPD (chronic obstructive pulmonary disease) model [39]. The same inhibitor also blocked anti-inflammatory actions of melatonin such as downregulation of TNFα and IL-1β in acute kidney injury of rats [36]. Several other results on sirtuin inhibition of melatonin treatment were obtained under conditions of more severe inflammation. This was observed in cardiac ischemia/reperfusion of normal [30] and diabetic rats [31], correspondingly in ER stress of H9C2 cardiomyocytes [31], in LPS-treated microglial cell lines [37], and in brain injury by cecal ligation/puncture in mice [32]. Further studies, without measurement of inflammatory parameters, in which melatonin effects were blunted by sirtuin inhibitors or *Sirt1* siRNA, are summarized elsewhere [5,17].

Numerous studies have demonstrated antioxidant and anti-inflammatory actions by SIRT1 that are also known from melatonin [17]. This concerns the suppression of NF-κB activation [153–156], the upregulation of Nrf2 [157–161], suppression of NLRP3 inflammasome activation [39,162–165], and inhibition of TLR4 signaling [166–168]. An important player in TLR4 activation is HMGB1 (high mobility group box-1), an inflammatory signaling molecule released by monocytes and macrophages, and also by other cells (e.g., endothelial). SIRT1 is known to deacetylate HMGB1 [169,170], to inhibit its nucleocytoplasmic transfer, and to prevent its release [167,171–176]. Importantly, HMGB1 also favors the polarization of macrophages and microglia towards the proinflammatory M1 type [177–181]. With regard to the melatonin–SIRT1 relationship, it is of interest that anti-inflammatory actions via HMGB1 inhibition have been also reported for melatonin, as recently summarized [17]. Again, SIRT1 may mediate melatonin effects in this case, an assumption to be experimentally confirmed.

SIRT1 also displays several additional anti-inflammatory effects, which cannot yet be matched with corresponding data from melatonin treatment. As mentioned in the previous section, some reports on melatonin effects on mTOR signaling have not been directly related to inflammation. Other studies on actions of melatonin on mTORC1 (mTOR complex 1) activity in cancer cells led to contradictory results [182–184] and, therefore, do not provide a reliable basis for comparison. However, SIRT1 has been shown to counteract adipose inflammation by suppressing mTORC1 signaling [185]. Another investigation on liver steatosis reported an increase of mTORC1 activity upon hepatocyte-specific deletion of SIRT1 [186]. Under conditions of sepsis, SIRT1 was also reported to deacetylate and, thereby, inhibit NICD (intracellular domain of Notch) [187,188]. Apart from its developmental roles, Notch is known to act in a proinflammatory way by promoting M1 polarization of macrophages or related cells. Conversely, SIRT1 knockdown upregulated Notch1 in stellate cells at both mRNA and protein levels [189]. However, the relationship to melatonin remains unclear because beneficial effects of Notch signaling were reported in studies on protection against ischemia/reperfusion and Aβ toxicity by melatonin [190–192], in spite of the fact that Notch activation is known to cause inflammatory responses. Either the endpoints studied were unrelated to macrophages or microglia, respectively, or Notch signaling may be a case in which SIRT1 does not mediate melatonin effects. This would be surprising, as melatonin was shown to upregulate SIRT1 under conditions of ischemia/reperfusion, and as some of the melatonin effects were blocked by EX527 or *Sirt1* siRNA [5,30,31].

Another recently reported anti-inflammatory action of SIRT1 has been observed in macrophages after treatment with LPS. The lncRNA-CCL2, which is related to the gene locus of the chemokine CCL2 and stimulates the release of proinflammatory cytokines, was downregulated by SIRT1 [193]. No corresponding data on changes of lncRNA-CCL2 expression by melatonin are actually available.

Despite the fact that many proinflammatory effects of melatonin are mediated by lymphocytes [15,44,49,54,56], such responses were also obtained in monocytes and monocyte-derived cells [11,15,44]. However, as macrophages and related cells represent major executive players in inflammation, their polarization into proinflammatory M1 or anti-inflammatory M2 phenotypes is of utmost importance for the pro-/anti-inflammatory balance. In fact, melatonin is capable of shifting this balance toward the anti-inflammatory side by favoring M2 and disfavoring M1 polarization, as recently reviewed [194]. One of the major anti-inflammatory effects in the inhibition of M1 function consists in the MT_1 receptor-mediated activation of JAK2 (Janus kinase 2), which phosphorylates STAT3 (signal transducer and activator of transcription 3) [194–197]. In the nucleus, pSTAT3 dimers activate SOCS1 (suppressor of cytokine signaling 1), which favors NF-κB degradation by virtue of its property as an E3 ubiquitin ligase and, thus, inhibits NF-κB actions at the chromatin [194,198–200]. Additionally, suppression of NF-κB actions has been reported for RORα [194,201]. This relationship is highly convincing, since RORα knockout causes strong upregulations of proinflammatory cytokines [202]. However, as RORα has been definitely shown to be incapable of binding melatonin [203,204], actions of melatonin via this transcription factor have to be of indirect nature [18,194]. A possibility of particular interest concerns an effect of SIRT1 on RORα, in its function as a partial mediator of melatonin effects. As far as melatonin upregulates SIRT1, a known mechanism can become effective, which comprises deacetylation of PGC-1α (peroxisome proliferator-activated receptor-γ coactivator-1α) and facilitates the binding of RORα to its response elements (ROREs) [41]. With regard to the number of documented anti-inflammatory actions of SIRT1 summarized above, this idea is insofar attractive as SIRT1 was shown to suppress NF-κB signaling, as likewise reported for RORα [153–155]. This does not exclude additional actions of SIRT1, which was also found to be involved in SOCS1 signaling [205].

M2 polarization of macrophages, and presumably in similar ways of microglial cells, is largely promoted by STAT6 phosphorylation, in pathways regulated by melatonin and anti-inflammatory cytokines, especially IL-4 and IL-13 [194]. A key step in these processes is tyrosine phosphorylation of IRS-2 (insulin receptor substrate 2), which can be achieved in multiple ways [194]: (1) via IL-4 or IL-13 binding to IL-4 receptor-α, phosphorylation of JAK1/3 or JAK1/Tyk2 (tyrosine kinase-2), which may cause tyrosine phosphorylation of STAT6 directly or, alternately, IRS-2, followed by GRB2 (growth factor receptor bound-2) activation and STAT6 phosphorylation, or (2) via melatonin and an $MT_{1/2}$-dependent cascade of pIRS-2, GRB2, and pSTAT6. pSTAT6 dimers are responsible for M2-specific gene expressions, partially in conjunction with KLF4 (krüppel-like factor 4), which interacts with pSTAT6. Additional effects of melatonin seem to contribute, such as $MT_{1/2}$-dependent inhibition of PI3K (phosphoinositide 3-kinase), thereby blocking the proinflammatory Akt/mTORC1 cascade [194]. Moreover, this inhibitory effect of melatonin can be assumed to prevent two other negative modulatory actions, (1) by the mTORC1 downstream factor GRB10 and (2) by p70S6K (p50S-6-kinase). Tyrosine phosphorylation of IRS-2 and expression of M2 genes were found to be substantially increased by GBR10 knockdown [206]. Serine phosphorylation of IRS-2 by p70S6K was shown to counteract M2 polarization [206].

4. Modulation of the Networks by Noncoding RNAs, an Emerging Field

The discovery of countless noncoding RNAs (ncRNAs) with regulatory properties has substantially changed our understanding of regulation. Many of these RNAs interfere with posttranscriptional processes. This is especially the case in most microRNAs (miRNAs), which target mRNAs, and snoRNAs (small nucleolar RNAs), which are involved in RNA processing [207,208]. Other categories of RNAs such as eRNAs (enhancer RNAs) or super-enhancer lncRNAs (super enhancer long noncoding RNAs) directly interact with the chromatin and even with DNA [209]. piRNAs (PIWI-interacting RNAs) silence transposable elements via both transcriptional and posttranscriptional mechanisms [210]. Multiple functions are known for lncRNAs, in addition to enhancer properties. They may either positively or negatively regulate gene expression, which has been even observed in their subgroup of asRNAs (antisense RNAs). Moreover, various lncRNAs are precursors of miRNAs or snoRNAs [209]. Moreover, many lncRNAs and circRNAs (circular RNAs) serve as miRNA sponges and gain particular relevance in the transmission of intercellular signals by exosomes and ectosomes [209,211,212]. With regard to melatonin, numerous ncRNAs of different categories were shown to be influenced by this pleiotropic regulator [209,213–215]. As ncRNAs also modulate inflammation, macrophage activities, and NF-κB signaling [216–219], the relationship to melatonin is of particular interest and will certainly gain increasing future importance.

A complete consideration of all ncRNAs with positive or negative actions on inflammation would go beyond the scope of this article. A comprehensive list of miRNAs that affect pro- or anti-inflammatory cytokines has been recently published [220]. With regard to the melatonin-related task of this review, only those ncRNAs shall be discussed that are influenced by melatonin or change its downstream factors or interfere with major regulatory parts of the networks, especially if they concern areas of action known to be a matter of beneficial actions by melatonin, such as sepsis, ischemia/reperfusion, aging, or neurodegeneration. Findings obtained in the field of microRNAs are summarized in Table 1. Only those miRNAs have been considered that are related to the field of inflammation regulation. Many additional studies have revealed effects of melatonin on other miRNAs [214,221,222]. However, the outcome in a non-inflammatory context cannot be expected to be the same. This concerns especially the situation in cancer, although cancer can also have an inflammatory aspect. The major difference between non-tumor and tumor cells concerns the inverse relationship between melatonin and SIRT1 and the pro-apoptotic activity of melatonin in tumor cells [5,16,17]. One of the most extensive studies on melatonin effects on miRNAs [221] that has been omitted from Table 1 has been conducted in breast cancer cell lines. The determinations revealed 12 miRNAs that were upregulated by melatonin and 10 others that were downregulated. The analysis of their 5'-utr sequences indicated that these 22 miRNAs might target 2029 mRNAs.

Table 1. Regulation of microRNA expression by melatonin and effects of microRNAs on SIRT1, Nrf2, and NF-κB in the context of inflammation.

miRNA	Change by Melatonin	Effect on Sirt1	Effect on Nrf2	Nrf2-Related Target	Effect on NF-κB	NF-κB-Related Target
miR-7		↓? [223]	↑ [224]	Keap1	↓ [225,226]	
miR-7-5p					↓ [227]	
miR-9		↓ [228–236]			↓ [236–242]	
miR-9					↑ [243,244]	MCPIP1, TRIM56
miR-20a					↑ [245,246]	CYLD
miR-21		↓ [247]				
miR-23a	↓ [194,248]	↓ [249–251]	↑ [252]	Keap1	↑ [194]	
miR-23a-3p		↓ [253]				
miR-23b		↓ [249]				
miR-23b-3p		↓ [155,254]				
miR-24	↓ [255]	↑ [256]				
miR-24-3p			↑ [257]	Keap1		
miR-26a					↓ [258]	
miR-27b					↓ [259]	
miR-29		↓ [260]	↑ [261,262]	Keap1		
miR-30a	↑ [263]	↓ [264]			↓ [265]	
miR-30a-3p		↓ [266,267]				
miR-30e-5p	↓ [268]					
miR-31					↓ [269]	TRADD
miR-34a	↓ [270]	↓ [271–275]				
miR-34a-5p		↓ [276]				
miR-101			↓ [277]			
miR-106a/b		↓ [278]				
miR-124a		↓ [279]				
miR-125a/b					↑ [280–284]	A20
miR-125b					↓ [285–289]	TRAF, MIP-1α
miR-126	↑ [270]	↑ [290]			↑ [291–293]	TOM1, IκB
miR-128		↓ [294–296]			↑ [297,298]	IκB
miR-132		↓ [249]				
miR-135a		↓ [228]				
miR-142-3p	↑ [299]					
miR-144			↓ [300]			
miR-145-5p		↓ [276]				
miR-146a	↑ [270]		↓ [301]		↑ [302–305]	TRAF6, IRAK1
miR-150					↓ [306–309]	
miR-152	↓ [310]	↓ [256]				
miR-153			↓ [300]			
miR-155	↓ [194,310]	↓ [279,311,312]	↑ [313,314]			
miR-181a-c		↓ [228,230,315]				
miR-182		↓ [316]			↑ [317–320]	CYLD, TCEAL7
miR-195-5p		↓ [276]				
miR-199b		↓ [228]				
miR-200a	↑ [321,322]	↓ [189,323–327]	↑ [328–336]	Keap1	↑ [337]	
miR-200a-3p		↓ [276,338,339]				
miR-204		↓ [228]				

Table 1. *Cont.*

miRNA	Change by Melatonin	Effect on Sirt1	Effect on Nrf2	Nrf2-Related Target	Effect on NF-κB	NF-κB-Related Target
miR-210					↓ [340,341]	DR6
miR-212	↓ [299]	↓ [249,267,342–344]				
miR-217		↓ [345]				
miR-301a					↑ [346–350]	NKRF
miR-326					↑ [351]	BCL2A1
miR-340			↓ [352]			
miR-340-5p			↓ [155,300,353]			
miR-495		↓ [354]				
miR-675	↑ [355]					
miR-675-3p	↑ [322]					
Let-7					↑ [356]	A20
Let-7a	↑ [357]					
Let-7e					↑ [358]	IκBβ
Let-7f					↑ [359]	A20
Let-7g					↑ [360]	
Let-7i		↓ [361]			↑ [361]	SIRT1

Abbreviations: A20, ubiquitin-editing enzyme A20 (=TNFAIP3, tumor necrosis factor alpha-induced protein 3); BCL2A1, B-cell lymphoma 2A1; CYLD, cylindromatosis; DR6, death receptor 6; IRAK1, interleukin-1 receptor-associated kinase 1; KEAP1, Kelch-like ECH-associated protein 1; MCPIP1, monocyte chemotactic protein-induced protein 1; MIP-1α, macrophage inflammatory protein-1α; NKRF, NF-κB repressing factor; TCEAL7, Transcription elongation factor A protein-like 7; TOM1, target of Myb1; TRADD, tumor necrosis factor receptor type 1-associated DEATH domain protein; TRAF6, tumor necrosis factor receptor-associated factor 6; TRIM56, tripartite-motif-containing protein 56. ↓↑ indicate down- or upregulations, respectively.

The pleiotropic targeting of various mRNAs by a single miRNA does not only open the possibility of jointly downregulating several mRNAs that are functionally connected, but may also reflect multiple functions in different contexts. This problem of overlapping actions of miRNAs with roles in both inflammation and cancer is also evident in the findings summarized in Table 1. Although the miRNAs were selected because of their modulation of inflammation, several of them exhibited other, perhaps independent, actions related to tumor promotion and progression. In fact, several molecules listed have been classified as oncomiRs, such as miR-21, miR-23a, miR-29, miR-106b, miR-125b, miR-155, and miR-182 [316,362–364].

The examples in Table 1, in which changes of miRNA expression by melatonin have been combined with actions of inflammation-related miRNAs on expression of SIRT1, Nrf2, and NF-κB, do not show a uniform picture. One of the reasons concerns the differences in context, cell types, and conditions, under which the studies have been performed. In many cases, changes by melatonin and effects on SIRT1, Nrf2, and NF-κB have been investigated in different systems. In a few cases, opposite effects have been obtained. In addition to organ specificity of melatonin actions, the difference between pro- and anti-inflammatory actions of melatonin has to be considered. The main purpose of this table is to provide information to investigators on which miRNAs they may focus on when studying melatonin effects in inflammation, especially in those cases where actions of melatonin on miRNAs have not yet been analyzed.

A few special types of connections between melatonin and the three possible downstream factors shall be briefly discussed. The levels of miR-7 have been shown to increase during aging, and it was assumed that it may downregulate SIRT1 [223]. On the other hand, the upregulation of Nrf2 and downregulation of NF-κB in the non-aging context of training athletes, which jointly indicate an antioxidant and anti-inflammatory action, is not easily compatible with reduced SIRT1 in senescence. In other cases of reduced SIRT1 expression, the reservation has to be made that decreased SIRT1 expression may be associated with elevated SIRT1 activity [23], as long as NAD⁺ levels or inhibition by sirtuin inhibitors have not been determined. In reports in which upregulation by melatonin was associated with downregulation of SIRT1 (miR-30a) or Nrf2 and upregulation of NF-κB (miR-146a), this combination may be interpreted as a route of melatonin's proinflammatory arm. The example of miR-24, which is suppressed by melatonin, but enhances SIRT1 expression, may be taken as a tumor-specific relationship, because this would conform to melatonin's known suppression of SIRT1 in cancer cells [5,365]. Therefore, this may not be applicable to inflammation control. The presence of miR-126, which is upregulated by melatonin and increases SIRT1 expression, indicates a role of this sirtuin as a mediator of melatonin and may appear, at first glance, as an example of the anti-inflammatory route. However, this contrasts strongly with the upregulation of NF-κB and would require clarification by studying all these factors in the same system. A number of SIRT1-targeting miRNAs is downregulated by melatonin, such as miR-23a, miR-34a, miR-152, miR-155, miR-212, and, according to some but not all data, miR-200a. This reduction of SIRT1-suppressing miRNAs would, again, be compatible with findings on SIRT1-mediated melatonin effects [17,18]. Unfortunately, the findings concerning miRNA-mediated changes of NF-κB are, to date, often not generally compatible with the alterations in the other factors. Further studies will be required for obtaining a coherent picture. However, this reservation concerns only the participation of miRNAs, whereas the effects of melatonin and SIRT1 on Nrf2 and NF-κB have been unequivocally demonstrated [17].

Available information concerning melatonin effects on other noncoding RNAs is, unfortunately, limited. A few publications have addressed such actions on lncRNAs. In the case of the lncRNA *H19*, a relationship to miR-675 (cf. Table 1) exists, which is a derivative of the former. The effects of melatonin on H_2O_2-exposed cardiac progenitor cells correspond to those of the microRNA and, consequently, *H19* knockdown abolishes the action of miR-675 [355]. The molecule miR-675 was shown to target *USP10* (ubiquitin carboxyl-terminal hydrolase 10) mRNA, which was concluded to cause downregulation of p53 and p21, thereby inhibiting premature senescence by oxidative stress [355]. Upregulation of *H19* and miR-675 by melatonin was also observed in protection experiments against brain injury after subarachnoid hemorrhage [357]. In a study on pulmonary hypertension, melatonin reduced *H19* expression and, therefore, also that of miR-675-3p, whereas *H19* was shown to suppress miR-200a [322]. Meanwhile, numerous other studies have confirmed the role of *H19* in inflammatory processes (not cited), however, they did not consider melatonin. Another type of interaction between lncRNA and miRNA was based on sponging. The lncRNA *MEG3* was reported to enhance pyroptosis in atherosclerotic models, such as human aortic endothelial cells exposed to oxidized LDL (low density lipoprotein). *MEG3* was found to sponge miR-223 and to activate NLRP3, whereas melatonin was shown to reduce pyroptosis by suppressing NLRP3 activation and downstream factors, such as NF-κB, gasdermin D, IL-1β, and IL-18 [366]. In hair follicle fibroblasts of cashmere goats, melatonin upregulated the lncRNA *MTC*, which was associated with proliferation and also NF-κB activation [367]. This finding was in contrast to many other results on NF-κB suppression by melatonin obtained in the context of antioxidant actions. Two other reports have dealt with melatonin's actions in hepatocellular carcinomas. In one study, melatonin was shown to reduce cancer progression by upregulating FoxA2, which induced the lncRNA *CPS1-IT1*, with a downstream effect of reduced HIF-1α activity [368]. In another investigation, melatonin suppressed DNA repair capacity by upregulating the antisense lncRNA *RAD51-AS1*, which interacts with *RAD51* mRNA, and the findings were interpreted as a sensitization to chemotherapeutic drugs [369]. A functionally different lncRNA, *TERRA* (telomeric repeat-containing RNA), has been investigated in the context of melatonin. This RNA interacts with PARP-1 (polyADP-ribose polymerase-1), a sensor of DNA damage. This interaction was shown to stimulate SASP, but melatonin suppressed SASP-related gene expression and, therefore, an important aging-promoting process [122].

Work on the mediation of melatonin effects on lncRNAs can be expected to become highly important in the future [198,370], although this line of investigation is still in its infancy. This assumption is also based on the increasing body of evidence in related fields, in particular, the circadian system [209,371] and gerontology [372,373]. Numerous lncRNAs, e.g., over 600 in murine liver, were found to vary in a circadian fashion, often at very high amplitudes [371]. It would be a surprise if the chronobiotic regulator melatonin would not modulate at least a fraction of them. Among the lncRNAs, particular attention should be paid to subforms of a different functionality, such as asRNAs, eRNAs and, importantly, super-enhancer lncRNAs, all of which contain numerous species under circadian control [209,371]. The prominent role of lncRNAs in the control of inflammation has also become evident and has been reviewed multiple times [374–380]. In this context, more research on melatonin is desired. A further category of RNAs that will gain importance in the melatonin field, but are still devoid of pertinent investigations, are the circRNAs, which are known to sponge miRNAs [211,381–383]. With regard to a miRNA discussed above, miR-7, which is sponged by the circRNA *CDR1as* (alias *ciRS-7*), a few recent publications indicate relationships to NF-κB signaling [384–386] and to Aβ secretion [384], and they also demonstrate the profoundness of *CDR1as* deficiency in the brain [387].

5. Conclusions

The decrease of melatonin in aging and in various aging-related diseases [1,2] has different, potentially opposite, consequences. The same is valid for melatonin replacement therapies. This results from the dual roles of melatonin as both a pro- and anti-inflammatory regulator [15–17]. The conditionality that determines whether melatonin promotes the one or the other mode of immunological response should be a precondition for appropriately judging what kind of change has to be expected. Unfortunately, to date, this conditionality is not completely understood. Tendentially, one can conclude that anti-inflammatory actions of melatonin are mostly prevailing [17], however, with the important exception of autoimmune diseases, in which the immune stimulatory properties of melatonin may turn out to be detrimental. This concerns especially rheumatoid arthritis [50–52] and multiple sclerosis [53]. Whether melatonin may be unfavorable in type 2 diabetes of humans, in spite of its clearly antidiabetic actions in rodents, remains to be clarified in detail and may partially depend on age [5]. The difference between humans and rodents has to be seen in the contrasting associations of melatonin with food consumption and related metabolic activities in diurnality vs. nocturnality [5,16]. The inflammatory aspect of type 2 diabetes extends to brain insulin resistance, which has been identified as a proinflammatory change in early AD [107–109]. Whether or not melatonin may be detrimental in PD has been a matter of controversy [388–390]. The reported improvements by melatonergic antagonists in PD [390] should at least be regarded as a caveat [391].

Under other conditions, melatonin has been shown to preferentially exert anti-inflammatory, antioxidant, and other beneficial actions in aging [10,26,70,74,93–98,392–394]. Concerning anti-inflammatory effects, melatonin suppresses various processes that lead to enhanced formation of reactive oxygen and nitrogen species and to proinflammatory signaling, as summarized in Section 2. Moreover, it also stimulates several anti-inflammatory pathways and cellular changes that favor this side of the immune system and promote healing, as outlined in Section 3. Among these activities, the promotion of macrophage polarization towards the M2 type may be of particular importance for the shift from pro- to anti-inflammatory behavior, an aspect that has recently received increased attention [137,194,197]. Another potentially decisive action of melatonin concerns the upregulation of SIRT1 in nontumor cells, especially observed in aging animals [3,5,17,18,28,70,395]. Apart from being a relevant factor in aging, SIRT1 displays antioxidant and anti-inflammatory properties (Section 3). As mentioned above, various actions of melatonin are abolished by sirtuin inhibitors or *Sirt1* siRNA. Moreover, some mitochondrial actions of melatonin were shown to be absent in Sirt1$^{-/-}$ mice [396]. The inclusion of SIRT1 and, perhaps, other sirtuins into the spectrum of melatonin's actions represents an important step forward in the understanding of its aging- and inflammation-related properties. However, one cannot expect that all SIRT1 actions mediate melatonergic regulation. SIRT1 is also controlled by various other factors, including accessory components of circadian oscillators that regulate NAMPT expression and NAD$^+$ levels, hormones such as triiodothyronine and glucocorticoids, oncogenes, lncRNAs such as *HOTAIR*, and various miRNAs not regulated by melatonin. The incoherence of melatonin effects on miRNAs and their targeting of typically otherwise melatonin-controlled transcription factors, such as Nrf2 and NF-κB, is evident from Table 1. Moreover, several miRNAs up- or downregulated by melatonin do not cause a rise in SIRT1. However, in the latter case, one has to remain aware that melatonin downregulates SIRT1 in cancer cells, and that several of the miRNAs have been studied in tumors. The selection of miRNAs in Table 1 was restricted to those with demonstrated functions in the control of inflammation. More extensive studies on melatonin, SIRT1, and miRNAs will presumably reveal additional cases that are related to the immune system and also to aging and age-related diseases. As examples, the associated roles of SIRT1 and miRNAs in SASP and in Aβ toxicity shall be briefly mentioned [125,397]. MicroRNAs and other noncoding RNAs have brought about a new, considerably expanded level of complexity into the relationships between melatonin, inflammation, and aging. There is considerable difficulty that arises from multiple mRNAs targeted by a single miRNA species. A further level of complexity in the regulatory networks can be expected as soon as the intercellular communication, via exosomal RNAs, is more profoundly understood in its details and its variations according to diseases and aging.

Conflicts of Interest: The author declares no conflict of interest.

References

1. Bubenik, G.A.; Konturek, S.J. Melatonin and aging: Prospects for human treatment. *J. Physiol. Pharmacol.* **2011**, *62*, 13–19. [PubMed]
2. Hardeland, R. Melatonin in aging and disease—Multiple consequences of reduced secretion, options and limits of treatment. *Aging Dis.* **2012**, *3*, 194–225. [PubMed]
3. Hardeland, R. Brain inflammaging: Roles of melatonin, circadian clocks and sirtuins. *J. Clin. Cell. Immunol.* **2018**, *9*, 543. [CrossRef]
4. Yamazaki, S.; Straume, M.; Tei, H.; Sakaki, Y.; Menaker, M.; Block, G.D. Effects of aging on central and peripheral mammalian clocks. *Proc. Natl. Acad. Sci. USA* **2002**, *99*, 10801–10806. [CrossRef] [PubMed]
5. Hardeland, R. Melatonin and the pathologies of weakened or dysregulated circadian oscillators. *J. Pineal Res.* **2017**, *62*, e12377. [CrossRef] [PubMed]
6. Ginaldi, L.; De Martinis, M.; D'Ostilio, A.; Marini, L.; Loreto, M.F.; Quaglino, D. Immunological changes in the elderly. *Aging* **1999**, *11*, 281–286. [CrossRef] [PubMed]
7. DelaRosa, O.; Pawelec, G.; Peralbo, E.; Wikby, A.; Mariani, E.; Mocchegiani, E.; Tarazona, R.; Solana, R. Immunological biomarkers of ageing in man: Changes in both innate and adaptive immunity are associated with health and longevity. *Biogerontology* **2006**, *7*, 471–481. [CrossRef] [PubMed]
8. Dewan, S.K.; Zheng, S.B.; Xia, S.J.; Bill, K. Senescent remodeling of the immune system and its contribution to the predisposition of the elderly to infections. *Chin. Med. J.* **2012**, *125*, 3325–3331. [PubMed]
9. Cardinali, D.P.; Esquifino, A.I.; Srinivasan, V.; Pandi-Perumal, S.R. Melatonin and the immune system in aging. *Neuroimmunomodulation* **2008**, *15*, 272–278. [CrossRef] [PubMed]
10. Hardeland, R. Melatonin and the theories of aging: A critical appraisal of melatonin's role in antiaging mechanisms. *J. Pineal Res.* **2013**, *55*, 325–356. [CrossRef] [PubMed]
11. Hardeland, R.; Cardinali, D.P.; Srinivasan, V.; Spence, D.W.; Brown, G.M.; Pandi-Perumal, S.R. Melatonin—A pleiotropic, orchestrating regulator molecule. *Prog. Neurobiol.* **2011**, *93*, 350–384. [CrossRef] [PubMed]
12. Gachon, F.; Nagoshi, E.; Brown, S.A.; Ripperger, J.; Schibler, U. The mammalian circadian timing system: From gene expression to physiology. *Chromosoma* **2004**, *113*, 103–112. [CrossRef] [PubMed]
13. Buijs, R.M.; Scheer, F.A.; Kreier, F.; Yi, C.; Bos, N.; Goncharuk, V.D.; Kalsbeek, A. Organization of circadian functions: Interaction with the body. *Prog. Brain Res.* **2006**, *153*, 341–360. [PubMed]
14. Hardeland, R. Melatonin and circadian oscillators in aging—A dynamic approach to the multiply connected players. *Interdisc. Top. Gerontol.* **2015**, *40*, 128–140. [PubMed]
15. Carrillo-Vico, A.; Lardone, P.J.; Álvarez-Sánchez, N.; Rodríguez-Rodríguez, A.; Guerrero, J.M. Melatonin: Buffering the immune system. *Int. J. Mol. Sci.* **2013**, *14*, 8638–8683. [CrossRef] [PubMed]
16. Hardeland, R. Opposite effects of melatonin in different systems and under different conditions. *Curr. Top. Biochem. Res.* **2016**, *17*, 57–69.

17.   Hardeland, R. Melatonin and inflammation—Story of a double-edged blade. *J. Pineal Res.* **2018**, *65*, e12525. [CrossRef] [PubMed]

18.   Hardeland, R. Melatonin and retinoid orphan receptors: Demand for new interpretations after their exclusion as nuclear melatonin receptors. *Melatonin Res.* **2018**, *1*, 77–92. [CrossRef]

19.   Nakahata, Y.; Sahar, S.; Astarita, G.; Kaluzova, M.; Sassone-Corsi, P. Circadian control of the NAD$^+$ salvage pathway by CLOCK-SIRT1. *Science* **2009**, *324*, 654–657. [CrossRef] [PubMed]

20.   Bellet, M.M.; Orozco-Solis, R.; Sahar, S.; Eckel-Mahan, K.; Sassone-Corsi, P. The time of metabolism: NAD$^+$, SIRT1, and the circadian clock. *Cold Spring Harb. Symp. Quant. Biol.* **2011**, *76*, 31–38. [CrossRef] [PubMed]

21.   Sahar, S.; Sassone-Corsi, P. The epigenetic language of circadian clocks. *Handb. Exp. Pharmacol.* **2013**, *217*, 29–44.

22.   Masri, S. Sirtuin-dependent clock control: New advances in metabolism, aging and cancer. *Curr. Opin. Clin. Nutr. Metab. Care* **2015**, *18*, 521–527. [CrossRef] [PubMed]

23.   Li, D.; Bi, F.F.; Chen, N.N.; Cao, J.M.; Sun, W.P.; Zhou, Y.M.; Li, C.Y.; Yang, Q. A novel crosstalk between BRCA1 and sirtuin 1 in ovarian cancer. *Sci. Rep.* **2014**, *4*, 6666. [CrossRef] [PubMed]

24.   Ramsey, K.M.; Mills, K.F.; Satoh, A.; Imai, S. Age-associated loss of sirt1-mediated enhancement of glucose-stimulated insulin secretion in beta cell-specific sirt1-overexpressing (besto) mice. *Aging Cell* **2008**, *7*, 78–88. [CrossRef] [PubMed]

25.   Elibol, B.; Kilic, U. High levels of SIRT1 expression as a protective mechanism against disease-related conditions. *Front Endocrinol.* **2018**, *9*, 614. [CrossRef] [PubMed]

26.   Gutierrez-Cuesta, J.; Tajes, M.; Jiménez, A.; Coto-Montes, A.; Camins, A.; Pallàs, M. Evaluation of potential pro-survival pathways regulated by melatonin in a murine senescence model. *J. Pineal Res.* **2008**, *45*, 497–505. [CrossRef] [PubMed]

27.   Tajes, M.; Gutierrez-Cuesta, J.; Ortuño-Sahagun, D.; Camins, A.; Pallàs, M. Anti-aging properties of melatonin in an in vitro murine senescence model: Involvement of the sirtuin 1 pathway. *J. Pineal Res.* **2009**, *47*, 228–237. [CrossRef] [PubMed]

28.   Cristòfol, R.; Porquet, D.; Corpas, R.; Coto-Montes, A.; Serret, J.; Camins, A.; Pallàs, M.; Sanfeliu, C. Neurons from senescence-accelerated SAMP8 mice are protected against frailty by the sirtuin 1 promoting agents melatonin and resveratrol. *J. Pineal Res.* **2012**, *52*, 271–281. [CrossRef] [PubMed]

29.   Guo, P.; Pi, H.; Xu, S.; Zhang, L.; Li, Y.; Li, M.; Cao, Z.; Tian, L.; Xie, J.; Li, R.; et al. Melatonin Improves mitochondrial function by promoting MT1/SIRT1/PGC-1 alpha-dependent mitochondrial biogenesis in cadmium-induced hepatotoxicity in vitro. *Toxicol. Sci.* **2014**, *142*, 182–195. [CrossRef] [PubMed]

30.   Yu, L.; Sun, Y.; Cheng, L.; Jin, Z.; Yang, Y.; Zhai, M.; Pei, H.; Wang, X.; Zhang, H.; Meng, Q.; et al. Melatonin receptor-mediated protection against myocardial ischemia/reperfusion injury: Role of SIRT1. *J. Pineal Res.* **2014**, *57*, 228–238. [CrossRef] [PubMed]

31.   Yu, L.; Liang, H.; Dong, X.; Zhao, G.; Jin, Z.; Zhai, M.; Yang, Y.; Chen, W.; Liu, J.; Yi, W.; et al. Reduced silent information regulator 1 signaling exacerbates myocardial ischemia-reperfusion injury in type 2 diabetic rats and the protective effect of melatonin. *J. Pineal Res.* **2015**, *59*, 376–390. [CrossRef] [PubMed]

32.   Zhao, L.; An, R.; Yang, Y.; Yang, X.; Liu, H.; Yue, L.; Li, X.; Lin, Y.; Reiter, R.J.; Qu, Y. Melatonin alleviates brain injury in mice subjected to cecal ligation and puncture via attenuating inflammation, apoptosis, and oxidative stress: The role of SIRT1 signaling. *J. Pineal Res.* **2015**, *59*, 230–239. [CrossRef] [PubMed]

33.   Zhou, L.; Chen, X.; Liu, T.; Gong, Y.; Chen, S.; Pan, G.; Cui, W.; Luo, Z.P.; Pei, M.; Yang, H.; et al. Melatonin reverses H$_2$O$_2$-induced premature senescence in mesenchymal stem cells via the SIRT1-dependent pathway. *J. Pineal Res.* **2015**, *59*, 190–205. [CrossRef] [PubMed]

34.   Yang, Y.; Jiang, S.; Dong, Y.; Fan, C.; Zhao, L.; Yang, X.; Li, J.; Di, S.; Yue, L.; Liang, G.; et al. Melatonin prevents cell death and mitochondrial dysfunction via a SIRT1-dependent mechanism during ischemic-stroke in mice. *J. Pineal Res.* **2015**, *58*, 61–70. [CrossRef] [PubMed]

35.   Lee, J.H.; Moon, J.H.; Nazim, U.M.; Lee, Y.J.; Seol, J.W.; Eo, S.K.; Lee, J.H.; Park, S.Y. Melatonin protects skin keratinocyte from hydrogen peroxide-mediated cell death via the SIRT1 pathway. *Oncotarget* **2016**, *7*, 12075–12088. [CrossRef] [PubMed]

36.   Bai, X.Z.; He, T.; Gao, J.X.; Liu, Y.; Liu, J.Q.; Han, S.C.; Li, Y.; Shi, J.H.; Han, J.T.; Tao, K.; et al. Melatonin prevents acute kidney injury in severely burned rats via the activation of SIRT1. *Sci. Rep.* **2016**, *6*, 32199. [CrossRef] [PubMed]

37. Shah, S.A.; Khan, M.; Jo, M.H.; Jo, M.G.; Amin, F.U.; Kim, M.O. Melatonin stimulates the SIRT1/Nrf2 signaling pathway counteracting lipopolysaccharide (LPS)-induced oxidative stress to rescue postnatal rat brain. *CNS Neurosci. Ther.* **2017**, *23*, 33–44. [CrossRef] [PubMed]

38. Yang, W.; Kang, X.; Qin, N.; Li, F.; Jin, X.; Ma, Z.; Qian, Z.; Wu, S. Melatonin protects chondrocytes from impairment induced by glucocorticoids via NAD$^+$-dependent SIRT1. *Steroids* **2017**, *126*, 24–29. [CrossRef] [PubMed]

39. Peng, Z.; Zhang, W.; Qiao, J.; He, B. Melatonin attenuates airway inflammation via SIRT1 dependent inhibition of NLRP3 inflammasome and IL-1β in rats with COPD. *Int. Immunopharmacol.* **2018**, *62*, 23–28. [CrossRef] [PubMed]

40. Hardeland, R.; Madrid, J.A.; Tan, D.X.; Reiter, R.J. Melatonin, the circadian multioscillator system and health: The need for detailed analyses of peripheral melatonin signaling. *J. Pineal Res.* **2012**, *52*, 139–166. [CrossRef] [PubMed]

41. Chang, H.C.; Guarente, L. SIRT1 mediates central circadian control in the SCN by a mechanism that decays with aging. *Cell* **2013**, *153*, 1448–1460. [CrossRef] [PubMed]

42. Hardeland, R. Extended signaling by melatonin. *Cell Cell. Life Sci. J.* **2018**, *3*, 000123.

43. Guerrero, J.M.; Reiter, R.J. Melatonin-immune system relationships. Curr. *Top. Med. Chem.* **2002**, *2*, 167–179. [CrossRef]

44. Carrillo-Vico, A.; Guerrero, J.M.; Lardone, P.J.; Reiter, R.J. A review of the multiple actions of melatonin on the immune system. *Endocrine* **2005**, *27*, 189–200. [CrossRef]

45. Carrillo-Vico, A.; Reiter, R.J.; Lardone, P.J.; Herrera, J.L.; Fernández-Montesinos, R.; Guerrero, J.M.; Pozo, D. The modulatory role of melatonin on immune responsiveness. *Curr. Opin. Investig. Drugs* **2006**, *7*, 423–431. [PubMed]

46. Miller, S.C.; Pandi-Perumal, S.R.; Esquifino, A.I.; Cardinali, D.P.; Maestroni, G.J.M. The role of melatonin in immuno-enhancement: potential application in cancer. *Int. J. Exp. Pathol.* **2006**, *87*, 81–87. [CrossRef] [PubMed]

47. Pandi-Perumal, S.R.; Srinivasan, V.; Maestroni, G.J.M.; Cardinali, D.P.; Poeggeler, B.; Hardeland, R. Melatonin: Nature's most versatile biological signal? *FEBS J.* **2006**, *273*, 2813–2838. [CrossRef] [PubMed]

48. Carrillo-Vico, A.; Calvo, J.R.; Abreu, P.; Lardone, P.J.; García-Mauriño, S.; Reiter, R.J.; Guerrero, J.M. Evidence of melatonin synthesis by human lymphocytes and its physiological significance: Possible role as intracrine, autocrine, and/or paracrine substance. *FASEB J.* **2004**, *18*, 537–539. [CrossRef] [PubMed]

49. Carrillo-Vico, A.; Lardone, P.J.; Fernandez-Santos, J.M.; Martín-Lacave, I.; Calvo, J.R.; Karasek, M.; Guerrero, J.M. Human lymphocyte-synthesized melatonin is involved in the regulation of the interleukin-2/interleukin-2 receptor system. *J. Clin. Endocrinol. MeTable* **2005**, *90*, 992–1000. [CrossRef] [PubMed]

50. Maestroni, G.J.M.; Cardinali, D.P.; Esquifino, A.I.; Pandi-Perumal, S.R. Does melatonin play a disease-promoting role in rheumatoid arthritis? *J. Neuroimmunol.* **2005**, *158*, 106–111. [CrossRef] [PubMed]

51. Cutolo, M.; Maestroni, G.J.M. The melatonin-cytokine connection in rheumatoid arthritis. *Ann. Rheum. Dis.* **2005**, *64*, 1109–1111. [CrossRef] [PubMed]

52. Maestroni, G.J.M.; Otsa, K.; Cutolo, M. Melatonin treatment does not improve rheumatoid arthritis. *Br. J. Clin. Pharmacol.* **2008**, *65*, 797–798. [CrossRef] [PubMed]

53. Ghareghani, M.; Dokoohaki, S.; Ghanbari, A.; Farhadi, N.; Zibara, K.; Khodadoust, S.; Parishani, M.; Ghavamizadeh, M.; Sadeghi, H. Melatonin exacerbates acute experimental autoimmune encephalomyelitis by enhancing the serum levels of lactate: A potential biomarker of multiple sclerosis progression. *Clin. Exp. Pharmacol. Physiol.* **2017**, *44*, 52–61. [CrossRef] [PubMed]

54. Carrillo-Vico, A.; García-Mauriño, S.; Calvo, J.R.; Guerrero, J.M. Melatonin counteracts the inhibitory effect of PGE2 on IL-2 production in human lymphocytes via its mt1 membrane receptor. *FASEB J.* **2003**, *17*, 755–757. [CrossRef] [PubMed]

55. Morrey, K.M.; McLachlan, J.A.; Serkin, C.D.; Bakouche, O. Activation of human monocytes by the pineal hormone melatonin. *J. Immunol.* **1994**, *153*, 2671–2680. [PubMed]

56. Kuklina, E.M.; Glebezdina, N.S.; Nekrasova, I.V. Role of melatonin in the regulation of differentiation of T cells producing interleukin-17 (Th17). *Bull. Exp. Biol. Med.* **2016**, *160*, 656–658. [CrossRef] [PubMed]

57. Jovanovic, D.V.; Di Battista, J.A.; Martel-Pelletier, J.; Jolicoeur, F.C.; He, Y.; Zhang, M.; Mineau, F.; Pelletier, J.P. IL-17 stimulates the production and expression of proinflammatory cytokines, IL-beta and TNF-alpha, by human macrophages. *J. Immunol.* **1998**, *160*, 3513–3521. [PubMed]

58. Shalom-Barak, T.; Quach, J.; Lotz, M. Interleukin-17-induced gene expression in articular chondrocytes is associated with activation of mitogen-activated protein kinases and NF-κB. *J. Biol. Chem.* **1998**, *273*, 27467–27473. [CrossRef] [PubMed]

59. Kawaguchi, M.; Kokubu, F.; Kuga, H.; Matsukura, S.; Hoshino, H.; Ieki, K.; Imai, T.; Adachi, M.; Huang, S.K. Modulation of bronchial epithelial cells by IL-17. *J. Allergy Clin. Immunol.* **2001**, *108*, 804–809. [CrossRef] [PubMed]

60. Molet, S.; Hamid, Q.; Davoine, F.; Nutku, E.; Taha, R.; Page, R.N.; Olivenstein, R.; Elias, J.; Chakir, J. IL-17 is increased in asthmatic airways and induces human bronchial fibroblasts to produce cytokines. *J. Allergy Clin. Immunol.* **2001**, *108*, 430–438. [CrossRef] [PubMed]

61. Paradowska, A.; Maśliński, W.; Grzybowska-Kowalczyk, A.; Łacki, J. The function of interleukin 17 in the pathogenesis of rheumatoid arthritis. *Arch. Immunol. Ther. Exp.* **2007**, *55*, 329–334. [CrossRef]

62. Ishigame, H.; Kakuta, S.; Nagai, T.; Kadoki, M.; Nambu, A.; Komiyama, Y.; Fujikado, N.; Tanahashi, Y.; Akitsu, A.; Kotaki, H.; et al. Differential roles of interleukin-17A and -17F in host defense against mucoepithelial bacterial infection and allergic responses. *Immunity* **2009**, *30*, 108–119. [CrossRef] [PubMed]

63. Zhang, Y.; Huang, D.; Gao, W.; Yan, J.; Zhou, W.; Hou, X.; Liu, M.; Ren, C.; Wang, S.; Shen, J. Lack of IL-17 signaling decreases liver fibrosis in murine schistosomiasis japonica. *Int. Immunol.* **2015**, *27*, 317–325. [CrossRef] [PubMed]

64. Shabgah, A.G.; Fattahi, E.; Shahneh, F.Z. Interleukin-17 in human inflammatory diseases. *Postepy Dermatol. Alergol.* **2014**, *31*, 256–261. [CrossRef] [PubMed]

65. Yang, Z.Y.; Yuan, C.X. IL-17A promotes the neuroinflammation and cognitive function in sevoflurane anesthetized aged rats via activation of NF-κB signaling pathway. *BMC Anesthesiol.* **2018**, *18*, 147. [CrossRef] [PubMed]

66. Liu, F.; Ng, T.B.; Fung, M.C. Pineal indoles stimulate the gene expression of immunomodulating cytokines. *J. Neural Transm.* **2001**, *108*, 397–405. [CrossRef] [PubMed]

67. Maestroni, G.J.M. The immunoneuroendocrine role of melatonin. *J. Pineal Res.* **1993**, *14*, 1–10. [CrossRef] [PubMed]

68. Maestroni, G.J.M.; Hertens, E.; Galli, P.; Conti, A.; Pedrinis, E. Melatonin-induced T-helper cell hematopoietic cytokines resembling both interleukin-4 and dynorphin. *J. Pineal Res.* **1996**, *21*, 131–139. [CrossRef] [PubMed]

69. Maestroni, G.J.M. κ-Opioid receptors in marrow stroma mediate the hematopoietic effects of melatonin-induced opioid cytokines. *Ann. N. Y. Acad. Sci.* **1998**, *840*, 411–419. [CrossRef] [PubMed]

70. Hardeland, R.; Cardinali, D.P.; Brown, G.M.; Pandi-Perumal, S.R. Melatonin and brain inflammaging. *Prog. Neurobiol.* **2015**, *127–128*, 46–63. [CrossRef] [PubMed]

71. Strindhall, J.; Nilsson, B.O.; Löfgren, S.; Ernerudh, J.; Pawelec, G.; Johansson, B.; Wikby, A. No Immune Risk Profile among individuals who reach 100 years of age: Findings from the Swedish NONA immune longitudinal study. *Exp. Gerontol.* **2007**, *42*, 753–761. [CrossRef] [PubMed]

72. Candore, G.; Caruso, C.; Colonna-Romano, G. Inflammation, genetic background and longevity. *Biogerontology* **2010**, *11*, 565–573. [CrossRef] [PubMed]

73. Ponnappan, S.; Ponnappan, U. Aging and immune function: Molecular mechanisms to interventions. *Antioxid. Redox Signal.* **2011**, *14*, 1551–1585. [CrossRef] [PubMed]

74. Hardeland, R.; Coto-Montes, A. New vistas on oxidative damage and aging. *Open Biol. J.* **2010**, *3*, 39–52. [CrossRef]

75. Hardeland, R. Melatonin and its metabolites as anti-nitrosating and anti-nitrating agents. *J. Exp. Integr. Med.* **2011**, *1*, 67–81. [CrossRef]

76. Hardeland, R. Neuroprotection by radical avoidance: Search for suitable agents. *Molecules* **2009**, *14*, 5054–5102. [CrossRef] [PubMed]

77. Hardeland, R. Melatonin and the electron transport chain. *Cell. Mol. Life Sci.* **2017**, *74*, 3883–3896. [CrossRef] [PubMed]

78. Reiter, R.J.; Rosales-Corral, S.; Tan, D.X.; Jou, M.J.; Galano, A.; Xu, B. Melatonin as a mitochondria-targeted antioxidant: One of evolution's best ideas. *Cell. Mol. Life Sci.* **2017**, *74*, 3863–3881. [CrossRef] [PubMed]

79. Paradies, G.; Paradies, V.; Ruggiero, F.M.; Petrosillo, G. Mitochondrial bioenergetics decay in aging: Beneficial effect of melatonin. *Cell. Mol. Life Sci.* **2017**, *74*, 3897–3911. [CrossRef] [PubMed]

80. Slominski, A.T.; Zmijewski, M.A.; Semak, I.; Kim, T.K.; Janjetovic, Z.; Slominski, R.M.; Zmijewski, J.W. Melatonin, mitochondria, and the skin. *Cell. Mol. Life Sci.* **2017**, *74*, 3913–3925. [CrossRef] [PubMed]

81. Cardinali, D.P.; Vigo, D.E. Melatonin, mitochondria, and the metabolic syndrome. *Cell. Mol. Life Sci.* **2017**, *74*, 3941–3954. [CrossRef] [PubMed]

82. Acuña-Castroviejo, D.; Rahim, I.; Acuña-Fernández, C.; Fernández-Ortiz, M.; Solera-Marín, J.; Sayed, R.K.A.; Díaz-Casado, M.E.; Rusanova, I.; López, L.C.; Escames, G. Melatonin, clock genes and mitochondria in sepsis. *Cell. Mol. Life Sci.* **2017**, *74*, 3965–3987. [CrossRef] [PubMed]

83. Wongprayoon, P.; Govitrapong, P. Melatonin as a mitochondrial protector in neurodegenerative diseases. *Cell. Mol. Life Sci.* **2017**, *74*, 3999–4014. [CrossRef] [PubMed]

84. Tan, B.; Choi, R.H.; Chin, T.J.; Kaur, C.; Ling, E.A. Manipulation of microglial activity as a therapy for Alzheimer's disease. *Front. Biosci. (Schol. Ed.)* **2012**, *4*, 1402–1412. [PubMed]

85. McLarnon, J.G. Correlated inflammatory responses and neurodegeneration in peptide-injected animal models of Alzheimer's disease. *Biomed. Res. Int.* **2014**, *2014*, 923670. [CrossRef] [PubMed]

86. Narayan, P.; Holmström, K.M.; Kim, D.H.; Whitcomb, D.J.; Wilson, M.R.; St George-Hyslop, P.; Wood, N.W.; Dobson, C.M.; Cho, K.; Abramov, A.Y.; et al. Rare individual amyloid-β oligomers act on astrocytes to initiate neuronal damage. *Biochemistry* **2014**, *53*, 2442–2453. [CrossRef] [PubMed]

87. Hanzel, C.E.; Pichet-Binette, A.; Pomentel, L.S.; Iulita, M.F.; Allard, S.; Ducatenzeiler, A.; Do Carmo, S.; Cuello, A.C. Neuronal driven pre-plaque inflammation in a transgenic rat model of Alzheimer's disease. *Neurobiol. Aging* **2014**, *35*, 2249–2262. [CrossRef] [PubMed]

88. Iliff, J.J.; Wang, M.; Liao, Y.; Plogg, B.A.; Peng, W.; Gundersen, G.A.; Benveniste, H.; Vates, G.E.; Deane, R.; Goldman, S.A.; et al. A paravascular pathway facilitates CSF flow through the brain parenchyma and the clearance of interstitial solutes, including amyloid β. *Sci. Transl. Med.* **2012**, *4*, 147ra111. [CrossRef] [PubMed]

89. Mendelsohn, A.R.; Larrick, J.W. Sleep facilitates clearance of metabolites from the brain: Glymphatic function in aging and neurodegenerative diseases. *Rejuvenation Res.* **2013**, *16*, 518–523. [CrossRef] [PubMed]

90. Plog, B.A.; Nedergaard, M. The glymphatic system in central nervous system health and disease: Past, present, and future. *Annu. Rev. Pathol.* **2018**, *13*, 379–394. [CrossRef] [PubMed]

91. Boespflug, E.L.; Iliff, J.J. The emerging relationship between interstitial fluid-cerebrospinal fluid exchange, amyloid-β, and sleep. *Biol. Psychiatry* **2018**, *83*, 328–336. [CrossRef] [PubMed]

92. Yulug, B.; Hanoglu, L.; Kilic, E. Does sleep disturbance affect the amyloid clearance mechanisms in Alzheimer's disease? *Psychiatry Clin. Neurosci.* **2017**, *71*, 673–677. [CrossRef] [PubMed]

93. Pappolla, M.A.; Matsubara, E.; Vidal, R.; Pacheco-Quinto, J.; Poeggeler, B.; Zagorski, M.; Sambamurti, K. Melatonin treatment enhances Aβ lymphatic clearance in a transgenic mouse model of amyloidosis. *Curr. Alzheimer Res.* **2018**, *15*, 637–642. [CrossRef] [PubMed]

94. Spinedi, E.; Cardinali, D.P. Neuroendocrine-metabolic dysfunction and sleep disturbances in neurodegenerative disorders: Focus on Alzheimer's Disease and melatonin. *Neuroendocrinology* **2018**. [CrossRef] [PubMed]

95. Pappolla, M.A.; Chyan, Y.; Poeggeler, B.; Frangione, B.; Wilson, G.; Ghiso, J.; Reiter, R.J. An assessment of the antioxidant and the antiamyloidogenic properties of melatonin: Implications for Alzheimer's disease. *J. Neural Transm.* **2000**, *107*, 203–231. [CrossRef] [PubMed]

96. Srinivasan, V.; Pandi-Perumal, S.R.; Cardinali, D.P.; Poeggeler, B.; Hardeland, R. Melatonin in Alzheimer's disease and other neurodegenerative disorders. *Behav. Brain Funct.* **2006**, *2*, 15. [CrossRef] [PubMed]

97. Rosales-Corral, S.A.; Acuña-Castroviejo, D.; Coto-Montes, A.; Boga, J.A.; Manchester, L.C.; Fuentes-Broto, L.; Korkmaz, A.; Ma, S.; Tan, D.X.; Reiter, R.J. Alzheimer's disease: Pathological mechanisms and the beneficial role of melatonin. *J. Pineal Res.* **2012**, *52*, 167–202. [CrossRef] [PubMed]

98. Lin, L.; Huang, Q.X.; Yang, S.S.; Chu, J.; Wang, J.Z.; Tian, Q. Melatonin in Alzheimer's disease. *Int. J. Mol. Sci.* **2013**, *14*, 14575–14593. [CrossRef] [PubMed]

99. Song, W.; Lahiri, D.K. Melatonin alters the metabolism of the β-amyloid precursor protein in the neuroendocrine cell line PC12. *J. Mol. Neurosci.* **1997**, *9*, 75–92. [CrossRef] [PubMed]

100. Panmanee, J.; Nopparat, C.; Chavanich, N.; Shukla, M.; Mukda, S.; Song, W.; Vincent, B.; Govitrapong, P. Melatonin regulates the transcription of βAPP-cleaving secretases mediated through melatonin receptors in human neuroblastoma SH-SY5Y cells. *J. Pineal Res.* **2015**, *58*, 151–165. [CrossRef] [PubMed]

101. Shukla, M.; Htoo, H.H.; Wintachai, P.; Hernandez, J.F.; Dubois, C.; Postina, R.; Xu, H.; Checler, F.; Smith, D.R.; Govitrapong, P.; et al. Melatonin stimulates the nonamyloidogenic processing of βAPP through the positive transcriptional regulation of ADAM10 and ADAM17. *J. Pineal Res.* **2015**, *58*, 151–165. [CrossRef] [PubMed]

102. Mukda, S.; Panmanee, J.; Boontem, P.; Govitrapong, P. Melatonin administration reverses the alteration of amyloid precursor protein-cleaving secretases expression in aged mouse hippocampus. *Neurosci. Lett.* **2016**, *621*, 39–46. [CrossRef] [PubMed]

103. Shukla, M.; Govitrapong, P.; Boontem, P.; Reiter, R.J.; Satayavivad, J. Mechanisms of melatonin in alleviating Alzheimer's disease. *Curr. Neuropharmacol.* **2017**, *15*, 1010–1031. [CrossRef] [PubMed]

104. Matsubara, E.; Bryant-Thomas, T.; Pacheco Quinto, J.; Henry, T.L.; Poeggeler, B.; Herbert, D.; Cruz-Sanchez, F.; Chyan, Y.J.; Smith, M.A.; Perry, G.; et al. Melatonin increases survival and inhibits oxidative and amyloid pathology in a transgenic model of Alzheimer's disease. *J. Neurochem.* **2003**, *85*, 1101–1108. [CrossRef] [PubMed]

105. Olcese, J.M.; Cao, C.; Mori, T.; Mamcarz, M.B.; Maxwell, A.; Runfeldt, M.J.; Wang, L.; Zhang, C.; Lin, X.; Zhang, G.; et al. Protection against cognitive deficits and markers of neurodegeneration by long-term oral administrations of melatonin in a transgenic model of Alzheimer disease. *J. Pineal Res.* **2009**, *47*, 82–96. [CrossRef] [PubMed]

106. Quinn, J.; Kulhanek, D.; Nowlin, J.; Jones, R.; Praticò, D.; Rokach, J.; Stackman, R. Chronic melatonin therapy fails to alter amyloid burden or oxidative damage in old Tg2576 mice: Implications for clinical trials. *Brain Res.* **2005**, *1037*, 209–213. [CrossRef] [PubMed]

107. Clark, I.A.; Vissel, B. Treatment implications of the altered cytokine-insulin axis in neurodegenerative disease. *Biochem. Pharmacol.* **2013**, *86*, 862–871. [CrossRef] [PubMed]

108. Jiang, T.; Yu, J.T.; Zhu, X.C.; Tan, L. TREM2 in Alzheimer's disease. *Mol. Neurobiol.* **2013**, *48*, 180–185. [CrossRef] [PubMed]

109. Ferreira, L.S.S.; Fernandes, C.S.; Vieira, M.N.N.; De Felice, F.G. Insulin resistance in Alzheimer's disease. *Front. Neurosci.* **2018**, *12*, 830. [CrossRef] [PubMed]

110. Hotamisligil, G.S.; Erbay, E. Nutrient sensing and inflammation in metabolic diseases. *Nat. Rev. Immunol.* **2008**, *8*, 923–934. [CrossRef] [PubMed]

111. Hotamisligil, G.S. Inflammation, metaflammation and immunometabolic disorders. *Nature* **2017**, *542*, 177–185. [CrossRef] [PubMed]

112. Franceschi, C.; Garagnani, P.; Parini, P.; Giuliani, C.; Santoro, A. Inflammaging: A new immune-metabolic viewpoint for age-related diseases. *Nat. Rev. Endocrinol.* **2018**, *14*, 576–590. [CrossRef] [PubMed]

113. Cardinali, D.P.; Hardeland, R. Inflammaging, metabolic syndrome and melatonin: A call for treatment studies. *Neuroendocrinology* **2017**, *104*, 382–397. [CrossRef] [PubMed]

114. Coppé, J.P.; Patil, C.K.; Rodier, F.; Sun, Y.; Muñoz, D.P.; Goldstein, J.; Nelson, P.S.; Desprez, P.Y.; Campisi, J. Senescence-associated secretory phenotypes reveal cell-nonautonomous functions of oncogenic RAS and p53 tumor suppressor. *PLoS Biol.* **2008**, *6*, 2853–2868. [CrossRef] [PubMed]

115. Young, A.R.J.; Narita, M. SASP reflects senescence. *EMBO Rep.* **2009**, *10*, 228–230. [CrossRef] [PubMed]

116. Fumagalli, M.; d'Adda di Fagagna, F. SASPense and DDRama in cancer and ageing. *Nat. Cell Biol.* **2009**, *11*, 921–923. [CrossRef] [PubMed]

117. Coppé, J.P.; Desprez, P.Y.; Krtolica, A.; Campisi, J. The senescence-associated secretory phenotype: The dark side of tumor suppression. *Annu. Rev. Pathol.* **2010**, *5*, 99–118. [CrossRef] [PubMed]

118. Pantsulaia, I.; Ciszewski, W.M.; Niewiarowska, J. Senescent endothelial cells: Potential modulators of immunosenescence and ageing. *Ageing Res. Rev.* **2016**, *29*, 13–25. [CrossRef] [PubMed]

119. Salminen, A.; Ojala, J.; Kaarniranta, K.; Haapasalo, A.; Hiltunen, M.; Soininen, H. Astrocytes in the aging brain express characteristics of senescence-associated secretory phenotype. *Eur. J. Neurosci.* **2011**, *34*, 3–11. [CrossRef] [PubMed]

120. Maciel-Barón, L.Á.; Morales-Rosales, S.L.; Silva-Palacios, A.; Rodríguez-Barrera, R.H.; García-Álvarez, J.A.; Luna-López, A.; Pérez, V.I.; Torres, C.; Königsberg, M. The secretory phenotype of senescent astrocytes isolated from Wistar newborn rats changes with anti-inflammatory drugs, but does not have a short-term effect on neuronal mitochondrial potential. *Biogerontology* **2018**, *19*, 415–433. [CrossRef] [PubMed]

121. Carroll, J.E.; Cole, S.W.; Seeman, T.E.; Breen, E.C.; Witarama, T.; Arevalo, J.M.; Ma, J.; Irwin, M.R. Partial sleep deprivation activates the DNA damage response (DDR) and the senescence-associated secretory phenotype (SASP) in aged adult humans. *Brain Behav. Immun.* **2016**, *51*, 223–229. [CrossRef] [PubMed]

122. Yu, S.; Wang, X.; Geng, P.; Tang, X.; Xiang, L.; Lu, X.; Li, J.; Ruan, Z.; Chen, J.; Xie, G.; et al. Melatonin regulates PARP1 to control the senescence-associated secretory phenotype (SASP) in human fetal lung fibroblast cells. *J. Pineal Res.* **2017**, *63*, e12405. [CrossRef] [PubMed]

123. Bae, W.J.; Park, J.S.; Kang, S.K.; Kwon, I.K.; Kim, E.C. Effects of melatonin and its underlying mechanism on ethanol-stimulated senescence and osteoclastic differentiation in human periodontal ligament cells and cementoblasts. *Int. J. Mol. Sci.* **2018**, *19*, 1742. [CrossRef] [PubMed]

124. Fang, J.; Yan, Y.; Teng, X.; Wen, X.; Li, N.; Peng, S.; Liu, W.; Donadeu, F.X.; Zhao, S.; Hua, J. Melatonin prevents senescence of canine adipose-derived mesenchymal stem cells through activating NRF2 and inhibiting ER stress. *Aging* **2018**, *10*, 2954–2972. [CrossRef] [PubMed]

125. Hekmatimoghaddam, S.; Dehghani Firoozabadi, A.; Zare-Khormizi, M.R.; Pourrajab, F. Sirt1 and Parp1 as epigenome safeguards and microRNAs as SASP-associated signals, in cellular senescence and aging. *Ageing Res. Rev.* **2017**, *40*, 120–141. [CrossRef] [PubMed]

126. Korkmaz, A.; Rosales-Corral, S.; Reiter, R.J. Gene regulation by melatonin linked to epigenetic phenomena. *Gene* **2012**, *503*, 1–11. [CrossRef] [PubMed]

127. Franceschi, C.; Garagnani, P.; Vitale, G.; Capri, M.; Salvioli, S. Inflammaging and 'garb-aging'. *Trends Endocrinol. MeTable* **2017**, *28*, 199–212. [CrossRef] [PubMed]

128. Gilthorpe, J.D.; Oozeer, F.; Nash, J.; Calvo, M.; Bennett, D.L.; Lumsden, A.; Pini, A. Extracellular histone H1 is neurotoxic and drives a pro-inflammatory response in microglia. *F1000Resarch* **2013**, *2*, 148. [CrossRef] [PubMed]

129. Shimada, K.; Crother, T.R.; Karlin, J.; Dagvadorj, J.; Chiba, N.; Chen, S.; Ramanujan, V.K.; Wolf, A.J.; Vergnes, L.; Ojcius, D.M.; et al. Oxidized mitochondrial DNA activates the NLRP3 inflammasome during apoptosis. *Immunity* **2012**, *36*, 401–414. [CrossRef] [PubMed]

130. Fang, C.; Wei, X.; Wei, Y. Mitochondrial DNA in the regulation of innate immune responses. *Protein Cell* **2016**, *7*, 11–16. [CrossRef] [PubMed]

131. Glück, S.; Ablasser, A. Innate immunosensing of DNA in cellular senescence. *Curr. Opin. Immunol.* **2018**, *56*, 31–36. [CrossRef] [PubMed]

132. Rongvaux, A. Innate immunity and tolerance toward mitochondria. *Mitochondrion* **2018**, *41*, 14–20. [CrossRef] [PubMed]

133. Escames, G.; López, L.C.; García, J.A.; García-Corzo, L.; Ortiz, F.; Acuña-Castroviejo, D. Mitochondrial DNA and inflammatory diseases. *Hum. Genet.* **2012**, *131*, 161–173. [CrossRef] [PubMed]

134. Rahim, I.; Djerdjouri, B.; Sayed, R.K.; Fernández-Ortiz, M.; Fernández-Gil, B.; Hidalgo-Gutiérrez, A.; López, L.C.; Escames, G.; Reiter, R.J.; Acuña-Castroviejo, D. Melatonin administration to wild-type mice and nontreated NLRP3 mutant mice share similar inhibition of the inflammatory response during sepsis. *J. Pineal Res.* **2017**, *63*, e12410. [CrossRef] [PubMed]

135. Glück, S.; Guey, B.; Gulen, M.F.; Wolter, K.; Kang, T.W.; Schmacke, N.A.; Bridgeman, A.; Rehwinkel, J.; Zender, L.; Ablasser, A. Innate immune sensing of cytosolic chromatin fragments through cGAS promotes senescence. *Nat. Cell Biol.* **2017**, *19*, 1061–1070. [CrossRef] [PubMed]

136. Hamann, L.; Ruiz-Moreno, J.S.; Szwed, M.; Mossakowska, M.; Lundvall, L.; Schumann, R.R.; Opitz, B.; Puzianowska-Kuznicka, M. STING SNP R293Q is associated with a decreased risk of aging-related diseases. *Gerontology* **2018**, *26*, 1–10. [CrossRef] [PubMed]

137. Liu, Z.; Gan, L.; Xu, Y.; Luo, D.; Ren, Q.; Wu, S.; Sun, C. Melatonin alleviates inflammasome-induced pyroptosis through inhibiting NF-κB/GSDMD signal in mice adipose tissue. *J. Pineal Res.* **2017**, *63*, e12414. [CrossRef] [PubMed]

138. Deng, W.G.; Tang, S.T.; Tseng, H.P.; Wu, K.K. Melatonin suppresses macrophage cyclooxygenase-2 and inducible nitric oxide synthase expression by inhibiting p52 acetylation and binding. *Blood* **2006**, *108*, 518–524. [CrossRef] [PubMed]

139. Negi, G.; Kumar, A.; Sharma, S.S. Melatonin modulates neuroinflammation and oxidative stress in experimental diabetic neuropathy: Effects on NF-κB and Nrf2 cascades. *J. Pineal Res.* **2011**, *50*, 124–131. [CrossRef] [PubMed]

140. Murakami, Y.; Yuhara, K.; Takada, N.; Arai, T.; Tsuda, S.; Takamatsu, S.; Machino, M.; Fujisawa, S. Effect of melatonin on cyclooxygenase-2 expression and nuclear factor-kappa B activation in RAW264. *7 macrophage-like cells stimulated with fimbriae of Porphyromonas gingivalis. In Vivo* **2011**, *25*, 641–647. [PubMed]

141. Shi, D.; Xiao, X.; Wang, J.; Liu, L.; Chen, W.; Fu, L.; Xie, F.; Huang, W.; Deng, W. Melatonin suppresses proinflammatory mediators in lipopolysaccharide-stimulated CRL1999 cells via targeting MAPK, NF-κB, c/EBPβ, and p300 signaling. *J. Pineal Res.* **2012**, *53*, 154–165. [CrossRef] [PubMed]

142. Chinchalongporn, V.; Shukla, M.; Govitrapong, P. Melatonin ameliorates Aβ42 -induced alteration of βAPP-processing secretases via the melatonin receptor through the Pin1/GSK3β/NF-κB pathway in SH-SY5Y cells. *J. Pineal Res.* **2018**, *64*, e12470. [CrossRef] [PubMed]

143. Xia, M.Z.; Liang, Y.L.; Wang, H.; Chen, X.; Huang, Y.Y.; Zhang, Z.H.; Chen, Y.H.; Zhang, C.; Zhao, M.; Xu, D.X.; et al. Melatonin modulates TLR4-mediated inflammatory genes through MyD88- and TRIF-dependent signaling pathways in lipopolysaccharide-stimulated RAW264.7 cells. *J. Pineal Res.* **2012**, *53*, 325–334. [CrossRef] [PubMed]

144. Chuffa, L.G.; Fioruci-Fontanelli, B.A.; Mendes, L.O.; Ferreira Seiva, F.R.; Martinez, M.; Fávaro, W.J.; Domeniconi, R.F.; Pinheiro, P.F.; Delazari Dos Santos, L.; Martinez, F.E. Melatonin attenuates the TLR4-mediated inflammatory response through MyD88- and TRIF-dependent signaling pathways in an in vivo model of ovarian cancer. *BMC Cancer* **2015**, *15*, 34. [CrossRef] [PubMed]

145. Kang, J.W.; Koh, E.J.; Lee, S.M. Melatonin protects liver against ischemia and reperfusion injury through inhibition of toll-like receptor signaling pathway. *J. Pineal Res.* **2011**, *50*, 403–411. [CrossRef] [PubMed]

146. Huang, S.H.; Cao, X.J.; Wei, W. Melatonin decreases TLR3-mediated inflammatory factor expression via inhibition of NF-κB activation in respiratory syncytial virus-infected RAW264.7 macrophages. *J. Pineal Res.* **2008**, *45*, 93–100. [CrossRef] [PubMed]

147. Ding, K.; Wang, H.; Xu, J.; Lu, X.; Zhang, L.; Zhu, L. Melatonin reduced microglial activation and alleviated neuroinflammation induced neuron degeneration in experimental traumatic brain injury: Possible involvement of mTOR pathway. *Neurochem. Int.* **2014**, *76*, 23–31. [CrossRef] [PubMed]

148. Jumnongprakhon, P.; Govitrapong, P.; Tocharus, C.; Pinkaew, D.; Tocharus, J. Melatonin protects methamphetamine-induced neuroinflammation through NF-κB and Nrf2 pathways in glioma cell line. *Neurochem. Res.* **2015**, *40*, 1448–1456. [CrossRef] [PubMed]

149. Ismail, I.A.; El-Bakry, H.A.; Soliman, S.S. Melatonin and tumeric ameliorate aging-induced changes: Implication of immunoglobulins, cytokines, DJ-1/NRF2 and apoptosis regulation. *Int. J. Physiol. Pathophysiol. Pharmacol.* **2018**, *10*, 70–82. [PubMed]

150. El-Bakry, H.A.; Ismail, I.A.; Soliman, S.S. Immunosenescence-like state is accelerated by constant light exposure and counteracted by melatonin or turmeric administration through DJ-1/Nrf2 and P53/Bax pathways. *J. Photochem. Photobiol. B* **2018**, *186*, 69–80. [CrossRef] [PubMed]

151. Wang, J.; Jiang, C.; Zhang, K.; Lan, X.; Chen, X.; Zang, W.; Wang, Z.; Guan, F.; Zhu, C.; Yang, X.; et al. Melatonin receptor activation provides cerebral protection after traumatic brain injury by mitigating oxidative stress and inflammation via the Nrf2 signaling pathway. *Free Radic. Biol. Med.* **2019**, *131*, 345–355. [CrossRef] [PubMed]

152. Ariga, H.; Takahashi-Niki, K.; Kato, I.; Maita, H.; Niki, T.; Iguchi-Ariga, S.M. Neuroprotective function of DJ-1 in Parkinson's disease. *Oxid. Med. Cell. Longev.* **2013**, *2013*, 683920. [CrossRef] [PubMed]

153. Poulose, N.; Raju, R. Sirtuin regulation in aging and injury. *Biochim. Biophys. Acta* **2015**, *1852*, 2442–2455. [CrossRef] [PubMed]

154. Zhang, W.; Huang, Q.; Zeng, Z.; Wu, J.; Zhang, Y.; Chen, Z. Sirt1 inhibits oxidative stress in vascular endothelial cells. *Oxid. Med. Cell. Longev.* **2017**, *2017*, 7543973. [CrossRef] [PubMed]

155. Karbasforooshan, H.; Karimi, G. The role of SIRT1 in diabetic cardiomyopathy. *Biomed. Pharmacother.* **2017**, *90*, 386–392. [CrossRef] [PubMed]

156. Mendes, K.L.; Lelis, D.F.; Santos, S.H.S. Nuclear sirtuins and inflammatory signaling pathways. *Cytokine Growth Factor Rev.* **2017**, *38*, 98–105. [CrossRef] [PubMed]

157. Chai, D.; Zhang, L.; Xi, S.; Cheng, Y.; Jiang, H.; Hu, R. Nrf2 activation induced by Sirt1 ameliorates acute lung injury after intestinal ischemia/reperfusion through NOX4-mediated gene regulation. *Cell. Physiol. Biochem.* **2018**, *46*, 781–792. [CrossRef] [PubMed]

158. Han, J.; Liu, X.; Li, Y.; Zhang, J.; Yu, H. Sirt1/Nrf2 signalling pathway prevents cognitive impairment in diabetic rats through anti-oxidative stress induced by miRNA-23b-3p expression. *Mol. Med. Rep.* **2018**, *17*, 8414–8422. [CrossRef] [PubMed]

159. Huang, K.; Li, R.; Wei, W. Sirt1 activation prevents anti-Thy 1.1 mesangial proliferative glomerulonephritis in the rat through the Nrf2/ARE pathway. *Eur. J. Pharmacol.* **2018**, *832*, 138–144. [CrossRef] [PubMed]

160. Yang, B.; Xu, B.; Zhao, H.; Wang, Y.B.; Zhang, J.; Li, C.W.; Wu, Q.; Cao, Y.K.; Li, Y.; Cao, F. Dioscin protects against coronary heart disease by reducing oxidative stress and inflammation via Sirt1/Nrf2 and p38 MAPK pathways. *Mol. Med. Rep.* **2018**, *18*, 973–980. [CrossRef] [PubMed]

161. Wang, X.; Yuan, B.; Cheng, B.; Liu, Y.; Zhang, B.; Wang, X.; Lin, X.; Yang, B.; Gong, G. Crocin alleviates myocardial ischemia/reperfusion-induced endoplasmic reticulum stress via regulation of miR-34a/Sirt1/Nrf2 pathway. *Shock* **2019**, *51*, 123–130. [CrossRef] [PubMed]

162. Li, Y.; Yang, G.; Yang, X.; Wang, W.; Zhang, J.; He, Y.; Zhang, W.; Jing, T.; Lin, R. Nicotinic acid inhibits NLRP3 inflammasome activation via SIRT1 in vascular endothelial cells. *Int. Immunopharmacol.* **2016**, *40*, 211–218. [CrossRef] [PubMed]

163. Li, Y.; Yang, X.; He, Y.; Wang, W.; Zhang, J.; Zhang, W.; Jing, T.; Wang, B.; Lin, R. Negative regulation of NLRP3 inflammasome by SIRT1 in vascular endothelial cells. *Immunobiology* **2017**, *222*, 552–561. [CrossRef] [PubMed]

164. Zhang, S.; Jiang, L.; Che, F.; Lu, Y.; Xie, Z.; Wang, H. Arctigenin attenuates ischemic stroke via SIRT1-dependent inhibition of NLRP3 inflammasome. *Biochem. Biophys. Res. Commun.* **2017**, *493*, 821–826. [CrossRef] [PubMed]

165. Zou, P.; Liu, X.; Li, G.; Wang, Y. Resveratrol pretreatment attenuates traumatic brain injury in rats by suppressing NLRP3 inflammasome activation via SIRT1. *Mol. Med. Rep.* **2018**, *17*, 3212–3217. [CrossRef] [PubMed]

166. Lin, Q.Q.; Geng, Y.W.; Jiang, Z.W.; Tian, Z.J. SIRT1 regulates lipopolysaccharide-induced CD40 expression in renal medullary collecting duct cells by suppressing the TLR4-NF-κB signaling pathway. *Life Sci.* **2017**, *170*, 100–107. [CrossRef] [PubMed]

167. Yuan, Y.; Liu, Q.; Zhao, J.; Tang, H.; Sun, J. SIRT1 attenuates murine allergic rhinitis by downregulated HMGB 1/TLR4 pathway. *Scand. J. Immunol.* **2018**, *87*, e12667. [CrossRef] [PubMed]

168. Li, K.; Lv, G.; Pan, L. Sirt1 alleviates LPS induced inflammation of periodontal ligament fibroblasts via downregulation of TLR4. *Int. J. Biol. Macromol.* **2018**, *119*, 249–254. [CrossRef] [PubMed]

169. Rabadi, M.M.; Xavier, S.; Vasko, R.; Kaur, K.; Goligorksy, M.S.; Ratliff, B.B. High-mobility group box 1 is a novel deacetylation target of Sirtuin1. *Kidney Int.* **2015**, *87*, 95–108. [CrossRef] [PubMed]

170. Hwang, J.S.; Choi, H.S.; Ham, S.A.; Yoo, T.; Lee, W.J.; Paek, K.S.; Seo, H.G. Deacetylation-mediated interaction of SIRT1-HMGB1 improves survival in a mouse model of endotoxemia. *Sci. Rep.* **2015**, *5*, 15971. [CrossRef] [PubMed]

171. Rickenbacher, A.; Jang, J.H.; Limani, P.; Ungethüm, U.; Lehmann, K.; Oberkofler, C.E.; Weber, A.; Graf, R.; Humar, B.; Clavien, P.A. Fasting protects liver from ischemic injury through Sirt1-mediated downregulation of circulating HMGB1 in mice. *J. Hepatol.* **2014**, *61*, 301–308. [CrossRef] [PubMed]

172. Xu, W.; Lu, Y.; Yao, J.; Li, Z.; Chen, Z.; Wang, G.; Jing, H.; Zhang, X.; Li, M.; Peng, J.; Tian, X. Novel role of resveratrol: Suppression of high-mobility group protein box 1 nucleocytoplasmic translocation by the upregulation of sirtuin 1 in sepsis-induced liver injury. *Shock* **2014**, *42*, 440–447. [CrossRef] [PubMed]

173. Hwang, J.S.; Lee, W.J.; Kang, E.S.; Ham, S.A.; Yoo, T.; Paek, K.S.; Lim, D.S.; Do, J.T.; Seo, H.G. Ligand-activated peroxisome proliferator-activated receptor-δ and -γ inhibit lipopolysaccharide-primed release of high mobility group box 1 through upregulation of SIRT1. *Cell Death Dis.* **2014**, *5*, e1432. [CrossRef] [PubMed]

174. Yin, Y.; Feng, Y.; Zhao, H.; Zhao, Z.; Yua, H.; Xu, J.; Che, H. SIRT1 inhibits releases of HMGB1 and HSP70 from human umbilical vein endothelial cells caused by IL-6 and the serum from a preeclampsia patient and protects the cells from death. *Biomed. Pharmacother.* **2017**, *88*, 449–458. [CrossRef] [PubMed]

175. Chen, X.; Wu, S.; Chen, C.; Wu, S.; Yang, F.; Fang, Z.; Fu, H.; Li, Y. Omega-3 polyunsaturated fatty acid supplementation attenuates microglial-induced inflammation by inhibiting the HMGB1/TLR4/NF-κB pathway following experimental traumatic brain injury. *J. Neuroinflamm.* **2017**, *14*, 143. [CrossRef] [PubMed]

176. Hwang, J.S.; Kang, E.S.; Han, S.G.; Lim, D.S.; Paek, K.S.; Lee, C.H.; Seo, H.G. Formononetin inhibits lipopolysaccharide-induced release of high mobility group box 1 by upregulating SIRT1 in a PPARδ-dependent manner. *Peer J.* **2018**, *6*, e4208. [CrossRef] [PubMed]

177. Su, Z.; Zhang, P.; Yu, Y.; Lu, H.; Liu, Y.; Ni, P.; Su, X.; Wang, D.; Liu, Y.; Wang, J.; et al. HMGB1 facilitated macrophage reprogramming towards a proinflammatory M1-like phenotype in experimental autoimmune myocarditis development. *Sci. Rep.* **2016**, *6*, 21884. [CrossRef] [PubMed]

178. Karuppagounder, V.; Giridharan, V.V.; Arumugam, S.; Sreedhar, R.; Palaniyandi, S.S.; Krishnamurthy, P.; Quevedo, J.; Watanabe, K.; Konishi, T.; Thandavarayan, R.A. Modulation of macrophage polarization and HMGB1-TLR2/TLR4 cascade plays a crucial role for cardiac remodeling in senescence-accelerated prone mice. *PLoS ONE* **2016**, *11*, e0152922. [CrossRef] [PubMed]

179. Son, M.; Porat, A.; He, M.; Suurmond, J.; Santiago-Schwarz, F.; Andersson, U.; Coleman, T.R.; Volpe, B.T.; Tracey, K.J.; Al-Abed, Y.; et al. C1q and HMGB1 reciprocally regulate human macrophage polarization. *Blood* **2016**, *128*, 2218–2228. [CrossRef] [PubMed]

180. Jiang, Y.; Chen, R.; Shao, X.; Ji, X.; Lu, H.; Zhou, S.; Zong, G.; Xu, H.; Su, Z. HMGB1 silencing in macrophages prevented their functional skewing and ameliorated EAM development: Nuclear HMGB1 may be a checkpoint molecule of macrophage reprogramming. *Int. Immunopharmacol.* **2018**, *56*, 277–284. [CrossRef] [PubMed]

181. Gao, T.; Chen, Z.; Chen, H.; Yuan, H.; Wang, Y.; Peng, X.; Wei, C.; Yang, J.; Xu, C. Inhibition of HMGB1 mediates neuroprotection of traumatic brain injury by modulating the microglia/macrophage polarization. *Biochem. Biophys. Res. Commun.* **2018**, *497*, 430–436. [CrossRef] [PubMed]

182. Prieto-Domínguez, N.; Méndez-Blanco, C.; Carbajo-Pescador, S.; Fondevila, F.; García-Palomo, A.; González-Gallego, J.; Mauriz, J.L. Melatonin enhances sorafenib actions in human hepatocarcinoma cells by inhibiting mTORC1/p70S6K/HIF-1α and hypoxia-mediated mitophagy. *Oncotarget* **2017**, *8*, 91402–91414. [CrossRef] [PubMed]

183. Fan, T.; Pi, H.; Li, M.; Ren, Z.; He, Z.; Zhu, F.; Tian, L.; Tu, M.; Xie, J.; Liu, M.; et al. Inhibiting MT2-TFE3-dependent autophagy enhances melatonin-induced apoptosis in tongue squamous cell carcinoma. *J. Pineal Res.* **2018**, *64*, e12457. [CrossRef] [PubMed]

184. Yun, S.M.; Woo, S.H.; Oh, S.T.; Hong, S.E.; Choe, T.B.; Ye, S.K.; Kim, E.K.; Seong, M.K.; Kim, H.A.; Noh, W.C.; et al. Melatonin enhances arsenic trioxide-induced cell death via sustained upregulation of Redd1 expression in breast cancer cells. *Mol. Cell. Endocrinol.* **2016**, *422*, 64–73. [CrossRef] [PubMed]

185. Liu, Z.; Gan, L.; Liu, G.; Chen, Y.; Wu, T.; Feng, F.; Sun, C. Sirt1 decreased adipose inflammation by interacting with Akt2 and inhibiting mTOR/S6K1 pathway in mice. *J. Lipid Res.* **2016**, *57*, 1373–1381. [CrossRef] [PubMed]

186. Chen, H.; Shen, F.; Sherban, A.; Nocon, A.; Li, Y.; Wang, H.; Xu, M.J.; Rui, X.; Han, J.; Jiang, B.; et al. DEP domain-containing mTOR-interacting protein suppresses lipogenesis and ameliorates hepatic steatosis and acute-on-chronic liver injury in alcoholic liver disease. *Hepatology* **2018**, *68*, 496–514. [CrossRef] [PubMed]

187. Guarani, V.; Deflorian, G.; Franco, C.A.; Krüger, M.; Phng, L.K.; Bentley, K.; Toussaint, L.; Dequiedt, F.; Mostoslavsky, R.; Schmidt, M.H.H.; et al. Acetylation-dependent regulation of endothelial Notch signalling by the SIRT1 deacetylase. *Nature* **2011**, *473*, 234–238. [CrossRef] [PubMed]

188. Bai, X.; He, T.; Liu, Y.; Li, X.; Shi, J.; Wang, K.; Han, F.; Zhang, W.; Zhang, Y.; Cai, W.; et al. Acetylation-dependent regulation of Notch signaling in macrophages by SIRT1 affects sepsis development. *Front. Immunol.* **2018**, *9*, 762. [CrossRef] [PubMed]

189. Yang, J.J.; Tao, H.; Liu, L.P.; Hu, W.; Deng, Z.Y.; Li, J. miR-200a controls hepatic stellate cell activation and fibrosis via SIRT1/Notch1 signal pathway. *Inflamm. Res.* **2017**, *66*, 341–352. [CrossRef] [PubMed]

190. Yu, L.; Liang, H.; Lu, Z.; Zhao, G.; Zhai, M.; Yang, Y.; Yang, J.; Yi, D.; Chen, W.; Wang, X.; et al. Membrane receptor-dependent Notch1/Hes1 activation by melatonin protects against myocardial ischemia-reperfusion injury: In vivo and in vitro studies. *J. Pineal Res.* **2015**, *59*, 420–433. [CrossRef] [PubMed]

191. Zhang, S.; Wang, P.; Ren, L.; Hu, C.; Bi, J. Protective effect of melatonin on soluble Aβ$_{1-42}$-induced memory impairment, astrogliosis, and synaptic dysfunction via the Musashi1/Notch1/Hes1 signaling pathway in the rat hippocampus. *Alzheimers Res. Ther.* **2016**, *8*, 40. [CrossRef] [PubMed]

192. Yu, L.; Fan, C.; Li, Z.; Zhang, J.; Xue, X.; Xu, Y.; Zhao, G.; Yang, Y.; Wang, H. Melatonin rescues cardiac thioredoxin system during ischemia-reperfusion injury in acute hyperglycemic state by restoring Notch1/Hes1/Akt signaling in a membrane receptor-dependent manner. *J. Pineal Res.* **2017**, *62*, e12375. [CrossRef] [PubMed]

193. Jia, Y.; Li, Z.; Cai, W.; Xiao, D.; Han, S.; Han, F.; Bai, X.; Wang, K.; Liu, Y.; Li, X.; et al. SIRT1 regulates inflammation response of macrophages in sepsis mediated by long noncoding RNA. *Biochim. Biophys. Acta* **2018**, *1864*, 784–792. [CrossRef] [PubMed]

194. Xia, Y.; Chen, S.; Zeng, S.; Zhao, Y.; Zhu, C.; Deng, B.; Zhu, G.; Yin, Y.; Wang, W.; Hardeland, R.; et al. Melatonin in macrophage biology: Current understanding and future perspectives. *J. Pineal Res.* **2018**, *66*, e12547. [CrossRef] [PubMed]

195. Lau, W.W.; Ng, J.K.; Lee, M.M.; Chan, A.S.; Wong, Y.H. Interleukin-6 autocrine signaling mediates melatonin $MT_{1/2}$ receptor-induced STAT3 Tyr_{705} phosphorylation. *J. Pineal Res.* **2012**, *52*, 477–489. [CrossRef] [PubMed]

196. Yang, Y.; Duan, W.; Jin, Z.; Yi, W.; Yan, J.; Zhang, S.; Wang, N.; Liang, Z.; Li, Y.; Chen, W.; et al. JAK2/STAT3 activation by melatonin attenuates the mitochondrial oxidative damage induced by myocardial ischemia/reperfusion injury. *J. Pineal Res.* **2013**, *55*, 275–286. [CrossRef] [PubMed]

197. Yi, W.J.; Kim, T.S. Melatonin protects mice against stress-induced inflammation through enhancement of M2 macrophage polarization. *Int. Immunopharmacol.* **2017**, *48*, 146–158. [CrossRef] [PubMed]

198. Ryo, A.; Suizu, F.; Yoshida, Y.; Perrem, K.; Liou, Y.C.; Wulf, G.; Rottapel, R.; Yamaoka, S.; Lu, K.P. Regulation of NF-κB signaling by Pin1-dependent prolyl isomerization and ubiquitin-mediated proteolysis of p65/RelA. *Mol. Cell* **2003**, *12*, 1413–1426. [CrossRef]

199. Strebovsky, J.; Walker, P.; Lang, R.; Dalpke, A.H. Suppressor of cytokine signaling 1 (SOCS1) limits NFκB signaling by decreasing p65 stability within the cell nucleus. *FASEB J.* **2011**, *25*, 863–874. [CrossRef] [PubMed]

200. Liu, Y.; Liu, X.; Hua, W.; Wei, Q.; Fang, X.; Zhao, Z.; Ge, C.; Liu, C.; Chen, C.; Tao, Y.; et al. Berberine inhibits macrophage M1 polarization via AKT1/SOCS1/NF-κB signaling pathway to protect against DSS-induced colitis. *Int. Immunopharmacol.* **2018**, *57*, 121–131. [CrossRef] [PubMed]

201. García, J.A.; Volt, H.; Venegas, C.; Doerrier, C.; Escames, G.; López, L.C.; Acuña-Castroviejo, D. Disruption of the NF-κB/NLRP3 connection by melatonin requires retinoid-related orphan receptor-α and blocks the septic response in mice. *FASEB J.* **2015**, *29*, 3863–3875. [CrossRef] [PubMed]

202. Nejati Moharrami, N.; Bjørkøy Tande, E.; Ryan, L.; Espevik, T.; Boyartchuk, V. RORα controls inflammatory state of human macrophages. *PLoS ONE* **2018**, *13*, e0207374. [CrossRef] [PubMed]

203. Slominski, A.T.; Kim, T.K.; Takeda, Y.; Janjetovic, Z.; Brozyna, A.A.; Skobowiat, C.; Wang, J.; Postlethwaite, A.; Li, W.; Tuckey, R.C.; et al. RORα and RORγ are expressed in human skin and serve as receptors for endogenously produced noncalcemic 20-hydroxy- and 20,23-dihydroxyvitamin D. *FASEB J.* **2014**, *28*, 2775–2789. [CrossRef] [PubMed]

204. Slominski, A.T.; Zmijewski, M.A.; Jetten, A.M. RORα is not a receptor for melatonin (response to DOI 10.1002/bies.201600018). *BioEssays* **2016**, *38*, 1193–1194. [CrossRef] [PubMed]

205. Zhang, S.; Gao, L.; Liu, X.; Lu, T.; Xie, C.; Jia, J. Resveratrol attenuates microglial activation via SIRT1-SOCS1 pathway. *Evid. Based Complement. Alternat. Med.* **2017**, *2017*, 8791832. [CrossRef] [PubMed]

206. Warren, K.J.; Fang, X.; Gowda, N.M.; Thompson, J.J.; Heller, N.M. The TORC1-activated proteins, p70S6K and GRB10, regulate IL-4 signaling and M2 macrophage polarization by modulating phosphorylation of insulin receptor substrate-2. *J. Biol. Chem.* **2016**, *291*, 24922–24930. [CrossRef] [PubMed]

207. Klinge, C.M. Non-coding RNAs in breast cancer: Intracellular and intercellular communication. *Noncoding RNA* **2018**, *4*, 40. [CrossRef] [PubMed]

208. Kufel, J.; Grzechnik, P. Small nucleolar RNAs tell a different tale. *Trends Genet.* **2019**, *35*, 104–117. [CrossRef] [PubMed]

209. Hardeland, R. On the relationships between lncRNAs and other orchestrating regulators: Role of the circadian system. *Epigenomes* **2018**, *2*, 9. [CrossRef]

210. Czech, B.; Munafò, M.; Ciabrelli, F.; Eastwood, E.L.; Fabry, M.H.; Kneuss, E.; Hannon, G.J. piRNA-guided genome defense: From biogenesis to silencing. *Annu. Rev. Genet.* **2018**, *52*, 131–157. [CrossRef] [PubMed]

211. Hardeland, R. Intercellular communication via exosomal and ectosomal microRNAs: Facing a jungle of countless microRNAs and targets. In *Mini-Reviews in Recent Melatonin Research*; Hardeland, R., Ed.; Cuvillier: Göttingen, Germany, 2017; pp. 109–122, ISBN 978-3-7369-9677-9.

212. Hardeland, R. Exosomal and ectosomal noncoding RNAs in cancer—More than diagnostic and prognostic markers. *Open Acc. J. Oncol. Med.* **2018**, *2*, 000144.

213. Hardeland, R. Melatonin, noncoding RNAs, messenger RNA stability and epigenetics—Evidence, hints, gaps and perspectives. *Int. J. Mol. Sci.* **2014**, *15*, 18221–18252. [CrossRef] [PubMed]

214. Hardeland, R. Interactions of melatonin and microRNAs. *Biochem. Mol. Biol. J.* **2018**, *4*, 7. [CrossRef]

215. Hardeland, R. Melatonin and chromatin. *Melatonin Res.* **2019**, *2*, 67–93. [CrossRef]

216. Mann, M.; Mehta, A.; Zhao, J.L.; Lee, K.; Marinov, G.K.; Garcia-Flores, Y.; Lu, L.F.; Rudensky, A.Y.; Baltimore, D. An NF-κB-microRNA regulatory network tunes macrophage inflammatory responses. *Nat. Commun.* **2017**, *8*, 851. [CrossRef] [PubMed]

217. Yang, Y.; Liu, D.; Xi, Y.; Li, J.; Liu, B.; Li, J. Upregulation of miRNA-140-5p inhibits inflammatory cytokines in acute lung injury through the MyD88/NF-κB signaling pathway by targeting TLR4. *Exp. Ther. Med.* **2018**, *16*, 3913–3920. [CrossRef] [PubMed]

218. Zhao, G.; Zhang, T.; Wu, H.; Jiang, K.; Qiu, C.; Deng, G. MicroRNA let-7c improves LPS-induced outcomes of endometritis by suppressing NF-κB signaling. *Inflammation* **2018**. [CrossRef] [PubMed]

219. Zheng, D.; Zang, Y.; Xu, H.; Wang, Y.; Cao, X.; Wang, T.; Pan, M.; Shi, J.; Li, X. MicroRNA-214 promotes the calcification of human aortic valve interstitial cells through the acceleration of inflammatory reactions with activated MyD88/NF-κB signaling. *Clin. Res. Cardiol.* **2018**. [CrossRef] [PubMed]

220. Garavelli, S.; De Rosa, V.; de Candia, P. The multifaceted interface between cytokines and microRNAs: An ancient mechanism to regulate the good and the bad of inflammation. *Front. Immunol.* **2018**, *9*, 3012. [CrossRef] [PubMed]

221. Lee, S.E.; Kim, S.J.; Youn, J.P.; Hwang, S.Y.; Park, C.S.; Park, Y.S. MicroRNA and gene expression analysis of melatonin-exposed breast cancer cell lines indicating involvement of the anticancer effect. *J. Pineal Res.* **2011**, *51*, 345–352. [CrossRef] [PubMed]

222. Hardeland, R. Melatonin and microRNAs: An emerging field. In *Mini-Reviews in Recent Melatonin Research*; Hardeland, R., Ed.; Cuvillier: Göttingen, Germany, 2017; pp. 105–108, ISBN 978-3-7369-9677-9.

223. Koltai, E.; Bori, Z.; Osvath, P.; Ihasz, F.; Peter, S.; Toth, G.; Degens, H.; Rittweger, J.; Boldogh, I.; Radak, Z. Master athletes have higher miR-7, SIRT3 and SOD2 expression in skeletal muscle than age-matched sedentary controls. *Redox Biol.* **2018**, *19*, 46–51. [CrossRef] [PubMed]

224. Kabaria, S.; Choi, D.C.; Chaudhuri, A.D.; Jain, M.R.; Li, H.; Junn, E. MicroRNA-7 activates Nrf2 pathway by targeting Keap1 expression. *Free Radic. Biol. Med.* **2015**, *89*, 548–556. [CrossRef] [PubMed]

225. Choi, D.C.; Chae, Y.J.; Kabaria, S.; Chaudhuri, A.D.; Jain, M.R.; Li, H.; Mouradian, M.M.; Junn, E. MicroRNA-7 protects against 1-methyl-4-phenylpyridinium-induced cell death by targeting RelA. *J. Neurosci.* **2014**, *34*, 12725–12737. [CrossRef] [PubMed]

226. Zhao, X.D.; Lu, Y.Y.; Guo, H.; Xie, H.H.; He, L.J.; Shen, G.F.; Zhou, J.F.; Li, T.; Hu, S.J.; Zhou, L.; et al. MicroRNA-7/NF-κB signaling regulatory feedback circuit regulates gastric carcinogenesis. *J. Cell Biol.* **2015**, *210*, 613–627. [CrossRef] [PubMed]

227. Giles, K.M.; Brown, R.A.; Ganda, C.; Podgorny, M.J.; Candy, P.A.; Wintle, L.C.; Richardson, K.L.; Kalinowski, F.C.; Stuart, L.M.; Epis, M.R.; et al. microRNA-7-5p inhibits melanoma cell proliferation and metastasis by suppressing RelA/NF-κB. *Oncotarget* **2016**, *7*, 31663–31680. [CrossRef] [PubMed]

228. Saunders, L.R.; Sharma, A.D.; Tawney, J.; Nakagawa, M.; Okita, K.; Yamanaka, S.; Willenbring, H.; Verdin, E. miRNAs regulate SIRT1 expression during mouse embryonic stem cell differentiation and in adult mouse tissues. *Aging* **2010**, *2*, 415–431. [CrossRef] [PubMed]

229. Ramachandran, D.; Roy, U.; Garg, S.; Ghosh, S.; Pathak, S.; Kolthur-Seetharam, U. Sirt1 and mir-9 expression is regulated during glucose-stimulated insulin secretion in pancreatic β-islets. *FEBS J.* **2011**, *278*, 1167–1174. [CrossRef] [PubMed]

230. Schonrock, N.; Humphreys, D.T.; Preiss, T.; Götz, J. Target gene repression mediated by miRNAs miR-181c and miR-9 both of which are down-regulated by amyloid-β. *J. Mol. Neurosci.* **2012**, *46*, 324–335. [CrossRef] [PubMed]

231. Ao, R.; Wang, Y.; Tong, J.; Wang, B.F. Altered microRNA-9 expression level is directly correlated with pathogenesis of nonalcoholic fatty liver disease by targeting Onecut2 and SIRT1. *Med. Sci. Monit.* **2016**, *22*, 3804–3819. [CrossRef] [PubMed]

232. D'Adamo, S.; Cetrullo, S.; Guidotti, S.; Borzì, R.M.; Flamigni, F. Hydroxytyrosol modulates the levels of microRNA-9 and its target sirtuin-1 thereby counteracting oxidative stress-induced chondrocyte death. *Osteoarthritis Cartilage* **2017**, *25*, 600–610. [CrossRef] [PubMed]

233. Zhou, L.; Fu, L.; Lv, N.; Chen, X.S.; Liu, J.; Li, Y.; Xu, Q.Y.; Huang, S.; Zhang, X.D.; Dou, L.P.; et al. A minicircuitry comprised of microRNA-9 and SIRT1 contributes to leukemogenesis in t(8;21) acute myeloid leukemia. *Eur. Rev. Med. Pharmacol. Sci.* **2017**, *21*, 786–794. [PubMed]

234. Owczarz, M.; Budzinska, M.; Domaszewska-Szostek, A.; Borkowska, J.; Polosak, J.; Gewartowska, M.; Slusarczyk, P.; Puzianowska-Kuznicka, M. miR-34a and miR-9 are overexpressed and SIRT genes are downregulated in peripheral blood mononuclear cells of aging humans. *Exp. Biol. Med.* **2017**, *242*, 1453–1461. [CrossRef] [PubMed]

235. Bu, P.; Luo, C.; He, Q.; Yang, P.; Li, X.; Xu, D. MicroRNA-9 inhibits the proliferation and migration of malignant melanoma cells via targeting sirtuin 1. *Exp. Ther. Med.* **2017**, *14*, 931–938. [CrossRef] [PubMed]

236. Khosravi, A.; Alizadeh, S.; Jalili, A.; Shirzad, R.; Saki, N. The impact of Mir-9 regulation in normal and malignant hematopoiesis. *Oncol. Rev.* **2018**, *12*, e348. [CrossRef] [PubMed]

237. Rushworth, S.A.; Murray, M.Y.; Barrera, L.N.; Heasman, S.A.; Zaitseva, L.; MacEwan, D.J. Understanding the role of miRNA in regulating NF-κB in blood cancer. *Am. J. Cancer Res.* **2012**, *2*, 65–74. [PubMed]

238. Wang, L.Q.; Kwong, Y.L.; Kho, C.S.; Wong, K.F.; Wong, K.Y.; Ferracin, M.; Calin, G.A.; Chim, C.S. Epigenetic inactivation of miR-9 family microRNAs in chronic lymphocytic leukemia—Implications on constitutive activation of NFκB pathway. *Mol. Cancer* **2013**, *12*, 173. [CrossRef] [PubMed]

239. Zhang, Z.X.; Liu, Z.Q.; Jiang, B.; Lu, X.Y.; Ning, X.F.; Yuan, C.T.; Wang, A.L. BRAF activated non-coding RNA (BANCR) promoting gastric cancer cells proliferation via regulation of NF-κB1. *Biochem. Biophys. Res. Commun.* **2015**, *465*, 225–231. [CrossRef] [PubMed]

240. Chakraborty, S.; Zawieja, D.C.; Davis, M.J.; Muthuchamy, M. MicroRNA signature of inflamed lymphatic endothelium and role of miR-9 in lymphangiogenesis and inflammation. *Am. J. Physiol. Cell Physiol.* **2015**, *309*, C680–C692. [CrossRef] [PubMed]

241. Gu, R.; Liu, N.; Luo, S.; Huang, W.; Zha, Z.; Yang, J. MicroRNA-9 regulates the development of knee osteoarthritis through the NF-kappaB1 pathway in chondrocytes. *Medicine* **2016**, *95*, e4315. [CrossRef] [PubMed]

242. Qian, D.; Wei, G.; Xu, C.; He, Z.; Hua, J.; Li, J.; Hu, Q.; Lin, S.; Gong, J.; Meng, H.; et al. Bone marrow-derived mesenchymal stem cells (BMSCs) repair acute necrotized pancreatitis by secreting microRNA-9 to target the NF-κB1/p50 gene in rats. *Sci. Rep.* **2017**, *7*, 581. [CrossRef] [PubMed]

243. Yao, H.; Ma, R.; Yang, L.; Hu, G.; Chen, X.; Duan, M.; Kook, Y.; Niu, F.; Liao, K.; Fu, M.; et al. MiR-9 promotes microglial activation by targeting MCPIP1. *Nat. Commun.* **2014**, *5*, 4386. [CrossRef] [PubMed]

244. Huang, G.; Liu, X.; Zhao, X.; Zhao, J.; Hao, J.; Ren, J.; Chen, Y. MiR-9 promotes multiple myeloma progression by regulating TRIM56/NF-κB pathway. *Cell Biol. Int.* **2019**. [CrossRef] [PubMed]

245. Zhu, M.; Zhou, X.; Du, Y.; Huang, Z.; Zhu, J.; Xu, J.; Cheng, G.; Shu, Y.; Liu, P.; Zhu, W.; et al. miR-20a induces cisplatin resistance of a human gastric cancer cell line via targeting CYLD. *Mol. Med. Rep.* **2016**, *14*, 1742–1750. [CrossRef] [PubMed]

246. Liu, Z.; Yu, H.; Guo, Q. MicroRNA-20a promotes inflammation via the nuclear factor-κB signaling pathway in pediatric pneumonia. *Mol. Med. Rep.* **2018**, *17*, 612–617. [CrossRef] [PubMed]

247. Lin, Q.; Geng, Y.; Zhao, M.; Lin, S.; Zhu, Q.; Tian, Z. MiR-21 regulates TNF-α-induced CD40 expression via the SIRT1-NF-κB pathway in renal inner medullary collecting duct cells. *Cell. Physiol. Biochem.* **2017**, *41*, 124–136. [CrossRef] [PubMed]

248. Kim, S.J.; Kang, H.S.; Lee, J.H.; Park, J.H.; Jung, C.H.; Bae, J.H.; Oh, B.C.; Song, D.K.; Baek, W.K.; Im, S.S. Melatonin ameliorates ER stress-mediated hepatic steatosis through miR-23a in the liver. *Biochem. Biophys. Res. Commun.* **2015**, *458*, 462–469. [CrossRef] [PubMed]

249. Weinberg, R.B.; Mufson, E.J.; Counts, S.E. Evidence for a neuroprotective microRNA pathway in amnestic mild cognitive impairment. *Front. Neurosci.* **2015**, *9*, 430. [CrossRef] [PubMed]

250. Sruthi, T.V.; Edatt, L.; Raji, G.R.; Kunhiraman, H.; Shankar, S.S.; Shankar, V.; Ramachandran, V.; Poyyakkara, A.; Kumar, S.V.B. Horizontal transfer of miR-23a from hypoxic tumor cell colonies can induce angiogenesis. *J. Cell. Physiol.* **2018**, *233*, 3498–3514. [CrossRef] [PubMed]

251. Luo, H.; Han, Y.; Liu, J.; Zhang, Y. Identification of microRNAs in granulosa cells from patients with different levels of ovarian reserve function and the potential regulatory function of miR-23a in granulosa cell apoptosis. *Gene* **2019**, *686*, 250–260. [CrossRef] [PubMed]

252. Khan, A.U.H.; Rathore, M.G.; Allende-Vega, N.; Vo, D.N.; Belkhala, S.; Orecchioni, S.; Talarico, G.; Bertolini, F.; Cartron, G.; Lecellier, C.H.; et al. Human leukemic cells performing oxidative phosphorylation (OXPHOS) generate an antioxidant response independently of reactive oxygen species (ROS) production. *EBioMedicine* **2015**, *3*, 43–53. [CrossRef] [PubMed]

253. Zhang, D.; Qiu, X.; Li, J.; Zheng, S.; Li, L.; Zhao, H. MiR-23a-3p-regulated abnormal acetylation of FOXP3 induces regulatory T cell function defect in Graves' disease. *Biol. Chem.* **2018**. [CrossRef] [PubMed]

254. Zhao, S.; Li, T.; Li, J.; Lu, Q.; Han, C.; Wang, N.; Qiu, Q.; Cao, H.; Xu, X.; Chen, H.; et al. miR-23b-3p induces the cellular metabolic memory of high glucose in diabetic retinopathy through a SIRT1-dependent signalling pathway. *Diabetologia* **2016**, *59*, 644–654. [CrossRef] [PubMed]

255. Mori, F.; Ferraiuolo, M.; Santoro, R.; Sacconi, A.; Goeman, F.; Pallocca, M.; Pulito, C.; Korita, E.; Fanciulli, M.; Muti, P.; et al. Multitargeting activity of miR-24 inhibits long-term melatonin anticancer effects. *Oncotarget* **2016**, *7*, 20532–20548. [CrossRef] [PubMed]

256. Tong, X.; Wang, X.; Wang, C.; Li, L. Elevated levels of serum MiR-152 and miR-24 in uterine sarcoma: Potential for inducing autophagy via SIRT1 and deacetylated LC3. *Br. J. Biomed. Sci.* **2018**, *75*, 7–12. [CrossRef] [PubMed]

257. Xiao, X.; Lu, Z.; Lin, V.; May, A.; Shaw, D.H.; Wang, Z.; Che, B.; Tran, K.; Du, H.; Shaw, P.X. MicroRNA miR-24-3p reduces apoptosis and regulates Keap1-Nrf2 pathway in mouse cardiomyocytes responding to ischemia/reperfusion injury. *Oxid. Med. Cell. Longev.* **2018**, *2018*, 7042105. [CrossRef] [PubMed]

258. Xie, Q.; Wei, M.; Kang, X.; Liu, D.; Quan, Y.; Pan, X.; Liu, X.; Liao, D.; Liu, J.; Zhang, B. Reciprocal inhibition between miR-26a and NF-κB regulates obesity-related chronic inflammation in chondrocytes. *Biosci. Rep.* **2015**, *35*, e00204. [CrossRef] [PubMed]

259. Thulasingam, S.; Massilamany, C.; Gangaplara, A.; Dai, H.; Yarbaeva, S.; Subramaniam, S.; Riethoven, J.J.; Eudy, J.; Lou, M.; Reddy, J. miR-27b*, an oxidative stress-responsive microRNA modulates nuclear factor-κB pathway in RAW 264.7 cells. *Mol. Cell. Biochem.* **2011**, *352*, 181–188. [CrossRef] [PubMed]

260. Kurtz, C.L.; Fannin, E.E.; Toth, C.L.; Pearson, D.S.; Vickers, K.C.; Sethupathy, P. Inhibition of miR-29 has a significant lipid-lowering benefit through suppression of lipogenic programs in liver. *Sci. Rep.* **2015**, *5*, 12911. [CrossRef] [PubMed]

261. Zhou, L.; Xu, D.Y.; Sha, W.G.; Shen, L.; Lu, G.Y.; Yin, X.; Wang, M.J. High glucose induces renal tubular epithelial injury via Sirt1/NF-kappaB/microR-29/Keap1 signal pathway. *J. Transl. Med.* **2015**, *13*, 352. [CrossRef] [PubMed]

262. Wei, R.; Zhang, R.; Li, H.; Li, H.; Zhang, S.; Xie, Y.; Shen, L.; Chen, F. MiR-29 Targets PUMA to suppress oxygen and glucose deprivation/reperfusion (OGD/R)-induced cell death in hippocampal neurons. *Curr. Neurovasc. Res.* **2018**, *15*, 47–54. [CrossRef] [PubMed]

263. Egan Benova, T.; Viczenczova, C.; Szeiffova Bacova, B.; Knezl, V.; Dosenko, V.; Rauchova, H.; Zeman, M.; Reiter, R.J.; Tribulova, N. Obesity-associated alterations in cardiac connexin-43 and PKC signaling are attenuated by melatonin and omega-3 fatty acids in female rats. *Mol. Cell. Biochem.* **2018**. [CrossRef] [PubMed]

264. Guan, Y.; Rao, Z.; Chen, C. miR-30a suppresses lung cancer progression by targeting SIRT1. *Oncotarget* **2017**, *9*, 4924–4934. [CrossRef] [PubMed]

265. Wan, Q.; Zhou, Z.; Ding, S.; He, J. The miR-30a negatively regulates IL-17-mediated signal transduction by targeting Traf3ip2. *J. Interferon Cytokine Res.* **2015**, *35*, 917–923. [CrossRef] [PubMed]

266. Volkmann, I.; Kumarswamy, R.; Pfaff, N.; Fiedler, J.; Dangwal, S.; Holzmann, A.; Batkai, S.; Geffers, R.; Lother, A.; Hein, L.; et al. MicroRNA-mediated epigenetic silencing of sirtuin1 contributes to impaired angiogenic responses. *Circ. Res.* **2013**, *113*, 997–1003. [CrossRef] [PubMed]

267. Kumarswamy, R.; Volkmann, I.; Beermann, J.; Napp, L.C.; Jabs, O.; Bhayadia, R.; Melk, A.; Ucar, A.; Chowdhury, K.; Lorenzen, J.M.; et al. Vascular importance of the miR-212/132 cluster. *Eur. Heart J.* **2014**, *35*, 3224–3231. [CrossRef] [PubMed]

268. Meng, F.; Dai, E.; Yu, X.; Zhang, Y.; Chen, X.; Liu, X.; Wang, S.; Wang, L.; Jiang, W. Constructing and characterizing a bioactive small molecule and microRNA association network for Alzheimer's disease. *J. R. Soc. Interface* **2013**, *11*, 20131057. [CrossRef] [PubMed]

269. Rajbhandari, R.; McFarland, B.C.; Patel, A.; Gerigk, M.; Gray, G.K.; Fehling, S.C.; Bredel, M.; Berbari, N.F.; Kim, H.; Marks, M.P.; et al. Loss of tumor suppressive microRNA-31 enhances TRADD/NF-κB signaling in glioblastoma. *Oncotarget* **2015**, *6*, 17805–17816. [CrossRef] [PubMed]

270. Carloni, S.; Favrais, G.; Saliba, E.; Albertini, M.C.; Chalon, S.; Longini, M.; Gressens, P.; Buonocore, G.; Balduini, W. Melatonin modulates neonatal brain inflammation through endoplasmic reticulum stress, autophagy, and miR-34a/silent information regulator 1 pathway. *J. Pineal Res.* **2016**, *61*, 370–830. [CrossRef] [PubMed]

271. Pogribny, I.P.; Muskhelishvili, L.; Tryndyak, V.P.; Beland, F.A. The tumor-promoting activity of 2-acetylaminofluorene is associated with disruption of the p53 signaling pathway and the balance between apoptosis and cell proliferation. *Toxicol. Appl. Pharmacol.* **2009**, *235*, 305–311. [CrossRef] [PubMed]

272. Tabuchi, T.; Satoh, M.; Itoh, T.; Nakamura, M. MicroRNA-34a regulates the longevity-associated protein SIRT1 in coronary artery disease: Effect of statins on SIRT1 and microRNA-34a expression. *Clin. Sci.* **2012**, *123*, 161–171. [CrossRef] [PubMed]

273. Roy, A.; Zhang, M.; Saad, Y.; Kolattukudy, P.E. Antidicer RNAse activity of monocyte chemotactic protein-induced protein-1 is critical for inducing angiogenesis. *Am. J. Physiol. Cell Physiol.* **2013**, *305*, C1021–C1032. [CrossRef] [PubMed]

274. Khee, S.G.; Yusof, Y.A.; Makpol, S. Expression of senescence-associated microRNAs and target genes in cellular aging and modulation by tocotrienol-rich fraction. *Oxid. Med. Cell. Longev.* **2014**, *2014*, 725929. [PubMed]

275. Karbasforooshan, H.; Karimi, G. The role of SIRT1 in diabetic retinopathy. *Biomed. Pharmacother.* **2018**, *97*, 190–194. [CrossRef] [PubMed]

276. Rossi, S.; Di Filippo, C.; Gesualdo, C.; Testa, F.; Trotta, M.C.; Maisto, R.; Ferraro, B.; Ferraraccio, F.; Accardo, M.; Simonelli, F.; et al. Interplay between intravitreal RvD1 and local endogenous sirtuin-1 in the protection from endotoxin-induced uveitis in rats. *Mediat. Inflamm.* **2015**, *2015*, 126408. [CrossRef] [PubMed]

277. Gao, A.M.; Zhang, X.Y.; Ke, Z.P. Apigenin sensitizes BEL-7402/ADM cells to doxorubicin through inhibiting miR-101/Nrf2 pathway. *Oncotarget* **2017**, *8*, 82085–82091. [CrossRef] [PubMed]

278. Raji, G.R.; Sruthi, T.V.; Edatt, L.; Haritha, K.; Sharath Shankar, S.; Sameer Kumar, V.B. Horizontal transfer of miR-106a/b from cisplatin resistant hepatocarcinoma cells can alter the sensitivity of cervical cancer cells to cisplatin. *Cell. Signal.* **2017**, *38*, 146–158. [CrossRef] [PubMed]

279. Heyn, J.; Luchting, B.; Hinske, L.C.; Hübner, M.; Azad, S.C.; Kreth, S. miR-124a and miR-155 enhance differentiation of regulatory T cells in patients with neuropathic pain. *J. Neuroinflamm.* **2016**, *13*, 248. [CrossRef] [PubMed]

280. Kim, S.W.; Ramasamy, K.; Bouamar, H.; Lin, A.P.; Jiang, D.; Aguiar, R.C. MicroRNAs miR-125a and miR-125b constitutively activate the NF-κB pathway by targeting the tumor necrosis factor alpha-induced protein 3 (TNFAIP3, A20). *Proc. Natl. Acad. Sci. USA* **2012**, *109*, 7865–7870. [CrossRef] [PubMed]

281. Haemmig, S.; Baumgartner, U.; Glück, A.; Zbinden, S.; Tschan, M.P.; Kappeler, A.; Mariani, L.; Vajtai, I.; Vassella, E. miR-125b controls apoptosis and temozolomide resistance by targeting TNFAIP3 and NKIRAS2 in glioblastomas. *Cell Death Dis.* **2014**, *5*, e1279. [CrossRef] [PubMed]

282. Parisi, C.; Napoli, G.; Amadio, S.; Spalloni, A.; Apolloni, S.; Longone, P.; Volonté, C. MicroRNA-125b regulates microglia activation and motor neuron death in ALS. *Cell Death Differ.* **2016**, *23*, 531–541. [CrossRef] [PubMed]

283. Zheng, Z.; Qu, J.Q.; Yi, H.M.; Ye, X.; Huang, W.; Xiao, T.; Li, J.Y.; Wang, Y.Y.; Feng, J.; Zhu, J.F.; et al. MiR-125b regulates proliferation and apoptosis of nasopharyngeal carcinoma by targeting A20/NF-κB signaling pathway. *Cell Death Dis.* **2017**, *8*, e2855. [CrossRef] [PubMed]

284. Xue, N.; Qi, L.; Zhang, G.; Zhang, Y. miRNA-125b regulates osteogenic differentiation of periodontal ligament cells through NKIRAS2/NF-κB pathway. *Cell. Physiol. Biochem.* **2018**, *48*, 1771–1781. [CrossRef] [PubMed]

285. Wang, D.; Cao, L.; Xu, Z.; Fang, L.; Zhong, Y.; Chen, Q.; Luo, R.; Chen, H.; Li, K.; Xiao, S. MiR-125b reduces porcine reproductive and respiratory syndrome virus replication by negatively regulating the NF-κB pathway. *PLoS ONE* **2013**, *8*, e55838. [CrossRef] [PubMed]

286. Wang, X.; Ha, T.; Zou, J.; Ren, D.; Liu, L.; Zhang, X.; Kalbfleisch, J.; Gao, X.; Williams, D.; Li, C. MicroRNA-125b protects against myocardial ischaemia/reperfusion injury via targeting p53-mediated apoptotic signalling and TRAF6. *Cardiovasc. Res.* **2014**, *102*, 385–395. [CrossRef] [PubMed]

287. Ma, H.; Wang, X.; Ha, T.; Gao, M.; Liu, L.; Wang, R.; Yu, K.; Kalbfleisch, J.H.; Kao, R.L.; Williams, D.L.; et al. MicroRNA-125b prevents cardiac dysfunction in polymicrobial sepsis by targeting TRAF6-mediated nuclear factor κB activation and p53-mediated apoptotic signaling. *J. Infect. Dis.* **2016**, *214*, 1773–1783. [CrossRef] [PubMed]

288. Wang, Y.; Tang, P.; Chen, Y.; Chen, J.; Ma, R.; Sun, L. Overexpression of microRNA-125b inhibits human acute myeloid leukemia cells invasion, proliferation and promotes cells apoptosis by targeting NF-κB signaling pathway. *Biochem. Biophys. Res. Commun.* **2017**, *488*, 60–66. [CrossRef] [PubMed]

289. Jia, J.; Wang, J.; Zhang, J.; Cui, M.; Sun, X.; Li, Q.; Zhao, B. MiR-125b inhibits LPS-induced inflammatory injury via targeting MIP-1α in chondrogenic cell ATDC5. *Cell. Physiol. Biochem.* **2018**, *45*, 2305–2316. [CrossRef] [PubMed]

290. Togliatto, G.; Trombetta, A.; Dentelli, P.; Gallo, S.; Rosso, A.; Cotogni, P.; Granata, R.; Falcioni, R.; Delale, T.; Ghigo, E.; et al. Unacylated ghrelin induces oxidative stress resistance in a glucose intolerance and peripheral artery disease mouse model by restoring endothelial cell miR-126 expression. *Diabetes* **2015**, *64*, 1370–1382. [CrossRef] [PubMed]

291. Oglesby, I.K.; Bray, I.M.; Chotirmall, S.H.; Stallings, R.L.; O'Neill, S.J.; McElvaney, N.G.; Greene, C.M. miR-126 is downregulated in cystic fibrosis airway epithelial cells and regulates TOM1 expression. *J. Immunol.* **2010**, *184*, 1702–1709. [CrossRef] [PubMed]

292. Feng, X.; Wang, H.; Ye, S.; Guan, J.; Tan, W.; Cheng, S.; Wei, G.; Wu, W.; Wu, F.; Zhou, Y. Up-regulation of microRNA-126 may contribute to pathogenesis of ulcerative colitis via regulating NF-kappaB inhibitor IκBα. *PLoS ONE* **2012**, *7*, e52782. [CrossRef] [PubMed]

293. Feng, X.; Tan, W.; Cheng, S.; Wang, H.; Ye, S.; Yu, C.; He, Y.; Zeng, J.; Cen, J.; Hu, J.; et al. Upregulation of microRNA-126 in hepatic stellate cells may affect pathogenesis of liver fibrosis through the NF-κB pathway. *DNA Cell Biol.* **2015**, *34*, 470–480. [CrossRef] [PubMed]

294. Adlakha, Y.K.; Saini, N. miR-128 exerts pro-apoptotic effect in a p53 transcription-dependent and -independent manner via PUMA-Bak axis. *Cell Death Dis.* **2013**, *4*, e542. [CrossRef] [PubMed]

295. Lian, B.; Yang, D.; Liu, Y.; Shi, G.; Li, J.; Yan, X.; Jin, K.; Liu, X.; Zhao, J.; Shang, W.; et al. miR-128 targets the SIRT1/ROS/DR5 pathway to sensitize colorectal cancer to TRAIL-induced apoptosis. *Cell. Physiol. Biochem.* **2018**, *49*, 2151–2162. [CrossRef] [PubMed]

296. Pan, S.; Cui, Y.; Fu, Z.; Zhang, L.; Xing, H. MicroRNA-128 is involved in dexamethasone-induced lipid accumulation via repressing SIRT1 expression in cultured pig preadipocytes. *J. Steroid Biochem. Mol. Biol.* **2019**, *186*, 185–195. [CrossRef] [PubMed]

297. Xia, Z.; Meng, F.; Liu, Y.; Fang, Y.; Wu, X.; Zhang, C.; Liu, D.; Li, G. Decreased MiR-128-3p alleviates the progression of rheumatoid arthritis by up-regulating the expression of TNFAIP3. *Biosci. Rep.* **2018**, *38*, BSR20180540. [CrossRef] [PubMed]

298. Geng, L.; Zhang, T.; Liu, W.; Chen, Y. Inhibition of miR-128 abates Aβ-mediated cytotoxicity by targeting PPAR-γ via NF-κB inactivation in primary mouse cortical neurons and Neuro2a cells. *Yonsei Med. J.* **2018**, *59*, 1096–1106. [CrossRef] [PubMed]

299. Ali, T.; Mushtaq, I.; Maryam, S.; Farhan, A.; Saba, K.; Jan, M.I.; Sultan, A.; Anees, M.; Duygu, B.; Hamera, S.; et al. Interplay of N-acetyl cysteine and melatonin in regulating oxidative stress-induced cardiac hypertrophic factors and microRNAs. *Arch. Biochem. Biophys.* **2019**, *661*, 56–65. [CrossRef] [PubMed]

300. Liu, L.; Sun, T.; Liu, Z.; Chen, X.; Zhao, L.; Qu, G.; Li, Q. Traumatic brain injury dysregulates microRNAs to modulate cell signaling in rat hippocampus. *PLoS ONE* **2014**, *9*, e103948. [CrossRef] [PubMed]

301. Smith, E.J.; Shay, K.P.; Thomas, N.O.; Butler, J.A.; Finlay, L.F.; Hagen, T.M. Age-related loss of hepatic Nrf2 protein homeostasis: Potential role for heightened expression of miR-146a. *Free Radic. Biol. Med.* **2015**, *89*, 1184–1191. [CrossRef] [PubMed]

302. Kamali, K.; Korjan, E.S.; Eftekhar, E.; Malekzadeh, K.; Soufi, F.G. The role of miR-146a on NF-κB expression level in human umbilical vein endothelial cells under hyperglycemic condition. *Bratisl. Lek. Listy* **2016**, *117*, 376–380. [CrossRef] [PubMed]

303. Loubaki, L.; Chabot, D.; Paré, I.; Drouin, M.; Bazin, R. MiR-146a potentially promotes IVIg-mediated inhibition of TLR4 signaling in LPS-activated human monocytes. *Immunol. Lett.* **2017**, *185*, 64–73. [CrossRef] [PubMed]

304. An, R.; Feng, J.; Xi, C.; Xu, J.; Sun, L. miR-146a attenuates sepsis-induced myocardial dysfunction by suppressing IRAK1 and TRAF6 via targeting ErbB4 expression. *Oxid. Med. Cell. Longev.* **2018**, *2018*, 7163057. [CrossRef] [PubMed]

305. Lindeløv Vestergaard, A.; Heiner Bang-Berthelsen, C.; Fløyel, T.; Lucien Stahl, J.; Christen, L.; Taheri Sotudeh, F.; de Hemmer Horskjær, P.; Stensgaard Frederiksen, K.; Greek Kofod, F.; Bruun, C.; et al. MicroRNAs and histone deacetylase inhibition-mediated protection against inflammatory β-cell damage. *PLoS ONE* **2018**, *13*, e0203713. [CrossRef] [PubMed]

306. Sang, W.; Wang, Y.; Zhang, C.; Zhang, D.; Sun, C.; Niu, M.; Zhang, Z.; Wei, X.; Pan, B.; Chen, W.; et al. MiR-150 impairs inflammatory cytokine production by targeting ARRB-2 after blocking CD28/B7 costimulatory pathway. *Immunol. Lett.* **2016**, *172*, 1–10. [CrossRef] [PubMed]

307. Xue, H.; Li, M.X. MicroRNA-150 protects against cigarette smoke-induced lung inflammation and airway epithelial cell apoptosis through repressing p53: MicroRNA-150 in CS-induced lung inflammation. *Hum. Exp. Toxicol.* **2018**, *37*, 920–928. [CrossRef] [PubMed]

308. Ma, Y.; Liu, Y.; Hou, H.; Yao, Y.; Meng, H. MiR-150 predicts survival in patients with sepsis and inhibits LPS-induced inflammatory factors and apoptosis by targeting NF-κB1 in human umbilical vein endothelial cells. *Biochem. Biophys. Res. Commun.* **2018**, *500*, 828–837. [CrossRef] [PubMed]

309. Luo, X.Y.; Zhu, X.Q.; Li, Y.; Wang, X.B.; Yin, W.; Ge, Y.S.; Ji, W.M. MicroRNA-150 restores endothelial cell function and attenuates vascular remodeling by targeting PTX3 through the NF-κB signaling pathway in mice with acute coronary syndrome. *Cell Biol. Int.* **2018**. [CrossRef] [PubMed]

310. Gu, J.; Lu, Z.; Ji, C.; Chen, Y.; Liu, Y.; Lei, Z.; Wang, L.; Zhang, H.T.; Li, X. Melatonin inhibits proliferation and invasion via repression of miRNA-155 in glioma cells. *Biomed. Pharmacother.* **2017**, *93*, 969–975. [CrossRef]

311. Wang, Y.; Zheng, Z.J.; Jia, Y.J.; Yang, Y.L.; Xue, Y.M. Role of p53/miR-155-5p/sirt1 loop in renal tubular injury of diabetic kidney disease. *J. Transl. Med.* **2018**, *16*, 146. [CrossRef] [PubMed]

312. Wang, X.; Wang, B.; Zhao, J.; Liu, C.; Qu, X.; Li, Y. MiR-155 is involved in major depression disorder and antidepressant treatment via targeting SIRT1. *Biosci. Rep.* **2018**, *38*, 20181139. [CrossRef] [PubMed]

313. Wan, C.; Han, R.; Liu, L.; Zhang, F.; Li, F.; Xiang, M.; Ding, W. Role of miR-155 in fluorooctane sulfonate-induced oxidative hepatic damage via the Nrf2-dependent pathway. *Toxicol. Appl. Pharmacol.* **2016**, *295*, 85–93. [CrossRef] [PubMed]

314. Gu, S.; Lai, Y.; Chen, H.; Liu, Y.; Zhang, Z. miR-155 mediates arsenic trioxide resistance by activating Nrf2 and suppressing apoptosis in lung cancer cells. *Sci. Rep.* **2017**, *7*, 12155. [CrossRef] [PubMed]

315. Rodriguez-Ortiz, C.J.; Baglietto-Vargas, D.; Martinez-Coria, H.; LaFerla, F.M.; Kitazawa, M. Upregulation of miR-181 decreases c-Fos and SIRT-1 in the hippocampus of 3xTg-AD mice. *J. Alzheimers Dis.* **2014**, *42*, 1229–1238. [CrossRef] [PubMed]

316. Chen, X.Y.; Zhang, H.S.; Wu, T.C.; Sang, W.W.; Ruan, Z. Down-regulation of NAMPT expression by miR-182 is involved in Tat-induced HIV-1 long terminal repeat (LTR) transactivation. *Int. J. Biochem. Cell Biol.* **2013**, *45*, 292–298. [CrossRef] [PubMed]

317. Eyler, C.E.; Rich, J.N. Looking in the miR-ror: TGF-β-mediated activation of NF-κB in glioma. *J. Clin. Investig.* **2012**, *122*, 3473–3475. [CrossRef] [PubMed]

318. Song, L.; Liu, L.; Wu, Z.; Li, Y.; Ying, Z.; Lin, C.; Wu, J.; Hu, B.; Cheng, S.Y.; Li, M.; et al. TGF-β induces miR-182 to sustain NF-κB activation in glioma subsets. *J. Clin. Investig.* **2012**, *122*, 3563–3578. [CrossRef] [PubMed]

319. Guo, Y.; Liao, Y.; Jia, C.; Ren, J.; Wang, J.; Li, T. MicroRNA-182 promotes tumor cell growth by targeting transcription elongation factor A-like 7 in endometrial carcinoma. *Cell. Physiol. Biochem.* **2013**, *32*, 581–590. [CrossRef] [PubMed]

320. Ling, T.; Yu, F.; Cao, H. miR-182 controls cell growth in gastrointestinal stromal tumors by negatively regulating CYLD expression. *Oncol. Rep.* **2018**, *40*, 3705–3713. [CrossRef] [PubMed]

321. Liu, Y.; Li, L.N.; Guo, S.; Zhao, X.Y.; Liu, Y.Z.; Liang, C.; Tu, S.; Wang, D.; Li, L.; Dong, J.Z.; et al. Melatonin improves cardiac function in a mouse model of heart failure with preserved ejection fraction. *Redox Biol.* **2018**, *18*, 211–221. [CrossRef] [PubMed]

322. Wang, R.; Zhou, S.; Wu, P.; Li, M.; Ding, X.; Sun, L.; Xu, X.; Zhou, X.; Zhou, L.; Cao, C.; et al. Identifying involvement of H19-miR-675-3p-IGF1R and H19-miR-200a-PDCD4 in treating pulmonary hypertension with melatonin. *Mol. Ther. Nucleic Acids* **2018**, *13*, 44–54. [CrossRef] [PubMed]

323. Eades, G.; Yao, Y.; Yang, M.; Zhang, Y.; Chumsri, S.; Zhou, Q. miR-200a regulates SIRT1 expression and epithelial to mesenchymal transition (EMT)-like transformation in mammary epithelial cells. *J. Biol. Chem.* **2011**, *286*, 25992–26002. [CrossRef] [PubMed]

324. Pan, F.; Qiu, X.F.; Yu, W.; Zhang, Q.P.; Chen, Q.; Zhang, C.Y.; Chen, Y.; Pan, L.J.; Zhang, A.X.; Dai, Y.T. MicroRNA-200a is up-regulated in aged rats with erectile dysfunction and could attenuate endothelial function via SIRT1 inhibition. *Asian J. Androl.* **2016**, *18*, 74–79. [PubMed]

325. Zhang, P.; Xu, L.; Guan, H.; Liu, L.; Liu, J.; Huang, Z.; Cao, X.; Liao, Z.; Xiao, H.; Li, Y. Beraprost sodium, a prostacyclin analogue, reduces fructose-induced hepatocellular steatosis in mice and in vitro via the microRNA-200a and SIRT1 signaling pathway. *Metabolism* **2017**, *73*, 9–21. [CrossRef] [PubMed]

326. Fu, H.; Song, W.; Chen, X.; Guo, T.; Duan, B.; Wang, X.; Tang, Y.; Huang, L.; Zhang, C. MiRNA-200a induce cell apoptosis in renal cell carcinoma by directly targeting SIRT1. *Mol. Cell. Biochem.* **2018**, *437*, 143–152. [CrossRef] [PubMed]

327. Salimian, N.; Peymani, M.; Ghaedi, K.; Nasr Esfahani, M.H. Modulation in miR-200a/SIRT1axis is associated with apoptosis in MPP⁺-induced SH-SY5Y cells. *Gene* **2018**, *674*, 25–30. [CrossRef] [PubMed]

328. Eades, G.; Yang, M.; Yao, Y.; Zhang, Y.; Zhou, Q. miR-200a regulates Nrf2 activation by targeting Keap1 mRNA in breast cancer cells. *J. Biol. Chem.* **2011**, *286*, 40725–40733. [CrossRef] [PubMed]

329. Murray-Stewart, T.; Hanigan, C.L.; Woster, P.M.; Marton, L.J.; Casero, R.A., Jr. Histone deacetylase inhibition overcomes drug resistance through a miRNA-dependent mechanism. *Mol. Cancer Ther.* **2013**, *12*, 2088–2099. [CrossRef] [PubMed]

330. Wei, J.; Zhang, Y.; Luo, Y.; Wang, Z.; Bi, S.; Song, D.; Dai, Y.; Wang, T.; Qiu, L.; Wen, L.; et al. Aldose reductase regulates miR-200a-3p/141-3p to coordinate Keap1-Nrf2, Tgfβ1/2, and Zeb1/2 signaling in renal mesangial cells and the renal cortex of diabetic mice. *Free Radic. Biol. Med.* **2014**, *67*, 91–102. [CrossRef] [PubMed]

331. Liu, Q.L.; Zhang, J.; Liu, X.; Gao, J.Y. Role of growth hormone in maturation and activation of dendritic cells via miR-200a and the Keap1/Nrf2 pathway. *Cell Prolif.* **2015**, *48*, 573–581. [CrossRef] [PubMed]

332. Liu, M.; Hu, C.; Xu, Q.; Chen, L.; Ma, K.; Xu, N.; Zhu, H. Methylseleninic acid activates Keap1/Nrf2 pathway via up-regulating miR-200a in human oesophageal squamous cell carcinoma cells. *Biosci. Rep.* **2015**, *35*, e00256. [CrossRef] [PubMed]

333. Sun, X.; Zuo, H.; Liu, C.; Yang, Y. Overexpression of miR-200a protects cardiomyocytes against hypoxia-induced apoptosis by modulating the kelch-like ECH-associated protein 1-nuclear factor erythroid 2-related factor 2 signaling axis. *Int. J. Mol. Med.* **2016**, *38*, 1303–1311. [CrossRef] [PubMed]

334. Wu, H.; Kong, L.; Tan, Y.; Epstein, P.N.; Zeng, J.; Gu, J.; Liang, G.; Kong, M.; Chen, X.; Miao, L.; et al. C66 ameliorates diabetic nephropathy in mice by both upregulating NRF2 function via increase in miR-200a and inhibiting miR-21. *Diabetologia* **2016**, *59*, 1558–1568. [CrossRef] [PubMed]

335. Zhao, S.; Mao, L.; Wang, S.G.; Chen, F.L.; Ji, F.; Fe, H.D. MicroRNA-200a activates Nrf2 signaling to protect osteoblasts from dexamethasone. *Oncotarget* **2017**, *8*, 104867–104876. [CrossRef] [PubMed]

336. Zhao, X.J.; Yu, H.W.; Yang, Y.Z.; Wu, W.Y.; Chen, T.Y.; Jia, K.K.; Kang, L.L.; Jiao, R.Q.; Kong, L.D. Polydatin prevents fructose-induced liver inflammation and lipid deposition through increasing miR-200a to regulate Keap1/Nrf2 pathway. *Redox Biol.* **2018**, *18*, 124–137. [CrossRef] [PubMed]

337. Shi, Z.; Hu, Z.; Chen, D.; Huang, J.; Fan, J.; Zhou, S.; Wang, X.; Hu, J.; Huang, F. MicroRNA-200a mediates nasopharyngeal carcinoma cell proliferation through the activation of nuclear factor-κB. *Mol. Med. Rep.* **2016**, *13*, 1732–1738. [CrossRef] [PubMed]

338. Zhang, Q.S.; Liu, W.; Lu, G.X. miR-200a-3p promotes β-Amyloid-induced neuronal apoptosis through down-regulation of SIRT1 in Alzheimer's disease. *J. Biosci.* **2017**, *42*, 397–404. [CrossRef] [PubMed]

339. Li, L.; Wang, Q.; Yuan, Z.; Chen, A.; Liu, Z.; Li, H.; Wang, Z. Long non-coding RNA H19 contributes to hypoxia-induced CPC injury by suppressing Sirt1 through miR-200a-3p. *Acta Biochim. Biophys. Sin.* **2018**, *50*, 950–959. [CrossRef] [PubMed]

340. Qi, J.; Qiao, Y.; Wang, P.; Li, S.; Zhao, W.; Gao, C. microRNA-210 negatively regulates LPS-induced production of proinflammatory cytokines by targeting NF-κB1 in murine macrophages. *FEBS Lett.* **2012**, *586*, 1201–1207. [CrossRef] [PubMed]

341. Zhang, D.; Cao, X.; Li, J.; Zhao, G. MiR-210 inhibits NF-κB signaling pathway by targeting DR6 in osteoarthritis. *Sci. Rep.* **2015**, *5*, 12775. [CrossRef] [PubMed]

342. Ramalinga, M.; Roy, A.; Srivastava, A.; Bhattarai, A.; Harish, V.; Suy, S.; Collins, S.; Kumar, D. MicroRNA-212 negatively regulates starvation induced autophagy in prostate cancer cells by inhibiting SIRT1 and is a modulator of angiogenesis and cellular senescence. *Oncotarget* **2015**, *6*, 34446–34457. [CrossRef] [PubMed]

343. Miao, H.; Zeng, H.; Gong, H. microRNA-212 promotes lipid accumulation and attenuates cholesterol efflux in THP-1 human macrophages by targeting SIRT1. *Gene* **2018**, *643*, 55–60. [CrossRef] [PubMed]

344. Li, D.; Bai, L.; Wang, T.; Xie, Q.; Chen, M.; Fu, Y.; Wen, Q. Function of miR-212 as a tumor suppressor in thyroid cancer by targeting SIRT1. *Oncol. Rep.* **2018**, *39*, 695–702. [CrossRef] [PubMed]

345. Fiorentino, L.; Cavalera, M.; Mavilio, M.; Conserva, F.; Menghini, R.; Gesualdo, L.; Federici, M. Regulation of TIMP3 in diabetic nephropathy: A role for microRNAs. *Acta Diabetol.* **2013**, *50*, 965–969. [CrossRef] [PubMed]

346. Lu, Z.; Li, Y.; Takwi, A.; Li, B.; Zhang, J.; Conklin, D.J.; Young, K.H.; Martin, R.; Li, Y. miR-301a as an NF-κB activator in pancreatic cancer cells. *EMBO J.* **2011**, *30*, 57–67. [CrossRef] [PubMed]

347. Zhou, P.; Jiang, W.; Wu, L.; Chang, R.; Wu, K.; Wang, Z. miR-301a is a candidate oncogene that targets the homeobox gene Gax in human hepatocellular carcinoma. *Dig. Dis. Sci.* **2012**, *57*, 1171–1180. [CrossRef] [PubMed]

348. Huang, L.; Liu, Y.; Wang, L.; Chen, R.; Ge, W.; Lin, Z.; Zhang, Y.; Liu, S.; Shan, Y.; Lin, Q.; et al. Down-regulation of miR-301a suppresses pro-inflammatory cytokines in Toll-like receptor-triggered macrophages. *Immunology* **2013**, *140*, 314–322. [PubMed]

349. Ma, X.; Yan, F.; Deng, Q.; Li, F.; Lu, Z.; Liu, M.; Wang, L.; Conklin, D.J.; McCracken, J.; Srivastava, S.; et al. Modulation of tumorigenesis by the pro-inflammatory microRNA miR-301a in mouse models of lung cancer and colorectal cancer. *Cell Discov.* **2015**, *1*, 15005. [CrossRef] [PubMed]

350. Tavakolpour, V.; Shokri, G.; Naser Moghadasi, A.; Mozafari Nahavandi, P.; Hashemi, M.; Kouhkan, F. Increased expression of mir-301a in PBMCs of patients with relapsing-remitting multiple sclerosis is associated with reduced NKRF and PIAS3 expression levels and disease activity. *J. Neuroimmunol.* **2018**, *325*, 79–86. [CrossRef] [PubMed]

351. Wu, C.T.; Huang, Y.; Pei, Z.Y.; Xi, X.; Zhu, G.F. MicroRNA-326 aggravates acute lung injury in septic shock by mediating the NF-κB signaling pathway. *Int. J. Biochem. Cell Biol.* **2018**, *101*, 1–11. [CrossRef] [PubMed]

352. Shi, L.; Chen, Z.G.; Wu, L.L.; Zheng, J.J.; Yang, J.R.; Chen, X.F.; Chen, Z.Q.; Liu, C.L.; Chi, S.Y.; Zheng, J.Y.; et al. miR-340 reverses cisplatin resistance of hepatocellular carcinoma cell lines by targeting Nrf2-dependent antioxidant pathway. *Asian Pac. J. Cancer Prev.* **2014**, *15*, 10439–10444. [CrossRef] [PubMed]

353. Wu, L.L.; Cai, W.P.; Lei, X.; Shi, Q.; Lin, X.Y.; Shi, L. NRAL mediates cisplatin resistance in hepatocellular carcinoma via miR-340-5p/Nrf2 axis. *J. Cell. Commun. Signal.* **2018**. [CrossRef] [PubMed]

354. Prins, S.A.; Przybycien-Szymanska, M.M.; Rao, Y.S.; Pak, T.R. Long-term effects of peripubertal binge EtOH exposure on hippocampal microRNA expression in the rat. *PLoS ONE* **2014**, *9*, e83166. [CrossRef] [PubMed]

355. Cai, B.; Ma, W.; Bi, C.; Yang, F.; Zhang, L.; Han, Z.; Huang, Q.; Ding, F.; Li, Y.; Yan, G.; et al. Long noncoding RNA H19 mediates melatonin inhibition of premature senescence of c-kit$^+$ cardiac progenitor cells by promoting miR-675. *J. Pineal Res.* **2016**, *61*, 82–95. [CrossRef] [PubMed]

356. Liu, J.; Zhu, L.; Xie, G.L.; Bao, J.F.; Yu, Q. Let-7 miRNAs modulate the activation of NF-κB by targeting TNFAIP3 and are involved in the pathogenesis of lupus nephritis. *PLoS ONE* **2015**, *10*, e0121256. [CrossRef] [PubMed]

357. Yang, S.; Tang, W.; He, Y.; Wen, L.; Sun, B.; Li, S. Long non-coding RNA and microRNA-675/let-7a mediates the protective effect of melatonin against early brain injury after subarachnoid hemorrhage via targeting TP53 and neural growth factor. *Cell Death Dis.* **2018**, *9*, 99. [CrossRef] [PubMed]

358. Lin, Z.; Ge, J.; Wang, Z.; Ren, J.; Wang, X.; Xiong, H.; Gao, J.; Zhang, Y.; Zhang, Q. Let-7e modulates the inflammatory response in vascular endothelial cells through ceRNA crosstalk. *Sci. Rep.* **2017**, *7*, 42498. [CrossRef] [PubMed]

359. Kumar, M.; Sahu, S.K.; Kumar, R.; Subuddhi, A.; Maji, R.K.; Jana, K.; Gupta, P.; Raffetseder, J.; Lerm, M.; Ghosh, Z.; et al. MicroRNA let-7 modulates the immune response to *Mycobacterium tuberculosis* infection via control of A20, an inhibitor of the NF-κB pathway. *Cell Host Microbe* **2015**, *17*, 345–356. [CrossRef] [PubMed]

360. Wang, Y.S.; His, E.; Cheng, H.Y.; Hsu, S.H.; Liao, Y.C.; Juo, S.H. Let-7g suppresses both canonical and non-canonical NF-κB pathways in macrophages leading to anti-atherosclerosis. *Oncotarget* **2017**, *8*, 101026–101041. [PubMed]

361. Xie, H.; Lei, N.; Gong, A.Y.; Chen, X.M.; Hu, G. *Cryptosporidium parvum* induces SIRT1 expression in host epithelial cells through downregulating let-7i. *Hum. Immunol.* **2014**, *75*, 760–765. [CrossRef] [PubMed]

362. Sarver, A.L.; Sarver, A.E.; Yuan, C.; Subramanian, S. OMCD: OncomiR Cancer Database. *BMC Cancer* **2018**, *18*, 1223. [CrossRef] [PubMed]

363. Ushio, N.; Rahman, M.M.; Maemura, T.; Lai, Y.C.; Iwanaga, T.; Kawaguchi, H.; Miyoshi, N.; Momoi, Y.; Miura, N. Identification of dysregulated microRNAs in canine malignant melanoma. *Oncol. Lett.* **2019**, *17*, 1080–1088. [CrossRef] [PubMed]

364. Zargar, S.; Tomar, V.; Shyamsundar, V.; Vijayalakshmi, R.; Somasundaram, K.; Karunagaran, D. A feedback loop between miRNA-155, Programmed cell death 4 and Activation Protein-1 modulates the expression of miR-155 and tumorigenesis in tongue cancer. *Mol. Cell. Biol.* **2019**. [CrossRef] [PubMed]

365. Jung-Hynes, B.; Schmit, T.L.; Reagan-Shaw, S.R.; Siddiqui, I.A.; Mukhtar, H.; Ahmad, N. Melatonin, a novel Sirt1 inhibitor, imparts proliferative effects against prostate cancer cells in vitro culture and in vivo in TRAMP model. *J. Pineal Res.* **2011**, *50*, 140–149. [PubMed]

366. Zhang, Y.; Liu, X.; Bai, X.; Lin, Y.; Li, Z.; Fu, J.; Li, M.; Zhao, T.; Yang, H.; Xu, R.; et al. Melatonin prevents endothelial cell pyroptosis via regulation of long noncoding RNA MEG3/miR-223/NLRP3 axis. *J. Pineal Res.* **2018**, *64*, e12449. [CrossRef] [PubMed]

367. Jin, M.; Cao, M.; Cao, Q.; Piao, J.; Zhao, F.; Piao, J. Long noncoding RNA and gene expression analysis of melatonin-exposed Liaoning cashmere goat fibroblasts indicating cashmere growth. *Naturwissenschaften* **2018**, *105*, 60. [CrossRef] [PubMed]

368. Wang, T.H.; Wu, C.H.; Yeh, C.T.; Su, S.C.; Hsia, S.M.; Liang, K.H.; Chen, C.C.; Hsueh, C.; Chen, C.Y. Melatonin suppresses hepatocellular carcinoma progression via lncRNA-CPS1-IT-mediated HIF-1α inactivation. *Oncotarget* **2017**, *8*, 82280–82293. [PubMed]

369. Chen, C.C.; Chen, C.Y.; Wang, S.H.; Yeh, C.T.; Su, S.C.; Ueng, S.H.; Chuang, W.Y.; Hsueh, C.; Wang, T.H. Melatonin sensitizes hepatocellular carcinoma cells to chemotherapy through long non-coding RNA RAD51-AS1-mediated suppression of DNA repair. *Cancers* **2018**, *10*, 320. [CrossRef] [PubMed]

370. Su, S.C.; Reiter, R.J.; Hsiao, H.Y.; Chung, W.H.; Yang, S.F. Functional interaction between melatonin signaling and noncoding RNAs. *Trends Endocrinol. MeTable* **2018**, *29*, 435–445. [CrossRef] [PubMed]

371. Fan, Z.; Zhao, M.; Joshi, P.D.; Li, P.; Zhang, Y.; Guo, W.; Xu, Y.; Wang, H.; Zhao, Z.; Yan, J. A class of circadian long non-coding RNAs mark enhancers modulating long-range circadian gene regulation. *Nucleic Acids Res.* **2017**, *45*, 5720–5738. [CrossRef] [PubMed]

372. Pereira Fernandes, D.; Bitar, M.; Jacobs, F.M.J.; Barry, G. Long non-coding RNAs in neuronal aging. *Noncoding RNA* **2018**, *4*, 12. [CrossRef] [PubMed]

373. Gomez-Verjan, J.C.; Vazquez-Martinez, E.R.; Rivero-Segura, N.A.; Medina-Campos, R.H. The RNA world of human ageing. *Hum. Genet.* **2018**, *137*, 865–879. [CrossRef] [PubMed]

374. Heward, J.A.; Lindsay, M.A. Long non-coding RNAs in the regulation of the immune response. *Trends Immunol.* **2014**, *35*, 408–419. [CrossRef] [PubMed]

375. Elling, R.; Chan, J.; Fitzgerald, K.A. Emerging role of long noncoding RNAs as regulators of innate immune cell development and inflammatory gene expression. *Eur. J. Immunol.* **2016**, *46*, 504–512. [CrossRef] [PubMed]

376. Murphy, M.B.; Medvedev, A.E. Long noncoding RNAs as regulators of Toll-like receptor signaling and innate immunity. *J. Leukoc. Biol.* **2016**, *99*, 839–850. [CrossRef] [PubMed]

377. Panda, A.C.; Abdelmohsen, K.; Gorospe, M. SASP regulation by noncoding RNA. *Mech. Ageing Dev.* **2017**, *168*, 37–43. [CrossRef] [PubMed]

378. Carpenter, S.; Fitzgerald, K.A. Cytokines and long noncoding RNAs. *Cold Spring Harb. Perspect. Biol.* **2018**, *10*, a028589. [CrossRef] [PubMed]

379. Wang, Z.; Zheng, Y. lncRNAs regulate innate immune responses and their roles in macrophage polarization. *Mediat. Inflamm.* **2018**, *2018*, 8050956. [CrossRef] [PubMed]

380. Yarani, R.; Mirza, A.H.; Kaur, S.; Pociot, F. The emerging role of lncRNAs in inflammatory bowel disease. *Exp. Mol. Med.* **2018**, *50*, 161. [CrossRef] [PubMed]

381. Memczak, S.; Jens, M.; Elefsinioti, A.; Torti, F.; Krueger, J.; Rybak, A.; Maier, L.; Mackowiak, S.D.; Gregersen, L.H.; Munschauer, M.; et al. Circular RNAs are a large class of animal RNAs with regulatory potency. *Nature* **2013**, *495*, 333–338. [CrossRef] [PubMed]

382. Hansen, T.B.; Jensen, T.I.; Clausen, B.H.; Bramsen, J.B.; Finsen, B.; Damgaard, C.K.; Kjems, J. Natural RNA circles function as efficient microRNA sponges. *Nature* **2013**, *495*, 384–388. [CrossRef] [PubMed]

383. Ebbesen, K.K.; Hansen, T.B.; Kjems, J. Insights into circular RNA biology. *RNA Biol.* **2017**, *14*, 1035–1045. [CrossRef] [PubMed]

384. Shi, Z.; Chen, T.; Yao, Q.; Zheng, L.; Zhang, Z.; Wang, J.; Hu, Z.; Cui, H.; Han, Y.; Han, X.; et al. The circular RNA ciRS-7 promotes APP and BACE1 degradation in an NF-κB-dependent manner. *FEBS J.* **2017**, *284*, 1096–1109. [CrossRef] [PubMed]

385. Su, C.; Han, Y.; Zhang, H.; Li, Y.; Yi, L.; Wang, X.; Zhou, S.; Yu, D.; Song, X.; Xiao, N.; et al. CiRS-7 targeting miR-7 modulates the progression of non-small cell lung cancer in a manner dependent on NF-κB signalling. *J. Cell. Mol. Med.* **2018**, *22*, 3097–3107. [CrossRef] [PubMed]

386. Huang, H.; Wei, L.; Qin, T.; Yang, N.; Li, Z.; Xu, Z. Circular RNA ciRS-7 triggers the migration and invasion of esophageal squamous cell carcinoma via miR-7/KLF4 and NF-κB signals. *Cancer Biol. Ther.* **2019**, *20*, 73–80. [CrossRef] [PubMed]

387. Piwecka, M.; Glažar, P.; Hernandez-Miranda, L.R.; Memczak, S.; Wolf, S.A.; Rybak-Wolf, A.; Filipchyk, A.; Klironomos, F.; Cerda Jara, C.A.; Fenske, P.; et al. Loss of a mammalian circular RNA locus causes miRNA deregulation and affects brain function. *Science* **2017**, *357*, eaam8526. [CrossRef] [PubMed]

388. Srinivasan, V.; Cardinali, D.P.; Srinivasan, U.S.; Kaur, C.; Brown, G.M.; Spence, D.W.; Hardeland, R.; Pandi-Perumal, S.R. Therapeutic potential of melatonin and its analogs in Parkinson's disease: Focus on sleep and neuroprotection. *Ther. Adv. Neurol. Disord.* **2011**, *4*, 297–317. [CrossRef] [PubMed]

389. Willis, G.L. Parkinson's disease as a neuroendocrine disorder of circadian function: Dopamine-melatonin imbalance and the visual system in the genesis and progression of the degenerative process. *Rev. Neurosci.* **2008**, *19*, 245–316. [CrossRef] [PubMed]

390. Willis, G.L. The role of ML-23 and other melatonin analogues in the treatment and management of Parkinson's disease. *Drug News Perspect.* **2005**, *18*, 437–444. [CrossRef] [PubMed]

391. Hardeland, R. Melatonin and synthetic melatoninergic agonists in psychiatric and age-associated disorders: Successful and unsuccessful approaches. *Curr. Pharm. Des.* **2016**, *22*, 1086–1101. [CrossRef] [PubMed]

392. Rodella, L.F.; Favero, G.; Rossini, C.; Foglio, E.; Bonomini, F.; Reiter, R.J.; Rezzani, R. Aging and vascular dysfunction: Beneficial melatonin effects. *Age* **2013**, *35*, 103–115. [CrossRef] [PubMed]

393. Favero, G.; Franceschetti, L.; Buffoli, B.; Moghadasian, M.H.; Reiter, R.J.; Rodella, L.F.; Rezzani, R. Melatonin: Protection against age-related cardiac pathology. *Ageing Res. Rev.* **2017**, *35*, 336–349. [CrossRef] [PubMed]

394. Majidinia, M.; Reiter, R.J.; Shakouri, S.K.; Yousefi, B. The role of melatonin, a multitasking molecule, in retarding the processes of ageing. *Ageing Res. Rev.* **2018**, *47*, 198–213. [CrossRef] [PubMed]

395. Kireev, R.A.; Vara, E.; Viña, J.; Tresguerres, J.A.F. Melatonin and oestrogen treatments were able to improve neuroinflammation and apoptotic processes in dentate gyrus of old ovariectomized female rats. *Age* **2014**, *36*, 9707. [CrossRef] [PubMed]

396. Ding, M.; Feng, N.; Tang, D.; Feng, J.; Li, Z.; Jia, M.; Liu, Z.; Gu, X.; Wang, Y.; Fu, F.; et al. Melatonin prevents Drp1-mediated mitochondrial fission in diabetic hearts through SIRT1-PGC1α pathway. *J. Pineal Res.* **2018**, *65*, e12491. [CrossRef] [PubMed]

397. Hadar, A.; Milanesi, E.; Walczak, M.; Puzianowska-Kuźnicka, M.; Kuźnicki, J.; Squassina, A.; Niola, P.; Chillotti, C.; Attems, J.; Gozes, I.; et al. SIRT1, miR-132 and miR-212 link human longevity to Alzheimer's Disease. *Sci. Rep.* **2018**, *8*, 8465. [CrossRef] [PubMed]

International Journal of
Molecular Sciences

MDPI

Review

Causes and Mechanisms of Hematopoietic Stem Cell Aging

Jungwoon Lee [1,†], Suk Ran Yoon [2,3,†], Inpyo Choi [2,3,*] and Haiyoung Jung [2,*]

[1]   Environmental Disease Research Center, Korea Research Institute of Bioscience and Biotechnology (KRIBB), 125 Gwahak-ro, Yuseong-gu, Daejeon 34141, Korea; jwlee821@kribb.re.kr
[2]   Immunotherapy Research Center, Korea Research Institute of Bioscience and Biotechnology (KRIBB), 125 Gwahak-ro, Yuseong-gu, Daejeon 34141, Korea; sryoon@kribb.re.kr
[3]   Department of Functional Genomics, University of Science and Technology (UST), 113 Gwahak-ro, Yuseong-gu, Daejeon 34113, Korea
*   Correspondence: ipchoi@kribb.re.kr (I.C.); haiyoung@kribb.re.kr (H.J.)
†   These authors contributed equally to this work.

Received: 15 February 2019; Accepted: 9 March 2019; Published: 13 March 2019

Abstract: Many elderly people suffer from hematological diseases known to be highly age-dependent. Hematopoietic stem cells (HSCs) maintain the immune system by producing all blood cells throughout the lifetime of an organism. Recent reports have suggested that HSCs are susceptible to age-related stress and gradually lose their self-renewal and regeneration capacity with aging. HSC aging is driven by cell-intrinsic and -extrinsic factors that result in the disruption of the immune system. Thus, the study of HSC aging is important to our understanding of age-related immune diseases and can also provide potential strategies to improve quality of life in the elderly. In this review, we delineate our understanding of the phenotypes, causes, and molecular mechanisms involved in HSC aging.

Keywords: hematopoietic stem cell aging; rejuvenation; self-renewal; differentiation

---

1. Introduction

In the hematopoietic system, hematopoietic stem cells (HSCs) continuously replenish the blood cells including B and T lymphocytes, erythrocytes, myeloid cells, platelets, natural killer (NK) cells, mast cells, and dendritic cells (DCs), throughout the lifetime of an organism [1–3]. HSCs were the first stem cells to be identified and isolated and remain the most-studied tissue-specific stem cells. HSCs constitute the pool of long-term HSCs (LT-HSCs), short-term HSCs (ST-HSCs), and multipotent progenitors (MPPs). They can be identified with specific cell-surface markers using fluorescence-activated cell sorting (FACS) technology. All murine HSCs are lineage (Lin$^-$), stem cell antigen-1 (Sca-1)$^+$, and cKit$^+$ (LSKs) that can be characterized as more or less primitive with CD150 (Slamf1), CD48 (Slamf2), CD34, and Flt3 [4,5]. Human HSCs can also be isolated and identified by the expression of cell surface markers such as Lin$^-$, CD34$^+$, CD38$^-$, Thy1.1$^+$, and CD45RA$^-$ [6–8] (Figure 1).

HSCs have the ability to self-renew and differentiate into immune cells; however, similar to other adult stem cells, HSCs are vulnerable to age-related stress [7,8]. With aging, HSCs gradually lose their self-renewal capacity and reconstitution potential and are therefore different from pluripotent embryonic stem cells (ESCs) and induced pluripotent stem cells (iPSCs) [9]. HSC aging is driven by both cell-intrinsic and -extrinsic factors. Although the functional change of HSCs with aging is mostly regulated by cell-intrinsic factors, including DNA damage, reactive oxygen species (ROS), epigenetic changes and polarity changes, it is also known that the hematopoietic niche or microenvironment-derived cell-extrinsic factors are essential for the maintenance of HSCs [7,10]. Hematopoietic aging alters the immune system, inducing many types of immune diseases including both myeloid and lymphoid leukemias, anemia, declining adaptive immunity, autoimmunity, increased susceptibility to infectious diseases, and vaccine failure [11,12]. Thus, the study of HSC aging is important to our understanding of age-related immune diseases and can also provide potential strategies to improve quality of life in the elderly. In this review, we summarized the hallmarks, causes and mechanisms, and rejuvenation of HSC aging, and also introduced recent emerging technologies for HSC study.

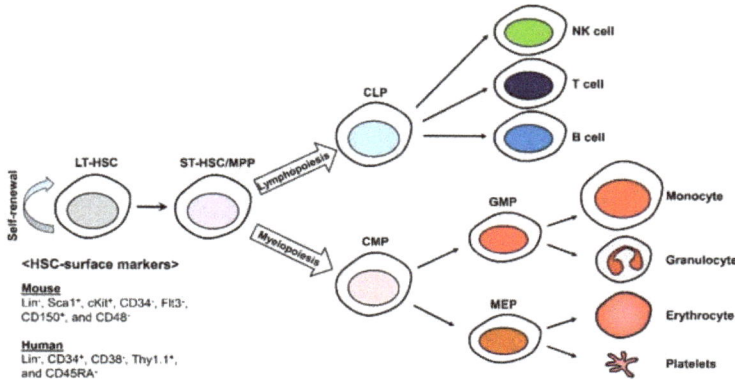

Figure 1. Differentiation of hematopoietic stem cells (HSCs). Long-term HSCs (LT-HSCs) are able to self-renew and are responsible for generating blood cells. CLP; the common lymphoid progenitor, CMP; the common myeloid progenitor, GMP; the granulocyte macrophage progenitor, and MEP; the megakaryocytic and erythroid progenitor.

2. Hallmarks of HSC Aging

2.1. Defect in Repopulating Capacity

Though we know that the number of HSCs in bone marrow (BM) increases 2- to 10-fold with aging in both mice and humans [12–14], the mechanisms by which such increases in HSC numbers occur in aged organisms remain elusive. One hypothesis proposes that the increase in the number of aged HSCs is a compensatory mechanism to overcome their loss of function in normal hematopoiesis [12]. Another study suggested that an increase in the frequency of self-renewing symmetric cell divisions might contribute to the increased numbers and impaired function of aged HSCs [15]. Recent reports show that aged HSCs are less quiescent, with larger fractions undergoing cell division than young HSCs [8,16], having accumulated more oxidative DNA damage [17]. These reports go on to suggest that the function of aged HSCs may be limited by these factors. To determine the functional differences between young and aged HSCs, a competitive transplantation assay was developed as a gold standard test to assess the long-term self-renewal and multi-lineage potential of HSCs. In this method, HSCs are mixed with young BM cells that are able to restore immunity in the recipient animal post-irradiation [3,5,11,12]. We and other research groups have reported that aged HSCs exhibited a functional decline in repopulation capacity [13,18,19]. These results imply that the increased number of aged HSCs does not compensate for their loss of function, thereby resulting in immune cell homeostasis defects in aged organisms [12,15].

2.2. Defect in Homing and Increase in Mobilization

HSCs reside in the marrow cavity of long bones during adult life, co-existing in close association with many other cell types in BM in a highly organized structure supported by stromal cells referred to as the niche [20–22]. The engraftment of HSCs into nonmyeloablative hosts resulted in spatial localization of stem cell "niches," while other transplanted BM cells were detected as flattened bone lining cells on the periosteal bone surface [23]. Osteoblastic cells constitute an important cellular component of the HSC niche [24,25]. Other groups assessed that different hematopoietic cell subsets are localized to distinct areas according to their stage of differentiation using live-imaging techniques [20,26,27]. Interestingly, transplanted HSCs tended to home to the endosteum in irradiated recipients but were randomly distributed and unstable in nonirradiated mice [27]. The homing ability of HSCs, involving the trafficking of a transplanted donor HSC to the BM of a recipient, is critical in BM transplantation procedures for the successful treatment of many blood disorders and malignancies including leukemia, lymphoma, and myeloma [28]. Liang and colleagues found that the BM homing efficiency of aged mouse HSCs was approximately three-fold lower than that of young HSCs; we also reported a reduction in the homing ability of aged HSCs in BM [1,19,29].

Systemic administration of cytokines and chemokines or cytotoxic agents mobilizes hematopoietic stem and progenitor cells (HSPCs) from the BM into peripheral blood, where they are collected in clinically useful quantities for stem cell therapies [22]. Granulocyte colony-stimulating factor (G-CSF) mobilizes hematopoietic cells from the marrow into circulation, with increased progenitor cells of all lineages detected in the spleens of G-CSF-treated mice [30,31]. In a G-CSF-receptor (G-CSFR)-deficient mice study, they found that expression of the G-CSFR on BM cells is required for mobilization of hematopoietic progenitor cells (HPCs), but this is not dependent on the expression of the G-CSFR on hematopoietic progenitor cells (HPCs). From these results, they suggested that the indirect effect of G-CSF on hematopoietic cells is essential for HSC mobilization [30,32]. G-CSF causes transient upregulation of stromal cell-derived factor-1 (SDF-1) and subsequently activates CXC chemokine receptor-4 (CXCR4) signaling for the production of hepatocyte growth factor (HGF). HGF binds to c-Met and thus activates c-Met signaling to regulate the mTOR/FOXO3a signaling pathway. Finally, G-CSF signaling causes ROS production and promotes the egress of HSCs out of the BM [31,33]. Xing and colleagues revealed that the mobilization of HSPCs from BM to peripheral blood in response to G-CSF requires the de-adhesion of HSPCs from the niche. This ability to mobilize HSCs is approximately five-fold greater in aged mice [22].

2.3. Lineage Skewing

Under normal conditions, HSCs differentiate into balanced myeloid and lymphoid lineages. However, aged adults show a higher prevalence of anemia and compromised adaptive immunity, due to reduced HSC number and function, caused by thymic involution and aged lymphoid progenitors [34,35]. Aging forces the differentiation of HSCs to myeloid bias, characterized by a higher percentage of myeloid cells in peripheral blood, both upon transplantation and at steady state [1,8,29,36]. Sudo and colleagues showed that although aged HSCs exhibit abnormal differentiation, they were able to regenerate blood cells over long periods of time and could self-renew. These aged HSCs exhibited myeloid biased differentiation [3,15]. This characteristic is also reflected in the relative expansion of myeloid progenitor numbers in aged compared to young mice—a feature of aged HSCs known to be a cell autonomous function [37]. The lineage skewing associated with HSC aging has been linked to an upregulation of myeloid-specific genes and a downregulation of lymphoid-specific genes [38,39].

3. Causes and Mechanisms of HSC Aging

3.1. DNA Damage

The DNA damage response (DDR) in cells is involved in cell cycle regulation, cell death and senescence, transcriptional regulation, as well as chromatin remodeling [40]. HSCs are also exposed to DDR stress due to their continuous self-renewal process as well as their microenvironment during aging. It has been proposed that DNA damage may be a principal mechanism regulating age-dependent stem cell decline; furthermore, the accumulation of genomic instability has been implicated in hematopoietic malignancy, possibly derived from transformed HSCs [40,41]. Researchers have reported that HSCs experience genomic instability during physiological aging. Rossi and colleagues tested these hypotheses in mice deficient in several genomic maintenance pathways including nucleotide excision repair, telomere maintenance, and non-homologous end-joining. DDR-deficient HSCs exhibited an accumulation of high levels of DNA double strand breaks (DSBs), reduced repopulation ability and self-renewal, as well as stem cell exhaustion, such as occurs in aged HSCs [1,40,42]. Similar to mouse HSCs, human CD34+ HSCs and hematopoietic progenitors showed a significant accumulation of DSBs during the normal aging process [43]. These reports strongly suggest that the appropriate DDR is important for the maintenance of HSCs and may prevent the functional decline of HSCs from aging.

3.2. ROS (Reactive Oxygen Species)

HSCs reside in hypoxic niches in the BM—an environment that presumably ensures HSCs are protected from oxidative stress and can maintain their self-renewal ability [31]. HSCs are quiescent, have a low metabolic rate, and therefore, generate low levels of reactive oxygen species (ROS). However, with aging, ROS levels accumulate and can result in ROS-induced oxidative stress in HSCs [1,44,45]. Aging increases ROS levels, which contribute to increased proliferation, senescence, or apoptosis. HSCs exposed to low ROS levels maintained a higher self-renewal potential in serial transplantation. However, HSCs exposed to high ROS levels failed to self-renew and showed higher levels of expression of activated p38 mitogen-activated protein kinase (p38) and mammalian target of rapamycin (mTOR) [46]. Porto and colleagues found that aged HSCs showed increased oxidative stress compared to young HSCs; mitochondria and NADPHox were the major sources of ROS production, whereas CYP450 contributed in middle and aged HSCs and xanthine oxidase only in aged HSCs [47]. Thus, oxidative stress might be considered an important cause of HSC dysfunction during the aging process [48]. We have reported the regulatory mechanisms of ROS in aged HSCs using specific gene knockout mice. We found that thioredoxin-interacting protein (TXNIP) regulated ROS levels in HSCs by regulating p53 activity via interference with p53-mouse double minute 2 (MDM2) interactions [5]. More recently, we revealed that TXNIP regulated p38 activity—a critical inducer of ROS in aged HSCs [19]. Recent reports have also demonstrated that ROS acted as critical regulators of HSC aging and go on to determine regulatory mechanisms. The Forkhead O (FOXO) subfamily of transcription factors—including Foxo1, Foxo3a, and Foxo4—regulated pools of HSCs and progenitors as well as ROS in HSCs [49,50]. Hypoxia-inducible factor-1α (HIF-1α) was highly expressed in HSCs. HIF-1α changed cellular metabolism of HSCs from mitochondrial respiration to glycolysis and thereby reduced ROS production. Deletion of HIF-1α in HSCs induced ROS levels and reduced long-term repopulation capacity [31,51]. Altogether, the importance of ROS as an inducer of aging in HSCs was determined by cellular regulatory mechanism studies.

3.3. Epigenetic Changes

Epigenetics refers to changes in gene expression that do not involve changes to the underlying DNA sequence, that is a change in phenotype without a change in genotype [52]. Analysis of gene transcriptional profiles indicated that myeloid differentiation-linked and inflammation- and stress response-related genes were upregulated, but lymphopoiesis, DNA repair, and chromatin silencing genes were downregulated in aged HSCs [13,53]. Aged HSCs exhibited broader H3K4me3 peaks across HSC identity and self-renewal genes, exhibiting increased DNA methylation at transcription factor binding sites associated with differentiation-promoting genes. There is a strong correlation between altered H3K4me3 levels and transcriptional activity, involving genes that undergo increased expression levels, most significantly with age [53]. These gene expression changes are in accordance with age-induced myeloid skewing, at the level of gene transcription. It also implies that HSC aging is related to epigenetic changes that impact transcriptional regulation. The function of DNA methylation includes the induction of physiological processes such as stem cell differentiation [52]. Genetic inactivation of DNA methyltransferase 1 (DNMT1), induced nearly immediate and complete loss of HSCs in vivo [54]. Reduced DNMT1 activity in HSCs restricted myeloerythroid differentiation due to both impaired silencing of key lineage determinant genes and an inability to prime master lymphoid regulators [55]. DNMT3a also regulates HSC fate decisions, as evidenced by the conditional inactivation of DNMT3a, which skews HSC divisions toward self-renewal at the expense of differentiation [56]. The expression of genes encoding DNA methyltransferase enzymes decreases in aged HSCs compared to young; moreover, aged HSCs are characterized by a gradual increase in global DNA methylation levels [53]. Corresponding with the DNA hypermethylation of aged HSCs, there is a concomitant reduction in 5-hmC in aged HSCs compared to young, with all three TET enzymes decreasing in abundance, with age [52,53]. Florian and colleagues demonstrated that young HSCs expressed higher levels of AcH4K16 and aged HSCs presented decreased levels of AcH4K16 [57]. However, these mechanisms are inadequate to explain the exact correlation between epigenetic changes and HSC aging; therefore, the functional consequences of epigenetic changes in HSCs remain to be determined.

3.4. Polarity Changes

The small RhoGTPase Cdc42 (Cdc42) cycles between an active (GTP-bound) and an inactive (GDP-bound) state and is known to regulate actin and tubulin organization, cell-cell and cell-extracellular matrix adhesion, and cell polarity in distinct cell types [57–59]. Florian and colleagues reported Cdc42 as a new aging marker for HSCs. They found that elevated Cdc42 activity in aged HSCs was causally linked to HSC aging and correlated with a loss of polarity in aged HSCs. Constitutively activated Cdc42 resulted in premature aging of HSCs and induced depolarization of Cdc42 and tubulin in aged HSCs. Pharmacological reduction of Cdc42 activity reversed the polarity of Cdc42 and tubulin and restored cellular function of aged HSCs [57]. We also demonstrated that aged HSCs showed depolarization of Cdc42 allowing us to use it as an aging marker for HSCs in our study [19].

4. Rejuvenation of Aged HSCs

4.1. Reduction of Nutrient Supply

Mutations in growth signaling pathways extend the life span while protecting against age-dependent DNA damage in yeast [60]. Prolonged fasting (PF) reduces progrowth signaling and activates pathways that enhance cellular resistance to toxins in mice and humans [60–62]. One group has suggested that caloric restriction by PF can rejuvenate aged HSCs. They reported that PF reduced circulating IGF-1 levels and PKA activity in various cell populations and rejuvenated the aging-associated phenotypes of aged HSCs including myeloid bias. PF also protected mice from chemotherapy-induced immunosuppression and mortality [63].

Reversible acetylation of metabolic enzymes regulates metabolic pathways in response to nutrient availability [64]. The sirtuin family are key regulators of the nutrient-sensitive metabolic regulatory circuit. Calorie restriction reduced ROS by inducing the activation of SIRT3-mediated superoxide dismutase 2 (SOD2) [65]. Brown and colleagues demonstrated that SIRT3 regulated the global acetylation of mitochondrial proteins, which are enriched in HSCs. The expression of SIRT3 decreased with aging; moreover, SIRT3 was dispensable for young HSCs but was essential under stress or for aged HSCs. Finally, the authors suggested that the plasticity of mitochondrial homeostasis controls HSC aging and showed that aged HSCs can be rejuvenated by SIRT3 expression [66]. Mohrin and colleagues have reported on the interaction between SIRT7 and nuclear respiratory factor 1 (NRF1). SIRT7-deficient HSCs showed a reduction in repopulating capacity and myeloid-biased differentiation in transplantation assays. The upregulation of SIRT7 improved the regenerative capacity of aged HSCs [67].

4.2. ROS Scavenging

HSCs exhibit low metabolic rates and produce less ROS. The ataxia telangiectasia mutated (ATM) gene is essential for HSC self-renewal and quiescence; ATM regulates ROS levels in HSCs. ATM-deficient mice showed a defect in HSC function and elevated ROS levels in HSCs. Treatment with the ROS scavenging agent, N-acetyl-L-cysteine (NAC), rescued the repopulating capacity of

ATM-deficient HSCs [68]. p38 limits the lifespan of HSCs in vivo by inducing ROS; the inhibition of p38 by treatment with SB203580, a p38 inhibitor, rescued ROS-induced defects in HSC repopulating capacity and HSC quiescence maintenance [69]. Recently, we developed a new p38 inhibitor—TN13—which is a cell-penetrating peptide-(CPP-), a conjugated peptide derived from the TXNIP-p38 interaction motif in TXNIP. TN13 inhibited p38 activity and rejuvenated aged HSCs by reducing ROS levels [19]. G protein-coupled receptor kinases (GRKs) regulate cytokine receptors in mature leukocytes. GRK6$^{-/-}$ mice exhibited HSC loss and impaired HSC self-renewal. GRK6 also regulates ROS response, and ROS scavenger α-lipoic acid treatment rescued HSC loss in GRK6$^{-/-}$ mice [70].

4.3. Epigenetic Modulation

Elevated Cdc42 activity is associated with HSC aging and induces decreased levels of H4K16Ac in aged HSCs. Treatment with a specific inhibitor of Cdc42 activity (CASIN) can restore aged HSC phenotypes by regulating both Cdc42 activity and epigenetic reprograming by elevating H4K16Ac levels to those of young cells [1,57]. Special AT-rich sequence binding 1 (Satb1), a nuclear architectural protein, regulates chromatin structure by epigenetic regulation and plays an important role in lymphoid lineage specification. The expression of Satb1 increased with early lymphoid differentiation and Satb1-deficient HSCs showed a reduction in lymphopoiesis, while induced Satb1 expression enhanced lymphocyte production. Exogenous Satb1 expression primed the lymphoid potential of aged HSCs [71].

4.4. Clearance of Senescent Cells

Aging and genotoxic stress induce cellular senescence of HSCs [72,73]. Senescent cells (SCs) accumulate in various tissues and organs with aging, thereby disrupting tissue structure and function by secreting cellular components. Clearance of SCs in a mouse model using a transgenic approach revealed delays in several age-associated disorders [74]. Chang and colleagues developed a potent senolytic drug ABT263—a specific inhibitor of BCL-2 and BCL-xL, that selectively kills SCs. Oral administration of ABT263 effectively depleted SCs—including senescent HSCs in both sub-lethally irradiated and normally aged mice. Depletion of SCs mitigated irradiation-induced premature aging of the hematopoietic system and rejuvenated aged HSCs in normally aged mice [75]. Given these results, Chang et al. determined that ABT263 may represent a new class of radiation mitigators and anti-aging agents.

5. Emerging Technologies for HSC Study

5.1. Single-Cell RNA-Sequencing (scRNA-Seq)

HSCs are heterogeneous and HSC fate decisions are performed at the individual cell level. Thus, single-cell profiling may be an essential tool for characterizing the heterogeneity of HSCs. Most recently, single-cell RNA-sequencing (scRNA-Seq) was developed to profile single cells and was applied in the studies of HSCs [76]. Kowalczyk and colleagues used scRNA-Seq to dissect variability in HSCs and HPCs from young and old mice. They found that transcriptional changes in HSCs during aging are inversely related to those upon HSC differentiation, and old ST-HSCs exist in a less differentiated state than young ST-HSCs [77]. One group have revealed that the megakaryocyte lineage is developed independently of other hematopoietic fates using scRNA-Seq. They have also found a functional hierarchy of unilineage- and oligolineage-producing clones within the MPPs [78]. Grover and colleagues identified an unrecognized class of HSCs that exclusively produce platelets with age. They suggested that increased platelet bias may contribute to the age-associated decrease in lymphopoiesis [79]. More recently, Baron and colleagues reported the genes and transcription factor networks activated during the endothelial-to-haematopoietic switch using scRNA-Seq. From this study, they provide an unprecedented complete resource to study in depth HSC generation in vivo [80].

5.2. Single-Cell Transplantation

HSC transplantation assays evaluate the capacity of HSCs to repopulate the hematopoietic system. HSC heterogeneity was initially identified by tracking of myeloid and lymphoid lineage output from transplanted single HSCs [81–83]. Single-cell transplantation has shown considerable heterogeneity of HSCs and different propensities of myeloid-, lymphoid- and platelet-biased HSCs in fate decision [84]. Carrelha and colleagues applied single-cell transplantation assays to investigate lineage-restricted fates of long-term self-renewing HSCs. They found that a distinct class of HSCs adopts a fate towards megakaryocyte/platelet-lineage tree, but not of other blood cell lineages. Finally, they suggested that a limited repertoire of distinct HSC subsets adopt a fate towards replenishment of a restricted set of blood lineages, before loss of self-renewal and multipotency [84]. Yamamoto and colleagues investigated the age-related functional changes in HSCs using single-cell transplantation assays. They found that myeloid-restricted repopulating progenitor (MyRPs) frequency increased dramatically with age. Interestingly, they identified new subsets, latent-HSCs, that were myeloid restricted in primary recipients but displayed multipotent (five blood-lineage) output in secondary recipients. From these results, they have raised a question about the traditional dogma of HSC aging and our current approaches to assay [85].

6. Conclusions

Here, we introduced the hallmarks of HSC aging including defects in repopulation capacity, homing, and an increase in mobilization and myeloid-biased skewing. HSC aging is driven by multiple cell-intrinsic factors such as DNA damage, ROS, epigenetic changes and polarity changes, and cell-extrinsic factors including cytokines and chemokines derived from the HSC niche. We also mentioned recent emerging technologies for single-HSC profiling such as scRNA-Seq and single-cell transplantation. These methods have proven highly valuable in the unravelling of HSC heterogeneity, as described above. Surprisingly, aged HSCs could be rejuvenated by reduction of the nutrient supply, ROS scavenging, epigenetic modulation or the clearance of senescent cells (Figure 2). Although we have described the defined mechanisms and the physiological or biological changes involved in HSC aging, this does not explain all the aspects of aging, many of which remain elusive. However, our accumulated knowledge may be helpful for understanding HSC aging and designing HSC aging studies. Fortunately, many HSC aging studies have suggested that the dysregulated functions of aged HSCs can be reversed or rescued by rejuvenating agents or the overexpression of specific genes. In the future, these approaches may shine a light on treating age-related immune diseases and can also provide a promising tool to improve quality of life for the elderly.

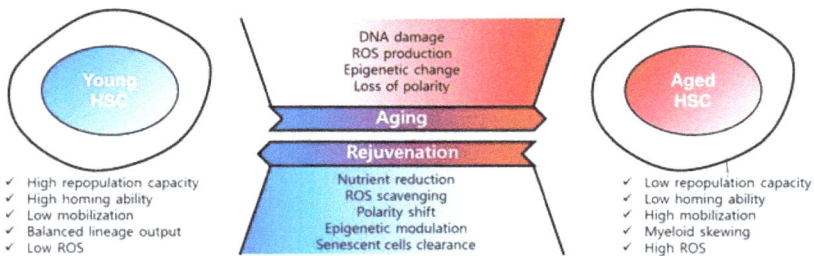

Figure 2. Regulation of HSC aging and rejuvenation. Aged HSCs have the hallmarks of low repopulation capacity, low homing ability, high mobility, myeloid skewing, and high ROS, among others. HSC aging is driven by DNA damage, ROS, epigenetic change, and loss of polarity. Aged HSCs can be rejuvenated by nutrient reduction, ROS scavenging, polarity shift, epigenetic modulation, and senescent cells clearance.

Int. J. Mol. Sci. **2019**, *20*, 1272

Author Contributions: J.L. and S.R.Y. wrote the manuscript. I.C. and H.J. wrote and supervised the manuscript.

Funding: This work was supported in part by the National Research Foundation of Korea (NRF) (2019R1A2C3002034 and 2019R1A2C1007906), and the KRIBB Research Initiative Program from the Korea government (MSIP).

Conflicts of Interest: The authors declare no conflict of interest.

References

1. Akunuru, S.; Geiger, H. Aging, Clonality, and Rejuvenation of Hematopoietic Stem Cells. *Trends Mol. Med.* **2016**, *22*, 701–712. [CrossRef]
2. Moehrle, B.M.; Geiger, H. Aging of hematopoietic stem cells: DNA damage and mutations? *Exp. Hematol.* **2016**, *44*, 895–901. [CrossRef] [PubMed]
3. Ergen, A.V.; Goodell, M.A. Mechanisms of hematopoietic stem cell aging. *Exp. Gerontol.* **2010**, *45*, 286–290. [CrossRef]
4. Kiel, M.J.; Yilmaz, O.H.; Iwashita, T.; Yilmaz, O.H.; Terhorst, C.; Morrison, S.J. SLAM family receptors distinguish hematopoietic stem and progenitor cells and reveal endothelial niches for stem cells. *Cell* **2005**, *121*, 1109–1121. [CrossRef]
5. Jung, H.; Kim, M.J.; Kim, D.O.; Kim, W.S.; Yoon, S.J.; Park, Y.J.; Yoon, S.R.; Kim, T.D.; Suh, H.W.; Yun, S.; et al. TXNIP maintains the hematopoietic cell pool by switching the function of p53 under oxidative stress. *Cell Metab.* **2013**, *18*, 75–85. [CrossRef]
6. Notta, F.; Doulatov, S.; Laurenti, E.; Poeppl, A.; Jurisica, I.; Dick, J.E. Isolation of single human hematopoietic stem cells capable of long-term multilineage engraftment. *Science* **2011**, *333*, 218–221. [CrossRef]
7. Latchney, S.E.; Calvi, L.M. The aging hematopoietic stem cell niche: Phenotypic and functional changes and mechanisms that contribute to hematopoietic aging. *Semin. Hematol.* **2017**, *54*, 25–32. [CrossRef] [PubMed]
8. Pang, W.W.; Schrier, S.L.; Weissman, I.L. Age-associated changes in human hematopoietic stem cells. *Semin. Hematol.* **2017**, *54*, 39–42. [CrossRef]
9. De Haan, G.; Lazare, S.S. Aging of hematopoietic stem cells. *Blood* **2018**, *131*, 479–487. [CrossRef] [PubMed]
10. Morrison, S.J.; Scadden, D.T. The bone marrow niche for haematopoietic stem cells. *Nature* **2014**, *505*, 327–334. [CrossRef]
11. Warren, L.A.; Rossi, D.J. Stem Cells and aging in the hematopoietic system. *Mech. Ageing Dev.* **2009**, *130*, 46–53. [CrossRef]
12. Geiger, H.; de Haan, G.; Florian, M.C. The ageing haematopoietic stem cell compartment. *Nat. Rev. Immunol.* **2013**, *13*, 376–389. [CrossRef]
13. Rossi, D.J.; Bryder, D.; Zahn, J.M.; Ahlenius, H.; Sonu, R.; Wagers, A.J.; Weissman, I.L. Cell intrinsic alterations underlie hematopoietic stem cell aging. *Proc. Natl. Acad. Sci. USA* **2005**, *102*, 9194–9199. [CrossRef] [PubMed]
14. De Haan, G.; Nijhof, W.; Van Zant, G. Mouse strain-dependent changes in frequency and proliferation of hematopoietic stem cells during aging: Correlation between lifespan and cycling activity. *Blood* **1997**, *89*, 1543–1550. [PubMed]
15. Sudo, K.; Ema, H.; Morita, Y.; Nakauchi, H. Age-associated characteristics of murine hematopoietic stem cells. *J. Exp. Med.* **2000**, *192*, 1273–1280. [CrossRef]
16. Pang, W.W.; Price, E.A.; Sahoo, D.; Beerman, I.; Maloney, W.J.; Rossi, D.J.; Schrier, S.L.; Weissman, I.L. Human bone marrow hematopoietic stem cells are increased in frequency and myeloid-biased with age. *Proc. Natl. Acad. Sci. USA* **2011**, *108*, 20012–20017. [CrossRef] [PubMed]
17. Yahata, T.; Takanashi, T.; Muguruma, Y.; Ibrahim, A.A.; Matsuzawa, H.; Uno, T.; Sheng, Y.; Onizuka, M.; Ito, M.; Kato, S.; et al. Accumulation of oxidative DNA damage restricts the self-renewal capacity of human hematopoietic stem cells. *Blood* **2011**, *118*, 2941–2950. [CrossRef] [PubMed]
18. Kim, M.; Moon, H.B.; Spangrude, G.J. Major age-related changes of mouse hematopoietic stem/progenitor cells. *Ann. N. Y. Acad. Sci.* **2003**, *996*, 195–208. [CrossRef]
19. Jung, H.; Kim, D.O.; Byun, J.E.; Kim, W.S.; Kim, M.J.; Song, H.Y.; Kim, Y.K.; Kang, D.K.; Park, Y.J.; Kim, T.D.; et al. Thioredoxin-interacting protein regulates haematopoietic stem cell ageing and rejuvenation by inhibiting p38 kinase activity. *Nat. Commun.* **2016**, *7*, 13674. [CrossRef]
20. Chen, J. Hematopoietic stem cell development, aging and functional failure. *Int. J. Hematol.* **2011**, *94*, 3–10. [CrossRef]

21. Spangrude, G.J.; Heimfeld, S.; Weissman, I.L. Purification and characterization of mouse hematopoietic stem cells. *Science* **1988**, *241*, 58–62. [CrossRef]

22. Xing, Z.; Ryan, M.A.; Daria, D.; Nattamai, K.J.; Van Zant, G.; Wang, L.; Zheng, Y.; Geiger, H. Increased hematopoietic stem cell mobilization in aged mice. *Blood* **2006**, *108*, 2190–2197. [CrossRef]

23. Nilsson, S.K.; Dooner, M.S.; Weier, H.U.; Frenkel, B.; Lian, J.B.; Stein, G.S.; Quesenberry, P.J. Cells capable of bone production engraft from whole bone marrow transplants in nonablated mice. *J. Exp. Med.* **1999**, *189*, 729–734. [CrossRef]

24. Calvi, L.M.; Adams, G.B.; Weibrecht, K.W.; Weber, J.M.; Olson, D.P.; Knight, M.C.; Martin, R.P.; Schipani, E.; Divieti, P.; Bringhurst, F.R.; et al. Osteoblastic cells regulate the haematopoietic stem cell niche. *Nature* **2003**, *425*, 841–846. [CrossRef]

25. Zhang, J.; Niu, C.; Ye, L.; Huang, H.; He, X.; Tong, W.G.; Ross, J.; Haug, J.; Johnson, T.; Feng, J.Q.; et al. Identification of the haematopoietic stem cell niche and control of the niche size. *Nature* **2003**, *425*, 836–841. [CrossRef] [PubMed]

26. Lo Celso, C.; Fleming, H.E.; Wu, J.W.; Zhao, C.X.; Miake-Lye, S.; Fujisaki, J.; Cote, D.; Rowe, D.W.; Lin, C.P.; Scadden, D.T. Live-animal tracking of individual haematopoietic stem/progenitor cells in their niche. *Nature* **2009**, *457*, 92–96. [CrossRef]

27. Xie, Y.; Yin, T.; Wiegraebe, W.; He, X.C.; Miller, D.; Stark, D.; Perko, K.; Alexander, R.; Schwartz, J.; Grindley, J.C.; et al. Detection of functional haematopoietic stem cell niche using real-time imaging. *Nature* **2009**, *457*, 97–101. [CrossRef]

28. Domingues, M.J.; Nilsson, S.K.; Cao, B. New agents in HSC mobilization. *Int. J. Hematol.* **2017**, *105*, 141–152. [CrossRef]

29. Liang, Y.; Van Zant, G.; Szilvassy, S.J. Effects of aging on the homing and engraftment of murine hematopoietic stem and progenitor cells. *Blood* **2005**, *106*, 1479–1487. [CrossRef]

30. Bendall, L.J.; Bradstock, K.F. G-CSF: From granulopoietic stimulant to bone marrow stem cell mobilizing agent. *Cytokine Growth Factor Rev.* **2014**, *25*, 355–367. [CrossRef]

31. Lee, J.; Cho, Y.S.; Jung, H.; Choi, I. Pharmacological Regulation of Oxidative Stress in Stem Cells. *Oxid. Med. Cell. Longev.* **2018**, *2018*, 4081890. [CrossRef]

32. Liu, F.; Poursine-Laurent, J.; Link, D.C. Expression of the G-CSF receptor on hematopoietic progenitor cells is not required for their mobilization by G-CSF. *Blood* **2000**, *95*, 3025–3031.

33. Tesio, M.; Golan, K.; Corso, S.; Giordano, S.; Schajnovitz, A.; Vagima, Y.; Shivtiel, S.; Kalinkovich, A.; Caione, L.; Gammaitoni, L.; et al. Enhanced c-Met activity promotes G-CSF-induced mobilization of hematopoietic progenitor cells via ROS signaling. *Blood* **2011**, *117*, 419–428. [CrossRef]

34. Dorshkind, K.; Montecino-Rodriguez, E.; Signer, R.A. The ageing immune system: Is it ever too old to become young again? *Nat. Rev. Immunol.* **2009**, *9*, 57–62. [CrossRef]

35. Geiger, H.; Rudolph, K.L. Aging in the lympho-hematopoietic stem cell compartment. *Trends Immunol.* **2009**, *30*, 360–365. [CrossRef]

36. Wahlestedt, M.; Erlandsson, E.; Kristiansen, T.; Lu, R.; Brakebusch, C.; Weissman, I.L.; Yuan, J.; Martin-Gonzalez, J.; Bryder, D. Clonal reversal of ageing-associated stem cell lineage bias via a pluripotent intermediate. *Nat. Commun.* **2017**, *8*, 14533. [CrossRef]

37. Beerman, I.; Bhattacharya, D.; Zandi, S.; Sigvardsson, M.; Weissman, I.L.; Bryder, D.; Rossi, D.J. Functionally distinct hematopoietic stem cells modulate hematopoietic lineage potential during aging by a mechanism of clonal expansion. *Proc. Natl. Acad. Sci. USA* **2010**, *107*, 5465–5470. [CrossRef]

38. Cho, R.H.; Sieburg, H.B.; Muller-Sieburg, C.E. A new mechanism for the aging of hematopoietic stem cells: Aging changes the clonal composition of the stem cell compartment but not individual stem cells. *Blood* **2008**, *111*, 5553–5561. [CrossRef]

39. Rundberg Nilsson, A.; Soneji, S.; Adolfsson, S.; Bryder, D.; Pronk, C.J. Human and Murine Hematopoietic Stem Cell Aging Is Associated with Functional Impairments and Intrinsic Megakaryocytic/Erythroid Bias. *PLoS ONE* **2016**, *11*, e0158369. [CrossRef]

40. Li, T.; Zhou, Z.W.; Ju, Z.; Wang, Z.Q. DNA Damage Response in Hematopoietic Stem Cell Ageing. *Genom. Proteom. Bioinform.* **2016**, *14*, 147–154. [CrossRef]

41. Park, Y.; Gerson, S.L. DNA repair defects in stem cell function and aging. *Annu. Rev. Med.* **2005**, *56*, 495–508. [CrossRef]

42. Rossi, D.J.; Bryder, D.; Seita, J.; Nussenzweig, A.; Hoeijmakers, J.; Weissman, I.L. Deficiencies in DNA damage repair limit the function of haematopoietic stem cells with age. *Nature* **2007**, *447*, 725–729. [CrossRef]

43. Rube, C.E.; Fricke, A.; Widmann, T.A.; Furst, T.; Madry, H.; Pfreundschuh, M.; Rube, C. Accumulation of DNA damage in hematopoietic stem and progenitor cells during human aging. *PLoS ONE* **2011**, *6*, e17487. [CrossRef]

44. Hamilton, M.L.; Van Remmen, H.; Drake, J.A.; Yang, H.; Guo, Z.M.; Kewitt, K.; Walter, C.A.; Richardson, A. Does oxidative damage to DNA increase with age? *Proc. Natl. Acad. Sci. USA* **2001**, *98*, 10469–10474. [CrossRef]

45. Simsek, T.; Kocabas, F.; Zheng, J.; Deberardinis, R.J.; Mahmoud, A.I.; Olson, E.N.; Schneider, J.W.; Zhang, C.C.; Sadek, H.A. The distinct metabolic profile of hematopoietic stem cells reflects their location in a hypoxic niche. *Cell Stem Cell* **2010**, *7*, 380–390. [CrossRef]

46. Jang, Y.Y.; Sharkis, S.J. A low level of reactive oxygen species selects for primitive hematopoietic stem cells that may reside in the low-oxygenic niche. *Blood* **2007**, *110*, 3056–3063. [CrossRef]

47. Porto, M.L.; Rodrigues, B.P.; Menezes, T.N.; Ceschim, S.L.; Casarini, D.E.; Gava, A.L.; Pereira, T.M.; Vasquez, E.C.; Campagnaro, B.P.; Meyrelles, S.S. Reactive oxygen species contribute to dysfunction of bone marrow hematopoietic stem cells in aged C57BL/6 J mice. *J. Biomed. Sci.* **2015**, *22*, 97. [CrossRef]

48. Urao, N.; Ushio-Fukai, M. Redox regulation of stem/progenitor cells and bone marrow niche. *Free Rad. Biol. Med.* **2013**, *54*, 26–39. [CrossRef]

49. Tothova, Z.; Kollipara, R.; Huntly, B.J.; Lee, B.H.; Castrillon, D.H.; Cullen, D.E.; McDowell, E.P.; Lazo-Kallanian, S.; Williams, I.R.; Sears, C.; et al. FoxOs are critical mediators of hematopoietic stem cell resistance to physiologic oxidative stress. *Cell* **2007**, *128*, 325–339. [CrossRef]

50. Miyamoto, K.; Araki, K.Y.; Naka, K.; Arai, F.; Takubo, K.; Yamazaki, S.; Matsuoka, S.; Miyamoto, T.; Ito, K.; Ohmura, M.; et al. Foxo3a is essential for maintenance of the hematopoietic stem cell pool. *Cell Stem Cell* **2007**, *1*, 101–112. [CrossRef]

51. Wheaton, W.W.; Chandel, N.S. Hypoxia. 2. Hypoxia regulates cellular metabolism. *Am. J. Physiol. Cell Physiol.* **2011**, *300*, C385–C393. [CrossRef]

52. Kramer, A.; Challen, G.A. The epigenetic basis of hematopoietic stem cell aging. *Semin. Hematol.* **2017**, *54*, 19–24. [CrossRef]

53. Sun, D.; Luo, M.; Jeong, M.; Rodriguez, B.; Xia, Z.; Hannah, R.; Wang, H.; Le, T.; Faull, K.F.; Chen, R.; et al. Epigenomic profiling of young and aged HSCs reveals concerted changes during aging that reinforce self-renewal. *Cell Stem Cell* **2014**, *14*, 673–688. [CrossRef]

54. Trowbridge, J.J.; Snow, J.W.; Kim, J.; Orkin, S.H. DNA methyltransferase 1 is essential for and uniquely regulates hematopoietic stem and progenitor cells. *Cell Stem Cell* **2009**, *5*, 442–449. [CrossRef]

55. Broske, A.M.; Vockentanz, L.; Kharazi, S.; Huska, M.R.; Mancini, E.; Scheller, M.; Kuhl, C.; Enns, A.; Prinz, M.; Jaenisch, R.; et al. DNA methylation protects hematopoietic stem cell multipotency from myeloerythroid restriction. *Nat. Genet.* **2009**, *41*, 1207–1215. [CrossRef]

56. Challen, G.A.; Sun, D.; Jeong, M.; Luo, M.; Jelinek, J.; Berg, J.S.; Bock, C.; Vasanthakumar, A.; Gu, H.; Xi, Y.; et al. Dnmt3a is essential for hematopoietic stem cell differentiation. *Nat. Genet.* **2011**, *44*, 23–31. [CrossRef]

57. Florian, M.C.; Dorr, K.; Niebel, A.; Daria, D.; Schrezenmeier, H.; Rojewski, M.; Filippi, M.D.; Hasenberg, A.; Gunzer, M.; Scharffetter-Kochanek, K.; et al. Cdc42 activity regulates hematopoietic stem cell aging and rejuvenation. *Cell Stem Cell* **2012**, *10*, 520–530. [CrossRef]

58. Cau, J.; Hall, A. Cdc42 controls the polarity of the actin and microtubule cytoskeletons through two distinct signal transduction pathways. *J. Cell Sci.* **2005**, *118*, 2579–2587. [CrossRef]

59. Florian, M.C.; Geiger, H. Concise review: Polarity in stem cells, disease, and aging. *Stem Cells* **2010**, *28*, 1623–1629. [CrossRef]

60. Guevara-Aguirre, J.; Balasubramanian, P.; Guevara-Aguirre, M.; Wei, M.; Madia, F.; Cheng, C.W.; Hwang, D.; Martin-Montalvo, A.; Saavedra, J.; Ingles, S.; et al. Growth hormone receptor deficiency is associated with a major reduction in pro-aging signaling, cancer, and diabetes in humans. *Sci. Transl. Med.* **2011**, *3*, 70ra13. [CrossRef]

61. Fontana, L.; Partridge, L.; Longo, V.D. Extending healthy life span–from yeast to humans. *Science* **2010**, *328*, 321–326. [CrossRef]

62. Holzenberger, M.; Dupont, J.; Ducos, B.; Leneuve, P.; Geloen, A.; Even, P.C.; Cervera, P.; Le Bouc, Y. IGF-1 receptor regulates lifespan and resistance to oxidative stress in mice. *Nature* **2003**, *421*, 182–187. [CrossRef]

63. Cheng, C.W.; Adams, G.B.; Perin, L.; Wei, M.; Zhou, X.; Lam, B.S.; Da Sacco, S.; Mirisola, M.; Quinn, D.I.; Dorff, T.B.; et al. Prolonged fasting reduces IGF-1/PKA to promote hematopoietic-stem-cell-based regeneration and reverse immunosuppression. *Cell Stem Cell* **2014**, *14*, 810–823. [CrossRef]

64. Shin, J.; Zhang, D.; Chen, D. Reversible acetylation of metabolic enzymes celebration: SIRT2 and p300 join the party. *Mol. Cell* **2011**, *43*, 3–5. [CrossRef]

65. Qiu, X.; Brown, K.; Hirschey, M.D.; Verdin, E.; Chen, D. Calorie restriction reduces oxidative stress by SIRT3-mediated SOD2 activation. *Cell Metab.* **2010**, *12*, 662–667. [CrossRef]

66. Brown, K.; Xie, S.; Qiu, X.; Mohrin, M.; Shin, J.; Liu, Y.; Zhang, D.; Scadden, D.T.; Chen, D. SIRT3 reverses aging-associated degeneration. *Cell Rep.* **2013**, *3*, 319–327. [CrossRef]

67. Mohrin, M.; Shin, J.; Liu, Y.; Brown, K.; Luo, H.; Xi, Y.; Haynes, C.M.; Chen, D. Stem cell aging. A mitochondrial UPR-mediated metabolic checkpoint regulates hematopoietic stem cell aging. *Science* **2015**, *347*, 1374–1377. [CrossRef]

68. Ito, K.; Hirao, A.; Arai, F.; Matsuoka, S.; Takubo, K.; Hamaguchi, I.; Nomiyama, K.; Hosokawa, K.; Sakurada, K.; Nakagata, N.; et al. Regulation of oxidative stress by ATM is required for self-renewal of haematopoietic stem cells. *Nature* **2004**, *431*, 997–1002. [CrossRef]

69. Ito, K.; Hirao, A.; Arai, F.; Takubo, K.; Matsuoka, S.; Miyamoto, K.; Ohmura, M.; Naka, K.; Hosokawa, K.; Ikeda, Y.; et al. Reactive oxygen species act through p38 MAPK to limit the lifespan of hematopoietic stem cells. *Nat. Med.* **2006**, *12*, 446–451. [CrossRef]

70. Le, Q.; Yao, W.; Chen, Y.; Yan, B.; Liu, C.; Yuan, M.; Zhou, Y.; Ma, L. GRK6 regulates ROS response and maintains hematopoietic stem cell self-renewal. *Cell Death Dis.* **2016**, *7*, e2478. [CrossRef]

71. Satoh, Y.; Yokota, T.; Sudo, T.; Kondo, M.; Lai, A.; Kincade, P.W.; Kouro, T.; Iida, R.; Kokame, K.; Miyata, T.; et al. The Satb1 protein directs hematopoietic stem cell differentiation toward lymphoid lineages. *Immunity* **2013**, *38*, 1105–1115. [CrossRef] [PubMed]

72. Shao, L.; Feng, W.; Li, H.; Gardner, D.; Luo, Y.; Wang, Y.; Liu, L.; Meng, A.; Sharpless, N.E.; Zhou, D. Total body irradiation causes long-term mouse BM injury via induction of HSC premature senescence in an Ink4a- and Arf-independent manner. *Blood* **2014**, *123*, 3105–3115. [CrossRef] [PubMed]

73. Le, O.N.; Rodier, F.; Fontaine, F.; Coppe, J.P.; Campisi, J.; DeGregori, J.; Laverdiere, C.; Kokta, V.; Haddad, E.; Beausejour, C.M. Ionizing radiation-induced long-term expression of senescence markers in mice is independent of p53 and immune status. *Aging Cell* **2010**, *9*, 398–409. [CrossRef]

74. Baker, D.J.; Wijshake, T.; Tchkonia, T.; LeBrasseur, N.K.; Childs, B.G.; van de Sluis, B.; Kirkland, J.L.; van Deursen, J.M. Clearance of p16Ink4a-positive senescent cells delays ageing-associated disorders. *Nature* **2011**, *479*, 232–236. [CrossRef]

75. Chang, J.; Wang, Y.; Shao, L.; Laberge, R.M.; Demaria, M.; Campisi, J.; Janakiraman, K.; Sharpless, N.E.; Ding, S.; Feng, W.; et al. Clearance of senescent cells by ABT263 rejuvenates aged hematopoietic stem cells in mice. *Nat. Med.* **2016**, *22*, 78–83. [CrossRef] [PubMed]

76. Papalexi, E.; Satija, R. Single-cell RNA sequencing to explore immune cell heterogeneity. *Nat. Rev. Immunol.* **2018**, *18*, 35–45. [CrossRef] [PubMed]

77. Kowalczyk, M.S.; Tirosh, I.; Heckl, D.; Rao, T.N.; Dixit, A.; Haas, B.J.; Schneider, R.K.; Wagers, A.J.; Ebert, B.L.; Regev, A. Single-cell RNA-seq reveals changes in cell cycle and differentiation programs upon aging of hematopoietic stem cells. *Genome Res.* **2015**, *25*, 1860–1872. [CrossRef]

78. Rodriguez-Fraticelli, A.E.; Wolock, S.L.; Weinreb, C.S.; Panero, R.; Patel, S.H.; Jankovic, M.; Sun, J.; Calogero, R.A.; Klein, A.M.; Camargo, F.D. Clonal analysis of lineage fate in native haematopoiesis. *Nature* **2018**, *553*, 212–216. [CrossRef] [PubMed]

79. Grover, A.; Sanjuan-Pla, A.; Thongjuea, S.; Carrelha, J.; Giustacchini, A.; Gambardella, A.; Macaulay, I.; Mancini, E.; Luis, T.C.; Mead, A.; et al. Single-cell RNA sequencing reveals molecular and functional platelet bias of aged haematopoietic stem cells. *Nat. Commun.* **2016**, *7*, 11075. [CrossRef] [PubMed]

80. Baron, C.S.; Kester, L.; Klaus, A.; Boisset, J.C.; Thambyrajah, R.; Yvernogeau, L.; Kouskoff, V.; Lacaud, G.; van Oudenaarden, A.; Robin, C. Single-cell transcriptomics reveal the dynamic of haematopoietic stem cell production in the aorta. *Nat. Commun.* **2018**, *9*, 2517. [CrossRef]

81. Muller-Sieburg, C.E.; Cho, R.H.; Thoman, M.; Adkins, B.; Sieburg, H.B. Deterministic regulation of hematopoietic stem cell self-renewal and differentiation. *Blood* **2002**, *100*, 1302–1309. [PubMed]

82. Dykstra, B.; Kent, D.; Bowie, M.; McCaffrey, L.; Hamilton, M.; Lyons, K.; Lee, S.J.; Brinkman, R.; Eaves, C. Long-term propagation of distinct hematopoietic differentiation programs in vivo. *Cell Stem Cell* **2007**, *1*, 218–229. [CrossRef] [PubMed]

83. Jacobsen, S.E.W.; Nerlov, C. Haematopoiesis in the era of advanced single-cell technologies. *Nat. Cell Biol.* **2019**, *21*, 2–8. [CrossRef] [PubMed]

84. Carrelha, J.; Meng, Y.; Kettyle, L.M.; Luis, T.C.; Norfo, R.; Alcolea, V.; Boukarabila, H.; Grasso, F.; Gambardella, A.; Grover, A.; et al. Hierarchically related lineage-restricted fates of multipotent haematopoietic stem cells. *Nature* **2018**, *554*, 106–111. [CrossRef] [PubMed]

85. Yamamoto, R.; Wilkinson, A.C.; Ooehara, J.; Lan, X.; Lai, C.Y.; Nakauchi, Y.; Pritchard, J.K.; Nakauchi, H. Large-Scale Clonal Analysis Resolves Aging of the Mouse Hematopoietic Stem Cell Compartment. *Cell Stem Cell* **2018**, *22*, 600–607e604. [CrossRef]

International Journal of
Molecular Sciences

MDPI

Review

Age-Associated Changes in the Immune System and Blood–Brain Barrier Functions

Michelle A. Erickson [1,2,]* and William A. Banks [1,2,]*

[1] VA Puget Sound Healthcare System, Geriatric Research Education and Clinical Center, Seattle, WA 98108, USA
[2] Division of Gerontology and Geriatric Medicine, Department of Medicine, University of Washington, Seattle, WA 98104, USA
* Correspondence: mericks9@uw.edu (M.A.E.); wabanks1@uw.edu (W.A.B.)

Received: 16 February 2019; Accepted: 29 March 2019; Published: 2 April 2019

Abstract: Age is associated with altered immune functions that may affect the brain. Brain barriers, including the blood–brain barrier (BBB) and blood–CSF barrier (BCSFB), are important interfaces for neuroimmune communication, and are affected by aging. In this review, we explore novel mechanisms by which the aging immune system alters central nervous system functions and neuroimmune responses, with a focus on brain barriers. Specific emphasis will be on recent works that have identified novel mechanisms by which BBB/BCSFB functions change with age, interactions of the BBB with age-associated immune factors, and contributions of the BBB to age-associated neurological disorders. Understanding how age alters BBB functions and responses to pathological insults could provide important insight on the role of the BBB in the progression of cognitive decline and neurodegenerative disease.

Keywords: blood–brain barrier; aging; inflammation

1. Introduction

Advances in modern medicine, nutrition, hygiene, and safety standards have doubled the life expectancy of humans worldwide over the last century and a half [1]. It has been estimated that in the next 50 years, the elderly will comprise approximately 20% of the world population [2]. Therefore, it is imperative that the scientific and medical communities investigate approaches that will minimize age-associated disease and maximize quality of life. Age-associated neurological and neurodegenerative diseases are especially debilitating to the afflicted and their families, having tremendous emotional and socioeconomic costs. Changes in the immune system have long been recognized to occur with aging, and it is now appreciated that neuroinflammation likely contributes to age-associated neurological diseases [3]. However, it is less well understood how specific changes in the immune system with aging may affect central nervous system (CNS) functions and contribute to neurological disease. We posit that brain barriers, especially the blood–brain barrier (BBB) and blood–CSF barrier (BCSFB), are important interfaces between CNS and peripheral tissues that are affected by age-associated changes in the immune system. The BBB/BCSFB may, in turn, affect homeostatic functions of the CNS, and/or exhibit more detrimental responses to pathological stimuli. In this review, we will provide a brief overview of changes known to occur in the peripheral immune system with aging, and then discuss recent works that have explored the relationships of BBB/BCSFB dysfunction, healthy aging, and the immune system. We will also briefly discuss how age might contribute to BBB/BCSFB dysfunction in different disease states.

2. Changes in the Immune System with Aging

Aging is associated with immune-related changes that come with clinical consequences. For example, as one ages, vulnerability to certain infections increases, and effectiveness of many vaccines decreases [4]. These clinical features of aging are attributed to an overall decline in protective immune responses, termed "immunosenescence" [1]. Aging is also associated with low-grade inflammation that occurs in the absence of overt infection, termed "inflammaging" [5]. Immunosenescence and inflammaging are interrelated processes [6], and may occur with age due to a number of factors that include latent infections, metabolic changes, and cell/tissue injury. Changes in the adaptive and innate immune systems, and related physiological processes that are detectable outside of the CNS are summarized below. Later sections will further discuss known relations of these changes to neuroimmune functions of the BBB and BCSFB.

2.1. Age-Associated Changes in the Adaptive Immune System

The main function of the adaptive immune system is to confer immunological memory to the organism, which facilitates the rapid recognition and neutralization of specific pathogens upon subsequent encounters. Changes in the cellular arm of the adaptive immune system with age have been described comprehensively by many groups [7–9]. One prominent feature of immunosenescence in the elderly is the change in T-cell composition. In particular, there is a decrease in the number of naïve T lymphocytes and an increase in memory and effector T cells with age, as well as a reduced diversity in T-cell receptors, and diminished functions of both naïve and memory T-cells [7,10]. The mechanistic underpinnings of these changes have been described elsewhere [7]. Changes in the B-cell compartment include reduced B-cell numbers, a reduced repertoire of B-cell receptors, reduced proliferative capacity, and decreases in immunoglobulin class-switch recombination [10,11]. The immunological consequences of reduced B-cell and T-cell functions include the reduced ability to generate immune memory to novel antigens, and thus, the reduced vaccine efficacy and increased vulnerability to certain infections in the elderly.

2.2. Age-Associated Changes in the Innate Immune System

The innate immune system is important for mounting initial protective responses against infections, and in sterile tissue injury and wound repair. The innate immune system also initiates cross-talk with the adaptive immune system through antigen presentation, co-stimulatory molecule expression, and cytokine production and so can contribute to adaptive immune responses [12]. The major changes in the innate immune system with aging include a heightened level of baseline inflammation, and an impaired ability to mount an efficient innate immune response against pathogenic stimuli. Specific changes in the function of innate immune cells have been comprehensively described elsewhere [13], and include impairments in phagocytosis, capacity to produce reactive oxygen and nitrogen species, T-cell priming, and signaling through pattern recognition receptors.

2.3. Age-Associated Changes in the Microbiome

A significant portion of the body's immune system resides in or near the gastrointestinal tract and can regulate the resident gut microbial populations. In young humans, the most numerous and diverse bacterial phylum is Firmicutes, with most in this phylum belonging to the Clostridia class. The second most abundant phylum is Bacteriodetes, which shows a high level of subject-to-subject variability in phylotypes detected [14]. In initial studies from the ELDERMET consortium that explored differences in gut microbial populations in young versus elderly subjects, it was found that elderly subjects had a lower proportion of Firmicutes, and atypical Bacteriodetes/Firmicutes ratios where Bacteriodetes predominated [15]. However, this study and others also demonstrated that elderly subjects show high variability in their microbiota profiles [15,16]. Notably, many of the Firmicutes are major producers of the short chain fatty acid, butyrate [17], which has histone deacetylase inhibitor activities and has been shown to protect against age-associated conditions such as sarcopenia and cognitive impairment in rodent models of neurodegenerative disease [18,19]. Increases in pathogenic bacteria that thrive in pro-inflammatory environments such as streptococci, staphylococci, enterococci, and enterobacteria have also been reported with aging [16], namely in centenarians [20].

Gut-associated lymphoid tissues (GALT) are the major immune interfaces of the gut that regulate the microbiome throughout lifespan. Age-associated changes in GALT include a reduction in antigen-specific IgA-immune responses which are, in part, mediated by aberrant cytokine responses of CD4+ T-cells [21]. It has also been shown that aging is associated with a reduced induction of immune tolerance to novel oral antigens [22]. Loss of the immunoregulatory environment within the gut with aging may have consequences, such as immune responses to novel antigens that would normally be tolerated, or a shift in gut microbial populations [16]. A pro-inflammatory environment within the gut could also result in microbial translocation and release of pathogenic microbes and/or their products (e.g., lipopolysaccharides) into the bloodstream, which could affect distal organs such as the CNS [23,24].

Dietary and environmental changes that are specifically associated with aging may also contribute to alterations in the microbiome. For example, dietary changes may occur upon new residence in assisted living institutions [16]. The elderly are also disproportionately affected by *Clostridium difficile* infection, and risk factors that may facilitate changes in the microbiome include increased use of antibiotics, and prior health care exposures where *C. difficile* may be contracted [25].

2.4. Age-Associated Changes in Peripheral Tissue Microenvironments and the Circulation

The transition of cells to a senescent phenotype is thought to be a protective mechanism against malignancies, and accumulation of senescent cells in multiple tissues occurs with aging [26]. Senescent cells are growth-arrested, but they remain metabolically active and undergo dramatic changes in protein expression and secretion, primarily in response to DNA damage [27]. The senescence-associated secretory phenotype (SASP) involves secretion of soluble cytokines, chemokines, and growth factors, proteases, extracellular matrix components, and reactive oxygen and nitrogen species which together modify the tissue microenvironment to promote local inflammation and tissue damage [26]. Therefore, cellular senescence may be one contributing factor to inflammaging. It has been proposed that cells of the CNS that have proliferative capacities such as endothelial cells and glia may also adopt a SASP, which could result in low-grade inflammation in the aging brain [28].

The BBB could also be affected by the accumulation of SASP cells in the periphery if exposed to pro-inflammatory secreted factors in the bloodstream. Many studies to date have demonstrated elevations in circulating inflammatory and acute phase proteins with aging [29–32]. We have also recently reviewed many aspects of neuroimmune interactions of the BBB and BCSFB with immune factors associated with SASP [33]. Recently, a novel aptamer-based proteomic approach was used to assess proteomic profiles in blood with healthy aging [34]. This study significantly detected an overall enrichment of SASP proteins in blood with aging, although some classical aging biomarkers such as interleukin-6 (IL-6), tumor necrosis factor-α (TNF-α), and insulin-like growth factor-1 (IGF-1) were not among the top-ranking age-associated proteins [34,35]. However, the proteins detected did reflect enriched signaling pathways such as cytokine/cytokine-receptor interactions, complement and coagulation cascades, and axon guidance. Notably, the protein that most strongly correlated with aging in this study was macrophage inhibitory cytokine-1 (MIC-1)/growth differentiation factor 15 (GDF15), which is a transforming growth factor-β (TGF-β) superfamily member that has anti-inflammatory activities in vitro [34,36]. Recent studies have also implicated GDF15 in obesity and the regulation of body weight, as well as frailty [37–39]. Overall, results from biomarker studies suggest that there may be differences in blood biomarkers that could signify healthy aging and predict SASP-associated disease.

3. Age-Associated Changes in Neuroimmune Functions

3.1. Altered Neuroimmune Phenotypes with Aging

Evidence supports that aging causes a pro-inflammatory environment in the CNS. Factors that have been proposed to contribute to increased baseline activation of inflammatory processes in the brain include reactive oxygen species, release of damage-associated molecular patterns (DAMPs) from injured or dying cells, increased abundance of cells with SASP phenotypes, and responses to peripheral inflammatory signals [40–44]. In humans, non-human primates, and rodents, aging is associated with increased numbers of reactive microglia and astrocytes [45]. The reactive phenotype of both astrocytes and microglia is typically determined by the expression levels of specific cell surface markers, as well as morphological changes of the cells. For example, there is an increased proportion of microglia in aged mice that stain positive for cell surface markers such as major histocompatibility complex II (MHCII), and cluster of differentiation (CD)11b, 86, and 68 [46]. Microglia in the healthy brain adopt a ramified morphology, characterized by long, branched extensions from the cell body that function in surveying the local environment [47]. With aging, microglia de-ramification is apparent: processes retract and thicken, and cell bodies enlarge [46,48]. Astrocytes also demonstrate morphological changes and increased expression of the inflammatory surface marker, glial fibrillary acidic protein (GFAP), with aging [46]. Aging also involves a shift in cytokine expression profiles, with increases in pro-inflammatory cytokines such as interleukin-1β (IL-1β) and IL-6, and decreases in anti-inflammatory cytokines such as interleukins 10 and 4 (IL-10 and IL-4) [46]. Overall, these pro-inflammatory phenotypes of glia at baseline are thought not only to reflect chronic, low-grade neuroinflammation, but also a "primed" phenotype whereby glia have more robust responses to immune stimuli [44].

3.2. Altered Neuroimmune Responses to Stimuli with Aging

It is appreciated that the aging brain may be more vulnerable to pathological changes in response to acute illness and infections. For example, urinary tract infections, which are not associated with cognitive symptoms in the young, can cause delirium and other neuropsychiatric conditions in the elderly [49]. Rodent models also suggest that aged mice have more severe neuroinflammatory responses and exacerbated behavioral outcomes following peripheral immune stimuli [50]. A prototypical stimulator of the innate immune system is lipopolysaccharide (LPS), which is a cell wall constituent of Gram-negative bacteria that activates inflammatory signaling cascades through the pattern recognition receptor Toll-like receptor 4 (TLR4) [51]. Young, healthy mice treated intraperitoneally with LPS exhibit a systemic cascade of cytokines and chemokines in the blood and brain [52], reactive gliosis, changes in body temperature and weight, and sickness behaviors. Intraperitoneal injection of LPS in aged mice causes increased pro-inflammatory cytokine responses and reactive microgliosis versus young mice [50,53,54]. Behavioral complications of peripheral infections and/or exposure to bacterial components is also more pronounced with age. For example, aged rodents are more vulnerable to cognitive impairment, sickness behavior, and depressive-like behavior following exposure to systemic inflammatory stimuli [50,54–56]. Neuroinflammatory stimuli, such as injection of the cytokines TNF-α and interferon-γ (IFN-γ) in the lateral ventricle, also result in increased reactive gliosis which occurs in the absence of apparent neurodegenerative changes [57].

4. The BBB as an Interface for Neuroimmune Communication

4.1. Anatomical, Cellular, and Subcellular Organization of Brain Barriers

4.1.1. The Vascular BBB

The primary anatomical unit of the vascular BBB is the brain endothelial cell (BEC). BECs have unique phenotypic properties that restrict the unregulated diffusion of molecules from blood into brain (barrier functions), and also those that regulate the passage of circulating nutrients, hormones, peptides, and proteins into and out of the brain (transport functions). In addition, the vascular BBB is an important signaling and secretory interface and uniquely regulates immune surveillance in the brain [33]. In Section 4.2, we will discuss how these functions of the BBB contribute to the immune-privileged status of the CNS and to unique aspects of neuroimmune communication.

Barrier functions of the vascular BBB are conferred by at least four distinct phenotypes of brain endothelial cells. These include expression of specialized tight junction proteins, reduced levels of pinocytosis, expression of efflux transporters, and expression of metabolic enzymes Tight junction protein complexes expressed by BECs localize to cell–cell junctions and prevent the diffusion of substances between cells (paracellular diffusion). Tight junctions are comprised of integral membrane proteins that include claudins (namely, claudin-5), occludin, and junctional adhesion molecules such as zonula occludens [58,59]. In addition to limiting paracellular diffusion, tight junctions can limit the lateral diffusion of membrane proteins and thus confer polarity to BECs. Tight junction proteins also interact with the cytoskeleton, adherens junctions, and the extracellular matrix, and are dynamically regulated by a range of stimuli at transcriptional and post-translational levels [58]. Relative reductions in fluid-phase pinocytosis also contribute to the BBB, and recent works have begun to elucidate molecular processes that are uniquely active in BECs and suppress formation of pinocytic vesicles. For example, the lipid transporter major facilitator superfamily domain containing 2A (Mfsd2a) confers a unique membrane lipid composition to brain endothelial cells that prevents assembly of caveolin-1 vesicles [60,61]. Finally, the BBB expresses specialized efflux transporters and metabolic enzymes that prevent the diffusion of circulating xenobiotics and other molecular substrates that would otherwise accumulate in the brain. Most efflux transporters at the BBB belong to the family of ATP-binding cassette transporters, and include P-glycoprotein (P-gp), multidrug resistance proteins (MRPs) and breast cancer resistance proteins (BCRPs) [62]. Examples of metabolic enzymes that contribute to BEC barrier functions include those that metabolize neurotransmitters (e.g., monoamine oxidases, cholinesterases, and aminopeptidases), and Phase I and II enzymes such as cytochrome P450s and transferases that are important for drug metabolism [63].

Like peripheral organs, the brain derives nutritive and trophic support from the circulation. However, energy and anabolic substrates such as glucose and amino acids that are derived from the circulation do not freely diffuse across BEC membranes, and so require transporters at the BBB to permit their passage from brain-to-blood in sufficient concentrations to support normal brain functions. Similarly, peptides and proteins such as insulin, leptin, ghrelin and some cytokines and chemokines can cross the intact BBB, and utilize specialized transport systems to do so. Transport systems at the BBB include solute carriers, which facilitate energy-independent transport down a concentration gradient, endocytic receptors, which bind ligands and transport them from one side of the membrane to the other in an energy-dependent process, and adsorptive endocytosis which involves interactions with the glycocalyx [33]. BBB transporters are important for conveying signals that relay aspects of metabolic status such as satiety and adipose mass, as well as inflammatory status which will be discussed in greater detail in Section 4.2. Also described in greater detail in Section 4.2 are the signaling and secretory interface functions of the BBB and their relevance to neuroimmune communication.

The specialized phenotype of BECs is greatly influenced by their local environment and closely associated supportive cells that are collectively termed the neurovascular unit (NVU). The most closely apposed cells to BECs are pericytes, which are found mostly around capillaries and post-capillary venules, and share a basement membrane with the brain endothelium. Pericytes are important for BBB induction and maintenance, as has been shown in mouse models with pericyte deficiencies [64]. Astrocyte end feet are also in very close proximity to the BBB, and ensheath the vessels. Astrocytes are also important for BBB induction and maintenance, as astrocyte conditioned medium is sufficient to promote BBB properties of BECs cultured in vitro [65]. Other components of the NVU include neurons, microglia, oligodendrocytes, and the extracellular matrix, which have been described for their contributions to BBB function under physiological and inflammatory states [33].

4.1.2. The Epithelial BCSFB

The BCSFB exists at the level of brain epithelial cells that comprise the choroid plexus (CP), which is located in each of the brain ventricles. Arachnoid epithelial cells also contribute to the BCSFB. Notably, endothelial cells comprising the vasculature in the CP do not have a BBB phenotype, and so permit leakage of serum components into the CP stroma [66]. The CP vasculature is also permissive to leukocyte trafficking, and so the stroma within the CP is a site where immune surveillance actively occurs. The CP epithelial cells of the BSCFB, similar to the BBB, express specialized tight junction proteins and efflux transporters that contribute to the barrier properties of the choroid plexus epithelium (CPE). The tight junction protein repertoire of the CPE is somewhat distinct from BECs in that they are comprised of distinct claudin proteins (1, 2, and 11) [59]. The CPE is the major site of cerebrospinal fluid (CSF) production, and CPE transporters are important for regulating CSF composition (reviewed in [67]). The arachnoid epithelium, while not a site for CSF production, does express the tight junction protein claudin 11, and efflux transporters such as P-gp and BCRP which may influence drug penetration in to the brain [33].

In contrast to the brain parenchyma, which has very low levels of blood-derived leukocytes under physiological conditions, the CSF and meninges do have resident populations of blood-derived leukocytes which must cross brain barriers to enter these compartments [68]. Indeed, the choroid plexus and arachnoid epithelial cells of the BCSFB are proposed to be major routes by which leukocytes gain entry to CSF under healthy conditions [68,69], and can also be a route of entry in injured states [70]. Aspects of leukocyte trafficking to the brain across brain barriers with aging will be discussed later in Section 5.1.4.

4.2. Neuroimmune Axes of the BBB

The BBB prevents the unregulated exchange of neuroimmune substances and cells between the CNS and blood. Hence, it is the BBB more than any other structure that secures the CNS as an immune-privileged tissue. However, the immune-privileged status of the CNS is relative as a number of mechanisms establish links between the peripheral components of the immune system and those of the CNS. These mechanisms are operational physiologically and, as discussed below, can be involved in aging and in aging-related diseases. Some of these mechanisms, such as vagal and other cranial nerve afferents, do not directly involve the BBB, whereas many others do. For convenience, mechanisms of BBB-neuroimmune interactions can be grouped into five categories or "axes" [33].

The first axis relates to the physiological regulation of the barrier properties that prevent leakage and is currently the least understood of the axes. As discussed above, much is known about how the barrier [71] is formed and even about how it can breakdown to once again become leaky [72–76]. However, there is some evidence that a degree of leakage may occur normally, if transiently. Hormones known to affect BBB tightness and that vary diurnally or with aging include insulin and dehydroepiandrosterone [77,78]. Such "physiological" leakage is probably at a very low level and its purpose is unknown.

A second axis is the alteration of other barrier functions, such as its transporter functions, by neuroimmune substances. There are many examples of these, such as TNF-α affecting the brain endothelial cell cytoskeleton [79], LPS increasing insulin transport [80], and granulocyte-macrophage colony-stimulating factor and IL-6 modulating BBB permeability to human immunodeficiency virus [81].

A third axis relates to the ability of the barriers to transport neuroimmune substances between the CNS and the blood. The best studied in this category are blood-to-brain transporters for cytokines, including IL-1α and β, IL-6, and TNF-α [82].

A fourth axis relates to immune cell trafficking as discussed above. This axis is clearly involved in both disease, as exemplified by multiple sclerosis, and in normal brain functioning [83]. The latter is illustrated by the belief that an impairment of immune cell surveillance in the brain can lead to progressive multifocal leukoencephalopathy [84].

The fifth axis relates to the ability of barrier cells themselves to secrete neuroimmune substances. For example, LPS acting at the luminal surface of brain endothelial cells induces release of prostaglandins into brain [85,86], resulting in fever. Barrier cells also secrete nitric oxide and cytokines [87]. Such release can be constitutive or induced. Secretion can be either from the same cell membrane surface (i.e., luminal–luminal or abluminal–abluminal) that receives the immune stimuli or, as in the case of LPS-prostaglandin-fever, from the opposite cell membrane surface [88].

These axes can interact in dynamic ways. As discussed below, they are known in some cases to be involved in aging and aging-related diseases.

5. Neuroimmune Mechanisms of Age-Associated Changes at the BBB

As was conveyed in Section 2, physiological aging is associated with changes in the immune system that may occur in response to the altered molecular environment of the aged organism. Although very few studies to date have explored direct relationships between BBB dysfunction and age-associated changes in peripheral components of the innate or adaptive immune systems (discussed in Section 5.1.5), emerging works have explored mechanistic changes at the BBB with aging that may contribute to altered neuroimmune functions. In this section, we will discuss changes at the BBB that are associated with aging in the absence of overt disease, and how physiological aging may affect BBB responses to immune stimuli. We will also consider activities of age-associated signaling pathways at the BBB and BCSFB, and how these might be affected using pharmacological approaches.

5.1. Changes in Brain Barrier Function with Aging

A challenge in the assessment of BBB dysfunction in healthy human aging is that many parameters can only be assessed in post-mortem tissues, and so it is difficult to distinguish changes at the BBB in humans that occur as a result of aging versus disease. Measurements of BBB dysfunction in living human subjects using imaging techniques such as PET, SPECT, and MRI are also becoming more robust with advances in instrumentation and analysis techniques, and have suggested that pathological changes at the BBB do occur progressively with aging, and predict clinical symptoms such as cognitive impairment. Findings in rodent models also corroborate general aging-associated phenotypes of the BBB and have elucidated possible mechanisms by which BBB functions are altered with age. These details are further described below.

5.1.1. Brain Barrier Disruption

One of the most-studied (and yet, poorly understood) aspects of BBB dysfunction is disruption [89], which is typically defined by the apparent leakage of normally BBB impenetrant molecules. Recent imaging results argue that BBB disruption does occur in healthy aging, and is worse in individuals with mild cognitive impairment, which is considered a prodrome of Alzheimer's disease (AD) [90,91]. One common approach to proxy BBB disruption in living humans is to measure the ratio of abundant, BBB-impermeant proteins such as albumin or immunoglobulin G (IgG) in CSF versus serum. However, these measures may be confounded by other known CNS deficits with aging, such as altered production and reabsorption of CSF, and inflammatory changes in the serum and CSF levels of these proteins, which have been discussed previously [57,92]. Further, there may be leakage of the BCSFB and altered protein synthesis at this site with age [93,94]. Recent studies have implemented advanced imaging technologies that can visualize leakage of intravenously injected tracers such as gadolinium via dynamic contrast MRI, and these have indicated that vascular BBB disruption does occur in the aging human brain, albeit at low levels [91].

In healthy aged mice (24 mo.), leakage of IgG into the parenchymal space of the cerebral cortex and hippocampus occurs when compared with young mice (3 mo.), suggesting that there is BBB disruption in this model. Increased IgG leakage in aged mice was associated with astrogliosis, endoplasmic reticulum (ER) stress, and increased endothelial cell levels of TNF-α; the latter measure significantly correlated with circulating levels of IL-6. In the same study, a significant reduction in occludin expression per brain endothelial cell was also observed in aged mice [95]. Other studies have corroborated findings of BBB disruption in aging mice [96]. Molecular mechanisms of BBB disruption in aging have been identified, and include reduced expression of sirtuin-1 [96], a de-acetylase enzyme which has been implicated in the regulation of lifespan, senescence, and inflammatory responses to environmental stress [97].

BBB disruption in the context of aging or disease could result in disease exacerbation through leakage of potentially harmful proteins into the brain [91]. However, it is not entirely clear that BBB disruption under any circumstance will always lead to brain damage. For example, certain therapeutic strategies for delivery of chemotherapeutics to the brain have relied on transiently disrupting the BBB, and are generally well-tolerated when brain cancers are the target [98]. Recent work has also indicated that repeated transient BBB disruption in humans with AD using focused ultrasound did not cause any serious clinical or radiological adverse events [99]. In contrast, healthy rodents with no prior brain abnormalities showed symptoms of reactive gliosis and neurodegeneration when transiently perfused with mannitol to cause widespread disruption of the BBB [100], and also had increased deposition of harmful serum proteins like fibrinogen in the CNS [101,102]. The apparent paradox in efforts to disrupt the BBB as a therapeutic strategy versus BBB disruption having known adverse consequences on the CNS and associations with many CNS diseases highlights the complexities of BEC barrier functions that are likely nuanced and context-specific. Why BBB disruption in and of itself is apparently innocuous under some conditions, but clearly detrimental in others remains to be understood in greater molecular detail.

5.1.2. Transporter Dysfunctions and Altered Signaling at Brain Barriers with Aging

Glucose transport: Glucose is the main energy source for the brain. The BBB regulates glucose uptake by the brain through expression of the glucose transporter GLUT1 on brain endothelial cells. GLUT1 is a uniporter that facilitates glucose diffusion from blood-to-brain. The amount of glucose uptake into brain is thus thought to depend on energy utilization by neurons which maintains a concentration gradient that drives glucose diffusion into the brain [103]. Under this assumption, neuronal dysfunction or neurodegeneration would result in reductions of glucose uptake by the brain due to reduced energy utilization and thus loss of the glucose concentration gradient. However, more recent works have suggested that reductions in brain glucose uptake could also reflect BBB dysfunction in glucose transport [103,104]. Brain glucose uptake can be measured in humans by imaging the uptake of ^{18}F-fludeoxyglucose into the brain with PET. Using this technique, it was shown that there is reduced glucose uptake into the brain in the frontal and temporal cortex with aging, even after correction for volume loss [105]. Aged rodents also show reduced brain glucose uptake, which is associated with cognitive impairment [106,107]. In mice, GLUT1 reductions at the BBB are apparent at 15 mo., and are even further reduced in an AD model of the same age [108].

Amyloid beta transport: Accumulation and deposition of the amyloid beta (Aβ) protein in the brain is a pathological hallmark of AD and contributes to neurodegeneration [109]. The BBB expresses transport systems for Aβ that mediate both transport into (influx) and transport out of (efflux) the CNS. Efflux transporters are thought to be important regulators of Aβ clearance from the brain, and these include the low-density lipoprotein receptor-related protein 1 (LRP-1) and P-gp [110–112]. The latter is also an important multidrug efflux transporter that can affect drug delivery to the brain. LRP-1 expression was shown to be decreased in brain microvessels with age, and in AD [110,113]. P-gp function is decreased in aged humans [114–116] as well as in aged mice [117]. Collectively, these changes at the BBB with age could contribute to Aβ accumulation in the brain with AD. It is also known that systemic inflammation in young mice can contribute to Aβ efflux deficits [118]; whether there is an inflammatory component to the Aβ efflux deficit in aging remains to be determined.

Insulin transport: Insulin is a trophic factor in the brain, and regulates critical functions such as feeding and learning and memory [119]. Brain insulin is not thought to be derived from CNS production, but rather from circulating insulin produced by the pancreas. Transport of insulin from blood-to-brain occurs through saturable transport mechanisms at the BBB [120–122]. In humans, it was recently shown that CSF/serum ratios of insulin decrease with aging [123], suggesting that BBB transport may be impaired. Reductions of insulin concentrations in brain tissues have also been reported with human aging [124]. In the senescence-accelerated mouse P8 (SAMP8) model of accelerated aging and AD-like cognitive decline, significant differences in the transport rate of insulin across the BBB were not observed in young versus aged mice. Increased insulin occupancy of vascular space was observed in aged SAMP8 mice in the parietal cortex, cerebellum, and thalamus, which indicates that there may be increased binding of insulin to brain endothelium with age in these regions [125]. It has not yet been determined whether commonly used mouse strains exhibit alterations in insulin transport across the BBB with age.

5.1.3. Interactions of Age-Associated Circulating Factors with Brain Barriers

Aging is associated with both increases in circulating factors that are harmful to the CNS, and decreases in circulating factors that are protective [29,126]. For example, circulating levels of growth differentiation factor 11 (GDF11, a member of the TGF-β superfamily), decline with age [127], and GDF11 treatments can stimulate vascular proliferation in vitro and in the subventricular zone of aged mice [128]. A circulating factor that increases in blood with aging is the chemokine CCL11, which has been shown through parabiosis studies to mediate cognitive impairment and to inhibit neurogenesis [29]. CCL11 in the circulation can access the CNS through a non-saturable or high capacity transport system at the BBB [129], indicating that increasing levels of circulating CCL11 in blood with age contribute to increased brain levels even when the BBB is intact. Another recent report has demonstrated that the enzyme acid sphingomyelinase (ASM) can contribute to BBB dysfunction [130]. This study showed that ASM concentrations increase in the circulation and in brain endothelial cells with aging. When compared with old mice that had reduced capillary density and evidence of BBB disruption, it was shown that mice heterozygous for sphingomyelin phosphodiesterase 1 (*Smpd1*) gene, which encodes ASM, were protected against these age-associated changes. It was further shown that ASM contributes to BBB disruption through induction of caveolae-cytoskeleton interactions that result in increased fluid-phase pinocytosis, but not through any apparent changes in paracellular/tight junction-regulated routes. ASM has enzymatic activity that facilitates the hydrolysis of sphingomyelin to ceramide and phosphorylcholine conversion [131], and thus altered membrane lipid composition could be contributing to the apparent changes in pinocytosis as well. Smpd1 heterozygosity also protected against age-associated deficits in learning and memory [130].

A summary of disruptive and non-disruptive changes at the vascular BBB with age and consequences to CNS function is depicted in Figure 1.

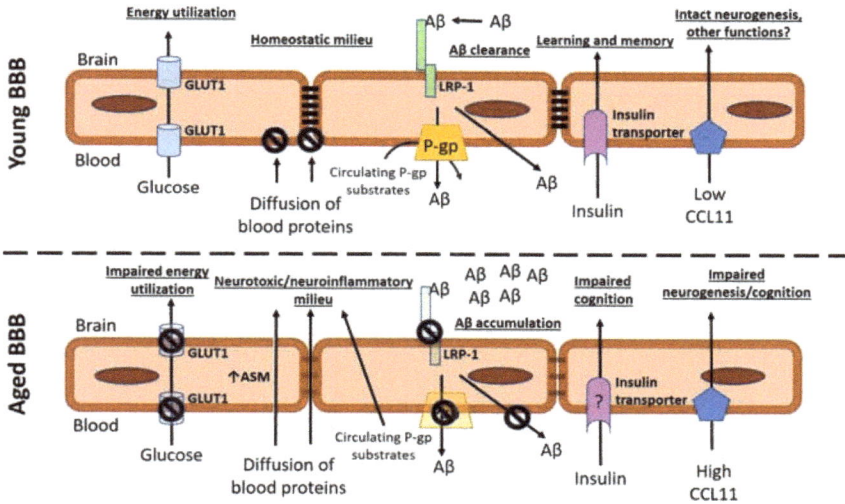

Figure 1. Changes in vascular blood–brain barrier (BBB) function that may lead to brain pathology with age. The upper panel depicts physiological functions of the BBB such as glucose transport, expression of intact tight junction complexes and suppression of vesicular processes that prevent the paracellular or transcellular leakage of blood proteins into the brain, intact functions of the efflux transporter P-gp, which contributes to barrier function by limiting diffusion of its substrates into the brain, and in concert with lipoprotein receptor-related protein 1 (LRP-1), facilitates amyloid beta (Aβ) clearance from the brain. Additionally, entry of insulin into the brain supports neuronal functions and contributes to learning and memory, and brain entry of circulating compounds with high-capacity transporters such as CCL11 is limited by low circulatory concentrations. The lower panel depicts aspects of BBB dysfunction that are either supported or suggested to occur with aging. Transparent appearance of transporters (GLUT1, LRP-1, and P-gp) indicates reduced protein expression levels at the BBB, and the interdictory circles over transporters indicate known functional impairments which may occur in the presence or absence of expression changes. The question mark on the insulin receptor suggests a possible mechanism by which aging influences brain insulin through altered function of the insulin transporter, which has not yet been definitively determined.

5.1.4. Age-Associated Changes in Inflammatory Signaling at the Choroid Plexus

The choroid plexus epithelium that comprises the BCSFB is an important immunological brain interface. The CSF is immunologically active, and contains cells of the adaptive immune system such as central memory T-cells which are thought to participate in CNS immune surveillance [132]. The BCSFB is an important site for leukocyte trafficking into CSF [132,133] and may regulate both protective and pathogenic types of immune cell recruitment to damaged tissues in the CNS and peripheral nervous system (PNS) [133]. Therefore, changes in BSCFB function with age may alter aspects of protective adaptive immunity in the CNS. We discuss this prospect in more detail in the following section.

Age-associated changes in gene expression profiles of the CPE have been found in both mice and humans, and some of the most robust changes in CPE gene expression were related to interferon (IFN)-associated pathways. With aging, there is an increased expression of type I IFN-response genes, and a decrease in type II IFN-response genes at the CP [134]. In the same study, it was found that circulating factors from aged mice reduced type II IFN gene expression, whereas factors in CSF increased type I IFN gene expression, suggesting that the aging brain and systemic compartments have distinct effects on CP gene expression. In aged mice that demonstrated deficits in spatial memory, it was shown that blocking signaling of the interferon α/β receptor, which binds type I IFN cytokines, could improve spatial memory. Aging also induces a shift in cytokine levels expressed at the CP, with increased expression of IL-4 and pro-inflammatory cytokines IL-1β and IL-6, and reduced expression of the type II interferon IFN-γ. It was also found that CPEs could express CCL11, which is induced by IL-4, when IFN-γ levels are low [134]. Therefore, the CP in addition to blood could be a source of CNS CCL11.

Another molecule thought to regulate the CPE with aging is the protein Klotho. Klotho is a transmembrane protein that facilitates signaling of fibroblast growth factor 23 (FGF23), and can also be secreted or cleaved from the membrane by a disintegrin and metalloproteinase domain-containing protein (ADAM) 10 or 17 and released as a soluble form to activate transient receptor potential cation channel subfamily V (TRPV5) signaling or inhibit IGF-1 and wnt signaling [134]. Mice lacking functional Klotho exhibit an accelerated aging phenotype which includes early thymic involution, osteopenia, skin atrophy, hearing loss, and neurodegeneration [135,136]. Klotho expression is not ubiquitous among tissues but is expressed at high levels in the CPE [136]. Expression of Klotho mRNA and protein is significantly reduced at the CPE with age [136]. Such reductions are also associated with increased expression of MHCII in CP stroma, increased levels of peripheral blood-derived macrophages in CP stroma and increased microglial activation, and NLRP3 inflammasome activation [137]. Therefore, Klotho may have important functions in suppressing activation of innate immunity in the CPE, and its reduction may be one mechanism by which neuroimmune functions change with age.

5.1.5. Immune Cell Trafficking

Immune cell trafficking in the brain under healthy conditions is largely compartmentalized to CSF and meningeal spaces, and is thought to be mediated in part through expression of P-selectins and intracellular adhesion molecule-1 (ICAM-1) expressed by the choroid plexus and arachnoid epithelium [68,138]. The post-capillary venules of the BBB are also interfaces for immune cell trafficking, particularly in disease states such as brain injury and multiple sclerosis. Detailed aspects of immune cell trafficking across brain barriers have been discussed by us in a recent review [33]. Relatively little is known about how immune cell populations in the CNS change with healthy aging, or how brain barriers may regulate such changes. However, it is plausible that relations do exist, since changes in peripheral leukocyte populations occur with aging (discussed in Section 2), and brain barriers are active sites of immune cell trafficking to the CNS in both healthy and injured/diseased states. Further, it should be considered whether peripheral changes in innate and adaptive immune cell composition with age could have important implications for CNS function. T-cells, which are the major blood-derived leukocyte population in the CNS (mostly residing in the CSF and meninges) have recently been shown to regulate aspects of learning and memory [138], but it is presently unclear whether age-related changes in T-cell subsets are associated with cognitive deficits.

Increases in T-cell and dendritic cell numbers have been observed in aged mice, starting at about 12 months [139]. One recent study in mice has explored relations among changes in leukocyte subsets in brain and blood [140]. The results of this study showed that numbers of T-cells, but not myeloid cells or other CD45+ cell types significantly increased in the brain with age. Further exploration of T-cell subsets demonstrated that in blood, the ratio of CD4+/CD8+ T-cells decreased with age, whereas age had no effect on the CD4+/CD8+ ratios in brain. In brain, the majority of T-cells detected were CD8+, and localized to perivascular spaces, brain parenchyma, and in the choroid plexus and meninges. Interestingly, the majority of T-cells in human CSF are CD4+ central memory T-cells [69], and so may reflect a different population than those found in brain parenchyma, although species differences may also explain the different abundances of CD4+ versus CD8+ T-cell subsets. The majority of the CD8+ T-cells in aged mouse brains had an effector memory phenotype, and the enrichment of these cells in the brain with age was not attributed to clonal expansion following exposure to brain antigens [140]. Age-associated increases in CD8+ T-cells positively correlated with microglia numbers, and phagocytosis, but negatively correlated with TNF-α positive microglia, suggesting that the CD8+ T-cells may be facilitating microglia polarization towards a phagocytic phenotype. However, it is also possible that the aged microglial phenotype could be driving T-cell recruitment. Finally, this study showed that CD8+ brain T-cells from aged mice produce greater levels of reactive oxygen species and pro-inflammatory cytokines following ischemic injury [140].

Overall, these data indicate that T-cell subsets in brain are distinct from those that predominate in blood, and that changes in T-cell subsets in aged blood are not necessarily reflected by the same population changes in the brain. Although it is presently not known which T-cell subsets in the circulation contribute to brain T-cell populations in parenchyma or CSF/meninges, findings from Ritzel et al. did indicate that T-cells from aged mice had elevated expression of adhesion molecules that are necessary for capture and diapedesis across brain barriers [140]. Future studies are needed to determine the contributions of brain barriers to age-associated increases in T-cell trafficking to the brain.

5.2. Effects of Aging on BBB Responses to Immune Stimuli

Dysfunction of the BBB can occur in concert with systemic and neuroinflammatory changes, however existing data suggest that the young, healthy BBB is relatively resistant to dysfunction caused by peripheral inflammatory insults, and relatively high doses of immune stimulators like LPS are required to elicit BBB disruption and dysfunction of transporters [118,141,142]. This is also supported in humans, where it was recently shown that in the absence of CSF abnormalities that would suggest disease, there were no correlations of systemic inflammatory markers with CSF/serum albumin ratios [143]. These findings further suggest that the healthy BBB of young adults is resistant to disruption induced by moderate systemic inflammation.

It is also understood that BBB dysfunction in response to neuroinflammatory stimuli may be regulated by the systemic inflammatory context. For example, intracerebroventricular (ICV) injection of IL-1β causes a robust influx of leukocytes into CNS parenchyma, but an intraperitoneal dose of LPS inhibits the ability of ICV IL-1β to recruit leukocytes to the brain [144]. The apparent resistance of the BBB to leukocyte trafficking in the presence of systemic inflammation in this context could be an adaptive advantage to protect the CNS from the systemic response to pathogen infections in the periphery.

As previously discussed, aging is associated with decreased BBB integrity and functional impairment of transporters. Aging may also exacerbate BBB responses to CNS injury and systemic inflammatory stimuli. In an LPS model that causes cerebral microhemorrhages (CMH), it was shown that mice aged 18 months had more numerous and severe CMH than young mice. This phenotype was associated with increased microgliosis and astrogliosis [145]. Aging also can cause dramatic changes in sleep, which is associated with increased production of pro-inflammatory cytokines such as TNF-α [146]. Cytokines such as TNF-α can cross the intact BBB, and so peripherally derived TNF-α may enter the brain to activate neuroinflammatory responses directly [147]. Sleep fragmentation in aged mice significantly increased the transport of TNF-α into the brain, but had no significant effect in young mice [146].

5.3. Interactions of the BBB with the Aging Microbiome

Interactions of the microbiome with the BBB have been reviewed recently [148]. Much of the current knowledge of these interactions is based on findings in germ-free mouse models, which exhibit increased BBB disruption that is apparently due to reduced tight junction protein expression and tight junction dysfunction [149]. It is therefore plausible that changes in the microbiome with age may affect BBB function. To date, no studies have directly tested this possibility, although emerging works suggest potential mechanisms. For example, short-chain fatty acids such as butyrate have been shown to protect against BBB disruption in germ-free mice [149]. Emerging evidence also supports that there is a reduced capacity of the microbiome to produce butyrate in the elderly [150], and so it may be that reduced butyrate levels contribute to age-associated BBB dysfunction. Recent studies have also begun to identify how age-associated changes in the microbiome might affect aspects of CNS function in which the BBB could be involved. For example, aged mice have altered cecal microbiota compositions, which is associated with increased gut permeability and higher levels of circulating pro-inflammatory cytokines in the periphery versus young mice. The same aged mouse cohort also demonstrated increased anxiety-like behaviors and impaired object–place recognition memory and social recognition [151]. Although altered BBB functions were not examined in this study, future works could further elucidate relationships of BBB dysfunction and aging with microbiome alterations. However, we also acknowledge that other plausible neuroimmune mechanisms, such as altered gut-to-brain signals as mediated by the vagus nerve or the BSCFB, could also be contributing to CNS changes caused by gut microbiome dyshomeostasis [152,153].

5.4. Effects of Aging on Non-Endothelial Cells of the Neurovascular Unit

Endothelial cells of the BBB develop and maintain their specialized phenotype through interactions with other associated cell types in the CNS that include pericytes, astrocytes, neurons, and also other glial cell types such as microglia and oligodendrocytes [33]. Pericytes and astrocytes are the most extensively studied for their roles in promoting and maintaining BBB functions and may contribute to BBB dysfunction with aging. Detailed aspects of these changes are discussed below.

Pericyte loss/dysfunction: Numerous functions have been ascribed to brain pericytes, including contractility, pluripotent stem cell-like properties, phagocytosis, and induction and maintenance of the BBB [154,155]. Platelet-derived growth factor receptor beta (PDGFRβ) heterozygous mice, which show an age-dependent loss of brain pericytes, also have increased evidence of BBB disruption with aging which coincides with pericyte loss and precedes associated neuroinflammation and learning and memory impairment in this model [156]. In studies of wild-type mice and humans, pericyte loss has been reported with age, but not consistently [156–159]. However, it is more clear that pericyte damage can occur with age, perhaps through phagocytosis of increasing amounts of cell debris [159], which also occurs under inflammatory conditions [160]. Recent studies have reported an increase of soluble PDGFRβ in CSF, a proposed marker of pericyte damage, with aging, BBB disruption, and cognitive impairment in humans [91,161,162].

Astrocyte changes: Astrocyte endfeet ensheath brain capillaries, and contribute to BBB maturation and maintenance [65]. Phenotypic changes have been observed in astrocytes with aging, such as reduced vascular coverage, increased GFAP expression, enlarged size, and reduced aquaporin-4 (AQP4) expression [157,158,163]. Such changes indicate increased reactive astrogliosis, which is also a process that occurs in response to pro-inflammatory stimuli. Given the important role of AQP4 in facilitating paravascular clearance of brain solutes [164], AQP4 reductions on astrocytes with aging could contribute to the neurotoxic accumulation of solutes in the brain.

6. The BBB in Age-Associated Neurological Diseases

Aging increases the risk of developing disease, and many neurological conditions in which the BBB has been implicated are also associated with aging. This section discusses some of these diseases in context of age-associated BBB dysfunction that may predispose or exacerbate the molecular mechanisms of disease.

6.1. Alzheimer's Disease

AD is the most common neurodegenerative disorder, and the greatest risk factor for AD is aging. There have been many recent reviews on the relations of BBB, inflammation, and AD that are beyond the scope of this review [57,165]. This section will focus on some recent conceptual advancements in the AD field that may relate inflammatory changes with aging and the BBB.

Evolving concepts in AD: Under a new research framework proposed by the National Institute on Aging and Alzheimer's Association, it was proposed that AD should be redefined by biological markers of disease, which include neurodegeneration and markers of deposition of two pathological proteins in the brain: amyloid beta and tau [166]. Whereas previous definitions of AD required a clinical diagnosis of dementia, it is now appreciated that pathological changes in AD precede onset of clinical symptoms by years, or even decades [167]. Although Aβ and tau are used to define AD as a unique neurodegenerative disease, it is now being considered that disease modifiers other than Aβ and tau may act in concert to regulate disease progression and manifestation of clinical symptoms. In previous sections, we have discussed aspects of BBB dysfunction that may be causal in AD, and here refer the reader to a recent detailed commentary on the importance of considering the neurovasculature as a possible driver of and therapeutic target for AD [104]. We also consider some additional timely findings that implicate interactions of the BBB, inflammation, and aging in AD.

ApoE isoform-dependent immunomodulatory activities: In humans, there are three major alleles of the apolipoprotein E (*APOE*) gene, which are *APOE2*, *APOE3*, and *APOE4*. *APOE4* is the strongest genetic risk factor for late-onset forms of Alzheimer's disease, which may be due to a number of distinct functions of ApoE4 protein versus the more prevalent ApoE3 protein. ApoE4 may be contributing to AD in part via limiting Aβ clearance from the brain [168], and also through tau-dependent effects [169]. ApoE4 also has diverse functions in regulating the immune system that may be independent or synergistic with Aβ and tau-driven brain pathology [170]. For example, transgenic ApoE4 mice have BBB disruption through the loss of interaction of ApoE4 with LRP-1 in pericytes, which is preserved in mice expressing ApoE3 or ApoE2. As a result, matrix metalloproteinase 9 (MMP9) activation occurring in brain endothelial cells contributes to BBB disruption in the model [74]. This molecular route of BBB disruption in APOE4 carriers has also been reported in human AD [171].

Low-grade CNS infections and AD: Since the discovery of AD by Alois Alzheimer, there have been speculations and a few studies supporting that CNS infections could be causal in AD [172,173]. Although this concept has been largely overshadowed by the amyloid cascade hypotheses and is still controversial, emerging studies have supported that CNS bacterial and viral infections may contribute to or exacerbate AD. Early works have shown that herpes simplex virus-1 (HSV-1) DNA is present in brains of humans with and without AD [174,175], but it was also questioned whether HSV-1 infection was directly involved in AD [176]. Subsequently, it was shown that HSV-1 infection of cultured neurons and glia and mouse brain can increase the production of Aβ [177], and induce cytoskeletal abnormalities in neurons that include tau hyperphosphorylation [178]. Recent works have indicated that Aβ has antimicrobial properties against bacteria and viruses [179,180], and have substantiated the associations of herpes virus infections and AD or dementia [181–185]. These findings suggest that Aβ and/or tau may be protective responses to CNS infections that would be more likely to occur with age-associated immunosenescence and a dysfunctional BBB. It remains to be determined whether antimicrobial strategies such as antibiotics, antivirals, or vaccines could protect the infected against AD progression.

Preclinical animal models of AD: Mouse models of AD have been used extensively to define mechanisms of disease pathology and therapeutic efficacy. Most of these models are based on genetic mutations which cause Aβ plaque deposition, and so are really models of Aβ-driven brain injury [186]. However, other factors such as extraphysiological expression of transgenes and individual or combinations of mutations that are not observed in sporadic AD could further confound these models. Additionally, the AD-like sequalae (plaques, neuroinflammation, cognitive deficits, and neuronal/synaptic loss) in most transgenic mouse models of AD occur when the mice are considered to be young (3–6 mo.) or middle-aged (10–14 mo.), and so exclude the aging component of AD. Accelerated aging models, such as SAMP8 mice which have modest increases in brain Aβ, deficits in Aβ clearance, and impaired learning and memory by 12 months of age [187,188] are less widely used, but have utility in studying the synergy of aging and AD. Along these lines, BBB disruption variably occurs in mouse models of Aβ-driven brain pathology and is not apparent in SAMP8 mice [125,189–192], and recent works have also indicated that tauopathies may also drive BBB disruption in rodents [193,194]. BBB efflux systems have also been identified for truncated forms of tau [195]. Inclusion of an aging component in preclinical AD models may reveal important therapeutic considerations of treatment, or novel aspects of disease progression that may improve the chances of success in drug development.

6.2. Depression

Depression affects individuals of all ages, but poses unique considerations in the aging population. Although depression is less prevalent in older versus younger adults, it is notable that over half of depression diagnoses in the aged are in those who have not previously been afflicted [196]. Depression in the aged is also associated with cognitive dysfunction, dementia risk, and vascular dysfunction [196]. Notably, cardiovascular disease (CVD) and depression are inter-related in that major depressive disorder (MDD) prevalence is more prevalent in individuals with CVD, and MDD increases CVD morbidity and mortality [197]. Systemic inflammation may also be a factor that drives MDD, with studies showing associations of MDD and cytokines and acute phase proteins in blood [198,199]. The BBB has recently been implicated as a possible mediator of depressive behaviors in mice. Mice that were vulnerable to depressive-like behaviors following chronic social defeat stress were shown to have reductions in the tight junction protein claudin-5 and BBB leakiness in the nucleus accumbens (NAc), as well as increased leukocyte trafficking and IL-6 accumulation in this region. Knock-down of claudin-5 in the NAc recapitulated depressive-like behaviors [200]. Notably, aged mice have increased inflammatory responses to social defeat stress [201], suggesting that synergy of BBB dysfunction, glial cell priming, and increased peripheral cytokine responses could all contribute to depressive responses to stress in the aged.

Another possible link between aging, the BBB, and depression is the microbiome. In humans with major depressive disorder, and in rodents subjected to a variety of stressors that can cause depressive-like behavior, composition of the gut microbiome is altered [202–204]. Humans with active MDD were shown to have an increase in Bacteroidetes and a reduction in Firmicutes, similar to age-associated microbiome changes that were discussed in Section 2.3. Recent work has also shown that the transplantation of microbiota from MDD patients into germ-free mice caused depressive-like behaviors, and altered metabolites of carbohydrates and amino acids [205]. Interestingly, the current data suggest that there is bidirectional regulation of the brain and gut microbiome in MDD. Future work is needed to determine how brain barriers may be contributing to gut-brain communication in MDD and other diseases.

6.3. Metabolic Syndrome

Metabolic syndrome is defined by a cluster of risk factors that increases risk of developing CVD, type II diabetes, stroke, and other co-morbid diseases [206,207]. These risk factors include insulin resistance, abdominal obesity, high serum triglycerides, high blood pressure, and hyperglycemia [208]. In the United States, metabolic syndrome is most prevalent in individuals aged 60 and older [209]. Age-associated factors such as low testosterone in males, and low levels of vitamin D may contribute to components of metabolic syndrome in the elderly, such as insulin resistance [209]. Testosterone depletion was recently linked to BBB dysfunction. Orchiectomized mice were shown to have increased BBB disruption, which was in part attributed to reduced expression of the tight junction proteins claudin-5 and zonula occludens-1. Castrated mice from this study also had evidence of reactive astrogliosis [210]. BBB deficits in castrated mice could be rescued by testosterone supplementation, but it was unclear whether the BBB effects occurred through direct actions of testosterone on the BBB, or indirect consequences of testosterone depletion. Notably, in mice, metabolic effects of orchidectomy are very minor or absent on a standard chow diet, but androgen deprivation can exacerbate adipose hypertrophy, glucose intolerance, insulin insensitivity, and systemic inflammation when fed a high-fat diet [211], suggesting that interactions between hormone changes with aging could synergize with diet and obesity to affect BBB disruption.

Obesity is associated with increased systemic inflammation with aging and can exacerbate autoimmune diseases such as rheumatoid arthritis [212]. Obesity can also contribute to BBB dysfunction. Mice that are made to become obese through high-fat diet feeding have evidence of BBB disruption in the hippocampus, which is also associated with learning and memory deficits. BBB disruption in mice fed a high-fat also have reduced levels of tight junction proteins at the BBB and BCSFB [213]. Obesity is also associated with reduced transport of proteins across the BBB that act on the CNS to regulate feeding, such as leptin, insulin, and ghrelin. Whereas insulin and leptin signal satiety to the brain and stimulate anorexia, ghrelin is an orexigenic signal. However, levels of circulating leptin increase with obesity, whereas ghrelin levels decrease with obesity and aging [214]. In the case of ghrelin, age and obesity were also shown to have synergistic suppressive effects on transport across the BBB [215]. Transport of lipids such as palmitate and free fatty acids across the BBB is increased with obesity [213]. Triglycerides, which are elevated in obesity, are known inhibitors of leptin transport, and so may contribute to a feed-forward cycle of leptin deficiency in the brain that leads to hyperphagia and further increases in triglycerides [216]. Obesity induced by a genetic mutation in the leptin receptor has also been associated with increased neuroinflammation, which included monocyte trafficking across the BBB [217].

Type II diabetes mellitus (T2DM), a consequence of metabolic syndrome, is also associated with aging and BBB dysfunction. Studies in rodents and monkeys have shown that increased BBB disruption occurs in T2DM, and is associated with reduced levels of tight junction proteins [218–220]. T2DM is associated with an increased risk of AD [221], and recent work has demonstrated a mechanistic link that implicates BBB dysfunction in this process. In a mouse model of type II diabetes with hyperinsulinemia, it was shown that BBB transport of Aβ into the brain was increased, and transport of Aβ out of the brain was decreased when compared with non-diabetic controls. Antidiabetic drugs reduced Aβ influx and increased Aβ efflux in diabetic mice, and these changes appeared to be mediated through decreases of the Aβ influx transporter, receptor for advanced glycation endproducts (RAGE), and increases in the Aβ efflux transporter, LRP-1 [222]. The BBB dysfunction that occurs as a result of obesity alone, or in combination with T2DM may be further exacerbated by aging, or vice-versa [213,223,224].

7. Conclusions

Inflammatory changes with aging are important drivers of CNS dysfunction, and we have described mechanisms by which BBB dysfunction in healthy aging could predispose to neurological disease. Clearly, more work is necessary to further explore how aging and associated inflammatory changes could affect brain barrier functions in health, infection, and injury. Developing a better understanding of the interactions of aging with known pathogenic mechanisms of disease is important in the development of novel therapies for neurological disorders.

Author Contributions: Writing—Original Draft Preparation, M.A.E. and W.A.B.; Writing—Review & Editing, M.A.E. and W.A.B.; Funding Acquisition, M.A.E. and W.A.B.

Funding: This research was funded by The Department of Veterans Affairs and the National Institutes of Health-National Institute of Environmental Health Sciences, grant number R21 ES029657.

Conflicts of Interest: The authors declare no conflict of interest.

References

1. Aw, D.; Silva, A.B.; Palmer, D.B. Immunosenescence: Emerging challenges for an ageing population. *Immunology* **2007**, *120*, 435–446. [CrossRef]
2. Ellison, D.; White, D.; Farrar, F.C. Aging population. *Nurs. Clin. North. Am.* **2015**, *50*, 185–213. [CrossRef]
3. Ransohoff, R.M. How neuroinflammation contributes to neurodegeneration. *Science* **2016**, *353*, 777–783. [CrossRef]
4. Weinberger, B.; Herndler-Brandstetter, D.; Schwanninger, A.; Weiskopf, D.; Grubeck-Loebenstein, B. Biology of immune responses to vaccines in elderly persons. *Clin. Infect. Dis.* **2008**, *46*, 1078–1084. [CrossRef] [PubMed]
5. Franceschi, C.; Bonafe, M.; Valensin, S.; Olivieri, F.; De Luca, M.; Ottaviani, E.; De Benedictis, G. Inflamm-aging. An evolutionary perspective on immunosenescence. *Ann. N. Y. Acad. Sci.* **2000**, *908*, 244–254. [CrossRef]

6. Fulop, T.; Larbi, A.; Dupuis, G.; Le Page, A.; Frost, E.H.; Cohen, A.A.; Witkowski, J.M.; Franceschi, C. Immunosenescence and Inflamm-Aging As Two Sides of the Same Coin: Friends or Foes? *Front. Immunol.* **2017**, *8*, 1960. [CrossRef]

7. Arnold, C.R.; Wolf, J.; Brunner, S.; Herndler-Brandstetter, D.; Grubeck-Loebenstein, B. Gain and loss of T cell subsets in old age–age-related reshaping of the T cell repertoire. *J. Clin. Immunol.* **2011**, *31*, 137–146. [CrossRef]

8. Nikolich-Zugich, J.; Li, G.; Uhrlaub, J.L.; Renkema, K.R.; Smithey, M.J. Age-related changes in CD8 T cell homeostasis and immunity to infection. *Semin. Immunol.* **2012**, *24*, 356–364. [CrossRef] [PubMed]

9. Frasca, D.; Diaz, A.; Romero, M.; Landin, A.M.; Blomberg, B.B. Age effects on B cells and humoral immunity in humans. *Ageing Res. Rev.* **2011**, *10*, 330–335. [CrossRef] [PubMed]

10. Poland, G.A.; Ovsyannikova, I.G.; Kennedy, R.B.; Lambert, N.D.; Kirkland, J.L. A systems biology approach to the effect of aging, immunosenescence and vaccine response. *Curr. Opin. Immunol.* **2014**, *29*, 62–68. [CrossRef] [PubMed]

11. Frasca, D.; Landin, A.M.; Lechner, S.C.; Ryan, J.G.; Schwartz, R.; Riley, R.L.; Blomberg, B.B. Aging down-regulates the transcription factor E2A, activation-induced cytidine deaminase, and Ig class switch in human B cells. *J. Immunol.* **2008**, *180*, 5283–5290. [CrossRef]

12. Jain, A.; Pasare, C. Innate Control of Adaptive Immunity: Beyond the Three-Signal Paradigm. *J. Immunol.* **2017**, *198*, 3791–3800. [CrossRef]

13. Shaw, A.C.; Goldstein, D.R.; Montgomery, R.R. Age-dependent dysregulation of innate immunity. *Nat. Rev. Immunol.* **2013**, *13*, 875–887. [CrossRef]

14. Eckburg, P.B.; Bik, E.M.; Bernstein, C.N.; Purdom, E.; Dethlefsen, L.; Sargent, M.; Gill, S.R.; Nelson, K.E.; Relman, D.A. Diversity of the human intestinal microbial flora. *Science* **2005**, *308*, 1635–1638. [CrossRef] [PubMed]

15. Claesson, M.J.; Cusack, S.; O'Sullivan, O.; Greene-Diniz, R.; de Weerd, H.; Flannery, E.; Marchesi, J.R.; Falush, D.; Dinan, T.; Fitzgerald, G.; et al. Composition, variability, and temporal stability of the intestinal microbiota of the elderly. *Proc. Natl. Acad. Sci. USA* **2011**, *108* (Suppl. 1), 4586–4591. [CrossRef] [PubMed]

16. Biagi, E.; Candela, M.; Turroni, S.; Garagnani, P.; Franceschi, C.; Brigidi, P. Ageing and gut microbes: Perspectives for health maintenance and longevity. *Pharmacol. Res.* **2013**, *69*, 11–20. [CrossRef] [PubMed]

17. Pryde, S.E.; Duncan, S.H.; Hold, G.L.; Stewart, C.S.; Flint, H.J. The microbiology of butyrate formation in the human colon. *FEMS Microbiol. Lett.* **2002**, *217*, 133–139. [CrossRef] [PubMed]

18. Walsh, M.E.; Bhattacharya, A.; Sataranatarajan, K.; Qaisar, R.; Sloane, L.; Rahman, M.M.; Kinter, M.; Van Remmen, H. The histone deacetylase inhibitor butyrate improves metabolism and reduces muscle atrophy during aging. *Aging Cell* **2015**, *14*, 957–970. [CrossRef] [PubMed]

19. Bourassa, M.W.; Alim, I.; Bultman, S.J.; Ratan, R.R. Butyrate, neuroepigenetics and the gut microbiome: Can a high fiber diet improve brain health? *Neurosci. Lett.* **2016**, *625*, 56–63. [CrossRef] [PubMed]

20. Biagi, E.; Nylund, L.; Candela, M.; Ostan, R.; Bucci, L.; Pini, E.; Nikkila, J.; Monti, D.; Satokari, R.; Franceschi, C.; et al. Through ageing, and beyond: Gut microbiota and inflammatory status in seniors and centenarians. *PLoS ONE* **2010**, *5*, e10667. [CrossRef]

21. Kawanishi, H.; Senda, S.; Ajitsu, S. Aging-associated intrinsic defects in IgA production by murine Peyer's patch B cells stimulated by autoreactive Peyer's patch T cell hybridoma-derived B cell stimulatory factors (BSF). *Mech. Ageing Dev.* **1989**, *49*, 61–78. [CrossRef]

22. Fujihashi, K.; McGhee, J.R. Mucosal immunity and tolerance in the elderly. *Mech. Ageing Dev.* **2004**, *125*, 889–898. [CrossRef] [PubMed]

23. D'Ettorre, G.; Douek, D.; Paiardini, M.; Ceccarelli, G.; Vullo, V. Microbial translocation and infectious diseases: What is the link? *Int. J. Microbiol.* **2012**, *2012*, 356981. [CrossRef] [PubMed]

24. Brenchley, J.M.; Price, D.A.; Schacker, T.W.; Asher, T.E.; Silvestri, G.; Rao, S.; Kazzaz, Z.; Bornstein, E.; Lambotte, O.; Altmann, D.; et al. Microbial translocation is a cause of systemic immune activation in chronic HIV infection. *Nat. Med.* **2006**, *12*, 1365–1371. [CrossRef] [PubMed]

25. Asempa, T.E.; Nicolau, D.P. Clostridium difficile infection in the elderly: An update on management. *Clin. Interv. Aging* **2017**, *12*, 1799–1809. [CrossRef] [PubMed]

26. Coppe, J.P.; Desprez, P.Y.; Krtolica, A.; Campisi, J. The senescence-associated secretory phenotype: The dark side of tumor suppression. *Annu. Rev. Pathol.* **2010**, *5*, 99–118. [CrossRef]

27. Tchkonia, T.; Zhu, Y.; van Deursen, J.; Campisi, J.; Kirkland, J.L. Cellular senescence and the senescent secretory phenotype: Therapeutic opportunities. *J. Clin. Investig.* **2013**, *123*, 966–972. [CrossRef]

28. Chinta, S.J.; Woods, G.; Rane, A.; Demaria, M.; Campisi, J.; Andersen, J.K. Cellular senescence and the aging brain. *Exp. Gerontol.* **2015**, *68*, 3–7. [CrossRef]

29. Villeda, S.A.; Luo, J.; Mosher, K.I.; Zou, B.; Britschgi, M.; Bieri, G.; Stan, T.M.; Fainberg, N.; Ding, Z.; Eggel, A.; et al. The ageing systemic milieu negatively regulates neurogenesis and cognitive function. *Nature* **2011**, *477*, 90–94. [CrossRef]

30. Byerley, L.O.; Leamy, L.; Tam, S.W.; Chou, C.W.; Ravussin, E.; Louisiana Healthy Aging Study. Development of a serum profile for healthy aging. *Age (Dordr)* **2010**, *32*, 497–507. [CrossRef] [PubMed]

31. Maggio, M.; Guralnik, J.M.; Longo, D.L.; Ferrucci, L. Interleukin-6 in aging and chronic disease: A magnificent pathway. *J. Gerontol. A Biol. Sci. Med. Sci.* **2006**, *61*, 575–584. [CrossRef]

32. Bruunsgaard, H.; Andersen-Ranberg, K.; Jeune, B.; Pedersen, A.N.; Skinhoj, P.; Pedersen, B.K. A high plasma concentration of TNF-alpha is associated with dementia in centenarians. *J. Gerontol. A Biol. Sci. Med. Sci.* **1999**, *54*, M357–M364. [CrossRef]

33. Erickson, M.A.; Banks, W.A. Neuroimmune Axes of the Blood-Brain Barriers and Blood-Brain Interfaces: Bases for Physiological Regulation, Disease States, and Pharmacological Interventions. *Pharmacol. Rev.* **2018**, *70*, 278–314. [CrossRef] [PubMed]

34. Tanaka, T.; Biancotto, A.; Moaddel, R.; Moore, A.Z.; Gonzalez-Freire, M.; Aon, M.A.; Candia, J.; Zhang, P.; Cheung, F.; Fantoni, G.; et al. Plasma proteomic signature of age in healthy humans. *Aging Cell* **2018**, *17*, e12799. [CrossRef] [PubMed]

35. Krabbe, K.S.; Pedersen, M.; Bruunsgaard, H. Inflammatory mediators in the elderly. *Exp. Gerontol.* **2004**, *39*, 687–699. [CrossRef] [PubMed]

36. Bootcov, M.R.; Bauskin, A.R.; Valenzuela, S.M.; Moore, A.G.; Bansal, M.; He, X.Y.; Zhang, H.P.; Donnellan, M.; Mahler, S.; Pryor, K.; et al. MIC-1, a novel macrophage inhibitory cytokine, is a divergent member of the TGF-beta superfamily. *Proc. Natl. Acad. Sci. USA* **1997**, *94*, 11514–11519. [CrossRef]

37. Villanueva, M.T. Obesity: GDF15 tells the brain to lose weight. *Nat. Rev. Drug Discov.* **2017**, *16*, 827. [CrossRef]

38. Xiong, Y.; Walker, K.; Min, X.; Hale, C.; Tran, T.; Komorowski, R.; Yang, J.; Davda, J.; Nuanmanee, N.; Kemp, D.; et al. Long-acting MIC-1/GDF15 molecules to treat obesity: Evidence from mice to monkeys. *Sci. Transl. Med.* **2017**, *9*. [CrossRef]

39. Cardoso, A.L.; Fernandes, A.; Aguilar-Pimentel, J.A.; de Angelis, M.H.; Guedes, J.R.; Brito, M.A.; Ortolano, S.; Pani, G.; Athanasopoulou, S.; Gonos, E.S.; et al. Towards frailty biomarkers: Candidates from genes and pathways regulated in aging and age-related diseases. *Ageing Res. Rev.* **2018**, *47*, 214–277. [CrossRef] [PubMed]

40. Floyd, R.A.; Hensley, K. Oxidative stress in brain aging. Implications for therapeutics of neurodegenerative diseases. *Neurobiol. Aging* **2002**, *23*, 795–807. [CrossRef]

41. Bhat, R.; Crowe, E.P.; Bitto, A.; Moh, M.; Katsetos, C.D.; Garcia, F.U.; Johnson, F.B.; Trojanowski, J.Q.; Sell, C.; Torres, C. Astrocyte senescence as a component of Alzheimer's disease. *PLoS ONE* **2012**, *7*, e45069. [CrossRef] [PubMed]

42. Fulop, G.A.; Kiss, T.; Tarantini, S.; Balasubramanian, P.; Yabluchanskiy, A.; Farkas, E.; Bari, F.; Ungvari, Z.; Csiszar, A. Nrf2 deficiency in aged mice exacerbates cellular senescence promoting cerebrovascular inflammation. *Geroscience* **2018**, *40*, 513–521. [CrossRef]

43. Wilhelm, I.; Nyul-Toth, A.; Kozma, M.; Farkas, A.E.; Krizbai, I.A. Role of pattern recognition receptors of the neurovascular unit in inflamm-aging. *Am. J. Physiol. Heart Circ. Physiol* **2017**, *313*, H1000–H1012. [CrossRef] [PubMed]

44. Perry, V.H.; Teeling, J. Microglia and macrophages of the central nervous system: The contribution of microglia priming and systemic inflammation to chronic neurodegeneration. *Semin. Immunopathol.* **2013**, *35*, 601–612. [CrossRef]

45. Godbout, J.P.; Johnson, R.W. Age and neuroinflammation: A lifetime of psychoneuroimmune consequences. *Immunol. Allergy Clin. N. Am.* **2009**, *29*, 321–337. [CrossRef]

46. Norden, D.M.; Godbout, J.P. Review: Microglia of the aged brain: Primed to be activated and resistant to regulation. *Neuropathol. Appl. Neurobiol.* **2013**, *39*, 19–34. [CrossRef] [PubMed]

47.	Nimmerjahn, A.; Kirchhoff, F.; Helmchen, F. Resting microglial cells are highly dynamic surveillants of brain parenchyma in vivo. *Science* **2005**, *308*, 1314–1318. [CrossRef] [PubMed]

48.	Karperien, A.; Ahammer, H.; Jelinek, H.F. Quantitating the subtleties of microglial morphology with fractal analysis. *Front. Cell Neurosci.* **2013**, *7*, 3. [CrossRef]

49.	Chae, J.H.; Miller, B.J. Beyond Urinary Tract Infections (UTIs) and Delirium: A Systematic Review of UTIs and Neuropsychiatric Disorders. *J. Psychiatr. Pract.* **2015**, *21*, 402–411. [CrossRef]

50.	Godbout, J.P.; Chen, J.; Abraham, J.; Richwine, A.F.; Berg, B.M.; Kelley, K.W.; Johnson, R.W. Exaggerated neuroinflammation and sickness behavior in aged mice following activation of the peripheral innate immune system. *FASEB J.* **2005**, *19*, 1329–1331. [CrossRef]

51.	Lu, Y.C.; Yeh, W.C.; Ohashi, P.S. LPS/TLR4 signal transduction pathway. *Cytokine* **2008**, *42*, 145–151. [CrossRef]

52.	Erickson, M.A.; Banks, W.A. Cytokine and chemokine responses in serum and brain after single and repeated injections of lipopolysaccharide: Multiplex quantification with path analysis. *Brain Behav. Immun.* **2011**, *25*, 1637–1648. [CrossRef] [PubMed]

53.	Henry, C.J.; Huang, Y.; Wynne, A.M.; Godbout, J.P. Peripheral lipopolysaccharide (LPS) challenge promotes microglial hyperactivity in aged mice that is associated with exaggerated induction of both pro-inflammatory IL-1beta and anti-inflammatory IL-10 cytokines. *Brain Behav. Immun.* **2009**, *23*, 309–317. [CrossRef] [PubMed]

54.	Chen, J.; Buchanan, J.B.; Sparkman, N.L.; Godbout, J.P.; Freund, G.G.; Johnson, R.W. Neuroinflammation and disruption in working memory in aged mice after acute stimulation of the peripheral innate immune system. *Brain Behav. Immun.* **2008**, *22*, 301–311. [CrossRef] [PubMed]

55.	Barrientos, R.M.; Higgins, E.A.; Biedenkapp, J.C.; Sprunger, D.B.; Wright-Hardesty, K.J.; Watkins, L.R.; Rudy, J.W.; Maier, S.F. Peripheral infection and aging interact to impair hippocampal memory consolidation. *Neurobiol. Aging* **2006**, *27*, 723–732. [CrossRef] [PubMed]

56.	Godbout, J.P.; Moreau, M.; Lestage, J.; Chen, J.; Sparkman, N.L.; O'Connor, J.; Castanon, N.; Kelley, K.W.; Dantzer, R.; Johnson, R.W. Aging exacerbates depressive-like behavior in mice in response to activation of the peripheral innate immune system. *Neuropsychopharmacology* **2008**, *33*, 2341–2351. [CrossRef]

57.	Erickson, M.A.; Banks, W.A. Blood-brain barrier dysfunction as a cause and consequence of Alzheimer's disease. *J. Cereb. Blood Flow Metab.* **2013**, *33*, 1500–1513. [CrossRef]

58.	Luissint, A.C.; Artus, C.; Glacial, F.; Ganeshamoorthy, K.; Couraud, P.O. Tight junctions at the blood brain barrier: Physiological architecture and disease-associated dysregulation. *Fluids Barriers CNS* **2012**, *9*, 23. [CrossRef]

59.	Tietz, S.; Engelhardt, B. Brain barriers: Crosstalk between complex tight junctions and adherens junctions. *J. Cell Biol.* **2015**, *209*, 493–506. [CrossRef] [PubMed]

60.	Andreone, B.J.; Chow, B.W.; Tata, A.; Lacoste, B.; Ben-Zvi, A.; Bullock, K.; Deik, A.A.; Ginty, D.D.; Clish, C.B.; Gu, C. Blood-Brain Barrier Permeability Is Regulated by Lipid Transport-Dependent Suppression of Caveolae-Mediated Transcytosis. *Neuron* **2017**, *94*, 581–594.e5. [CrossRef]

61.	Ben-Zvi, A.; Lacoste, B.; Kur, E.; Andreone, B.J.; Mayshar, Y.; Yan, H.; Gu, C. Mfsd2a is critical for the formation and function of the blood-brain barrier. *Nature* **2014**, *509*, 507–511. [CrossRef]

62.	Begley, D.J. ABC transporters and the blood-brain barrier. *Curr. Pharm. Des.* **2004**, *10*, 1295–1312. [CrossRef] [PubMed]

63.	Agundez, J.A.; Jimenez-Jimenez, F.J.; Alonso-Navarro, H.; Garcia-Martin, E. Drug and xenobiotic biotransformation in the blood-brain barrier: A neglected issue. *Front. Cell Neurosci.* **2014**, *8*, 335. [CrossRef] [PubMed]

64.	Sweeney, M.D.; Ayyadurai, S.; Zlokovic, B.V. Pericytes of the neurovascular unit: Key functions and signaling pathways. *Nat. Neurosci.* **2016**, *19*, 771–783. [CrossRef] [PubMed]

65.	Abbott, N.J.; Ronnback, L.; Hansson, E. Astrocyte-endothelial interactions at the blood-brain barrier. *Nat. Rev. Neurosci.* **2006**, *7*, 41–53. [CrossRef] [PubMed]

66.	Johanson, C.E.; Stopa, E.G.; McMillan, P.N. The blood-cerebrospinal fluid barrier: Structure and functional significance. *Methods Mol. Biol.* **2011**, *686*, 101–131. [CrossRef]

67.	Damkier, H.H.; Brown, P.D.; Praetorius, J. Cerebrospinal fluid secretion by the choroid plexus. *Physiol. Rev.* **2013**, *93*, 1847–1892. [CrossRef] [PubMed]

68.	Engelhardt, B.; Ransohoff, R.M. The ins and outs of T-lymphocyte trafficking to the CNS: Anatomical sites and molecular mechanisms. *Trends Immunol.* **2005**, *26*, 485–495. [CrossRef] [PubMed]

69. Kivisakk, P.; Mahad, D.J.; Callahan, M.K.; Trebst, C.; Tucky, B.; Wei, T.; Wu, L.; Baekkevold, E.S.; Lassmann, H.; Staugaitis, S.M.; et al. Human cerebrospinal fluid central memory CD4+ T cells: Evidence for trafficking through choroid plexus and meninges via P-selectin. *Proc. Natl. Acad. Sci. USA* **2003**, *100*, 8389–8394. [CrossRef] [PubMed]

70. Llovera, G.; Benakis, C.; Enzmann, G.; Cai, R.; Arzberger, T.; Ghasemigharagoz, A.; Mao, X.; Malik, R.; Lazarevic, I.; Liebscher, S.; et al. The choroid plexus is a key cerebral invasion route for T cells after stroke. *Acta Neuropathol.* **2017**, *134*, 851–868. [CrossRef]

71. Daneman, R.; Zhou, L.; Agalliu, D.; Cahoy, J.D.; Kaushal, A.; Barres, B.A. The mouse blood-brain barrier transcriptome: A new resource for understanding the development and function of brain endothelial cells. *PLoS ONE* **2010**, *5*, e13741. [CrossRef]

72. Stamatovic, S.M.; Shakui, P.; Keep, R.F.; Moore, B.B.; Kunkel, S.L.; Van Rooijen, N.; Andjelkovic, A.V. Monocyte chemoattractant protein-1 regulation of blood-brain barrier permeability. *J. Cereb. Blood Flow Metab.* **2005**, *25*, 593–606. [CrossRef]

73. Higashida, T.; Kreipke, C.W.; Rafols, J.A.; Peng, C.; Schafer, S.; Schafer, P.; Ding, J.Y.; Dornbos, D.; LI, X.; Guthikonda, M.; et al. The role of hypoxia-inducible factor-1alpha, aquaporin-4, and matrix metalloproteinase-9 in blood-brain barrier disruption and brain edema after traumatic brain injury. *J. Neurosurg.* **2011**, *114*, 92–101. [CrossRef]

74. Bell, R.D.; Winkler, E.A.; Singh, I.; Sagare, A.P.; Deane, R.; Wu, Z.; Holtzman, D.M.; Betsholtz, C.; Armulik, A.; Sallstrom, J.; et al. Apolipoprotein E controls cerebrovascular integrity via cyclophilin A. *Nature* **2012**, *485*, 512–516. [CrossRef] [PubMed]

75. Persidsky, Y.; Dykstra, F.S.; Reichenbach, N.L.; Rom, S.; Rameriz, S.H. Activation of cannabinoid type two receptors (CB2) diminish inflammatory responses in macrophages and brain endothelium. *J. Neuroimmune Pharmacol.* **2015**, *10*, 302–308. [CrossRef]

76. Zhang, J.; Sadowska, G.B.; Chen, X.; Park, S.Y.; Kim, J.E.; Bodge, C.A.; Cummings, E.E.; Lim, Y.P.; Makeyev, O.; Besio, W.G.; et al. Anti-IL-6 neutralizing antibody modulates blood-brain barrier function in the ovine fetus. *FASEB J.* **2015**, *29*, 1739–1753. [CrossRef] [PubMed]

77. Papadopoulos, D.; Scheiner-Bobis, G. Dehydroepiandrosterone sulfate augments blood-brain barrier and tight junction protein expression in brain endothelial cells. *Biochim. Biophys. Acta Mol. Cell Res.* **2017**, *8*, 1382–1392. [CrossRef] [PubMed]

78. Ito, S.; Yanai, M.; Yamaguchi, S.; Couraud, P.O.; Ohtsuki, S. Regulation of tight-junction integrity by insulin in an in vitro model of human blood-brain barrier. *J. Pharm. Sci.* **2017**, *106*, 2599–2605. [CrossRef] [PubMed]

79. Megyeri, P.; Abraham, C.S.; Temesvari, P.; Kovacs, J.; Vas, T.; Speer, C.P. Recombinant human tumor necrosis factor alpha constricts pial arterioles and increases blood-brain barrier permeability in newborn piglets. *Neurosci. Lett.* **1992**, *148*, 137–140. [CrossRef]

80. Xaio, H.; Banks, W.A.; Niehoff, M.L.; Morley, J.E. Effect of LPS on the permeability of the blood-brain barrier to insulin. *Brain Res.* **2001**, *896*, 36–42. [CrossRef]

81. Dohgu, S.; Fleegal-DeMotta, M.A.; Banks, W.A. Lipopolysaccharide-enhanced transcellular transport of HIV-1 across the blood-brain barrier is mediated by luminal microvessel IL-6 and GM-CSF. *J. Neuroinflamm.* **2011**, *8*, 167. [CrossRef] [PubMed]

82. Banks, W.A.; Kastin, A.J.; Broadwell, R.D. Passage of cytokines across the blood-brain barrier. *Neuroimmunomodulation* **1995**, *2*, 241–248. [CrossRef]

83. Greenwood, J.; Heasman, S.J.; Alvarez, J.I.; Pratt, A.; Lyck, R.; Engelhardt, B. Review: Leukocyte-endothelial cell crosstalk at the blood-brain barrier: A prerequisite for successful immune cell entry to the brain. *Neuropathol. Appl. Neurobiol.* **2011**, *37*, 24–39. [CrossRef]

84. Warnke, C.; Menge, T.; Hartung, H.-P.; Racke, M.K.; Cravens, P.D.; Bennett, J.L.; Frohman, E.M.; Greenberg, B.M.; Zamvil, S.S.; Gold, R.; et al. Natalizumab and progressive mutifocal leukoencephalopathy. *JAMA Neurol.* **2010**, *67*, 923–930.

85. Engstrom, L.; Ruud, J.; Eskilsson, A.; Larsson, A.; Mackerlova, L.; Kugelberg, U.; Qian, H.; Vasilache, A.M.; Larsson, P.; Engblom, D.; et al. Lipopolysaccharide-induced fever depends on prostaglandin E2 production specifically in brain endothelial cells. *Endocrinology* **2012**, *153*, 4849–4861. [CrossRef]

86. Nilsson, A.; Wilhelms, D.B.; Mirrasekhian, E.; Jaarola, M.; Blomqvist, A.; Engblom, D. Inflammation-induced anorexia and fever are elicited by distinct prostaglandin dependent mechanisms, whereas conditioned taste aversion is prostaglandin independent. *Brain Behav. Immun.* **2017**, *61*, 236–243. [CrossRef]

87. Banks, W.A.; Kovac, A.; Morofuji, Y. Neurovascular unit crosstalk: Pericytes and astrocytes modify cytokine secretion patterns of brain endothelial cells. *J. Cereb. Blood Flow Metab.* **2017**, *38*, 1104–1118. [CrossRef] [PubMed]

88. Verma, S.; Nakaoke, R.; Dohgu, S.; Banks, W.A. Release of cytokines by brain endothelial cells: A polarized response to lipopolysaccharide. *Brain Behav. Immun.* **2006**, *20*, 449–455. [CrossRef]

89. Varatharaj, A.; Galea, I. The blood-brain barrier in systemic inflammation. *Brain Behav. Immun.* **2017**, *60*, 1–12. [CrossRef] [PubMed]

90. Farrall, A.J.; Wardlaw, J.M. Blood-brain barrier: Ageing and microvascular disease–systematic review and meta-analysis. *Neurobiol. Aging* **2009**, *30*, 337–352. [CrossRef]

91. Montagne, A.; Barnes, S.R.; Sweeney, M.D.; Halliday, M.R.; Sagare, A.P.; Zhao, Z.; Toga, A.W.; Jacobs, R.E.; Liu, C.Y.; Amezcua, L.; et al. Blood-brain barrier breakdown in the aging human hippocampus. *Neuron* **2015**, *85*, 296–302. [CrossRef] [PubMed]

92. Chen, R.L. Is it appropriate to use albumin CSF/plasma ratio to assess blood brain barrier permeability? *Neurobiol. Aging* **2011**, *32*, 1338–1339. [CrossRef] [PubMed]

93. Chen, R.L.; Kassem, N.A.; Redzic, Z.B.; Chen, C.P.; Segal, M.B.; Preston, J.E. Age-related changes in choroid plexus and blood-cerebrospinal fluid barrier function in the sheep. *Exp. Gerontol.* **2009**, *44*, 289–296. [CrossRef]

94. Chen, R.L.; Athauda, S.B.; Kassem, N.A.; Zhang, Y.; Segal, M.B.; Preston, J.E. Decrease of transthyretin synthesis at the blood-cerebrospinal fluid barrier of old sheep. *J. Gerontol. A Biol. Sci. Med. Sci.* **2005**, *60*, 852–858. [CrossRef]

95. Elahy, M.; Jackaman, C.; Mamo, J.C.; Lam, V.; Dhaliwal, S.S.; Giles, C.; Nelson, D.; Takechi, R. Blood-brain barrier dysfunction developed during normal aging is associated with inflammation and loss of tight junctions but not with leukocyte recruitment. *Immun. Ageing* **2015**, *12*, 2. [CrossRef]

96. Stamatovic, S.M.; Martinez-Revollar, G.; Hu, A.; Choi, J.; Keep, R.F.; Andjelkovic, A.V. Decline in Sirtuin-1 expression and activity plays a critical role in blood-brain barrier permeability in aging. *Neurobiol. Dis.* **2018**. [CrossRef] [PubMed]

97. Rahman, S.; Islam, R. Mammalian Sirt1: Insights on its biological functions. *Cell Commun. Signal.* **2011**, *9*, 11. [CrossRef] [PubMed]

98. Doolittle, N.D.; Muldoon, L.L.; Culp, A.Y.; Neuwelt, E.A. Delivery of chemotherapeutics across the blood-brain barrier: Challenges and advances. *Adv. Pharmacol.* **2014**, *71*, 203–243. [CrossRef] [PubMed]

99. Lipsman, N.; Meng, Y.; Bethune, A.J.; Huang, Y.; Lam, B.; Masellis, M.; Herrmann, N.; Heyn, C.; Aubert, I.; Boutet, A.; et al. Blood-brain barrier opening in Alzheimer's disease using MR-guided focused ultrasound. *Nat. Commun.* **2018**, *9*, 2336. [CrossRef] [PubMed]

100. Salahuddin, T.S.; Johansson, B.B.; Kalimo, H.; Olsson, Y. Structural changes in the rat brain after carotid infusions of hyperosmolar solutions: A light microscopic and immunohistochemical study. *Neuropathol. Appl. Neurobiol.* **1988**, *14*, 467–482. [CrossRef]

101. Salahuddin, T.S.; Kalimo, H.; Johansson, B.B.; Olsson, Y. Observations on exsudation of fibronectin, fibrinogen and albumin in the brain after carotid infusion of hyperosmolar solutions. An immunohistochemical study in the rat indicating longlasting changes in the brain microenvironment and multifocal nerve cell injuries. *Acta Neuropathol.* **1988**, *76*, 1–10.

102. Merlini, M.; Rafalski, V.A.; Rios Coronado, P.E.; Gill, T.M.; Ellisman, M.; Muthukumar, G.; Subramanian, K.S.; Ryu, J.K.; Syme, C.A.; Davalos, D.; et al. Fibrinogen Induces Microglia-Mediated Spine Elimination and Cognitive Impairment in an Alzheimer's Disease Model. *Neuron* **2019**. [CrossRef] [PubMed]

103. Patching, S.G. Glucose Transporters at the Blood-Brain Barrier: Function, Regulation and Gateways for Drug Delivery. *Mol. Neurobiol.* **2017**, *54*, 1046–1077. [CrossRef]

104. Sweeney, M.D.; Montagne, A.; Sagare, A.P.; Nation, D.A.; Schneider, L.S.; Chui, H.C.; Harrington, M.G.; Pa, J.; Law, M.; Wang, D.J.J.; et al. Vascular dysfunction-The disregarded partner of Alzheimer's disease. *Alzheimers Dement.* **2019**, *15*, 158–167. [CrossRef] [PubMed]

105. Bonte, S.; Vandemaele, P.; Verleden, S.; Audenaert, K.; Deblaere, K.; Goethals, I.; Van Holen, R. Healthy brain ageing assessed with 18F-FDG PET and age-dependent recovery factors after partial volume effect correction. *Eur. J. Nucl. Med. Mol. Imaging* **2017**, *44*, 838–849. [CrossRef]

106. Jiang, T.; Yin, F.; Yao, J.; Brinton, R.D.; Cadenas, E. Lipoic acid restores age-associated impairment of brain energy metabolism through the modulation of Akt/JNK signaling and PGC1alpha transcriptional pathway. *Aging Cell* **2013**, *12*, 1021–1031. [CrossRef] [PubMed]

107. Gage, F.H.; Kelly, P.A.; Bjorklund, A. Regional changes in brain glucose metabolism reflect cognitive impairments in aged rats. *J. Neurosci.* **1984**, *4*, 2856–2865. [CrossRef] [PubMed]

108. Ding, F.; Yao, J.; Rettberg, J.R.; Chen, S.; Brinton, R.D. Early decline in glucose transport and metabolism precedes shift to ketogenic system in female aging and Alzheimer's mouse brain: Implication for bioenergetic intervention. *PLoS ONE* **2013**, *8*, e79977. [CrossRef]

109. Selkoe, D.J.; Hardy, J. The amyloid hypothesis of Alzheimer's disease at 25 years. *EMBO Mol. Med.* **2016**, *8*, 595–608. [CrossRef]

110. Shibata, M.; Yamada, S.; Kumar, S.R.; Calero, M.; Bading, J.; Frangione, B.; Holtzman, D.M.; Miller, C.A.; Strickland, D.K.; Ghiso, J.; et al. Clearance of Alzheimer's amyloid-ss(1-40) peptide from brain by LDL receptor-related protein-1 at the blood-brain barrier. *J. Clin. Investig.* **2000**, *106*, 1489–1499. [CrossRef] [PubMed]

111. Storck, S.E.; Hartz, A.M.S.; Bernard, J.; Wolf, A.; Kachlmeier, A.; Mahringer, A.; Weggen, S.; Pahnke, J.; Pietrzik, C.U. The concerted amyloid-beta clearance of LRP1 and ABCB1/P-gp across the blood-brain barrier is linked by PICALM. *Brain Behav. Immun.* **2018**, *73*, 21–33. [CrossRef] [PubMed]

112. Hartz, A.M.; Miller, D.S.; Bauer, B. Restoring blood-brain barrier P-glycoprotein reduces brain amyloid-beta in a mouse model of Alzheimer's disease. *Mol. Pharmacol.* **2010**, *77*, 715–723. [CrossRef] [PubMed]

113. Ramanathan, A.; Nelson, A.R.; Sagare, A.P.; Zlokovic, B.V. Impaired vascular-mediated clearance of brain amyloid beta in Alzheimer's disease: The role, regulation and restoration of LRP1. *Front. Aging Neurosci.* **2015**, *7*, 136. [CrossRef]

114. Toornvliet, R.; van Berckel, B.N.; Luurtsema, G.; Lubberink, M.; Geldof, A.A.; Bosch, T.M.; Oerlemans, R.; Lammertsma, A.A.; Franssen, E.J. Effect of age on functional P-glycoprotein in the blood-brain barrier measured by use of (R)-[(11)C]verapamil and positron emission tomography. *Clin. Pharmacol. Ther.* **2006**, *79*, 540–548. [CrossRef]

115. van Assema, D.M.; Lubberink, M.; Boellaard, R.; Schuit, R.C.; Windhorst, A.D.; Scheltens, P.; Lammertsma, A.A.; van Berckel, B.N. P-glycoprotein function at the blood-brain barrier: Effects of age and gender. *Mol. Imaging Biol.* **2012**, *14*, 771–776. [CrossRef] [PubMed]

116. Chiu, C.; Miller, M.C.; Monahan, R.; Osgood, D.P.; Stopa, E.G.; Silverberg, G.D. P-glycoprotein expression and amyloid accumulation in human aging and Alzheimer's disease: Preliminary observations. *Neurobiol. Aging* **2015**, *36*, 2475–2482. [CrossRef]

117. Hoffman, J.D.; Parikh, I.; Green, S.J.; Chlipala, G.; Mohney, R.P.; Keaton, M.; Bauer, B.; Hartz, A.M.S.; Lin, A.L. Age Drives Distortion of Brain Metabolic, Vascular and Cognitive Functions, and the Gut Microbiome. *Front. Aging Neurosci.* **2017**, *9*, 298. [CrossRef]

118. Jaeger, L.B.; Dohgu, S.; Sultana, R.; Lynch, J.L.; Owen, J.B.; Erickson, M.A.; Shah, G.N.; Price, T.O.; Fleegal-Demotta, M.A.; Butterfield, D.A.; et al. Lipopolysaccharide alters the blood-brain barrier transport of amyloid beta protein: A mechanism for inflammation in the progression of Alzheimer's disease. *Brain Behav. Immun.* **2009**, *23*, 507–517. [CrossRef] [PubMed]

119. Banks, W.A.; Owen, J.B.; Erickson, M.A. Insulin in the brain: There and back again. *Pharmacol. Ther.* **2012**, *136*, 82–93. [CrossRef] [PubMed]

120. Duffy, K.R.; Pardridge, W.M. Blood-brain barrier transcytosis of insulin in developing rabbits. *Brain Res.* **1987**, *420*, 32–38. [CrossRef]

121. Banks, W.A.; Jaspan, J.B.; Huang, W.; Kastin, A.J. Transport of insulin across the blood-brain barrier: Saturability at euglycemic doses of insulin. *Peptides* **1997**, *18*, 1423–1429. [CrossRef]

122. Baura, G.D.; Foster, D.M.; Porte, D., Jr.; Kahn, S.E.; Bergman, R.N.; Cobelli, C.; Schwartz, M.W. Saturable transport of insulin from plasma into the central nervous system of dogs in vivo. A mechanism for regulated insulin delivery to the brain. *J. Clin. Investig.* **1993**, *92*, 1824–1830. [CrossRef] [PubMed]

123. Sartorius, T.; Peter, A.; Heni, M.; Maetzler, W.; Fritsche, A.; Haring, H.U.; Hennige, A.M. The brain response to peripheral insulin declines with age: A contribution of the blood-brain barrier? *PLoS ONE* **2015**, *10*, e0126804. [CrossRef] [PubMed]

124. Frolich, L.; Blum-Degen, D.; Bernstein, H.G.; Engelsberger, S.; Humrich, J.; Laufer, S.; Muschner, D.; Thalheimer, A.; Turk, A.; Hoyer, S.; et al. Brain insulin and insulin receptors in aging and sporadic Alzheimer's disease. *J. Neural. Transm (Vienna)* **1998**, *105*, 423–438. [CrossRef] [PubMed]

125. Banks, W.A.; Farr, S.A.; Morley, J.E. Permeability of the blood-brain barrier to albumin and insulin in the young and aged SAMP8 mouse. *J. Gerontol. A Biol. Sci. Med. Sci.* **2000**, *55*, B601–B606. [CrossRef] [PubMed]

126. Yates, D. Brain ageing: Blood-derived rejuvenation. *Nat. Rev. Neurosci.* **2014**, *15*, 352–353. [CrossRef] [PubMed]

127. Poggioli, T.; Vujic, A.; Yang, P.; Macias-Trevino, C.; Uygur, A.; Loffredo, F.S.; Pancoast, J.R.; Cho, M.; Goldstein, J.; Tandias, R.M.; et al. Circulating Growth Differentiation Factor 11/8 Levels Decline With Age. *Circ. Res.* **2016**, *118*, 29–37. [CrossRef]

128. Katsimpardi, L.; Litterman, N.K.; Schein, P.A.; Miller, C.M.; Loffredo, F.S.; Wojtkiewicz, G.R.; Chen, J.W.; Lee, R.T.; Wagers, A.J.; Rubin, L.L. Vascular and neurogenic rejuvenation of the aging mouse brain by young systemic factors. *Science* **2014**, *344*, 630–634. [CrossRef]

129. Erickson, M.A.; Morofuji, Y.; Owen, J.B.; Banks, W.A. Rapid transport of CCL11 across the blood-brain barrier: Regional variation and importance of blood cells. *J. Pharmacol. Exp. Ther.* **2014**, *349*, 497–507. [CrossRef]

130. Park, M.H.; Lee, J.Y.; Park, K.H.; Jung, I.K.; Kim, K.T.; Lee, Y.S.; Ryu, H.H.; Jeong, Y.; Kang, M.; Schwaninger, M.; et al. Vascular and Neurogenic Rejuvenation in Aging Mice by Modulation of ASM. *Neuron* **2018**, *100*, 167–182. [CrossRef] [PubMed]

131. Li, C.; Wang, A.; Wu, Y.; Gulbins, E.; Grassme, H.; Zhao, Z. Acid Sphingomyelinase-Ceramide System in Bacterial Infections. *Cell Physiol. Biochem.* **2019**, *52*, 280–301. [CrossRef]

132. Ransohoff, R.M.; Engelhardt, B. The anatomical and cellular basis of immune surveillance in the central nervous system. *Nat. Rev. Immunol.* **2012**, *12*, 623–635. [CrossRef] [PubMed]

133. Baruch, K.; Schwartz, M. CNS-specific T cells shape brain function via the choroid plexus. *Brain Behav. Immun.* **2013**, *34*, 11–16. [CrossRef] [PubMed]

134. Baruch, K.; Deczkowska, A.; David, E.; Castellano, J.M.; Miller, O.; Kertser, A.; Berkutzki, T.; Barnett-Itzhaki, Z.; Bezalel, D.; Wyss-Coray, T.; et al. Aging. Aging-induced type I interferon response at the choroid plexus negatively affects brain function. *Science* **2014**, *346*, 89–93. [CrossRef] [PubMed]

135. Kuro-o, M. Klotho and aging. *Biochim. Biophys. Acta* **2009**, *1790*, 1049–1058. [CrossRef] [PubMed]

136. Kuro-o, M.; Matsumura, Y.; Aizawa, H.; Kawaguchi, H.; Suga, T.; Utsugi, T.; Ohyama, Y.; Kurabayashi, M.; Kaname, T.; Kume, E.; et al. Mutation of the mouse klotho gene leads to a syndrome resembling ageing. *Nature* **1997**, *390*, 45–51. [CrossRef] [PubMed]

137. Zhu, L.; Stein, L.R.; Kim, D.; Ho, K.; Yu, G.Q.; Zhan, L.; Larsson, T.E.; Mucke, L. Klotho controls the brain-immune system interface in the choroid plexus. *Proc. Natl. Acad. Sci. USA* **2018**, *115*, E11388–E11396. [CrossRef] [PubMed]

138. Schwartz, M.; Baruch, K. Breaking peripheral immune tolerance to CNS antigens in neurodegenerative diseases: Boosting autoimmunity to fight-off chronic neuroinflammation. *J. Autoimmun.* **2014**, *54*, 8–14. [CrossRef]

139. Stichel, C.C.; Luebbert, H. Inflammatory processes in the aging mouse brain: Participation of dendritic cells and T-cells. *Neurobiol. Aging* **2007**, *28*, 1507–1521. [CrossRef]

140. Ritzel, R.M.; Crapser, J.; Patel, A.R.; Verma, R.; Grenier, J.M.; Chauhan, A.; Jellison, E.R.; McCullough, L.D. Age-Associated Resident Memory CD8 T Cells in the Central Nervous System Are Primed to Potentiate Inflammation after Ischemic Brain Injury. *J. Immunol.* **2016**, *196*, 3318–3330. [CrossRef]

141. Erickson, M.A.; Liang, W.S.; Fernandez, E.G.; Bullock, K.M.; Thysell, J.A.; Banks, W.A. Genetics and sex influence peripheral and central innate immune responses and blood-brain barrier integrity. *PLoS ONE* **2018**, *13*, e0205769. [CrossRef] [PubMed]

142. Banks, W.A.; Gray, A.M.; Erickson, M.A.; Salameh, T.S.; Damodarasamy, M.; Sheibani, N.; Meabon, J.S.; Wing, E.E.; Morofuji, Y.; Cook, D.G.; et al. Lipopolysaccharide-induced blood-brain barrier disruption: Roles of cyclooxygenase, oxidative stress, neuroinflammation, and elements of the neurovascular unit. *J. Neuroinflamm.* **2015**, *12*, 223. [CrossRef] [PubMed]

143. Elwood, E.; Lim, Z.; Naveed, H.; Galea, I. The effect of systemic inflammation on human brain barrier function. *Brain Behav. Immun.* **2017**, *62*, 35–40. [CrossRef] [PubMed]

144. Ching, S.; Zhang, H.; Lai, W.; Quan, N. Peripheral injection of lipopolysaccharide prevents brain recruitment of leukocytes induced by central injection of interleukin-1. *Neuroscience* **2006**, *137*, 717–726. [CrossRef]

145. Sumbria, R.K.; Grigoryan, M.M.; Vasilevko, V.; Paganini-Hill, A.; Kilday, K.; Kim, R.; Cribbs, D.H.; Fisher, M.J. Aging exacerbates development of cerebral microbleeds in a mouse model. *J. Neuroinflamm.* **2018**, *15*, 69. [CrossRef]

146. Opp, M.R.; George, A.; Ringgold, K.M.; Hansen, K.M.; Bullock, K.M.; Banks, W.A. Sleep fragmentation and sepsis differentially impact blood-brain barrier integrity and transport of tumor necrosis factor-alpha in aging. *Brain Behav. Immun.* **2015**, *50*, 259–265. [CrossRef]

147. Gutierrez, E.G.; Banks, W.A.; Kastin, A.J. Murine tumor necrosis factor alpha is transported from blood to brain in the mouse. *J. Neuroimmunol* **1993**, *47*, 169–176. [CrossRef]

148. Logsdon, A.F.; Erickson, M.A.; Rhea, E.M.; Salameh, T.S.; Banks, W.A. Gut reactions: How the blood-brain barrier connects the microbiome and the brain. *Exp. Biol. Med. (Maywood)* **2018**, *243*, 159–165. [CrossRef] [PubMed]

149. Braniste, V.; Al-Asmakh, M.; Kowal, C.; Anuar, F.; Abbaspour, A.; Toth, M.; Korecka, A.; Bakocevic, N.; Ng, L.G.; Kundu, P.; et al. The gut microbiota influences blood-brain barrier permeability in mice. *Sci. Transl. Med.* **2014**, *6*, 263ra158. [CrossRef]

150. Hippe, B.; Zwielehner, J.; Liszt, K.; Lassl, C.; Unger, F.; Haslberger, A.G. Quantification of butyryl CoA:acetate CoA-transferase genes reveals different butyrate production capacity in individuals according to diet and age. *FEMS Microbiol. Lett.* **2011**, *316*, 130–135. [CrossRef] [PubMed]

151. Scott, K.A.; Ida, M.; Peterson, V.L.; Prenderville, J.A.; Moloney, G.M.; Izumo, T.; Murphy, K.; Murphy, A.; Ross, R.P.; Stanton, C.; et al. Revisiting Metchnikoff: Age-related alterations in microbiota-gut-brain axis in the mouse. *Brain Behav. Immun.* **2017**, *65*, 20–32. [CrossRef] [PubMed]

152. Gorle, N.; Blaecher, C.; Bauwens, E.; Vandendriessche, C.; Balusu, S.; Vandewalle, J.; Van Cauwenberghe, C.; Van Wonterghem, E.; Van Imschoot, G.; Liu, C.; et al. The choroid plexus epithelium as a novel player in the stomach-brain axis during Helicobacter infection. *Brain Behav. Immun.* **2018**, *69*, 35–47. [CrossRef] [PubMed]

153. Dinan, T.G.; Cryan, J.F. The Microbiome-Gut-Brain Axis in Health and Disease. *Gastroenterol. Clin. N. Am.* **2017**, *46*, 77–89. [CrossRef] [PubMed]

154. Dore-Duffy, P. Pericytes: Pluripotent cells of the blood brain barrier. *Curr. Pharm. Des.* **2008**, *14*, 1581–1593. [CrossRef]

155. Winkler, E.A.; Sagare, A.P.; Zlokovic, B.V. The pericyte: A forgotten cell type with important implications for Alzheimer's disease? *Brain Pathol.* **2014**, *24*, 371–386. [CrossRef] [PubMed]

156. Bell, R.D.; Winkler, E.A.; Sagare, A.P.; Singh, I.; LaRue, B.; Deane, R.; Zlokovic, B.V. Pericytes control key neurovascular functions and neuronal phenotype in the adult brain and during brain aging. *Neuron* **2010**, *68*, 409–427. [CrossRef]

157. Duncombe, J.; Lennen, R.J.; Jansen, M.A.; Marshall, I.; Wardlaw, J.M.; Horsburgh, K. Ageing causes prominent neurovascular dysfunction associated with loss of astrocytic contacts and gliosis. *Neuropathol. Appl. Neurobiol.* **2017**, *43*, 477–491. [CrossRef]

158. Goodall, E.F.; Wang, C.; Simpson, J.E.; Baker, D.J.; Drew, D.R.; Heath, P.R.; Saffrey, M.J.; Romero, I.A.; Wharton, S.B. Age-associated changes in the blood-brain barrier: Comparative studies in human and mouse. *Neuropathol. Appl. Neurobiol.* **2018**, *44*, 328–340. [CrossRef]

159. Erdo, F.; Denes, L.; de Lange, E. Age-associated physiological and pathological changes at the blood-brain barrier: A review. *J. Cereb. Blood Flow Metab.* **2017**, *37*, 4–24. [CrossRef]

160. Pieper, C.; Marek, J.J.; Unterberg, M.; Schwerdtle, T.; Galla, H.J. Brain capillary pericytes contribute to the immune defense in response to cytokines or LPS in vitro. *Brain Res.* **2014**, *1550*, 1–8. [CrossRef] [PubMed]

161. Sagare, A.P.; Sweeney, M.D.; Makshanoff, J.; Zlokovic, B.V. Shedding of soluble platelet-derived growth factor receptor-beta from human brain pericytes. *Neurosci. Lett.* **2015**, *607*, 97–101. [CrossRef] [PubMed]

162. Nation, D.A.; Sweeney, M.D.; Montagne, A.; Sagare, A.P.; D'Orazio, L.M.; Pachicano, M.; Sepehrband, F.; Nelson, A.R.; Buennagel, D.P.; Harrington, M.G.; et al. Blood-brain barrier breakdown is an early biomarker of human cognitive dysfunction. *Nat. Med.* **2019**, *25*, 270–276. [CrossRef]

163. Middeldorp, J.; Hol, E.M. GFAP in health and disease. *Prog. Neurobiol.* **2011**, *93*, 421–443. [CrossRef] [PubMed]

164. Iliff, J.J.; Wang, M.; Liao, Y.; Plogg, B.A.; Peng, W.; Gundersen, G.A.; Benveniste, H.; Vates, G.E.; Deane, R.; Goldman, S.A.; et al. A paravascular pathway facilitates CSF flow through the brain parenchyma and the clearance of interstitial solutes, including amyloid beta. *Sci. Transl. Med.* **2012**, *4*, 147ra111. [CrossRef]

165. Zenaro, E.; Piacentino, G.; Constantin, G. The blood-brain barrier in Alzheimer's disease. *Neurobiol. Dis.* **2017**, *107*, 41–56. [CrossRef] [PubMed]

166. Jack, C.R., Jr.; Bennett, D.A.; Blennow, K.; Carrillo, M.C.; Dunn, B.; Haeberlein, S.B.; Holtzman, D.M.; Jagust, W.; Jessen, F.; Karlawish, J.; et al. NIA-AA Research Framework: Toward a biological definition of Alzheimer's disease. *Alzheimers Dement.* **2018**, *14*, 535–562. [CrossRef] [PubMed]

167. Jack, C.R., Jr.; Knopman, D.S.; Jagust, W.J.; Petersen, R.C.; Weiner, M.W.; Aisen, P.S.; Shaw, L.M.; Vemuri, P.; Wiste, H.J.; Weigand, S.D.; et al. Tracking pathophysiological processes in Alzheimer's disease: An updated hypothetical model of dynamic biomarkers. *Lancet Neurol.* **2013**, *12*, 207–216. [CrossRef]

168. Castellano, J.M.; Kim, J.; Stewart, F.R.; Jiang, H.; DeMattos, R.B.; Patterson, B.W.; Fagan, A.M.; Morris, J.C.; Mawuenyega, K.G.; Cruchaga, C.; et al. Human apoE isoforms differentially regulate brain amyloid-beta peptide clearance. *Sci. Transl. Med.* **2011**, *3*, 89ra57. [CrossRef] [PubMed]

169. Shi, Y.; Yamada, K.; Liddelow, S.A.; Smith, S.T.; Zhao, L.; Luo, W.; Tsai, R.M.; Spina, S.; Grinberg, L.T.; Rojas, J.C.; et al. ApoE4 markedly exacerbates tau-mediated neurodegeneration in a mouse model of tauopathy. *Nature* **2017**, *549*, 523–527. [CrossRef] [PubMed]

170. Tai, L.M.; Ghura, S.; Koster, K.P.; Liakaite, V.; Maienschein-Cline, M.; Kanabar, P.; Collins, N.; Ben-Aissa, M.; Lei, A.Z.; Bahroos, N.; et al. APOE-modulated Abeta-induced neuroinflammation in Alzheimer's disease: Current landscape, novel data, and future perspective. *J. Neurochem.* **2015**, *133*, 465–488. [CrossRef]

171. Halliday, M.R.; Rege, S.V.; Ma, Q.; Zhao, Z.; Miller, C.A.; Winkler, E.A.; Zlokovic, B.V. Accelerated pericyte degeneration and blood-brain barrier breakdown in apolipoprotein E4 carriers with Alzheimer's disease. *J. Cereb. Blood Flow Metab.* **2016**, *36*, 216–227. [CrossRef]

172. Fulop, T.; Witkowski, J.M.; Bourgade, K.; Khalil, A.; Zerif, E.; Larbi, A.; Hirokawa, K.; Pawelec, G.; Bocti, C.; Lacombe, G.; et al. Can an Infection Hypothesis Explain the Beta Amyloid Hypothesis of Alzheimer's Disease? *Front. Aging Neurosci.* **2018**, *10*, 224. [CrossRef]

173. Itzhaki, R.F.; Lathe, R.; Balin, B.J.; Ball, M.J.; Bearer, E.L.; Braak, H.; Bullido, M.J.; Carter, C.; Clerici, M.; Cosby, S.L.; et al. Microbes and Alzheimer's Disease. *J. Alzheimers Dis.* **2016**, *51*, 979–984. [CrossRef] [PubMed]

174. Deatly, A.M.; Haase, A.T.; Fewster, P.H.; Lewis, E.; Ball, M.J. Human herpes virus infections and Alzheimer's disease. *Neuropathol. Appl. Neurobiol.* **1990**, *16*, 213–223. [CrossRef]

175. Ball, M.J.; Lewis, E.; Haase, A.T. Detection of herpes virus genome in Alzheimer's disease by in situ hybridization: A preliminary study. *J. Neural. Transm. Suppl.* **1987**, *24*, 219–225.

176. Walker, D.G.; O'Kusky, J.R.; McGeer, P.L. In situ hybridization analysis for herpes simplex virus nucleic acids in Alzheimer disease. *Alzheimer Dis. Assoc. Disord.* **1989**, *3*, 123–131. [CrossRef] [PubMed]

177. Wozniak, M.A.; Itzhaki, R.F.; Shipley, S.J.; Dobson, C.B. Herpes simplex virus infection causes cellular beta-amyloid accumulation and secretase upregulation. *Neurosci. Lett.* **2007**, *429*, 95–100. [CrossRef]

178. Zambrano, A.; Solis, L.; Salvadores, N.; Cortes, M.; Lerchundi, R.; Otth, C. Neuronal cytoskeletal dynamic modification and neurodegeneration induced by infection with herpes simplex virus type 1. *J. Alzheimers Dis.* **2008**, *14*, 259–269. [CrossRef]

179. Eimer, W.A.; Vijaya Kumar, D.K.; Navalpur Shanmugam, N.K.; Rodriguez, A.S.; Mitchell, T.; Washicosky, K.J.; Gyorgy, B.; Breakefield, X.O.; Tanzi, R.E.; Moir, R.D. Alzheimer's Disease-Associated beta-Amyloid Is Rapidly Seeded by Herpesviridae to Protect against Brain Infection. *Neuron* **2018**, *99*, 56–63. [CrossRef] [PubMed]

180. Kumar, D.K.; Choi, S.H.; Washicosky, K.J.; Eimer, W.A.; Tucker, S.; Ghofrani, J.; Lefkowitz, A.; McColl, G.; Goldstein, L.E.; Tanzi, R.E.; et al. Amyloid-beta peptide protects against microbial infection in mouse and worm models of Alzheimer's disease. *Sci. Transl. Med.* **2016**, *8*, 340ra372. [CrossRef]

181. Readhead, B.; Haure-Mirande, J.V.; Funk, C.C.; Richards, M.A.; Shannon, P.; Haroutunian, V.; Sano, M.; Liang, W.S.; Beckmann, N.D.; Price, N.D.; et al. Multiscale Analysis of Independent Alzheimer's Cohorts Finds Disruption of Molecular, Genetic, and Clinical Networks by Human Herpesvirus. *Neuron* **2018**, *99*, 64–82. [CrossRef] [PubMed]

182. Wozniak, M.A.; Mee, A.P.; Itzhaki, R.F. Herpes simplex virus type 1 DNA is located within Alzheimer's disease amyloid plaques. *J. Pathol.* **2009**, *217*, 131–138. [CrossRef]

183. Tsai, M.C.; Cheng, W.L.; Sheu, J.J.; Huang, C.C.; Shia, B.C.; Kao, L.T.; Lin, H.C. Increased risk of dementia following herpes zoster ophthalmicus. *PLoS ONE* **2017**, *12*, e0188490. [CrossRef]

184. Chen, V.C.; Wu, S.I.; Huang, K.Y.; Yang, Y.H.; Kuo, T.Y.; Liang, H.Y.; Huang, K.L.; Gossop, M. Herpes Zoster and Dementia: A Nationwide Population-Based Cohort Study. *J. Clin. Psychiatry* **2018**, *79*. [CrossRef] [PubMed]

185. Tzeng, N.S.; Chung, C.H.; Lin, F.H.; Chiang, C.P.; Yeh, C.B.; Huang, S.Y.; Lu, R.B.; Chang, H.A.; Kao, Y.C.; Yeh, H.W.; et al. Anti-herpetic Medications and Reduced Risk of Dementia in Patients with Herpes Simplex Virus Infections-a Nationwide, Population-Based Cohort Study in Taiwan. *Neurotherapeutics* **2018**, *15*, 417–429. [CrossRef] [PubMed]

186. Elder, G.A.; Gama Sosa, M.A.; De Gasperi, R. Transgenic mouse models of Alzheimer's disease. *Mt. Sinai. J. Med.* **2010**, *77*, 69–81. [CrossRef] [PubMed]

187. Banks, W.A.; Robinson, S.M.; Verma, S.; Morley, J.E. Efflux of human and mouse amyloid beta proteins 1-40 and 1-42 from brain: Impairment in a mouse model of Alzheimer's disease. *Neuroscience* **2003**, *121*, 487–492. [CrossRef]

188. Kumar, V.B.; Farr, S.A.; Flood, J.F.; Kamlesh, V.; Franko, M.; Banks, W.A.; Morley, J.E. Site-directed antisense oligonucleotide decreases the expression of amyloid precursor protein and reverses deficits in learning and memory in aged SAMP8 mice. *Peptides* **2000**, *21*, 1769–1775. [CrossRef]

189. Bien-Ly, N.; Boswell, C.A.; Jeet, S.; Beach, T.G.; Hoyte, K.; Luk, W.; Shihadeh, V.; Ulufatu, S.; Foreman, O.; Lu, Y.; et al. Lack of Widespread BBB Disruption in Alzheimer's Disease Models: Focus on Therapeutic Antibodies. *Neuron* **2015**, *88*, 289–297. [CrossRef]

190. Gustafsson, S.; Gustavsson, T.; Roshanbin, S.; Hultqvist, G.; Hammarlund-Udenaes, M.; Sehlin, D.; Syvanen, S. Blood-brain barrier integrity in a mouse model of Alzheimer's disease with or without acute 3D6 immunotherapy. *Neuropharmacology* **2018**, *143*, 1–9. [CrossRef] [PubMed]

191. Dickstein, D.L.; Biron, K.E.; Ujiie, M.; Pfeifer, C.G.; Jeffries, A.R.; Jefferies, W.A. Abeta peptide immunization restores blood-brain barrier integrity in Alzheimer disease. *FASEB J.* **2006**, *20*, 426–433. [CrossRef]

192. Montagne, A.; Zhao, Z.; Zlokovic, B.V. Alzheimer's disease: A matter of blood-brain barrier dysfunction? *J. Exp. Med.* **2017**, *214*, 3151–3169. [CrossRef] [PubMed]

193. Kovac, A.; Zilkova, M.; Deli, M.A.; Zilka, N.; Novak, M. Human truncated tau is using a different mechanism from amyloid-beta to damage the blood-brain barrier. *J. Alzheimers Dis.* **2009**, *18*, 897–906. [CrossRef] [PubMed]

194. Blair, L.J.; Frauen, H.D.; Zhang, B.; Nordhues, B.A.; Bijan, S.; Lin, Y.C.; Zamudio, F.; Hernandez, L.D.; Sabbagh, J.J.; Selenica, M.L.; et al. Tau depletion prevents progressive blood-brain barrier damage in a mouse model of tauopathy. *Acta Neuropathol. Commun.* **2015**, *3*, 8. [CrossRef] [PubMed]

195. Banks, W.A.; Kovac, A.; Majerova, P.; Bullock, K.M.; Shi, M.; Zhang, J. Tau Proteins Cross the Blood-Brain Barrier. *J. Alzheimers Dis.* **2017**, *55*, 411–419. [CrossRef] [PubMed]

196. Fiske, A.; Wetherell, J.L.; Gatz, M. Depression in older adults. *Annu. Rev. Clin. Psychol.* **2009**, *5*, 363–389. [CrossRef]

197. Seligman, F.; Nemeroff, C.B. The interface of depression and cardiovascular disease: Therapeutic implications. *Ann. N. Y. Acad. Sci.* **2015**, *1345*, 25–35. [CrossRef]

198. Chamberlain, S.R.; Cavanagh, J.; de Boer, P.; Mondelli, V.; Jones, D.N.C.; Drevets, W.C.; Cowen, P.J.; Harrison, N.A.; Pointon, L.; Pariante, C.M.; et al. Treatment-resistant depression and peripheral C-reactive protein. *Br. J. Psychiatry* **2019**, *214*, 11–19. [CrossRef]

199. Hodes, G.E.; Kana, V.; Menard, C.; Merad, M.; Russo, S.J. Neuroimmune mechanisms of depression. *Nat. Neurosci.* **2015**, *18*, 1386–1393. [CrossRef]

200. Menard, C.; Pfau, M.L.; Hodes, G.E.; Kana, V.; Wang, V.X.; Bouchard, S.; Takahashi, A.; Flanigan, M.E.; Aleyasin, H.; LeClair, K.B.; et al. Social stress induces neurovascular pathology promoting depression. *Nat. Neurosci.* **2017**, *20*, 1752–1760. [CrossRef] [PubMed]

201. Kinsey, S.G.; Bailey, M.T.; Sheridan, J.F.; Padgett, D.A. The inflammatory response to social defeat is increased in older mice. *Physiol. Behav.* **2008**, *93*, 628–636. [CrossRef]

202. Jiang, H.; Ling, Z.; Zhang, Y.; Mao, H.; Ma, Z.; Yin, Y.; Wang, W.; Tang, W.; Tan, Z.; Shi, J.; et al. Altered fecal microbiota composition in patients with major depressive disorder. *Brain Behav. Immun.* **2015**, *48*, 186–194. [CrossRef] [PubMed]

203. O'Mahony, S.M.; Marchesi, J.R.; Scully, P.; Codling, C.; Ceolho, A.M.; Quigley, E.M.; Cryan, J.F.; Dinan, T.G. Early life stress alters behavior, immunity, and microbiota in rats: Implications for irritable bowel syndrome and psychiatric illnesses. *Biol. Psychiatry* **2009**, *65*, 263–267. [CrossRef]

204. Bailey, M.T.; Dowd, S.E.; Galley, J.D.; Hufnagle, A.R.; Allen, R.G.; Lyte, M. Exposure to a social stressor alters the structure of the intestinal microbiota: Implications for stressor-induced immunomodulation. *Brain Behav. Immun.* **2011**, *25*, 397–407. [CrossRef]

205. Zheng, P.; Zeng, B.; Zhou, C.; Liu, M.; Fang, Z.; Xu, X.; Zeng, L.; Chen, J.; Fan, S.; Du, X.; et al. Gut microbiome remodeling induces depressive-like behaviors through a pathway mediated by the host's metabolism. *Mol. Psychiatry* **2016**, *21*, 786–796. [CrossRef] [PubMed]

206. Mauro, C.; De Rosa, V.; Marelli-Berg, F.; Solito, E. Metabolic syndrome and the immunological affair with the blood-brain barrier. *Front. Immunol.* **2014**, *5*, 677. [CrossRef]

207. Dominguez, L.J.; Barbagallo, M. The biology of the metabolic syndrome and aging. *Curr. Opin. Clin. Nutr. Metab. Care* **2016**, *19*, 5–11. [CrossRef] [PubMed]

208. Kaur, J. A comprehensive review on metabolic syndrome. *Cardiol Res Pract* **2014**, *2014*, 943162. [CrossRef] [PubMed]

209. Morley, J.E.; Sinclair, A. The metabolic syndrome in older persons: A loosely defined constellation of symptoms or a distinct entity? *Age Ageing* **2009**, *38*, 494–497. [CrossRef] [PubMed]

210. Atallah, A.; Mhaouty-Kodja, S.; Grange-Messent, V. Chronic depletion of gonadal testosterone leads to blood-brain barrier dysfunction and inflammation in male mice. *J. Cereb. Blood Flow Metab.* **2017**, *37*, 3161–3175. [CrossRef]

211. Dubois, V.; Laurent, M.R.; Jardi, F.; Antonio, L.; Lemaire, K.; Goyvaerts, L.; Deldicque, L.; Carmeliet, G.; Decallonne, B.; Vanderschueren, D.; et al. Androgen Deficiency Exacerbates High-Fat Diet-Induced Metabolic Alterations in Male Mice. *Endocrinology* **2016**, *157*, 648–665. [CrossRef]

212. Frasca, D.; Blomberg, B.B.; Paganelli, R. Aging, Obesity, and Inflammatory Age-Related Diseases. *Front. Immunol.* **2017**, *8*, 1745. [CrossRef]

213. Rhea, E.M.; Salameh, T.S.; Logsdon, A.F.; Hanson, A.J.; Erickson, M.A.; Banks, W.A. Blood-Brain Barriers in Obesity. *AAPS J.* **2017**, *19*, 921–930. [CrossRef]

214. Klok, M.D.; Jakobsdottir, S.; Drent, M.L. The role of leptin and ghrelin in the regulation of food intake and body weight in humans: A review. *Obes. Rev.* **2007**, *8*, 21–34. [CrossRef] [PubMed]

215. Banks, W.A.; Burney, B.O.; Robinson, S.M. Effects of triglycerides, obesity, and starvation on ghrelin transport across the blood-brain barrier. *Peptides* **2008**, *29*, 2061–2065. [CrossRef]

216. Banks, W.A.; Farr, S.A.; Morley, J.E. The effects of high fat diets on the blood-brain barrier transport of leptin: Failure or adaptation? *Physiol. Behav.* **2006**, *88*, 244–248. [CrossRef]

217. Stranahan, A.M.; Hao, S.; Dey, A.; Yu, X.; Baban, B. Blood-brain barrier breakdown promotes macrophage infiltration and cognitive impairment in leptin receptor-deficient mice. *J. Cereb. Blood Flow Metab.* **2016**, *36*, 2108–2121. [CrossRef]

218. Fujihara, R.; Chiba, Y.; Nakagawa, T.; Nishi, N.; Murakami, R.; Matsumoto, K.; Kawauchi, M.; Yamamoto, T.; Ueno, M. Albumin microvascular leakage in brains with diabetes mellitus. *Microsc. Res. Tech.* **2016**, *79*, 833–837. [CrossRef]

219. Xu, Z.; Zeng, W.; Sun, J.; Chen, W.; Zhang, R.; Yang, Z.; Yao, Z.; Wang, L.; Song, L.; Chen, Y.; et al. The quantification of blood-brain barrier disruption using dynamic contrast-enhanced magnetic resonance imaging in aging rhesus monkeys with spontaneous type 2 diabetes mellitus. *Neuroimage* **2017**, *158*, 480–487. [CrossRef] [PubMed]

220. Salameh, T.S.; Mortell, W.G.; Logsdon, A.F.; Butterfield, D.A.; Banks, W.A. Disruption of the hippocampal and hypothalamic blood-brain barrier in a diet-induced obese model of type II diabetes: Prevention and treatment by the mitochondrial carbonic anhydrase inhibitor, topiramate. *Fluids Barriers CNS* **2019**, *16*, 1. [CrossRef] [PubMed]

221. Arnold, S.E.; Arvanitakis, Z.; Macauley-Rambach, S.L.; Koenig, A.M.; Wang, H.Y.; Ahima, R.S.; Craft, S.; Gandy, S.; Buettner, C.; Stoeckel, L.E.; et al. Brain insulin resistance in type 2 diabetes and Alzheimer disease: Concepts and conundrums. *Nat. Rev. Neurol.* **2018**, *14*, 168–181. [CrossRef] [PubMed]

222. Chen, F.; Dong, R.R.; Zhong, K.L.; Ghosh, A.; Tang, S.S.; Long, Y.; Hu, M.; Miao, M.X.; Liao, J.M.; Sun, H.B.; et al. Antidiabetic drugs restore abnormal transport of amyloid-beta across the blood-brain barrier and memory impairment in db/db mice. *Neuropharmacology* **2016**, *101*, 123–136. [CrossRef] [PubMed]

Int. J. Mol. Sci. **2019**, *20*, 1632

223. Tucsek, Z.; Toth, P.; Sosnowska, D.; Gautam, T.; Mitschelen, M.; Koller, A.; Szalai, G.; Sonntag, W.E.; Ungvari, Z.; Csiszar, A. Obesity in aging exacerbates blood-brain barrier disruption, neuroinflammation, and oxidative stress in the mouse hippocampus: Effects on expression of genes involved in beta-amyloid generation and Alzheimer's disease. *J. Gerontol. A Biol. Sci. Med. Sci.* **2014**, *69*, 1212–1226. [CrossRef] [PubMed]

224. Goldwaser, E.L.; Acharya, N.K.; Sarkar, A.; Godsey, G.; Nagele, R.G. Breakdown of the Cerebrovasculature and Blood-Brain Barrier: A Mechanistic Link Between Diabetes Mellitus and Alzheimer's Disease. *J. Alzheimers Dis.* **2016**, *54*, 445–456. [CrossRef] [PubMed]

International Journal of
Molecular Sciences

MDPI

Communication

Expression of Longevity Genes Induced by a Low-Dose Fluvastatin and Valsartan Combination with the Potential to Prevent/Treat "Aging-Related Disorders"

Miodrag Janić [1] , Mojca Lunder [2], Srdjan Novaković [3], Petra Škerl [3] and Mišo Šabovič [1,*]

[1]   Department of Vascular Diseases, Ljubljana University Medical Centre, Zaloška cesta 7,
      SI-1000 Ljubljana, Slovenia; miodrag.janic@kclj.si
[2]   Department of Endocrinology, Diabetes and Metabolic Diseases, Ljubljana University Medical Centre,
      Zaloška cesta 7, SI-1000 Ljubljana, Slovenia; mojca.lunder@kclj.si
[3]   Department of Molecular Diagnostics, Institute of Oncology Ljubljana, SI-1000 Ljubljana, Slovenia;
      snovakovic@onko-i.si (S.N.); pskerl@onko-i.si (P.Š.)
*   Correspondence: miso.sabovic@kclj.si; Tel.: +386-15228032; Fax: +386-15228070

Received: 13 March 2019; Accepted: 12 April 2019; Published: 14 April 2019

Abstract: The incidence of aging-related disorders may be decreased through strategies influencing the expression of longevity genes. Although numerous approaches have been suggested, no effective, safe, and easily applicable approach is yet available. Efficacy of low-dose fluvastatin and valsartan, separately or in combination, on the expression of the longevity genes in middle-aged males, was assessed. Stored blood samples from 130 apparently healthy middle-aged males treated with fluvastatin (10 mg daily), valsartan (20 mg daily), fluvastatin-valsartan combination (10 and 20 mg, respectively), and placebo (control) were analyzed. They were taken before and after 30 days of treatment and, additionally, five months after treatment discontinuation. The expression of the following longevity genes was assessed: *SIRT1*, *PRKAA*, *KLOTHO*, *NFE2L2*, *mTOR*, and *NF-κB*. Treatment with fluvastatin and valsartan in combination significantly increased the expression of *SIRT1* (1.8-fold; $p < 0.0001$), *PRKAA* (1.5-fold; $p = 0.262$) and *KLOTHO* (1.7-fold; $p < 0.0001$), but not *NFE2L2*, *mTOR* and *NF-κB*. Both fluvastatin and valsartan alone significantly, but to a lesser extent, increased the expression of *SIRT1*, and did not influence the expression of other genes. Five months after treatment discontinuation, genes expression decreased to the basal levels. In addition, analysis with previously obtained results revealed significant correlation between *SIRT1* and both increased telomerase activity and improved arterial wall characteristics. We showed that low-dose fluvastatin and valsartan, separately and in combination, substantially increase expression of *SIRT1*, *PRKAA*, and *KLOTHO* genes, which may be attributed to their so far unreported pleiotropic beneficial effects. This approach could be used for prevention of ageing (and longevity genes)–related disorders.

Keywords: aging-related disorders; longevity genes; arterial aging; low-dose fluvastatin and valsartan combination

---

1. Introduction

An aging population, along with increased life expectancy and the prevalence of associated chronic diseases, has become an important medical and economic issue. Consequently, the burden of so-called "aging-related disorders," such as associated cardiovascular diseases, degenerative diseases of the central nervous system, and malignant diseases, is also increasing [1,2]. These diseases represent one of the leading problems of the healthcare systems in developed countries around the globe. Effective strategies for their prevention are therefore needed. Aging-related disorders have been found to be causally associated with the altered expression of aging-related or so-called longevity genes [3]. In addition, telomere length per se [4], as well as telomerase expression [5], are also associated with aging-related disorders. In a narrower context, it seems logical that these genes represent mechanistic intracellular targets that could be altered to change the intracellular pro-aging milieu. It is expected that with the induction of expression of protective and suppression of expression of harmful genes, a new rejuvenated cellular phenotype could be reached that could influence the occurrence and course of aging-related disorders. In summary, so-called rejuvenating strategies, focused on modifying such genes' expression, could possibly importantly impact the prevalence of aging-related disorders [3].

In our prior studies, we explored the functional and structural characteristics of the arterial wall, some of which are also characteristic for arterial aging, and were particularly interested in the improvement of altered characteristics of the arterial wall by low doses of fluvastatin and valsartan [6]. We found significant improvement of arterial wall characteristics after 30 days of treatment in middle-aged males, with the beneficial effect slowly declining within nine months of treatment discontinuation [7–9]. Importantly, for age-related changes, the improvement of arterial wall characteristics was associated with increased telomerase expression and reduced inflammation as well as oxidative stress parameters [10,11]. Both decreased telomerase expression along with decreased telomeres' length and increased activation of inflammation and oxidative stress are characteristic of aging, arterial aging, and aging-related disorders. Therefore, we hypothesized that "anti-aging" or "rejuvenation" effects could be achieved through treatment impacting the longevity genes. This impact should have the capacity to influence "aging-related" disorders. Considering our previous results, this approach could be particularly effective for decreasing aging-related changes of the arterial system.

In the present study, we explored the efficacy of an approach consisting of (short-term) treatment with low-dose fluvastatin and valsartan alone or in combination on the expression of longevity genes in middle-aged males who had already (sub-clinically) impaired functional and structural arterial wall characteristics which could at least partially be attributed to the initial processes of aging.

2. Results

2.1. Expression of Longevity Genes

We analyzed the expression of longevity genes in the treatment and control groups. At the beginning of the study, there was no difference in longevity genes expression in the four different groups (low-dose fluvastatin, low-dose valsartan, low-dose fluvastatin and valsartan combination, and the control group). Differences between the groups were only observed after 30 days of treatment. Five months after treatment discontinuation, the expression of longevity genes in all treatment groups decreased almost to initial values. In the control group, the expression of longevity genes did not change during the study (Figure 1A–F).

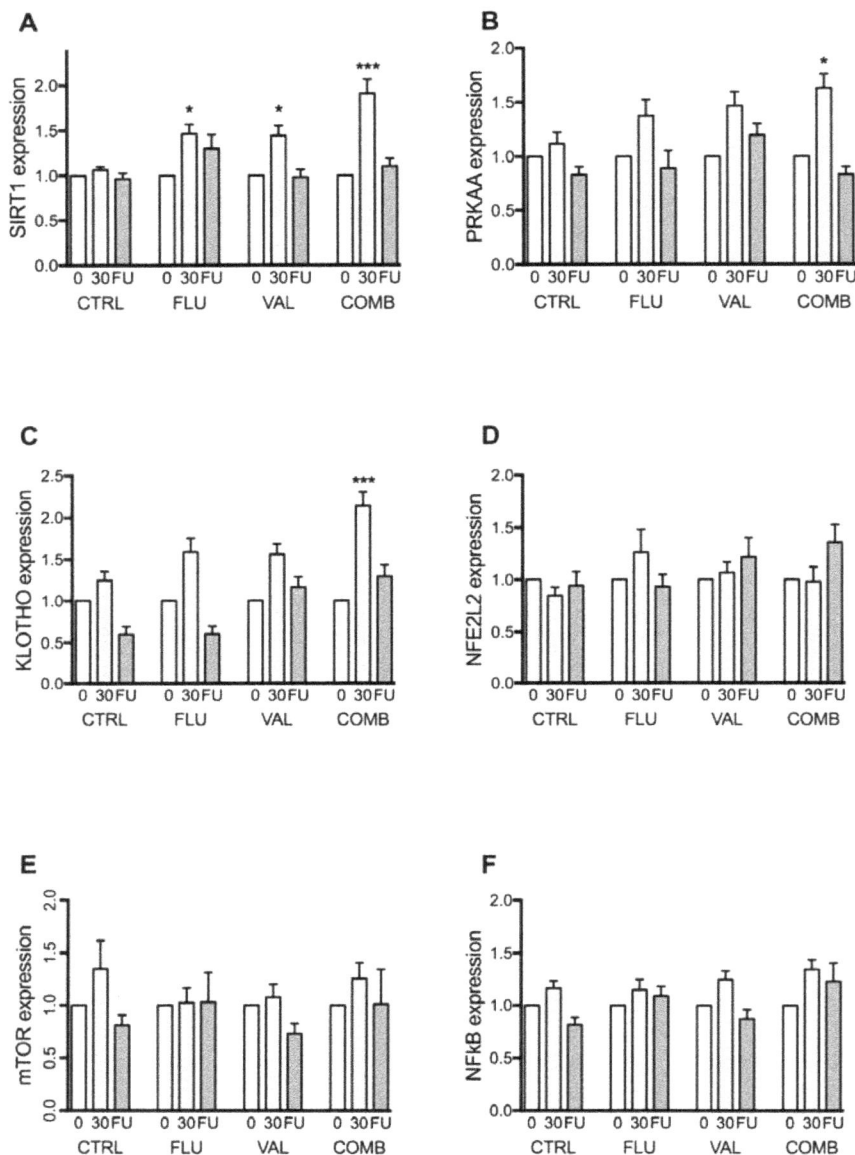

Figure 1. The expression of longevity genes: (**A**) *SIRT1*, (**B**) *PRKAA*, (**C**) *KLOTHO*, (**D**) *NFE2L2*, (**E**) *mTOR*, and (**F**) *NF-κB* in placebo (CTRL), low-dose fluvastatin (FLU), low-dose valsartan (VAL), or low-dose fluvastatin and valsartan combination (COMB). "0" represents the time before treatment, "30" represents the end of treatment, i.e., after 30 days (30) and "FU" represents five months after treatment discontinuation. Values are presented as means ± SEM. Presented p values are after Benjamini-Hochberg false discovery rate (FDR) correction, significance threshold set at $p < 0.05$. * signifies $p < 0.05$ and *** $p < 0.001$, vs. control group. *SIRT1*—sirtuin 1 gene; *PRKAA*—5'-AMP-activated protein kinase catalytic subunit α-2 gene; *NFE2L2*—nuclear factor (erythroid-derived 2)-like 2 gene; *mTOR*—mechanistic target of rapamycin (mTOR) gene; *NF-κB1*—nuclear factor κB gene.

2.1.1. Sirtuin 1 (*SIRT1*) Gene Expression

Both low-dose fluvastatin and valsartan separately increased the expression of the *SIRT1* gene after 30 days of treatment up to 1.4-fold ($p = 0.0165$ and $p = 0.0229$, respectively), while their low-dose combination increased its expression up to 1.8-fold ($p < 0.0001$) compared to the control group. Five months after treatment discontinuation, no significant effects of the treatment on the *SIRT1* gene expression were observed (Figure 1A).

2.1.2. 5′-AMP-Activated Protein Kinase Catalytic Subunit α-2 (*PRKAA*) Gene Expression

After 30 days of treatment, only the low-dose fluvastatin and valsartan combination significantly increased the expression of the *PRKAA* gene, up to 1.5-fold ($p = 0.0262$) compared to the control group. Separate drugs had no influence on the *PRKAA* gene expression (Figure 1B).

2.1.3. *KLOTHO* Gene Expression

After 30 days of treatment, only the low-dose fluvastatin and valsartan combination significantly increased the expression of the *KLOTHO* gene, up to 1.7-fold ($p < 0.0001$) compared to the control group. Separate drugs had no influence on the *KLOTHO* gene expression (Figure 1C).

2.1.4. Nuclear Factor (Erythroid-Derived 2)-Like 2 Gene Expression (*NFE2L2*)

No significant changes of the *NFE2L2* gene expression were observed in any of the study groups (Figure 1D).

2.1.5. Mechanistic Target of Rapamycin (*mTOR*) Gene Expression

No significant changes were observed in the *mTOR* gene expression in any of the study groups (Figure 1E).

2.1.6. Nuclear Factor κB (*NF-κB*) Gene Expression

No statistically significant changes in the expression of the *NF-κB* gene were observed either (Figure 1F).

2.2. *Correlations between the Expression of Longevity Genes, Telomerase Activity and Arterial Wall Properties*

The correlations between longevity genes expression and previously described telomerase activity and arterial wall properties [7–10], all measured after 30 days of treatment, were calculated. In the separate, low-dose fluvastatin and valsartan groups, the expression of the *SIRT1* gene positively correlated with telomerase activity ($r = 0.42$; $p = 0.04$ and $r = 0.39$; $p = 0.03$, respectively). Importantly, in the low-dose combination group, the expression of the *SIRT1* gene positively correlated with telomerase activity ($r = 0.62$; $p = 0.01$) and with brachial artery flow-mediated dilation (FMD) ($r = 0.52$; $p = 0.05$) while negatively correlated with carotid artery beta stiffness ($r = -0.45$; $p = 0.02$) and pulse wave velocity (PWV) ($r = -0.56$; $p = 0.05$).

3. Discussion

In the present study, we showed that low-dose fluvastatin and valsartan in combination significantly increased the expression of several important longevity genes (*SIRT1*, *PRKAA*, *KLOTHO*). Moreover, these changes correlated with an improvement of functional and structural arterial wall characteristics as well as with telomerase activity, both assessed previously [7–10]. Overall, the results revealed increased expression of several longevity genes that seems to be causally associated with increased telomerase activity and improvement of arterial wall (initial aging-related) characteristics. The results are very promising and indicate that our relatively simple but innovative approach could have potential efficacy as a "rejuvenating agent," particularly in the efforts to decrease the occurrence of aging-related disorders.

The present study was designed as a two-part study: The first part comprised the measurement of longevity genes in relation to treatment with low-dose fluvastatin and valsartan, and the second part comprised of a correlation analysis with previously measured relevant parameters (telomerase activity and functional/structural arterial wall characteristics). The studied population was a group of middle-aged males with already present aging-related changes. Since we found that a low-dose combination of fluvastatin and valsartan improved arterial wall characteristics, we aimed to further explore the mechanism behind these beneficial effects. Thus, from the previously obtained samples, we assessed the expression of longevity genes to explore the potential rejuvenating effect of our approach. We found that the low-dose fluvastatin and valsartan combination increased the expression of the *SIRT1* (1.8-fold; $p < 0.0001$), *PRKAA* (1.5-fold; $p = 0.0262$) and *KLOTHO* (1.7-fold; $p < 0.0001$) genes after 30 days, whereas no differences in the expression of the *NFE2L2*, *mTOR*, and *NF-κB* genes were observed. Fluvastatin or valsartan alone were less effective, increasing only the expression of the *SIRT1* gene to a lesser extent. Moreover, the expression of the *SIRT1* gene in the combination group positively correlated with telomerase activity and improvement of the arterial wall characteristics.

Importantly, the FDA recently approved the first interventional "anti-aging" study (MILES—Metformin in Longevity Study). Metformin, which, for this purpose, was repositioned from a solely antidiabetic drug to an "anti-aging" drug, is the interventional drug. This is based on a wealth of data indicating that metformin could influence the aging process, or more importantly, aging-related disorders. Interestingly, one of the most prominent of the several hypotheses underlying the "anti-aging" effects of metformin is the activation of longevity genes with consequent effects on energy metabolism, inflammation and oxidative stress. Several other currently ongoing studies with metformin are focusing on its "anti-aging" effects, such as VA-IMPACT, TAME and ePREDICE. In any event, a new period in which specific treatments of "aging-related disorders" are being studied, has already begun.

Sirtuins are a family of nicotinamide adenine dinucleotide (NAD)-dependent deacetylases, and according to some studies, are one of the key molecules involved in the regulation of aging and aging-related disorders [12]. *SIRT1* regulates DNA transcription and repair as well as cell survival, thus also inducing longevity. It was shown to have an important role in aging-related disorders of the cardiovascular [5] and nervous systems [13]. Some statins in therapeutic doses were shown to induce *SIRT1* expression, for example, simvastatin in endothelial progenitor cells [14]. On the other hand, atorvastatin and rosuvastatin reduced its expression in patients with coronary artery disease [15]. Until now, the effect of fluvastatin in therapeutic or low doses on *SIRT1* expression has not been assessed. To the best of our knowledge and according to the literature, no studies have assessed the effect of valsartan on the *SIRT1* gene in humans, either. A few studies were performed on rats or mice, in which valsartan and other sartans increased the expression of the *SIRT1* gene [16–18]. The *PRKAA* gene encodes the catalytic subunit of the AMPK. AMPK is the primary regulator of cellular responses and acts as a sensor to maintain energy balance within the cell [19]. In animal and cell culture studies, several statins were shown to induce the AMPK and eNOS pathway, acting in a vasoprotective manner [20–22]. Valsartan was also shown to act protectively through activation of the AMPK pathway in diabetic rats [23]; similar effects have been shown for telmisartan in human coronary artery cells [24]. The consequences of AMPK activations were divergent: In acute activation, it caused cell protection, whereas in chronic activation, it might have activated the pro-aging pathways and progressive degeneration during cellular senescence. There are various interactions between sirtuin and AMPK pathways [19]. The expression of *KLOTHO* decreased in aging-related disorders [25]. Valsartan in therapeutic doses increased the amount of plasma-soluble *KLOTHO* and consequently induced cardiorenal beneficial effects in patients with diabetes mellitus and diabetic kidney disease [26]. The effects of valsartan in low doses have not yet been studied in such a setting. The potential beneficial effects of statins on *KLOTHO* expression was only shown in animal studies [27]. The *NFE2L2* gene encodes a transcription factor, which regulates the proteins involved in responding to injury and inflammation. According to some studies, enhancing *NFE2L2* activity may be beneficial in diabetic cardiomyopathy, mitochondrial dysfunction, and as an anti-aging agent, but further studies are needed [28]. Fluvastatin was shown to induce *NFE2L2* in the vascular smooth muscle cells [29,30]. Mammalian target of Rapamycin (mTOR) was shown to have an important role in cardiovascular diseases, oxidative stress and longevity [31]. In animal or cell studies, statins influenced the *mTOR*, which was proven for fluvastatin in rats [32] and for lovastatin in vascular smooth muscle cells [33]. In one study on rats, valsartan induced cardio protection against ischemic-reperfusion injury through the mTOR [34]. *NF-κB* gene expression mediated vascular and myocardial inflammation and was additionally associated with impaired endothelial function [35]. There is evidence that both statins and sartans could reduce the *NF-κB* gene activation in various animal and cell line models [36,37].

To the best of our knowledge, no study like the present one, has been published. In our review of the literature, we found several different studies that assessed the effects of statins or sartans on longevity genes, but most of those studies were performed either on cell lines or animals. Therefore, the present study is the first to show that our new preventive cardiovascular approach, which was proven to induce the improvement of arterial wall functional and structural characteristics, and consequently decrease arterial age, additionally acted through the expression of several longevity genes. This could be one of the mechanisms lying behind the beneficial effects observed in our previous clinical studies [7–9]. Nevertheless, one of the limitations of the present study is that we used only qPCR, but validation by Western blotting would be of added value.

The results of the present study indicate that our innovative approach using short-term low-dose fluvastatin and valsartan has a potential in inducing the expression of certain longevity genes. These effects could be anti-aging or rejuvenating as well as act as a potential specific prevention/treatment for "aging-related" disorders. It can be speculated that using cycling, intermittent treatment with low-dose combination (every 6–12 years) starting at middle-age could postpone the occurrence of aging-related disorders. On the other hand, this approach could be used in the same population and with the same aim as metformin in the MILES trial. One of the major advantages of this approach is its cyclic, intermittent character, which, as previously described, could potentially activate the beneficial longevity genes for a time short enough not activate their contra regulatory mechanisms as well. With intermittently repeating cycles, this could lead to repetitive beneficial activations of the protective longevity genes. Thus, it could be speculated that the cumulative effect of these repeating cycles of treatment might eventually lead to successful specific prevention/treatment of "aging-related" disorders, most likely cardiovascular "aging-related" disorders.

4. Materials and Methods

4.1. Participants and Study Design

The stored blood samples from our three prior studies were used together for the present longevity gene expression study. Overall, 130 middle-aged, apparently healthy male participants were recruited and treated for one month (30 days): 25 persons with fluvastatin 10 mg daily, 20 persons with valsartan 20 mg daily, and 20 persons with a low-dose combination of fluvastatin (10 mg daily) and valsartan (20 mg daily). Accordingly, 65 participants received placebo. All the participants were blindly randomized into the relevant group, as in our previous studies, which are described in more detail elsewhere [7–9]. Blood samples were collected and ultrasound measurements of the arterial wall properties (endothelial function, arterial stiffness) were performed at the beginning and at the end of the treatment period (day 0 and day 30). The measurements were also repeated five months after treatment discontinuation.

The National Medical Ethics Committee of Slovenia approved the studies (Approval date 3 July 2009, Approval No.: 21k/05/09) and informed consent was obtained from all participants. Inclusion criteria were: age between 30 and 50 years, non-smoking status, normal blood pressure values, body mass index values below 30 kg/m^2, no clinical cardiovascular disease, no history of any other chronic disease, and no regular medication therapy. The characteristics of the subjects were already extensively described in our previous publications [7–9].

4.2. Blood Sampling

Three samples of whole peripheral blood were collected from each participant: before treatment (day 0), after treatment conclusion (day 30), and five months after treatment discontinuation (follow-up). The whole blood samples were collected in 10 mL EDTA tubes and stored at −80 °C. Prior to RNA extraction, samples were centrifuged at 4000 rpm for 25 min to obtain the pellet of cells and cell debris. The pellets were then used for RNA extraction.

4.3. RNA Isolation

Total RNA was isolated using a miRNeasy Mini kit (Qiagen, Hilden, Germany) according to the manufacturer's instructions. RNA was quantified using the NanoDrop, and cDNA was synthesized from 300 ng of the total RNA using the High-Capacity cDNA Reverse Transcription Kit with RNase Inhibitor (Applied Biosystems, Foster City, CA, USA) according to the manufacturer's protocol.

4.4. Quantitative Real-Time PCR (qPCR) for Human Telomerase Reverse Transcriptase (hTERT) Expression

The expression of target genes in the tested samples was performed using TaqMan Gene Expression Assays (Applied Biosystems) according to the manufacturer's instructions: Assay Hs00183100_m1 for the *KLOTHO* gene; assay Hs01009006_m1 for the Sirtuin 1 (*SIRT1*) gene; assay Hs01562315_m1 for the 5′AMP-activated protein kinase (*AMPK*) gene; assay Hs00765730_m1 for the Nuclear factor κB (*NF-κB1*) gene; assay Hs00234522_m1 for the Mechanistic target of rapamycin (*mTOR*) gene and assay Hs00975961_g1 for the Nuclear factor (erythroid-derived 2)-like 2 (*NFE2L2*) gene. The housekeeping gene glyceraldehyde 3-phoshate dehydrogenase (*GAPDH*) was used as an endogenous control. Described briefly, qPCR was performed using the ABI 7900 instrument (Applied Biosystems). Individual qPCR reactions were carried out in 10 μL reaction mix with 2xTaqMan Universal PCR Master Mix (Applied Biosystems), 1× TaqMan Gene Expression Assay (Applied Biosystems) and 200 ng cDNA. Each sample was analyzed in triplicate. RNA isolated from healthy volunteers ($n = 5$) was used as a positive control for target genes expression. In each run, the dilutions of control RNA (pool of RNA from healthy volunteers) was included. The data were analyzed by the SDS2.4 software and Ct values were extracted. Fold-differences in *hTERT* expression were calculated using the comparative Ct method as described previously [38], where data were normalized to day 0 for each participant.

4.5. Data Analysis

All values were expressed as means ± SEM. Differences between values were assessed by one-way analysis of variance (ANOVA). When significant interaction was present, the Bonferroni post-test was performed. Benjamini-Hochberg's correction method was used to control false discovery rate (FDR), with significance threshold set at $p < 0.05$. Correlations between arterial wall properties and telomerase activity that were described previously [7–9,11] and longevity genes expression assessed in the present study were calculated after 30 days treatment period using Pearson correlation coefficients. A $p < 0.05$ was considered significant. All statistical analyses were performed using the Graph Pad Prism 5.0 software.

5. Conclusions

In conclusion, the present study has shown that low-dose fluvastatin and valsartan treatment increased the expression of beneficial longevity genes (*SIRT1*, *PRKAA*, and *KLOTHO*) and could therefore represent a promising new treatment approach for "aging-related" disorders. Additional, population-based research is needed to additionally prove the proposed concept.

Author Contributions: Conceptualization, M.Š.; Data curation, M.J., M.L. and M.Š.; Formal analysis, M.J., M.L. and M.Š.; Investigation, M.J., M.L., S.N. and P.Š.; Methodology, S.N. and P. Š.; Supervision, M.Š.; Writing—original draft, M.J., M.L. and M.Š.; Writing—review & editing, M.J., M.L. and M.Š.

Funding: This research received no external funding.

Conflicts of Interest: The authors declare no conflict of interest.

Abbreviations

AMPK	AMP-activated protein kinase
ANOVA	Analysis of variance
COMB	Combination group
CTRL	Placebo group
FDA	Food and Drug Administration
FLU	Fluvastatin group
FMD	Flow mediated dilation
hTERT	human telomerase reverse transcriptase
MILES	Metformin in Longevity Study
mTOR	Mechanistic target of rapamycin
NAD	Nicotinamide adenine dinucleotide
NF-κB1	Nuclear factor κB
NFE2L2	Nuclear factor (erythroid-derived 2)-like 2 gene
PRKAA	5′-AMP-activated protein kinase catalytic subunit α-2 gene
PWV	Pulse wave velocity
qPCR	Quantitative real-time polymerase chain reaction
SIRT1	Sirtuin 1 gene
VAL	Valsartan group

References

1. Stevens, W.; Peneva, D.; Li, J.Z.; Liu, L.Z.; Liu, G.; Gao, R.; Lakdawalla, D.N. Estimating the future burden of cardiovascular disease and the value of lipid and blood pressure control therapies in China. *BMC Health Serv. Res.* **2016**, *16*, 175. [CrossRef] [PubMed]
2. Global Burden of Cardiovascular Diseases Collaboration; Roth, G.A.; Johnson, C.O.; Abate, K.H.; Abd-Allah, F.; Ahmed, M.; Alam, K.; Alam, T.; Alvis-Guzman, N.; Ansari, H.; et al. The Burden of Cardiovascular Diseases Among US States, 1990–2016. *JAMA Cardiol.* **2018**. [CrossRef] [PubMed]
3. Van Raamsdonk, J.M. Mechanisms underlying longevity: A genetic switch model of aging. *Exp. Gerontol.* **2017**. [CrossRef]
4. Rizvi, S.; Raza, S.T.; Mahdi, F. Telomere length variations in aging and age-related diseases. *Curr. Aging Sci.* **2014**, *7*, 161–167. [CrossRef]
5. Yeh, J.K.; Wang, C.Y. Telomeres and Telomerase in Cardiovascular Diseases. *Genes* **2016**, *7*, 58. [CrossRef] [PubMed]
6. Janic, M.; Lunder, M.; Sabovic, M. A new anti-ageing strategy focused on prevention of arterial ageing in the middle-aged population. *Med. Hypotheses* **2013**, *80*, 837–840. [CrossRef] [PubMed]
7. Lunder, M.; Janic, M.; Habjan, S.; Sabovic, M. Subtherapeutic, low-dose fluvastatin improves functional and morphological arterial wall properties in apparently healthy, middle-aged males—A pilot study. *Atherosclerosis* **2011**, *215*, 446–451. [CrossRef] [PubMed]
8. Lunder, M.; Janic, M.; Sabovic, M. Reduction of age-associated arterial wall changes by low-dose valsartan. *Eur. J. Prev. Cardiol.* **2012**, *19*, 1243–1249. [CrossRef]
9. Lunder, M.; Janic, M.; Jug, B.; Sabovic, M. The effects of low-dose fluvastatin and valsartan combination on arterial function: A randomized clinical trial. *Eur. J. Int. Med.* **2012**, *23*, 261–266. [CrossRef]
10. Janic, M.; Lunder, M.; Cerkovnik, P.; Prosenc Zmrzljak, U.; Novakovic, S.; Sabovic, M. Low-Dose Fluvastatin and Valsartan Rejuvenate the Arterial Wall Through Telomerase Activity Increase in Middle-Aged Men. *Rejuvenation Res.* **2016**, *19*, 115–119. [CrossRef] [PubMed]

11. Janic, M.; Lunder, M.; Prezelj, M.; Sabovic, M. A combination of low-dose fluvastatin and valsartan decreases inflammation and oxidative stress in apparently healthy middle-aged males. *J. Cardiopulm. Rehabil. Prev.* **2014**, *34*, 208–212. [CrossRef] [PubMed]

12. Park, S.; Mori, R.; Shimokawa, I. Do sirtuins promote mammalian longevity? A critical review on its relevance to the longevity effect induced by calorie restriction. *Mol. Cells* **2013**, *35*, 474–480. [CrossRef] [PubMed]

13. Braidy, N.; Jayasena, T.; Poljak, A.; Sachdev, P.S. Sirtuins in cognitive ageing and Alzheimer's disease. *Curr. Opin. Psychiatry* **2012**, *25*, 226–230. [CrossRef] [PubMed]

14. Du, G.; Song, Y.; Zhang, T.; Ma, L.; Bian, N.; Chen, X.; Feng, J.; Chang, Q.; Li, Z. Simvastatin attenuates TNFα-induced apoptosis in endothelial progenitor cells via the upregulation of *SIRT1. Int. J. Mol. Med.* **2014**, *34*, 177–182. [CrossRef] [PubMed]

15. Kilic, U.; Gok, O.; Elibol-Can, B.; Uysal, O.; Bacaksiz, A. Efficacy of statins on sirtuin 1 and endothelial nitric oxide synthase expression: The role of sirtuin 1 gene variants in human coronary atherosclerosis. *Clin. Exp. Pharmacol. Physiol.* **2015**, *42*, 321–330. [CrossRef] [PubMed]

16. Ye, Y.; Qian, J.; Castillo, A.C.; Perez-Polo, J.R.; Birnbaum, Y. Aliskiren and Valsartan reduce myocardial AT1 receptor expression and limit myocardial infarct size in diabetic mice. *Cardiovasc. Drugs Ther.* **2011**, *25*, 505–515. [CrossRef]

17. Shiota, A.; Shimabukuro, M.; Fukuda, D.; Soeki, T.; Sato, H.; Uematsu, E.; Hirata, Y.; Kurobe, H.; Maeda, N.; Sakaue, H.; et al. Telmisartan ameliorates insulin sensitivity by activating the *AMPK/SIRT1* pathway in skeletal muscle of obese db/db mice. *Cardiovasc. Diabetol.* **2012**, *11*, 139. [CrossRef] [PubMed]

18. Pantazi, E.; Bejaoui, M.; Zaouali, M.A.; Folch-Puy, E.; Pinto Rolo, A.; Panisello, A.; Palmeira, C.M.; Rosello-Catafau, J. Losartan activates sirtuin 1 in rat reduced-size orthotopic liver transplantation. *World J. Gastroenterol.* **2015**, *21*, 8021–8031. [CrossRef]

19. Wang, Y.; Liang, Y.; Vanhoutte, P.M. *SIRT1* and *AMPK* in regulating mammalian senescence: A critical review and a working model. *FEBS Lett.* **2011**, *585*, 986–994. [CrossRef] [PubMed]

20. Jia, F.; Wu, C.; Chen, Z.; Lu, G. Atorvastatin inhibits homocysteine-induced endoplasmic reticulum stress through activation of AMP-activated protein kinase. *Cardiovasc. Ther.* **2012**, *30*, 317–325. [CrossRef]

21. Sun, W.; Lee, T.S.; Zhu, M.; Gu, C.; Wang, Y.; Zhu, Y.; Shyy, J.Y. Statins activate AMP-activated protein kinase in vitro and in vivo. *Circulation* **2006**, *114*, 2655–2662. [CrossRef] [PubMed]

22. Hermida, N.; Markl, A.; Hamelet, J.; Van Assche, T.; Vanderper, A.; Herijgers, P.; van Bilsen, M.; Hilfiker-Kleiner, D.; Noppe, G.; Beauloye, C.; et al. HMGCoA reductase inhibition reverses myocardial fibrosis and diastolic dysfunction through AMP-activated protein kinase activation in a mouse model of metabolic syndrome. *Cardiovasc. Res.* **2013**, *99*, 44–54. [CrossRef] [PubMed]

23. Ha, Y.M.; Park, E.J.; Kang, Y.J.; Park, S.W.; Kim, H.J.; Chang, K.C. Valsartan independent of AT(1) receptor inhibits tissue factor, TLR-2 and -4 expression by regulation of Egr-1 through activation of AMPK in diabetic conditions. *J. Cell. Mol. Med.* **2014**, *18*, 2031–2043. [CrossRef]

24. Kurokawa, H.; Sugiyama, S.; Nozaki, T.; Sugamura, K.; Toyama, K.; Matsubara, J.; Fujisue, K.; Ohba, K.; Maeda, H.; Konishi, M.; et al. Telmisartan enhances mitochondrial activity and alters cellular functions in human coronary artery endothelial cells via AMP-activated protein kinase pathway. *Atherosclerosis* **2015**, *239*, 375–385. [CrossRef] [PubMed]

25. Donate-Correa, J.; Martin-Nunez, E.; Mora-Fernandez, C.; Muros-de-Fuentes, M.; Perez-Delgado, N.; Navarro-Gonzalez, J.F. Klotho in cardiovascular disease: Current and future perspectives. *World J. Biol. Chem.* **2015**, *6*, 351–357. [CrossRef] [PubMed]

26. Karalliedde, J.; Maltese, G.; Hill, B.; Viberti, G.; Gnudi, L. Effect of renin-angiotensin system blockade on soluble Klotho in patients with type 2 diabetes, systolic hypertension, and albuminuria. *Clin. J. Am. Soc. Nephrol.* **2013**, *8*, 1899–1905. [CrossRef] [PubMed]

27. Yoon, H.E.; Lim, S.W.; Piao, S.G.; Song, J.H.; Kim, J.; Yang, C.W. Statin upregulates the expression of *KLOTHO*, an anti-aging gene, in experimental cyclosporine nephropathy. *Nephron. Exp. Nephrol.* **2012**, *120*, e123–e133. [CrossRef] [PubMed]

28. Satta, S.; Mahmoud, A.M.; Wilkinson, F.L.; Yvonne Alexander, M.; White, S.J. The Role of Nrf2 in Cardiovascular Function and Disease. *Oxid. Med. Cell. Longev.* **2017**, *2017*, 9237263. [CrossRef]

29. Hwang, A.R.; Han, J.H.; Lim, J.H.; Kang, Y.J.; Woo, C.H. Fluvastatin inhibits AGE-induced cell proliferation and migration via an ERK5-dependent Nrf2 pathway in vascular smooth muscle cells. *PLoS ONE* **2017**, *12*, e0178278. [CrossRef]

30. Makabe, S.; Takahashi, Y.; Watanabe, H.; Murakami, M.; Ohba, T.; Ito, H. Fluvastatin protects vascular smooth muscle cells against oxidative stress through the Nrf2-dependent antioxidant pathway. *Atherosclerosis* **2010**, *213*, 377–384. [CrossRef]

31. Sciarretta, S.; Volpe, M.; Sadoshima, J. Mammalian target of rapamycin signaling in cardiac physiology and disease. *Circ. Res.* **2014**, *114*, 549–564. [CrossRef] [PubMed]

32. Cui, X.; Long, C.; Zhu, J.; Tian, J. Protective Effects of Fluvastatin on Reproductive Function in Obese Male Rats Induced by High-Fat Diet through Enhanced Signaling of mTOR. *Cell. Physiol. Biochem.* **2017**, *41*, 598–608. [CrossRef]

33. Wagner, R.J.; Martin, K.A.; Powell, R.J.; Rzucidlo, E.M. Lovastatin induces VSMC differentiation through inhibition of Rheb and mTOR. *Am. J. Physiol. Cell Physiol.* **2010**, *299*, C119–C127. [CrossRef] [PubMed]

34. Wu, X.; He, L.; Cai, Y.; Zhang, G.; He, Y.; Zhang, Z.; He, X.; He, Y.; Zhang, G.; Luo, J. Induction of autophagy contributes to the myocardial protection of valsartan against ischemiareperfusion injury. *Mol. Med. Rep.* **2013**, *8*, 1824–1830. [CrossRef] [PubMed]

35. Costantino, S.; Paneni, F.; Cosentino, F. Ageing, metabolism and cardiovascular disease. *J. Physiol.* **2016**, *594*, 2061–2073. [CrossRef] [PubMed]

36. Lenglet, S.; Quercioli, A.; Fabre, M.; Galan, K.; Pelli, G.; Nencioni, A.; Bauer, I.; Pende, A.; Python, M.; Bertolotto, M.; et al. Statin treatment is associated with reduction in serum levels of receptor activator of NF-κB ligand and neutrophil activation in patients with severe carotid stenosis. *Mediators Inflamm.* **2014**, *2014*, 720987. [CrossRef]

37. Wu, B.; Lin, R.; Dai, R.; Chen, C.; Wu, H.; Hong, M. Valsartan attenuates oxidative stress and NF-κB activation and reduces myocardial apoptosis after ischemia and reperfusion. *Eur. J. Pharmacol.* **2013**, *705*, 140–147. [CrossRef]

38. Livak, K.J.; Schmittgen, T.D. Analysis of relative gene expression data using real-time quantitative PCR and the $2^{-\Delta\Delta Ct}$ Method. *Methods* **2001**, *25*, 402–408. [CrossRef]

International Journal of
Molecular Sciences

MDPI

Review

Quiescence Entry, Maintenance, and Exit in Adult Stem Cells

Karamat Mohammad, Paméla Dakik, Younes Medkour, Darya Mitrofanova and Vladimir I. Titorenko *

Department of Biology, Concordia University, 7141 Sherbrooke Street, West, SP Building, Room 501-13, Montreal, QC H4B 1R6, Canada; karamat.mohammad@concordia.ca (K.M.); pameladakik@gmail.com (P.D.); writetoyounes@gmail.com (Y.M.); mitrofanova_darya@hotmail.com (D.M.)
* Correspondence: vladimir.titorenko@concordia.ca; Tel.: +1-514-848-2424 (ext. 3424)

Received: 12 March 2019; Accepted: 28 April 2019; Published: 1 May 2019

Abstract: Cells of unicellular and multicellular eukaryotes can respond to certain environmental cues by arresting the cell cycle and entering a reversible state of quiescence. Quiescent cells do not divide, but can re-enter the cell cycle and resume proliferation if exposed to some signals from the environment. Quiescent cells in mammals and humans include adult stem cells. These cells exhibit improved stress resistance and enhanced survival ability. In response to certain extrinsic signals, adult stem cells can self-renew by dividing asymmetrically. Such asymmetric divisions not only allow the maintenance of a population of quiescent cells, but also yield daughter progenitor cells. A multistep process of the controlled proliferation of these progenitor cells leads to the formation of one or more types of fully differentiated cells. An age-related decline in the ability of adult stem cells to balance quiescence maintenance and regulated proliferation has been implicated in many aging-associated diseases. In this review, we describe many traits shared by different types of quiescent adult stem cells. We discuss how these traits contribute to the quiescence, self-renewal, and proliferation of adult stem cells. We examine the cell-intrinsic mechanisms that allow establishing and sustaining the characteristic traits of adult stem cells, thereby regulating quiescence entry, maintenance, and exit.

Keywords: cell cycle; cellular quiescence; mechanisms of quiescence maintenance; mechanisms of quiescence entry and exit; adult stem cells; metabolism; mitochondria; reactive oxygen species; cell signaling; proteostasis

---

1. Introduction

Cellular quiescence is a reversible state of a temporary cell cycle arrest that can be induced in both metazoans and unicellular eukaryotes as a response to some anti-mitogenic factors [1–4]. These factors include cell-nonautonomous, extrinsic environmental cues and cell-autonomous, intrinsic regulatory mechanisms [1–4]. In mammals, the temporary cell cycle arrest and quiescence entry occur before cells reach the growth factor-dependent "restriction" (R) point of the G_1 phase [5,6]. In the budding yeast *Saccharomyces cerevisiae*, the nutrient-dependent "START A" point at the G_1 phase of the cell cycle is believed to be evolutionarily related to the R point in mammals [6–8]. Notably, under certain conditions some unicellular and multicellular cells, eukaryotic organisms can undergo a temporary cell cycle arrest and enter the quiescent state not only from the G_1 phase of the cell cycle, but also from the S, G_2, or M phase [9–19]. Studies in budding yeast suggest that this is because the entry into quiescence is controlled not by (or not only by) the cell cycle regulation machinery, but by (or also by) the metabolic status of the cell at a certain cell cycle phase [17,18]. Once the cell cycle is arrested at the R or START A point, cells enter a reversible G_0 phase of the cell cycle and become quiescent. In budding yeast and mammals, this reversible G_0 state of quiescence is also called the "quiescence cycle" of cell oscillation between at least two functional states [1–4,20].

The entry of cells into the reversible G_0 state of quiescence prevents their entry into the irreversible G_0 state of senescence or the irreversible G_0 state of terminal differentiation [2–4]. Of note, some "irreversibly arrested" senescent or terminally differentiated cells retain an intact (although silenced) mechanism for cell cycle re-entry, as they can resume proliferation in response to certain cell-extrinsic and cell-intrinsic factors [21–25]. Quiescent cells in the reversible G_0 state do not divide, but rather retain the ability to re-enter the cell cycle and resume proliferation in response to certain pro-mitogenic factors, which include cell-extrinsic environmental signals and cell-intrinsic regulatory mechanisms [2,4]. Cellular quiescence is actively maintained by complex multiprotein networks and represents a collection of heterogeneous states in both multicellular and unicellular eukaryotes [2–4,26–32].

Populations of unicellular eukaryotic organisms (such as various yeast species) in the wild are always able to undergo a reversible switch between the states of cellular quiescence and proliferation; such a switch is controlled by nutrient availability and some other environmental factors [3,8,19,33–36]. Adult organisms in "lower" metazoan organisms (such as nematodes and fruit flies) and in "higher" metazoans (such as plants, mammals, and humans) contain several distinct types of quiescent cells; adult stem cells are among these quiescent cells in mammals and humans [2,37–44].

Quiescent adult stem cells in different mammalian tissues are long-lived [2,29,45,46]. This is because they can actively support their resistance to various stresses and toxicities [2,29,45,46]. This is also because, when stimulated, quiescent adult stem cells can often self-renew by dividing infrequently and asymmetrically to form a new quiescent stem cell and an actively dividing daughter progenitor cell; then, the daughter progenitor cell can advance through a hierarchically organized and tightly controlled series of events that yield one or more types of terminally differentiated cells [2,47–51]. Of note, quiescent adult stem cells can also sometime undergo two types of symmetric divisions, either a proliferation division (which yields two identical quiescent stem cells) or a differentiation division (which yields two differentiated cells) [49,52–60].

A body of evidence indicates that the abilities of quiescent adult stem cells to resist stresses, self-renew, and produce fully differentiated cells are crucial for tissue repair and regeneration, and are vital for the growth, development, and health of the adult body [2,19,48,61–64]. The number of quiescent adult stem cells and the efficiencies with which they resist stresses, self-renew, and produce fully differentiated cells declines with age [29,31,32,45,46,61,62,64–81]. Such an age-related numerical and functional decline of quiescent adult stem cells impairs their ability to balance quiescence with proliferation activity, and has been implicated in the pathophysiology of cancer and other aging-associated diseases in mammals and humans [19,29,31,32,45,46,48,61,62,64,66,67,69,70,72–77,79–84]. Some genetic, dietary, and pharmacological interventions can delay cellular and organismal aging and postpone the onset of aging-associated diseases by decelerating an age-related decline in the number and/or functionality of quiescent adult stem cells [2,19,29,31,32,45,46,48,61,62,64–84]. Aging-associated changes in the specialized cellular neighborhoods of adult stem cells, which are known as the stem cell niches, provide an essential contribution to an age-related decline in the abilities of adult stem cells to sustain quiescence, proliferation capacity, and differentiation potential [85–91]. This is because the stem cell niches in diverse tissues produce and release certain short-range molecular signals that are indispensable for maintaining the quiescence, self-renewal, proliferation capacity, differentiation potential, and functionality of neighboring adult stem cells [2,91–93]. Since the efficiencies with which the stem cell niches produce and release such transmissible molecular signals are either enhanced or weakened with age, the cell-nonautonomous mechanisms orchestrated by these niches critically contribute to the age-related weakening of various aspects of stem cell functionality [85–91].

Here, we examine the characteristic metabolic, signal transduction, gene expression, epigenetic, stress survival, and cell cycle regulation features of quiescent adult stem cells in mammals and humans. We explore cell-intrinsic mechanisms regulating quiescence entry, maintenance, and exit in these cells.

2. Common Traits of Quiescent Adult Stem Cells and Cell-Intrinsic Mechanisms That in These Cells Control Quiescence Entry, Maintenance, and Exit

Quiescent adult stem cells residing in different tissues share a discrete set of metabolic, signal transduction, gene expression, epigenetic, stress survival, and cell cycle regulation traits, all of which are distinct from those of the fully differentiated progeny of such cells [94–97]. Many of these common traits are actively maintained by quiescent adult stem cells and define their stress resistance, self-renewal potential, and regulated proliferation and differentiation routes; thus, adult stem cells have developed cell-autonomous, intrinsic regulatory mechanisms for quiescence entry, maintenance, and exit [28,94–99]. These traits and mechanisms are outlined below and schematically depicted in Figures 1–6.

2.1. Common Metabolic Traits of Quiescent Adult Stem Cells Define Their Fate

Quiescent adult stem cells metabolize glucose and other carbohydrates mainly through aerobic glycolysis that yields pyruvate; instead of being transported to mitochondria, converted to acetyl-CoA, and incorporated into the mitochondrial tricarboxylic acid (TCA) cycle, this pyruvate is then transformed to lactate in the cytosol of these cells [28,43,97,98,100–103]. This metabolic signature of quiescent adult stem cells is actively sustained by the following processes and features that are characteristic of these cells: (1) the transcription factor hypoxia-inducible factor 1α (HIF-1α)-dependent upregulation of levels of many glycolytic enzymes (such as hexokinase, phosphofructokinase, glyceraldehyde 3-phosphate dehydrogenase, phosphoglycerate kinase, and enolase) and lactate dehydrogenase A (an enzyme involved in the formation of lactate from pyruvate) in the cytosol; (2) a downregulation of the MPC1 subunit of the pyruvate carrier complex in mitochondria; (3) a HIF-1α-dependent upregulation of the mitochondrial pyruvate dehydrogenase kinases PDK2 and PDK4, both of which inhibit the pyruvate dehydrogenase complex-driven conversion of pyruvate to acetyl CoA in mitochondria; (4) a dynamin-related protein 1 (DRP1)-dependent fragmentation of the mitochondrial network into globular and immature mitochondria with underdeveloped cristae; (5) an increased abundance of the mitochondrial membrane uncoupling protein UCP2 that uncouples and lowers oxidative phosphorylation (OXPHOS) in mitochondria; (6) an elevated concentration of the ATPase inhibitory factor 1 (IF1) that suppresses mitochondrial adenosine triphosphate (ATP) synthase activity; (7) an upregulation of the mitochondrial carrier homolog 2 (MTCH2), a negative regulator of mitochondrial OXPHOS that induces mitochondrial depolarization; (8) an HIF-2α-dependent upregulation of the primary antioxidant enzymes involved in the detoxification of reactive oxygen species (ROS); among these enzymes are catalase (CAT) in peroxisomes and mitochondria, glutathione peroxidase type 1 (GPX1) in peroxisomes and mitochondria, copper/zinc superoxide dismutase (SOD1) in the cytosol, mitochondria, and peroxisomes, and manganese superoxide dismutase (SOD2) in mitochondria; and (9) a decline in the extent of ROS-inflicted apoptotic cell death (Figure 1) [27,43,94,96,97,100,103–112].

A body of evidence supports the notion that the common features of adult stem cells are to metabolize carbohydrates mainly through aerobic glycolysis in the cytosol, to suppress carbohydrate oxidation in mitochondria, to fragment mitochondrial network into globular and immature mitochondria with underdeveloped cristae, and to stimulate ROS detoxification in several cellular locations are essential for the maintenance of quiescence, identity, high number, regulated proliferation, and controlled differentiation of these cells [43,71,94–97,100,101,110,113]. Specifically, the HIF-1α, HIF-2α, and transcription factor Meis1 (myeloid ecotropic viral insertion site 1)-dependent program of an intensified glycolytic flow, a weakened mitochondrial OXPHOS, and an enhanced ROS detoxification within hematopoietic stem cells residing in a hypoxic niche is essential for the maintenance of their quiescent state and number (Figure 1) [28,109,110,114–121]. An increased glycolytic flow and a decreased mitochondrial OXPHOS are also required for the quiescent state maintenance and number preservation in mesenchymal stromal cells that are known to reside within the hypoxic environment of the bone marrow niche [28,122].

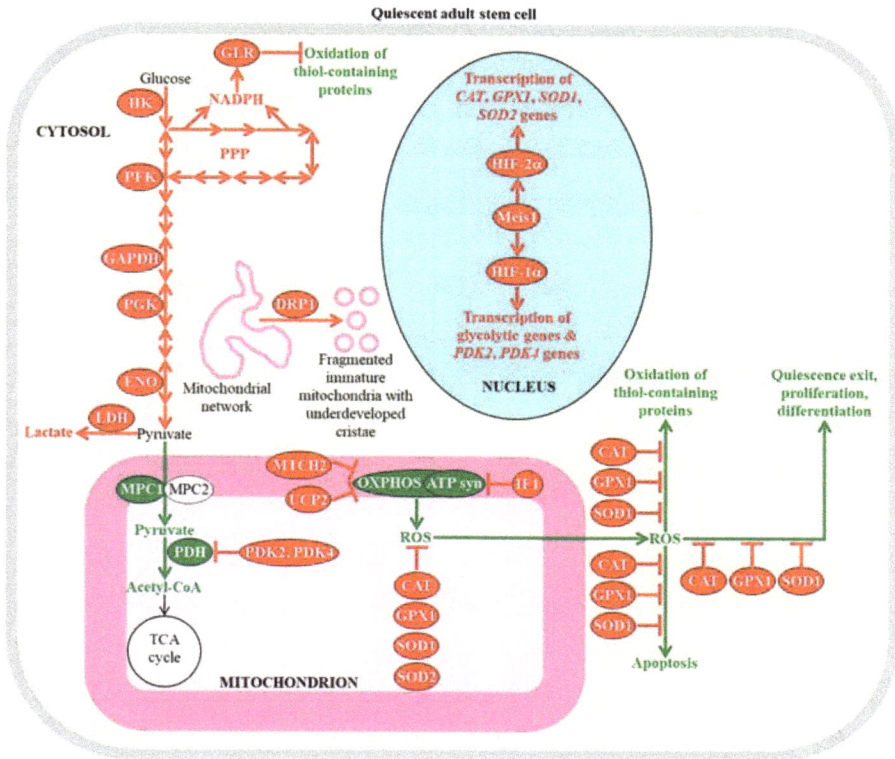

Figure 1. Some common metabolic features of adult stem cells are essential for the maintenance of their quiescence, number, proliferation potential, and differentiation ability. Among these metabolic features are carbohydrate metabolism mainly through aerobic glycolysis in the cytosol, suppressed carbohydrate oxidation in mitochondria, mitochondrial network fragmentation into globular and immature mitochondria with underdeveloped cristae, and stimulated ROS detoxification in several cellular locations. Enzymes, metabolites and processes whose activities, concentrations and rates are increased or decreased in quiescent adult stem cells (as compared to their fully differentiated progeny) are displayed in red or green color, respectively. The red one-way arrows and the red two-way arrows define irreversible and reversible (respectively) chemical reactions whose rates are increased in quiescent adult stem cells. The red inhibitory bars define inhibitory effects whose intensities are increased in quiescent adult stem cells. The green arrows define chemical reactions or processes whose rates or intensities are decreased in quiescent adult stem cells. The black arrow defines the irreversible chemical reaction whose rate is not changed in quiescent adult stem cells. See text for more details. Abbreviations: ATP syn, ATP synthase; CAT, catalase; DRP1, dynamin-related protein 1; ENO, enolase; GAPDH, glyceraldehyde 3-phosphate dehydrogenase; GLR, glutathione reductase; GPX1, glutathione peroxidase type 1; HIF-1α and HIF-2α, transcription factor hypoxia-inducible factors 1α and 2α, respectively; HK, hexokinase; IF1, inhibitory factor 1; LDH, lactate dehydrogenase; Meis1, myeloid ecotropic viral insertion site 1; MPC1 and MPC2, mitochondrial pyruvate carrier subunits 1 and 2; MTCH2, mitochondrial carrier homolog 2; OXPHOS, oxidative phosphorylation; PDH, pyruvate dehydrogenase; PDK2 and PDK4, pyruvate dehydrogenase kinases 2 and 4 (respectively); PFK, phosphofructokinase; PGK, phosphoglycerate kinase; PPP, pentose phosphate pathway; ROS, reactive oxygen species; SOD1 and SOD2, superoxide dismutases 1 and 2 (respectively); UCP2, uncoupling protein 2.

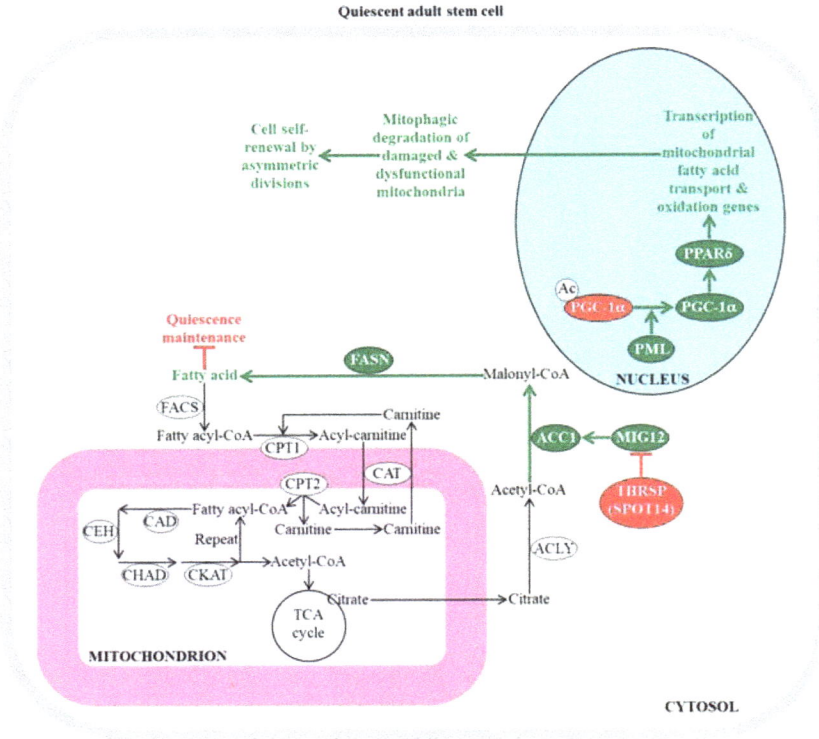

Figure 2. Mitochondrial β-oxidation of fatty acids and their synthesis in the cytosol of adult stem cells control the efficiency with which these cells can sustain quiescence or self-renew by asymmetric divisions. An inhibition of fatty acid synthesis in the cytosol is needed to sustain the quiescent state of adult neural stem cells. An activation of the transcription of nuclear genes that are involved in mitochondrial fatty acid transport and β-oxidation within adult stem cells is essential for the self-renewal of these cells by asymmetric divisions; such divisions lead to the formation of a new quiescent stem cell and an actively dividing daughter progenitor cell. Enzymes, metabolites, and processes whose activities, concentrations, and rates must be increased to maintain the quiescence of adult stem cells are displayed in red. Enzymes, metabolites, and processes whose activities, concentrations, and rates need be decreased to promote the self-renewal of adult stem cells by asymmetric divisions are displayed in green. The back arrows define chemical reactions whose rates are not essential for the maintenance of quiescence by adult stem cells. The green arrows define chemical reactions or processes whose rates or intensities must be decreased to maintain the quiescence of adult stem cells. The red inhibitory bars define inhibitory effects whose intensities must be increased to maintain the quiescence of adult stem cells. See text for more details. Abbreviations: Ac, acetyl group; ACC1, acetyl-CoA carboxylase 1; ACLY, ATP citrate lyase; CAD, acyl-CoA dehydrogenase; CAT, carnitine acylcarnitine translocase; CEH, enoyl-CoA hydratase; CHAD, hydroxyacyl-CoA dehydrogenase; CKAT, ketothiolase; CPT1 and CPT2, carnitine palmitoyltransferases 1 and 2 (respectively); FACS, fatty acyl-CoA synthase; FASN, fatty acid synthase; MIG12, midline-1-interacting G12-like protein; PGC-1α, peroxisome proliferator-activated receptor-gamma coactivator 1α; PML, promyelocytic leukaemia protein; PPARδ, peroxisome proliferator-activating receptor type δ; SPOT14, the 14th spot of proteins; THRSP, thyroid hormone-inducible hepatic protein.

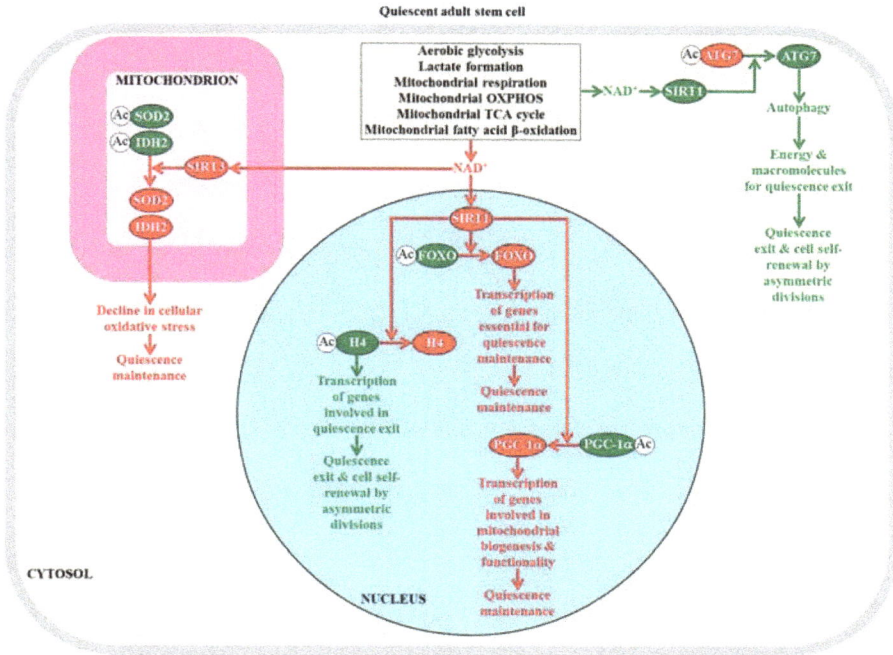

Figure 3. NAD$^+$ concentration within adult stem cells defines how efficiently these cells can maintain quiescence or self-renew by undergoing asymmetric divisions. The NAD$^+$-activated sirtuin SIRT1 in the nucleus is required for the maintenance of quiescence in adult stem cells, because this SIRT1 deacetylates histone H4 and the transcriptional factors FOXO and PGC-1α. The NAD$^+$-activated sirtuin SIRT3 in mitochondria is essential for quiescence maintenance by adult stem cells, because this SIRT3 deacetylates and activates superoxide dismutase SOD2 and isocitrate dehydrogenase IDH2, both of which weaken cellular oxidative stress. The NAD$^+$-activated sirtuin SIRT1 in the cytosol is required for the transition from quiescence to self-renewal by asymmetric divisions, because this SIRT1 deacetylates and activates the autophagy-related protein ATG7 to promote autophagy, which provides the energy and macromolecules required for such transitions. Proteins, metabolites, and processes whose activities, concentrations, and rates must be increased to maintain the quiescence of adult stem cells are displayed in red. Proteins, metabolites, and processes whose activities, concentrations, and rates must be increased to promote the self-renewal of adult stem cells by asymmetric divisions are displayed in green. The red arrows define processes whose intensities must be increased to maintain the quiescence of adult stem cells. The green arrows define processes whose intensities must be decreased to maintain the quiescence of adult stem cells. See text for more details. Abbreviations: Ac, acetyl group; ATG7, the autophagy-related protein 7; FOXO, transcriptional factors of the Forkhead family; H4, histone H4; IDH2, mitochondrial isocitrate dehydrogenase 2; OXPHOS, oxidative phosphorylation; PGC-1α, peroxisome proliferator-activated receptor-gamma coactivator 1α; SIRT1 and SIRT3, NAD$^+$-dependent protein deacetylases sirtuin 1 and sirtuin 3 (respectively); SOD2, mitochondrial manganese superoxide dismutase.

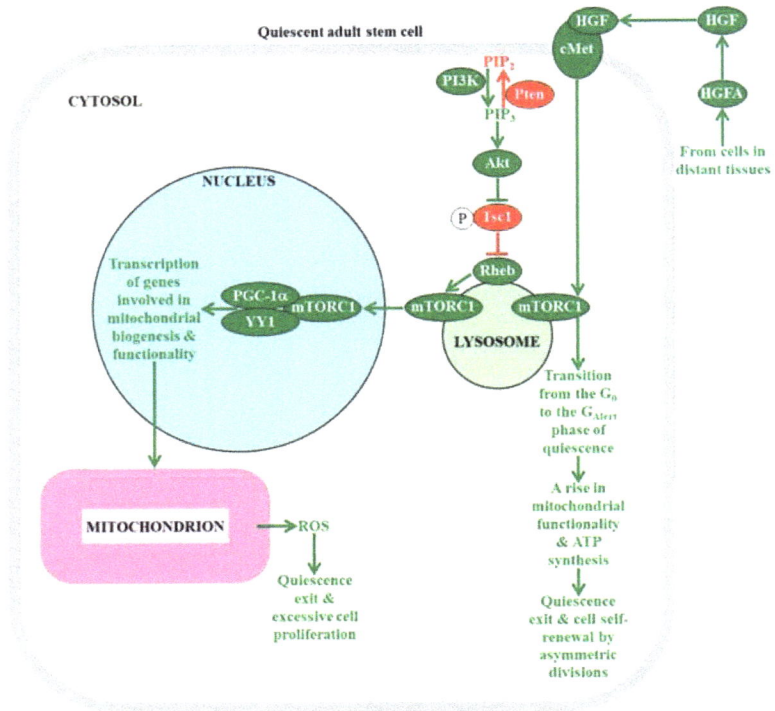

Figure 4. The intensity of information flow through the mTORC1-signaling pathway defines the fate of quiescent adult stem cells. A low intensity of mTORC1 signaling in adult stem cells decreases the extent of mitochondrial ROS production and release, thereby preventing the exit of these cells from the quiescent state and their excessive proliferation. An activation of mTORC1 signaling in response to hepatocyte growth factor, a cell-extrinsic pro-mitogenic signal, commits quiescent adult stem cells to cell cycle entry, proliferation, and differentiation. Proteins, metabolites, and processes whose activities, concentrations, and rates must be increased to maintain the quiescence of adult stem cells are displayed in red. Proteins, metabolites, and processes whose activities, concentrations, and rates need to be increased to promote the exit of adult stem cells from the state of quiescence are displayed in green. The red arrow defines the chemical reaction whose rate must be increased to maintain the quiescence of adult stem cells. The red inhibitory bar defines the inhibitory effect whose intensity must be increased to maintain the quiescence of adult stem cells. The green arrows define chemical reactions or processes whose rates or intensities must be increased to promote the exit of adult stem cells from the state of quiescence. See the text for more details. Abbreviations: Akt1, RAC-alpha serine/threonine protein kinase 1 or v-akt murine thymoma viral oncogene homolog 1; cMet, mesenchymal–epithelial transition factor; G_{Alert}, the "alert" phase of quiescence; HGF, hepatocyte growth factor; HGFA, hepatocyte growth factor activator; mTORC1, the mammalian (or mechanistic) target of rapamycin complex 1; PGC-1α, peroxisome proliferator-activated receptor-gamma coactivator 1α; PIP_2, phosphatidylinositol 4,5-bisphosphate; PIP_3, phosphatidylinositol (3,4,5)-trisphosphate; PI3K, phosphoinositide 3-kinase; Pten, phosphatase and tensin homologue; Rheb, Ras homolog enriched in brain; ROS, reactive oxygen species; Tsc1, tuberous sclerosis 1; YY1, yin-yang 1 transcription factor.

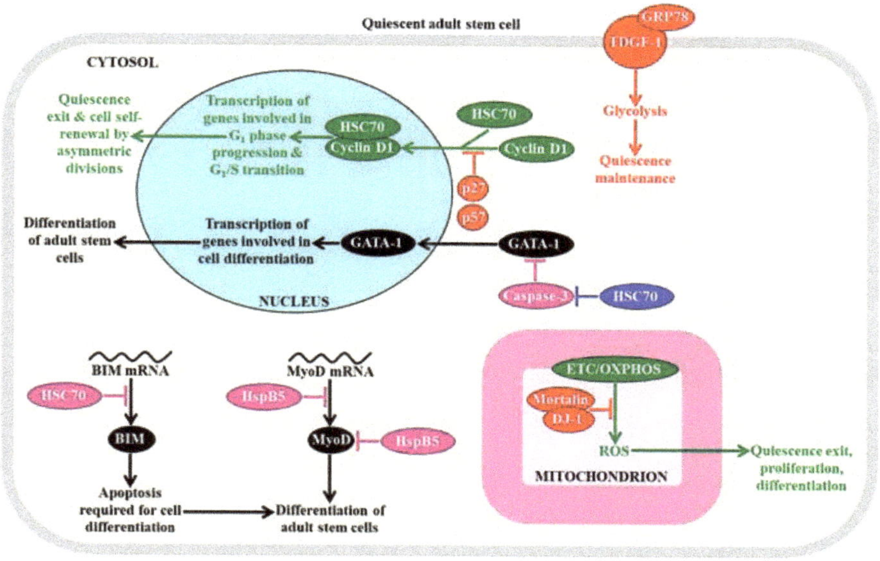

Figure 5. Several heat shock proteins (HSPs) play essential roles in preserving the quiescence, affecting the proliferation, and influencing the differentiation of adult stem cells. These HSPs control the nuclear import and degradation of some transcription factors, as well as the translation and stability of certain proteins that promote cell differentiation, glycolysis, and concentrations of mitochondrially-generated reactive oxygen species (ROS). Proteins, metabolites, and processes whose activities, concentrations, and rates must be increased to maintain the quiescence of adult stem cells are displayed in red. Proteins, metabolites, and processes whose activities, concentrations, and rates must be increased to promote the exit of adult stem cells from the state of quiescence are displayed in green. Proteins and processes whose activities and rates must be increased to promote the differentiation of adult stem cells are displayed in black. Proteins whose activities must be increased to suppress the differentiation of adult stem cells are displayed in pink. See text for more details. Abbreviations: BIM, Bcl-2-like protein 11; GATA-1, GATA-binding factor 1; GRP78, immunoglobulin heavy chain-binding protein homolog; HSC70, heat shock cognate 71 kDa protein; HspB5, alpha-crystallin B chain; MyoD, myoblast determination protein D; TDGF-1, teratocarcinoma-derived growth factor 1.

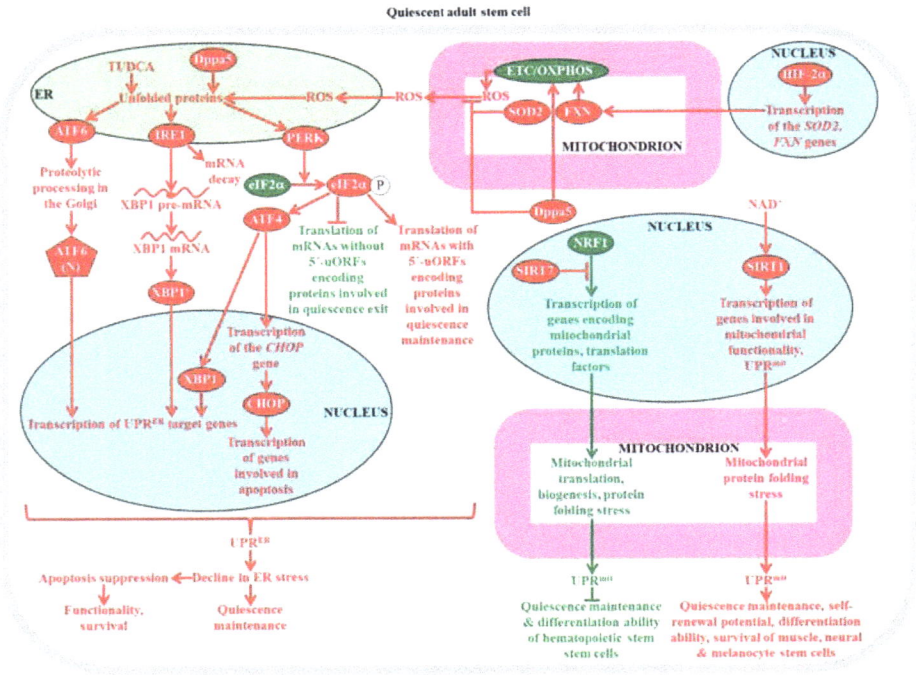

Figure 6. The unfolded protein response (UPR^ER) and (UPR^mit) systems of proteostasis restoration in adult stem cells are required for the maintenance of their quiescence, self-renewal proficiency, proliferation potential, differentiation competence, functionality, and viability. A stimulation of the ATF6 (activating transcription factor 6), IRE1 (inositol requiring enzyme 1), and PERK (PKR-like endoplasmic reticulum kinase) branches of the UPR^ER system is essential for the maintenance, self-renewal, functionality, and survival of adult stem cells. A SIRT7-dependent inhibition of the UPR^mit system is indispensable for sustaining the quiescence and differentiation capability of hematopoietic stem cells. A NAD^+/SIRT1-dependent stimulation of the UPR^mit system is required for the maintenance of the quiescence, self-renewal potential, differentiation ability, and viability of muscle stem cells, neural stem cells, and melanocyte stem cells. Proteins, metabolites, and processes whose activities, concentrations, and rates must be increased to maintain the quiescence of adult stem cells are displayed in red. Proteins, metabolites, and processes whose activities, concentrations, and rates need to be increased to promote the exit of adult stem cells from the state of quiescence are displayed in green. See text for more details. Abbreviations: ATF4, activating transcription factor 4; ATF6(N), N-terminal cytosolic fragment; CHOP, C/EBP homologous protein; Dppa5, developmental pluripotency-associated 5 protein; eIF2α, eukaryotic translation initiation factor 2α; ER, the endoplasmic reticulum; ETC, electron transport chain; FXN, frataxin; HIF-2α, hypoxia-inducible factor 2α; NRF1, nuclear respiratory factor 1; OXPHOS, oxidative phosphorylation; ROS, reactive oxygen species; SIRT1 and SIRT7, sirtuin 1 and sirtuin 7 (respectively); SOD2, mitochondrial manganese superoxide dismutase; TUDCA, tauroursodeoxycholic acid; uORFs, upstream open reading frames; XBP1, X-box binding protein 1; XBP1^s, X-box binding protein 1 translated as a protein product the spliced mRNA for XBP1.

Mechanisms through which the augmented glycolytic flow in adult stem cells can contribute to the maintenance of the quiescent state and number of these cells by regulating some downstream cellular processes require further investigation. One possibility is that such flow allows the glycolytic intermediate glucose-6-phosphate to enter the pentose phosphate pathway (PPP), which generates nicotinamide adenine dinucleotide phosphate (NADPH) (Figure 1) [123]. NADPH is not only the source of cellular reducing equivalents required for the synthesis of nucleic acids and lipids; it is also the electron donor that is essential for sustaining cellular redox homeostasis via the glutathione reductase system [123]. Since this system protects cellular macromolecules from oxidative damage inflicted by ROS, the intensified glycolytic flow in adult stem cells can defend macromolecules in these cells against ROS-dependent oxidative damage (Figure 1) [96,124–126]. The decline in mitochondrial OXPHOS, which is also characteristic of adult stem cells and generates the bulk of ROS, can contribute to the maintenance of the quiescent state and number of these cells through the two mechanisms described in Section 2.3.

2.2. Fatty Acid Oxidation and Synthesis Define the Fate of Quiescent Adult Stem Cells

Since mitochondrial transport and the oxidation of pyruvate are actively suppressed in adult stem cells (see above), the transport of fatty acid to mitochondria and the β-oxidation of fatty acids in mitochondria define the rate of mitochondrial respiration in these quiescent cells [43,97].

The efficiencies of mitochondrial fatty acid transport and β-oxidation regulate the self-renewal potential and proliferation capacity of adult stem cells [28,95,97]. A promyelocytic leukaemia protein (PML)/peroxisome proliferator-activated receptor-gamma coactivator 1α (PGC-1α)/peroxisome proliferator-activating receptor type δ (PPARδ)-dependent activation of transcription of nuclear genes involved in mitochondrial fatty acid transport and β-oxidation within hematopoietic stem cells is essential for the self-renewal of these cells by asymmetric divisions, in favor of symmetric differentiating divisions [127–129]; this is likely because such activation promotes the mitophagic degradation of damaged and dysfunctional mitochondria (Figure 2) [130].

The self-renewal of neural stem cells by asymmetric divisions also requires the efficiency of mitochondrial fatty acid β-oxidation to reach a certain threshold; it remains unknown why and how such optimal efficiency of mitochondrial fatty acid β-oxidation can fuel asymmetric divisions of neural stem cells but not their symmetric differentiating divisions [95,131,132].

Moreover, an optimal efficiency of mitochondrial fatty acid β-oxidation is needed for the self-renewal of quiescent skeletal muscle stem cells; however, the functionality and viability of these cells are impaired if such efficiency exceeds a certain, oxidative damage-inducing level [95,133,134].

Taken together, these findings indicate that there is a certain optimal efficiency of mitochondrial fatty acid β-oxidation at which this metabolic pathway actively supports the self-renewal potential and proliferation capacity of adult stem cells [95]. Mechanisms underlying this effect of mitochondrial fatty acid β-oxidation are presently unknown. One possible mechanism may involve the demonstrated ability of mitochondrial fatty acid β-oxidation to generate the bulk of NADPH [28,135], which is essential for sustaining cellular redox homeostasis via the glutathione reductase system in adult stem cells (see Section 2.1).

The self-renewal potential and proliferation capacity of adult stem cells can be controlled not only by the efficiencies of mitochondrial fatty acid transport and β-oxidation, but also by the efficiency with which fatty acids are synthesized in the cytosol. Specifically, the acetyl-CoA carboxylase (ACC) and fatty acid synthase (FASN)-dependent synthesis of fatty acids in the cytosol must be inhibited by thyroid hormone-inducible hepatic protein (THRSP; also known as the 14th spot of proteins [SPOT14] protein) to sustain the quiescent state of adult neural stem cells (Figure 2) [136]. This is most likely because certain concentrations of endogenous fatty acids can stimulate the exit of these cells from the quiescent state and their subsequent proliferation and differentiation [28,136]. However, some exogenously added acids and a high-fat diet rich in fatty acids exhibit the opposite effect on the quiescence and proliferation of a different kind of adult stem cell, specifically of intestinal stem cells [137]. In fact, endogenous fatty acids promote the quiescent state of these stem cells, and suppress their proliferation and differentiation [137].

In sum, these findings support the notion that a balance between the mitochondrial β-oxidation of fatty acids and their synthesis in the cytosol defines the fate of adult stem cells, because this balance controls the concentration of fatty acids within these cells. A regulated change of this balance can increase or decrease the intracellular concentration of fatty acids, thereby altering the efficiency with which these cells can sustain quiescence, self-renew by asymmetric divisions, or differentiate. It is also conceivable that different kinds of adult stem cells may differ in the threshold level of fatty acid concentrations that are capable of regulating the fate of these cells.

2.3. A Dual Role of ROS in Defining the Homeostasis and Functionality of Quiescent Adult Stem Cells

Since mitochondrial respiration is the major source of ROS within mammalian cells [138], the suppression of mitochondrial respiration and OXPHOS observed in adult stem cells elicits a significant decline in ROS within these cells [96,139,140].

The low concentration of ROS is a hallmark of adult stem cells, and it plays a dual essential role in sustaining the homeostasis, functionality, and long-term survival of these cells [2,71,95–97,124–126,141–145].

One role of such regulated decline in ROS is to protect the molecules of proteins, lipids, and nucleic acids within adult stem cells (including hematopoietic stems cells, neural stem cells, spermatogonial stem cells, and tracheal stem cells) against oxidative damage inflicted by ROS concentrations exceeding a toxic threshold [71,96,105,146–150]. If ROS concentrations within adult stem cells exceed such a toxic threshold, the DNA and other macromolecules in these cells are oxidatively damaged; this causes cell death, a decline in the number of adult stem cells, and ultimately leads to the exhaustion of the pool of these cells [71,151,152]. The protection of cellular macromolecules against ROS-dependent oxidative damage within adult stem cells is further enhanced by the high efficiencies of the antioxidant defense and DNA repair systems that are characteristic of these cells [96,124–126,153–155]. The antioxidant defense and DNA repair systems within adult stem cells include the following: (1) the NADPH-dependent glutathione reductase system for sustaining cellular redox homeostasis, which is fueled by the intensified glycolytic flow within these cells, as described in Section 2.1 (Figure 1); (2) the antioxidant system driven by the transcriptional activator FoxO3 of the forkhead family, which is activated by the NAD^+-dependent protein deacetylase sirtuin 1 (SIRT1); this system includes peroxisomal and mitochondrial CAT, mitochondrial SOD2, and mitochondrial thioredoxin-dependent peroxide reductase (PRDX3) [156–162]; (3) the antioxidant system orchestrated by the nuclear respiratory factor 2 (NRF2); this system includes enzymes involved in the synthesis of reduced glutathione and thioredoxin, NADPH-generating enzymes of the TCA cycle and pentose phosphate pathway, several forms of PRDX, and cytosolic thioredoxin reductase 1 (TXNRD1) [163,164]; and 4) the non-homologous end-joining (NHEJ) pathway for repairing DNA double-strand breaks [2,165]. In sum, the intensity of oxidative stress in adult stem cells is low. Such intensity is known to be defined by a balance between the efficiencies of cellular ROS production, ROS-inflicted oxidative damage, ROS detoxification, and oxidative damage repair [166–169]. Thus, the low concentration of ROS within adult stem cells, their high antioxidant capacity, and their enhanced ability to repair oxidatively damaged DNA protect these cells from excessive oxidative stress and damage.

It needs to be emphasized that the FoxO3-dependent and NRF2-dependent antioxidant defense systems, as well as the NHEJ pathway of DNA repair, are indispensable for the quiescence and/or long-term survival of adult stem cells. In fact, the FoxO3-dependent antioxidant defense system is essential for the self-renewal of neural stem cells by asymmetric divisions that allow maintaining a pool of these cells in a quiescent state [157,159,170,171]. Furthermore, the NRF2-dependent antioxidant defense system is required to preserve a pool of intestinal stem cells in the state of quiescence [163,164,172,173]. Moreover, the NHEJ pathway of DNA repair is needed for assuring the long-term survival of adult stem cells [2,174].

Another role of the regulated decline in ROS within quiescent adult stem cells is to suppress their symmetric differentiating divisions. These divisions: 1) are stimulated by ROS concentrations that exceed a certain threshold but remain non-toxic; and 2) cause proliferation and then the terminal differentiation of the cells, thereby exhausting the pool of quiescent adult stem cells [71,96,99,139,140,147,148,150,175–178]. Such stem cell proliferation-based and differentiation-based exhaustion of the pool of quiescent adult stem cells in response to a rise in ROS concentrations to a level below the toxic threshold has been reported for hematopoietic stems cells, intestinal stem cells, and skeletal muscle stem cells [71,157,159,163,164,170,171,175,179–182].

The downstream targets of ROS concentrations that elicit the cell proliferation-based and differentiation-based exhaustion of the pool of quiescent adult stem cells remain to be characterized. One of these targets of ROS in hematopoietic stems cells can be the p38α mitogen-activated protein kinase signaling pathway [183,184]. In response to increased ROS concentrations, this pathway stimulates the transcription factor Mitf (microphthalmia-associated transcription factor) that then activates the transcription of a gene for a rate-limiting enzyme of purine metabolism [184–186]. The resulting remodeling of purine metabolism alters the levels of amino acids and purine nucleotides, thus promoting cell cycle progression and cell proliferation in response to increased ROS concentrations [184–186].

2.4. Some Mitochondrial TCA Cycle Intermediates Contribute to Quiescence Maintenance by Adult Stem Cells

Recent findings suggest that the suppression of mitochondrial respiration and OXPHOS is not the only mitochondria-related factor involved in sustaining the homeostasis and functionality of adult stem cells. Specifically, it has been shown that some intermediates of the mitochondrial TCA cycle play essential roles in determining the fate of these cells; these intermediates include acetyl CoA, oxaloacetate, citrate, and α-ketoglutarate [31,77,94,187–191].

The essential roles of these TCA cycle intermediates in the fate of adult stem cells is due to their abilities to be cofactors or substrates for chromatin-modifying enzymes that catalyze some epigenetic modifications in the nucleus, such as histone acetylation/deacetylation and histone methylation, as well as DNA methylation and hypermethylation [2,31,71,77,94,187–196]. Some of these mitochondrial metabolism-driven modifications of histones and DNA have been observed only in neural stem cells, skeletal muscle stem cells, adipose-derived mesenchymal stem cells, muscle-derived mesenchymal stem cells, and bone marrow-derived mesenchymal stem cells, but not in their fully differentiated progeny; such stem cell-specific epigenetic modifications in the nucleus play essential roles in maintaining quiescence, regulating self-renewal and proliferation, and/or suppressing the differentiation of these adult stem cells [2,31,77,133,197–212].

Many of these stem cell-specific epigenetic modifications in the nucleus are known to activate or repress the transcription of genes implicated in different aspects of fate programming in adult stem cells, including the maintenance of their quiescence and self-renewal ability as well as their regulated proliferation and differentiation [2,31,77,188,189,197–203,206,207,210]. Nuclear genes whose transcription is activated in quiescent neural stem cells, muscle stem cells, hair follicle stem cells, umbilical cord-derived mesenchymal stem cells, and bone marrow-derived mesenchymal stem cells include genes that encode proteins that are involved in the G1/S cell cycle transition, epigenetic modifications in the nucleus, transcription of genes implicated in stem cell fate decisions, stemness, and microRNA formation [2,189,213–216]. Nuclear genes whose transcription is repressed in quiescent neural stem cells, muscle stem cells, hair follicle stem cells, adipose-derived mesenchymal stem cells, muscle-derived mesenchymal stem cells, and bone marrow-derived mesenchymal stem cells include genes that encode proteins implicated in DNA replication, chromosome segregation, cell cycle delay at several checkpoints, cell cycle progression, lineage specificity, and mitochondrial functionality [2,188,213–216].

Taken together, these findings suggest that certain mitochondrial metabolism-driven modifications of the epigenetic landscape within the nucleus of adult stem cells can alter the efficiencies with which some transcription factors activate or repress the expression of nuclear genes that are involved in the quiescent state maintenance by adult stem cells and/or in their response to environmental cues, proliferation, and differentiation [2,31,77,78,94,96].

Further investigation is required on the mechanisms through which the TCA cycle intermediates acetyl CoA, oxaloacetate, citrate, and α-ketoglutarate can maintain the homeostasis and functionality of adult stem cells by promoting epigenetic modifications of nuclear genes and altering the epigenetic landscape of these cells. In hematopoietic stems cells, skeletal muscle stem cells, and hair follicle stem cells, the histone 3 methyltransferases EZH1 and EZH2 (enhancer of zeste proteins 1 and 2, respectively) are essential components of such mechanisms; in fact, the downregulation of both proteins leads to a depletion of the pool of these quiescent adult stem cells, whereas the upregulation of EZH2 decelerates the exhaustion of hematopoietic stems cells [217–220].

2.5. NAD$^+$ Concentration within Adult Stem Cells Defines Their Fate

Relative rates of aerobic glycolysis, lactate formation, mitochondrial respiration, OXPHOS, mitochondrial TCA cycle, and mitochondrial fatty acid β-oxidation control the cellular concentrations of the oxidized and reduced forms of nicotinamide dinucleotide (NAD$^+$ and NADH, respectively) and, thus, the NAD$^+$/NADH ratio in adult stem cells [94,96,221]. The NAD$^+$/NADH ratio plays an essential role in defining the fate of adult stem cells not only because NAD$^+$ is a cofactor of all these metabolic processes, but also because NAD$^+$ is an activator of SIRT1 in the cytosol and SIRT3 in mitochondria [222–224].

A body of evidence supports the essential role of NAD$^+$-activated SIRT1 in defining the fate of adult neural stem cells. However, there are two conflicting views on how exactly NAD$^+$-activated SIRT1 can influence the fate of these cells. On the one hand, some studies suggest that NAD$^+$-activated SIRT1 promotes the quiescence of adult neural stem cells and suppresses their proliferation and differentiation. These studies have shown that (1) the genetic inactivation or pharmacological inhibition of SIRT1 in adult neural stem cells promotes their exit from quiescence and stimulates their proliferation and differentiation [225–229], and (2) SIRT1 overexpression or its resveratrol-driven activation preserves the quiescence of adult neural stem cells and suppresses their proliferation and differentiation [226,229]. On the other hand, other studies have suggested that NAD$^+$-activated SIRT1 suppresses the quiescence of adult neural stem cells and promotes their proliferation and differentiation. These other studies have demonstrated that (1) SIRT1 genetic downregulation or pharmacological inhibition promotes the quiescence of these cells and suppresses their proliferation and differentiation; and (2) SIRT1 overexpression stimulates the quiescence exit of these cells and accelerates their proliferation and differentiation in a mechanism that involves a transient translocation of SIRT1 from the cytosol into the nucleus [230–232]. In sum, the exact role of NAD$^+$-activated SIRT1 in defining the fate of adult neural stem cells remains controversial and requires further investigation.

The essential role of NAD$^+$ concentration and SIRT1 activity in defining the fate of adult stem cells is underscored by the finding that a decline in NAD$^+$ concentration within skeletal muscle stem cells lowers SIRT1 activity; this causes an increase in the extent of histone H4 acetylation and ultimately activates the transcription of nuclear genes involved in the transition from quiescence to the proliferation and differentiation of these stem cells [133,233]. Thus, one role of the NAD$^+$-dependent protein deacetylase SIRT1 in maintaining the quiescence of adult skeletal muscle stem cells is linked to its known ability to suppress the transcription of certain genes by specifically remodeling their epigenetic landscape through histone H4 deacetylation (Figure 3) [234–236].

Another mechanism through which NAD$^+$-activated SIRT1 contributes to quiescence maintenance in adult neural stem cells and adult skeletal muscle stem cells consists of the SIRT1-dependent deacetylation and activation of PGC-1α (a transcriptional activator of mitochondrial biogenesis) to promote mitochondrial functionality (Figure 3) [75,237–239].

It needs to be emphasized that the exact role of NAD$^+$-activated SIRT1 in the quiescence of adult skeletal muscle stem cells requires further investigation. Specifically, it has been demonstrated that when SIRT1 is activated by NAD$^+$ in these stem cells, it promotes autophagy by deacetylating and stimulating the autophagy-related protein 7 (ATG7) to provide the energy and macromolecules required for the transition from quiescence to proliferation followed by differentiation (Figure 3) [240,241]. This finding suggests that under certain conditions, NAD$^+$-activated SIRT1 can act via an autophagy-based mechanism that stimulates the exit from quiescence and promotes the proliferation and differentiation of adult skeletal muscle stem cells.

Of note, it has been proposed that NAD$^+$-activated SIRT1 in hematopoietic stem cells and neural stem cells may help to sustain their quiescent state by deacetylating another quiescence-related target of SIRT1, namely the DNA-binding forkhead box O (FOXO) transcription factors [48]. The SIRT1-dependent deacetylation of the FOXO transcriptional factors is known to enhance their ability to activate the transcription of many genes that are essential for preserving the quiescence of hematopoietic stem cells and neural stem cells, preventing their premature differentiation and maintaining their viability (Figure 3) [48,157,159,170,171,242,243].

SIRT3, a mitochondrial form of the NAD$^+$-dependent sirtuins, is essential for the maintenance of quiescence in adult hematopoietic stem cells [46,244,245]. Since NAD$^+$-stimulated SIRT3 deacetylates and activates some mitochondrial proteins involved in oxidative stress protection, it contributes to the quiescence maintenance of these cells by diminishing oxidative stress inside and outside of their mitochondria [46,244,245]. Specifically, when activated by NAD$^+$, SIRT3 deacetylates and activates SOD2, which is a major mitochondrial antioxidant enzyme; this reduces mitochondrial superoxide and weakens cellular oxidative stress (Figure 3) [245–247]. NAD$^+$-stimulated SIRT3 also deacetylates and activates isocitrate dehydrogenase 2 (IDH2) [244,248]. Mitochondrial IDH2 catalyzes a reaction that yields NADPH, the electron donor that is essential for sustaining cellular redox homeostasis via the glutathione reductase and thioredoxin reductase systems [123]. Thus, the SIRT3-driven deacetylation and activation of IDH2 also lessens cellular oxidative stress (Figure 3) [244,248].

2.6. A Low-Energy Status of Adult Stem Cells Helps Sustain Their Quiescence

As noted above, low efficiencies of OXPHOS and ATP synthesis in mitochondria are characteristic features of the metabolic signature of quiescent adult stem cells [94,96,97]. This causes a rise in the intracellular concentrations of adenosine monophosphate (AMP) and adenosine diphosphate (ADP), thereby creating a low-energy stress and activating the liver kinase B1 (LKB1; a master energy-sensing protein kinase) and the AMP-activated protein kinase (AMPK; a downstream phosphorylation target of LKB1) [249–252].

LKB1 is essential for the quiescence maintenance, functionality, and survival of hematopoietic stem cells in mice. In fact, a depletion of LKB has the following two temporally separated effects on hematopoietic stem cells and on the multipotent progenitors formed from these adult stem cells: (1) an initial loss of stem cell quiescence and the resulting rise in the number of cells comprising each of these two cell populations; and (2) a subsequent decline in the number, and the ultimate depletion, of both cell populations [253–255]. The hematopoietic stem cells of mice depleted of LKB1 lost the ability to regenerate the hematopoietic system of control mice, and this decline in the functionality of LKB1-depleted hematopoietic stem cells is followed by their apoptotic death [253–255]. A downstream target of LKB1 in hematopoietic stem cells is mitochondrial functionality, as LKB1 depletion in these cells decreases the expression of the transcriptional activators of mitochondrial biogenesis PGC-1α and PGC-1β, lowers mitochondrial ATP synthesis, and impairs mitochondrial integrity by decreasing mitochondrial membrane potential [253–255]. The decline in the integrity of mitochondria observed within LKB1-depleted hematopoietic stem cells is likely to be responsible for their accelerated apoptotic death [254]. These cells also undergo an induction of autophagy, which plays an essential protective role in the survival of LKB1-depleted hematopoietic stem cells [254]. It is conceivable that the ability of LKB1 to sustain the functionality and integrity of mitochondria, thereby decelerating apoptotic cell death, is linked to the essential role of this master energy-sensing protein kinase in preserving the quiescence, functionality, and survival of hematopoietic stem cells [253–256]. The involvement of LKB1 in preserving the quiescence of hematopoietic stem cells is complemented by the essential role of autophagy—a degradative elimination of dysfunctional macromolecules and organelles—in sustaining the quiescent state of these adult stem cells [257]. Although AMPK—the downstream phosphorylation and activation target of LKB1—is known to promote autophagy because it phosphorylates the autophagy-initiating protein kinase ULK1 [252]—a mechanism through which autophagic degradation is actively stimulated and sustained in hematopoietic stem cells to preserve their quiescence—it also requires further investigation.

The essential role of LKB1 in sustaining the quiescence, functionality, and survival of hematopoietic stem cells is not due to its activating effect on AMPK in these cells [253–255]. AMPK is a sensor of energetic stress and mitochondrial dysfunction that does not influence the fate of hematopoietic stem cells, neither before nor after being phosphorylated and activated by LKB1 [253–255,258]. Phosphorylated and activated AMPK is known to suppress the mammalian (or mechanistic) target of rapamycin complex 1 (mTORC1) signaling pathway [48,252], whose genetic enhancement elicits a depletion of hematopoietic stem cells [259–264] (see Section 2.7 for more details). Despite such an essential role of a low-intensity mTORC1 signaling in maintaining hematopoietic stem cell quiescence, the indispensable role of LKB1 in sustaining the quiescence, functionality, and survival of hematopoietic stem cells is not caused by the LKB1-driven and AMPK-dependent activation of the mTORC1 signaling pathway [253–255,258].

Although LKB1 controls the fates of hematopoietic stem cells and the multipotent progenitors formed from these cells, it does not affect the abundance or functionality of the fully differentiated progeny of such cells [253–255]. In contrast, LKB1 is not required for sustaining the quiescence, functionality, and survival of a different type of adult stem cells, namely intestinal stem cells [265]. However, LKB1 is essential for the normal differentiation and maturation of intestinal stem cells; mechanisms underlying such effects of LKB1 are presently unknown [265–267]. Thus, LKB1 can differently influence the self-renewal, proliferation, and differentiation of different types of adult stem cells.

2.7. The Maintenance of mTORC1 Signaling at Low Intensity in Adult Stem Cells Is Essential for Sustaining Their Quiescence, Self-Renewal, and Functionality

The mTORC1 signaling pathway is a key signaling hub that—in response to changes in extracellular and intracellular nutrients, pro-mitogenic stimuli, and stresses—alters the efficiencies of many anabolic and catabolic cellular processes to ensure a proper adaptation of the cell to such changes [268,269]. The anabolic processes controlled by mTORC1 include glycolysis and the pentose phosphate pathway, lipid and nucleotide synthesis, and protein synthesis [268,269]. Among the catabolic processes controlled by mTORC1 are autophagy and lysosome biogenesis, and the ubiquitin-dependent proteasomal degradation of proteins [268,269]. mTORC1 also controls cellular energy homeostasis by activating the YY1 (yin-yang 1)/PGC-1α-dependent transcription of nuclear genes involved in mitochondrial biogenesis and functionality [270–272].

Since some of the cellular processes regulated by mTORC1 can be involved in creating the metabolic signature of adult stem cells that is essential for their quiescence (see Sections 2.1–2.6), and because some of the processes creating such signatures are known to control mTORC1 via feedback regulation mechanisms [28,270–272], it is conceivable that mTORC1 may play an essential role in defining the fate of quiescent adult stem cells. In fact, an overactivation of the mTORC1 pathway by a mutation that depletes the protein inhibitor Pten (phosphatase and tensin homologue) of this pathway causes an exit of hematopoietic stem cells from quiescence, leads to a decline in functionality of these cells, decreases their self-renewal potential, alters their differentiation pattern, and eventually results in an exhaustion of the hematopoietic stem cell population pool in a cell-autonomous, mTORC1-dependent fashion (Figure 4) [259,260]. Furthermore, an enhancement of the mTORC1 pathway by a constitutive expression of the upstream protein activator Akt1 (RAC-alpha serine/threonine protein kinase 1 or v-akt murine thymoma viral oncogene homolog 1) of mTORC1 transiently increases the abundance of hematopoietic stem cells; then, it causes their excessive proliferation, and ultimately elicits the apoptotic death of hematopoietic stem cells and the resulting depletion of their pool in an mTORC1-dependent manner (Figure 4) [263,264]. Moreover, an overstimulation of the information flow through the mTORC1 pathway by a mutation that conditionally deletes the upstream protein inhibitor Tsc1 (tuberous sclerosis

1) of mTORC1 promotes an exit of hematopoietic stem cells from the quiescent state, impairs their functionality and self-renewal ability, promotes their apoptotic death, and eventually depletes the pool of hematopoietic stem cells in an mTORC1-dependent fashion (Figure 4) [261,262]. These effects of Tsc1 deletion on the fate of hematopoietic stem cells (1) coincide with a rise in the abundance of mitochondria and a rise in cellular ROS; and (2) can be reversed with the help of the antioxidant *N*-acetylcysteine, a ROS antagonist [261]. Thus, the maintenance of mTORC1 signaling at low intensity in hematopoietic stem cells is essential for sustaining their quiescence, because it allows the suppression of the YY1/PGC-1α-dependent transcription of nuclear genes involved in mitochondrial biogenesis and functionality [270]. Such suppression prevents a rise of mitochondrially-produced cellular ROS above a threshold level that is capable of inducing the quiescence exit and proliferation of these adult stem cells (Figure 4) [261].

The intensity of information flow through the mTORC1 signaling pathway also defines the fate of a different type of quiescent adult stem cells, namely neural stem cells, as well as the neural progenitor cells that are formed from them. In fact, a mutation that conditionally deletes Tsc1 to enhance this pathway impairs the migration patterns of both neural stem cells and neural progenitor cells in the subventricular zone of the brain [273]. Moreover, a downregulation of the mTORC1 signaling pathway by caloric restriction in Paneth cells of the intestinal stem cell niche triggers a cell non-autonomous mechanism that promotes the self-renewal of intestinal stem cells and improves their functionality [274]. Thus, the fate of intestinal stem cells also depends on the intensity of information flow through the mTORC1 signaling pathway. Downstream cellular processes whose regulation by mTORC1 controls quiescence maintenance, self-renewal, and functionality of neural stem cells and intestinal stem cells are presently unknown.

The nature of the intracellular pro-mitogenic and/or anti-mitogenic signals whose ability to control mTORC1 in a cell-autonomous manner defines the fate of quiescent adult stem cells requires further investigation. Recent studies uncovered a cell-nonautonomous mechanism of such mTORC1 control in muscle stem cells. Specifically, it was demonstrated that the abilities of muscle stem cells to respond to injury-induced systemic signals from distant tissues by retaining a pool of quiescent stem cells and by creating a population of differentiated cells for distant tissue repair are under tight control of the mTORC1 signaling within these adult stem cells [20,275,276]. Tissue injury stimulates the circulating protease hepatocyte growth factor activator (HGFA); then, HGFA relays a signal to muscle stem cells in tissues distant to the zone of injury by proteolytically activating HGF to stimulate the HGF receptor cMet (mesenchymal–epithelial transition factor) and promote mTORC1 signaling in these stem cells via a presently unknown mechanism (Figure 4) [20,275,276]. Once promoted, mTORC1 orchestrates a transition of adult stem cells from the G_0 phase to the "alert" (G_{Alert}) phase of quiescence; this increases the cell size, mitochondrial activity, and ATP concentration and, ultimately commits these stem cells to cell cycle entry, proliferation, and differentiation (Figure 4) [20,275,276].

2.8. Several Mechanisms of Proteostasis Maintenance in Quiescent Adult Stem Cells Define Their Fate

As discussed in Section 2.3, an essential role of ROS in defining the homeostasis and functionality of quiescent adult stem cells consists in the ability of ROS concentrations that exceed a toxic threshold to inflict oxidative damage to cellular proteins. Several mechanisms for maintaining protein homeostasis (proteostasis) within these cells are known to play essential roles in preserving the quiescence, self-renewal, proliferation capacity, differentiation potential, and long-term survival of adult stem cells. Some of these mechanisms prevent a collapse of cellular proteostasis by regulating protein synthesis and folding, whereas other mechanisms are involved in the proteasomal or autophagic degradation of damaged and dysfunctional proteins and organelles [29,277–280]. These mechanisms are integrated into a proteostatic network orchestrated by several signaling pathways and transcriptional factors that define the fate of adult stem cells [29,278,279]. The common traits of the proteostatic network operating in adult stem cells include the following: (1) a suppression of protein synthesis on free ribosomes in the cytosol; (2) a rise in the concentrations of heat shock proteins (HSPs) that act as chaperones or co-chaperones to assist the proper folding and stability of newly synthesized proteins; (3) an activation of the unfolded protein response (UPR) systems in the endoplasmic reticulum (ER) and mitochondria (UPR^{ER} and UPR^{mit}, respectively); (4) an enhancement of the ubiquitin system for the ubiquitin/proteasome-dependent proteolytic clearance of improperly folded proteins, cell-cycle regulators, and transcriptional factors, as well as for the proteasome-independent regulation of surface protein receptors, histones, ribosome assembly factors, and vesicle transport proteins; and (5) a stimulation of autophagy, a quality control mechanism for the regulated lysosomal degradation of dysfunctional or excessive proteins and organelles, in some types of adult stem cells or its maintenance at a basal level in other types of such cells (Figures 5 and 6) [29,79,277–284]. The involvement of each of these proteostatic mechanisms in the homeostasis and functionality of adult stem cells is outlined below in this section.

2.8.1. Protein Synthesis on Free Ribosomes in the Cytosol

The rate of protein synthesis in hematopoietic stem cells is significantly lower than that in differentiating progenitor cells formed from them, which is likely because the repressor 4EBP1 (eukaryotic translation initiation factor 4E-binding protein 1) of cap-dependent translation is less phosphorylated and more active in these adult stem cells [285]. The protein synthesis rate that is maintained in hematopoietic stem cells is optimal for sustaining their quiescence and self-renewal potential. In support of this notion, it was shown that (1) a mutation that increases such a rate within hematopoietic stem cells in an mTORC1-dependent manner depletes their pool; (2) a mutation that decreases the protein synthesis rate within hematopoietic stem cells weakens their self-renewal ability; and (3) a combination of these two mutations restores the "optimal" protein synthesis rate within hematopoietic stem cells and salvages the above effects of both mutations on stem cell quiescence and self-renewal [259,285]. Therefore, the 4EBP1-dependent decline in protein synthesis rate observed in hematopoietic stem cells is required for the maintenance of their homeostasis and functionality. The efficiency of ribosome assembly in hematopoietic stem cells is an essential contributing factor to such an indispensable role of the "optimal" protein synthesis rate in defining the fate of these adult stem cells. In fact, a mutation in the ribosome protein Rpl22l impairs the emergence of hematopoietic stem cells [286]. Furthermore, a depletion of the transcription factor Runx1 in hematopoietic stem cells weakens ribosome biogenesis, decelerates protein synthesis, decreases cell susceptibility to apoptotic death, and makes these adult stem cells more resistant to genotoxic and ER stresses [287]. Moreover, a mutation that impairs the maturation and nuclear export of the pre-60S ribosomal subunit in hematopoietic stem cells causes an exhaustion of the pools of quiescent cells and the multipotent progenitors formed from them, but not of the pool of mature hematopoietic cells [288]. In addition, a decline in ribosome abundance within hematopoietic stem cells suppresses the translation of a specific set of mRNAs that are essential for erythroid lineage commitment after these stem cells exit the quiescent state [289].

Akin to hematopoietic stem cell quiescence, self-renewal, and differentiation, the fate of neural stem cells also depends on the rate of mTORC1-regulated protein synthesis. Specifically, the constitutive activation of 4EBP1 (which represses cap-dependent translation) in neural stem cells with hyperactive mTORC1 and protein synthesis exceeding the optimal rate has been shown to restore the optimal rate of protein synthesis, increase self-renewal potential, and slow down the accelerated differentiation of these adult stem cells [290].

Protein synthesis on free ribosomes in the cytosol also defines the homeostasis and functionality of epidermal stem cells, as a mutation that impairs the recycling and rescue of ribosomes stalled before protein synthesis completion has been demonstrated to increase the synthesis rates of all the cellular proteins in an mTORC1-dependent manner, cause the excessive proliferation of these adult stem cells, and alter their normal differentiation pattern [291]. Since a suppression of mTORC1 signaling by rapamycin in epidermal stem cells carrying this mutation decreases the global protein synthesis rate and salvages the above effects of the mutation of the fate of epidermal stem cells, the maintenance of a protein synthesis rate below a certain threshold is essential for sustaining the homeostasis and functionality of these adult stem cells [291].

The fate of skeletal muscle stem cells depends on the synthesis rates of myogenic factor 5 (Myf5) and myoblast determination protein (MyoD) that are translated from a distinct set of mRNA templates. In quiescent skeletal muscle stem cells, the translation of these mRNAs is selectively suppressed either by a protein kinase R (PKR)-like endoplasmic reticulum kinase (PERK)-dependent phosphorylation of translation initiation factor eIF2α [292] or by their association with certain microRNAs [293,294]. This causes an accumulation of such mRNAs in cytoplasmic mRNP (messenger ribonucleoprotein) granules, thus silencing their translation [292–294]. Such silencing is essential for the maintenance of a reversible quiescent state by skeletal muscle stem cells [292–294]. Furthermore, the dissociation of the cytoplasmic mRNP granules caused either by an eIF2α dephosphorylation or by a downregulation of the microRNAs allows the translation of these mRNAs in skeletal muscle stem cells, and is required for both their exit from quiescence and their entry into a differentiation program [292–294].

2.8.2. HSP Concentrations

Many HSPs are chaperones or co-chaperones that assist the proper folding and stability of newly synthesized proteins [295,296]. The concentrations of some HSPs in various types of quiescent adult stem cells exceed those in the fully differentiated progeny of such cells [279,297]. Specifically, the abundance of HSP70 protein 5 (HSPA5), HSP70 protein 8 (HSPA8), and HOP (an HSP70-HSP90 organizing protein) in neural stem cells and mesenchymal stem cells is higher than in their differentiated progeny [298]. Adipose-derived stem cells exhibit higher concentrations of HSP27 (HSPB1), αB-crystallin (HSPB5), HSP20 (HSPB6), and HSP60 than the fully differentiated progeny of such cells [299].

Some of these HSPs play essential roles in preserving the quiescence, abundance, survival, proliferation capacity, and differentiation potential of adult stem cells, as outlined below.

Mechanisms underlying such effects of some HSPs have begun to emerge. HSC70 (HSPA8) is essential for the cytokine-mediated survival of hematopoietic stem cells and prevents their differentiation; this is because this HSP decreases the stability of mRNA encoding BIM, which is a BH3-only pro-apoptotic factor that compromises stem cell viability but is required for apoptosis during hematopoiesis and leukemogenesis (Figure 5) [300]. HSC70 (HSPA8) also interacts with cyclin D1 and cyclin-dependent kinase inhibitors p27 and p57 in the cytosol of hematopoietic stem cells [301]. Both p27 and p57 prevent the nuclear import of the HSC70/cyclin D1 complex, thereby allowing the maintenance of quiescence of hematopoietic stem cells (Figure 5) [301]. The GRP78 member of the HSP70 protein family binds to teratocarcinoma-derived growth factor 1 (TDGF-1) on the surface of hematopoietic stem cells, and this binding allows maintaining the quiescent state of these adult cells via the TDGF-1-dependent induction of glycolysis (Figure 5) [302]. HSP70 (HSPA5) indirectly stimulates the erythroid differentiation of hematopoietic stem cells because it prevents the caspase-3-mediated proteolysis of GATA sequence protein 1 (GATA-1), which is a transcriptional factor that is essential for the terminal differentiation of erythroid progenitors formed from these adult stem cells (Figure 5) [303]. The HSP mortalin in the mitochondria of hematopoietic stem cells interacts with the antioxidant deglycase protein 1(DJ-1), which is implicated in Parkinson's disease, and the mortalin/DJ-1 complex allows sustaining hematopoietic stem cell quiescence, abundance, and self-renewal capacity, because it protects the mitochondria of these cells from ROS accumulation and oxidative macromolecular damage (Figure 5) [304].

An overexpression of αB-crystallin (HspB5), a small HSP that is essential for muscle development and homeostasis [305,306], slows down the differentiation of skeletal muscle stem cells; this effect of αB-crystallin (HspB5) is due to its ability to decelerate synthesis and accelerate the degradation of MyoD, which is a master protein regulator of such differentiation (Figure 5) [307]. Moreover, HSP70 (HSPA5) is indispensable for the osteogenic and chondrogenic differentiation of human mesenchymal stem cells, perhaps because it enhances the expression of the bone morphogenetic protein 2 member of the transforming growth factor β (TGF-β) superfamily of secreted polypeptide factors [308,309].

2.8.3. The UPR^{ER} and UPR^{mit} Systems

If the HSPs-assisted folding of newly synthesized proteins is not enough to sustain their proper conformations, the build-up of misfolded and unfolded proteins in the ER and mitochondria activates the UPR^{ER} and UPR^{mit} systems (respectively) to refold or degrade these proteins and restore cellular proteostasis [310–313].

When activated, the UPR^{ER} system allows restoring proteostasis in the ER, because it decelerates protein synthesis to decrease protein flow to the ER, promotes a refolding of some improperly folded proteins accumulated in this organelle, and directs other improperly folded proteins amassed in the ER for degradation by ER-associated degradation (ERAD) or autophagy [310,311,314]. If these protective processes that are integrated into the UPR^{ER} system are unable to reinstate proteostasis in the ER, a coordinated action of the ER and mitochondria triggers the mitochondria-controlled apoptotic death of the entire cell [315].

When the UPR^{mit} system is activated, it enables proteostasis reinstatement in mitochondria because it stimulates the refolding of some improperly folded proteins that accumulated in these organelles with the help of mitochondrial chaperone systems and because it also promotes the proteolytic degradation of other improperly folded proteins with the help of proteolytic systems in mitochondria [312,313,316,317].

As outlined below, both the UPR^{ER} and UPR^{mit} systems of proteostasis restoration are indispensable for maintaining the quiescence, self-renewal, proliferation capacity, differentiation potential, and functionality of adult stem cells.

The RNA binding protein Dppa5 (developmental pluripotency-associated 5 protein), the tauroursodeoxycholic bile acid, and hypoxia-inducible factor 2α (HIF-2α) stimulate the UPR^{ER} system to elicit a decline in ER stress within hematopoietic stem cells and protect these cells from apoptotic death caused by excessive ER stress [318–320]. Such stimulation of the UPR^{ER} system is required for the maintenance, self-renewal, functionality, and survival of hematopoietic stem cells (Figure 6) [318–320]. If ER stress in the quiescent hematopoietic stem cells of humans exceeds a certain threshold and cannot be resolved by the UPR^{ER} system, these cells commit suicide by undergoing apoptotic death [321]. This prevents the propagation of dysfunctional quiescent hematopoietic stem cells that amass improperly folded and aggregated proteins in the ER [321]. Progenitor cell populations that are formed from these human quiescent hematopoietic stem cells following their activation can avoid an ER stress-induced apoptotic death, because they undergo an adaptive enhancement of the UPR^{ER} system and can resolve ER stress [321]. The UPR^{ER} system is also indispensable for the lymphopoietic and erythropoietic differentiation programs of primed mouse hematopoietic stem cells; however, mechanisms underlying such effects of the UPR^{ER} system require further investigation [322–324].

Of note, the UPR^{ER} system does not always define the fate of hematopoietic stem cells, because under certain conditions, ER stress can be resolved with the help of chemical compounds assisting in maintaining proteostasis within the ER. Specifically, bile acids secreted from maternal and fetal liver act as chemical chaperones that prevent the accumulation of improperly folded and aggregated proteins in the ER, thereby decreasing ER stress and allowing the expansion of a population of hematopoietic stem cells during hematopoiesis in the fetal liver of mice [325].

The protein kinase PERK of the UPR^{ER} phosphorylates the eukaryotic translation initiation factor 2α (eIF2α) in mouse skeletal muscle stem cells, thus suppressing the translation of many mRNAs that lack upstream open reading frames (uORFs) in their 5′ untranslated regions [292]. mRNAs whose translation is suppressed by eIF2α phosphorylation in these adult stem cells encode the proteins that are needed for quiescence exit followed by proliferation and myogenic differentiation, whereas mRNAs with uORFs whose translation is not suppressed by such phosphorylation encode proteins involved in quiescence maintenance (Figure 6) [292]. Mutations that prevent the PERK-dependent eIF2α phosphorylation impair the abilities of skeletal muscle stem cells to maintain quiescence and self-renew, while an inhibitor of eIF2a dephosphorylation increases the self-renewal potential of these cells and improves their functionality [292]. Thus, the PERK arm of the UPR^{ER} system controls the quiescence, self-renewal, differentiation, and functionality of skeletal muscle stem cells in mice. The activation of this arm of the UPR^{ER} system following muscle injury in mice is also essential for the survival of progenitor cells formed from primed skeletal muscle stem cells and for the differentiation of these progenitor cells during regenerative myogenesis and muscle formation [326].

In the nucleus of mouse quiescent hematopoietic stem cells, the histone deacetylase SIRT7 is very abundant and interacts with the transcription factor nuclear respiratory factor 1 (NRF1), which in its free form activates the transcription of genes encoding mitochondrial ribosomal proteins and translation factors [327]. The interaction between NRF1 and SIRT7 impairs the ability of NRF1 to activate the transcription of these genes, thus inhibiting mitochondrial translation and biogenesis, lowering mitochondrial protein stress, and suppressing the UPR^{mit} system (Figure 6) [327]. Such SIRT7-driven suppression of the UPR^{mit} system is essential for maintaining the quiescence and differentiation potential of hematopoietic stem cells [327].

Pharmacological interventions that rise the intracellular concentration of NAD$^+$ in mice have been shown to stimulate the protein deacetylase SIRT1, which then increases mitochondrial functionality and activates the UPR^{mit} system by increasing the abundance of two members of the prohibitin protein family of mitochondrial stress sensors and effectors (Figure 6) [75,239]. Such NAD$^+$/SIRT1-dependent activation of the UPR^{mit} system (including prohibitins) is indispensable for the quiescence, self-renewal, differentiation, and viability of muscle stem cells, neural stem cells, and melanocyte stem cells in mice (Figure 6) [75,239].

2.8.4. The Ubiquitin System

The enhancement of the ubiquitin system characteristic of quiescent adult stem cells is sometimes considered only as a mechanism for the ubiquitin/proteasome-dependent proteolytic clearance of improperly folded proteins whose proper conformations in these stem cells cannot be restored by their HSPs-assisted folding and/or by their UPR^{ER} and UPR^{mit}-dependent refolding or degradation [29,278,279]. However, the ubiquitin system is known to play essential proteolysis-related and proteolysis-unrelated roles in preserving the quiescence, self-renewal, proliferation capacity, and differentiation potential of adult stem cells, not because it helps to clear improperly folded proteins, but mainly because it regulates the stability, conformation, activity, localization, protein-binding specificity, or vesicular trafficking of properly folded cell-cycle regulators, surface protein receptors, transcription factors, histones, ribosome assembly factors, and vesicle transport proteins that define the fate of these stem cells [281,283,284].

The specificity with which ubiquitin is covalently attached to certain lysine residues of the proteins that define the fate of adult stem cells depends on many E3 ubiquitin ligases, each ubiquitinating a distinct set of protein targets [284,328]. The ubiquitin E3 ligases c-Cbl, Itch, Fbw7, Skp2, and Huwe1 are indispensable for sustaining quiescence and preventing the excessive self-renewal and overproliferation of hematopoietic stem cells, thus precluding a depletion of this adult stem cell population and impeding an exhaustion of hematopoiesis [29,281,283,284,329–346]. This is because these ubiquitin E3 ligases ubiquitinate and prime for proteasome degradation a distinct group of proteins that are essential for maintaining hematopoietic stem cell homeostasis by regulating signal transduction, the transcription of many nuclear genes, cell growth and proliferation, cell cycle progression, and mitochondria-controlled apoptosis; among these proteins are Notch1, c-Kit, STAT5, c-Myc, n-Myc, cyclin E, p27, and Mcl-1 [29,281,283,284,329–346]. A depletion of the ubiquitin E3 ligases c-Cbl, Itch, Fbw7, Skp2 and Huwe1 in hematopoietic stem cells elicits their loss of quiescence, enhances their self-renewal, increases their abundance, prompts their overproliferation, decelerates their differentiation, and eventually causes a depletion of the hematopoietic stem cell population [29,281,283,284,340,343,347–351].

When hematopoietic stem cells move out from their hypoxic niche in the low-oxygen environment of the bone marrow to enter the differentiation program, the ubiquitin E3 ligase VHL (von Hippel–Lindau) ubiquitinates the hypoxia-inducible transcription factor HIF-1α, thus promoting its proteasomal degradation [116,352]. HIF-1α is a master controller of the transcription program for sustaining the quiescence, functionality, and survival of the hematopoietic stem cells sustained within the low-tension oxygen niche [117,120,353]. The VHL-driven proteasomal degradation of HIF-1α in the hematopoietic stem cell niche is required for the migration of these cells from the niche and for the ensuing exit of these cells from quiescence and entry into differentiation [116]. In fact, VHL depletion increases the stability of HIF-1α, promotes the quiescence of hematopoietic stem cells and their early progenitors, and impairs the proliferation and differentiation of these cells during hematopoiesis [116].

In adult neural stem cells, the ubiquitin E3 ligase APC/C (anaphase-promoting complex) controls a balance between the cell quiescence maintenance program and the programs for cell quiescence exit, proliferation, and differentiation, because it primes for proteasome degradation or non-proteolytically activates certain cell-cycle regulation proteins, transcription factors, translation factors, and cell surface receptor proteins [284,354–359]. HUWE1 (HECT, UBA and WWE domain containing protein 1), another ubiquitin E3 ligase in adult neural stem cells, stimulates a re-entry of early neural progenitor cells into the quiescent state, because it ubiquitinates and primes some transcription factors for proteasome degradation, cell-cycle regulation proteins, and apoptotic death regulators [284,360–364].

In sum, the above findings indicate that the ability of ubiquitin E3 ligases to prime for proteasome degradation (or, in some cases, to activate in a non-proteolytic manner) a distinct set of proteins in different types of adult stem cells is essential for controlling a balance between the processes of cell quiescence maintenance, exit, and re-entry. As it has been reviewed elsewhere, the ubiquitin E3 ligases-dependent proteasome degradation is also indispensable for regulated self-renewal, which is a response to signals emerging from the niche, proliferation, differentiation, functionality and survival of various adult stem cells [281,283,284].

2.8.5. Autophagy

Autophagy, a quality control mechanism for regulated lysosomal degradation of damaged or dysfunctional proteins and organelles, is required for sustaining the quiescent state, self-renewal ability, proliferation capacity, differentiation potential, fitness, functionality, and/or long-term survival of quiescent adult stem cells [29,78,79,278,279,282,365,366].

In different types of adult stem cells, the activity of autophagy is either higher or lower than that in the differentiating progenitor cells formed from them [29,78,79,278,279,282,365,367].

Autophagy activity in hematopoietic, dermal, epidermal, and mesenchymal stem cells exceeds that in progenitor cells derived from these adult stem cells [368,369]. Such a rise in autophagic activity within adult hematopoietic stem cells is due to the induced expression of many pro-autophagy genes, which is driven by the forkhead transcription factor FoxO3 [369–371]. The increase of autophagic activity within hematopoietic stem cells is essential for sustaining their quiescent state, self-renewal ability, differentiation potential, high resistance to stresses, fitness, functionality, and viability [257,368,369,371–375]. In fact, mutations eliminating several key protein components of autophagic machinery impair these essential features of hematopoietic stem cells [257,368,369,371–375]. These mutations increase the abundance of mitochondria, stimulate mitochondrial ROS production, enhance ROS-inflicted oxidative damage to cellular components, and accelerate mitochondria-controlled apoptotic cell death [257,371–375]. Therefore, it is believed that the FoxO3-induced autophagy is essential for the quiescence, self-renewal, differentiation, stress resistance, fitness, functionality, and viability of hematopoietic stem cells, because it selectively eliminates functional and dysfunctional mitochondria to suppress the excessive formation of ROS in these organelles, lower oxidative cellular damage, and prevent mitochondria-controlled cell death [257,371–375].

Autophagy activity in neural stem cells and cardiac stem cells is lower than that in the progenitor cells formed from these adult stem cells; such activity further rises during the differentiation of both types of progenitor cells [376–378]. Mutations eliminating the Atg5 and Ambra1 protein components of autophagic machinery in neural stem cells and the pharmacological interventions that suppress autophagy in both these types of adult stem cells impair early steps of stem and progenitor cell differentiation [376–378]. In contrast, pharmacological interventions that stimulate autophagy in cardiac stem cells promote the differentiation of these stem cells and their early progenitors [377,378]. It is conceivable that autophagy activation in progenitor cells formed from neural stem cells and cardiac stem cells is required to fulfill the high-energy demands of the differentiation process [376–378]. Of note, akin to Atg5 and Ambra1 depletions, a depletion of the autophagy-inducing protein FIP200 impairs the differentiation of early progenitors of neural stem cells [379]. Since such an FIP200-dependent impairment of neural stem cell progenitors can be salvaged with the help of an anti-oxidant chemical compound, autophagy activation in these progenitor cells may not only provide energy to fuel the differentiation process, but may also suppress excessive oxidative damage to cellular components during differentiation [379]. FIP200 depletion also causes a progressive depletion of the pool of neural stem cells because it triggers their apoptotic death [379]. Thus, the basal level of autophagy activity observed in these adult stem cells plays an essential role in protecting them from the apoptotic death caused by excessive cellular stress.

Similar to the autophagy activity in neural stem cells and cardiac stem cells, such activity in the muscle stem cells of young mice is lower than the autophagy activity in the progenitor cells that are derived from them [365,367]. The basal level of autophagy activity in these muscle stem cells is essential for the establishment and maintenance of their quiescence state, as a depletion of the Atg7 protein component of autophagic machinery in these cells impairs the reversible G_0 state of quiescence, decreases stem cell number and functionality, and promotes entry into the irreversible G_0 state of senescence [365,367]. The entry of autophagy-deficient muscle stem cells into the irreversible senescent state is caused by a build-up of dysfunctional mitochondria, an accumulation of ROS in excessive concentrations, a rise in oxidative damage to cellular components, and a resulting decline of proteostasis in these cells [365,367]. The accumulation of excessive ROS in autophagy-deficient muscle stem cells is responsible for their entry into the irreversible senescent state; in fact, an anti-oxidant chemical compound that decreases cellular ROS has been shown to prevent such senescence entry and restore the self-renewal of autophagy-deficient muscle stem cells [365,367].

In addition to the essential role of autophagy in establishing and maintaining the quiescence of muscle stem cells, autophagy is also indispensable for the exit of muscle stem cells from the quiescent state, the activation of their myogenic differentiation, and the progression through several stages of such differentiation and muscle regeneration. Specifically, during the transition of muscle stem cells from the quiescent state to the activation state, autophagy activity rises in a SIRT1-dependent manner to provide the nutrients and ATP needed for the energy-demanding process of entry into myogenic differentiation [240]. Furthermore, in response to muscle injury, autophagy is stimulated in early progenitors of muscle stem cells, and the extent of such autophagy stimulation defines the efficiency of muscle regeneration via a presently unknown mechanism [380]. Moreover, autophagy is also involved in myotube formation during muscle regeneration, but the mechanism of such involvement remains to be investigated [381].

2.9. Cell Cycle Regulatory Proteins Are Essential for Sustaining the Quiescence, Self-Renewal, Functionality, and Differentiation Potential of Adult Stem Cells

Several cell cycle regulatory proteins are integrated into a network of cell-intrinsic mechanisms that define the fate of quiescent adult stem cells [2,4,382–387].

One of these proteins is the tumor suppressor protein p53, which is a transcription factor that can arrest the cell cycle at the G_1/S checkpoint because in response to DNA damage, it activates the expression of the cyclin-dependent kinase Cdk2 inhibitor p21 [388–390]. A depletion of p53 in hematopoietic stem cells, which promotes quiescence exit and cell cycle entry, enhances self-renewal and raises the number of fully functional hematopoietic stem cells [391–394]. Thus, p53 is essential for sustaining the quiescent state of hematopoietic stem cells and inhibiting their proliferation and self-renewal. The essential role of p53 in sustaining the pool of hematopoietic stem cells at a certain level is independent of p21 [394] but requires Necdin, which is a growth-suppressing protein whose expression is activated by p53 at the transcriptional level [395]. The ability of p53 to fully sustain the functional hematopoietic stem cells by promoting the apoptotic death of those of them that build-up excessive DNA damage is under the control of Aspp1, which is a member of the family of apoptosis-stimulating proteins of p53 [396].

p53 also suppresses the proliferation and self-renewal of neural stem cells and promotes the apoptotic death of those of them that are dysfunctional (likely because they amass DNA damage). In fact, p53 depletion in neural stem cells stimulates their proliferation and mitigates their apoptotic death [397]. The mechanisms underlying these effects of p53 in neural stem cells are presently unknown.

Three members of the retinoblastoma protein family (pRBs) are tumor suppressor proteins that, in their hypophosphorylated forms, can arrest the cell cycle at the G_1/S checkpoint, because they are transcriptional repressors of genes that are essential for G_1/S transition [398,399]. If pRBs are hyperphosphorylated by different cyclin/cyclin-dependent kinase complexes in response to certain pro-mitogenic stimuli, they unable to suppress cell cycle entry at the G_1/S checkpoint [399,400]. As outlined below, similar to p53, pRBs also define the fate of different types of quiescent adult stem cells.

A simultaneous depletion of all the pRBs in adult hematopoietic stem cells elicits the exit of these stem cells from quiescence, impairs the reconstitution ability of these stem cells in a transplantation assay, causes an excessive proliferation of both stem cells and their early hematopoietic progenitors, and promotes apoptosis in lymphoid (but not in myeloid) progenitor populations [401]. Hence, a collective action of pRBs is essential for sustaining the quiescence of hematopoietic stem cells, preventing their unbalanced proliferation, maintaining their transplantation functionality, and retaining an unbiased differentiation of the lymphoid and myeloid cell lineages in the hematopoietic system.

A depletion of a single member of the pRBs family in muscle stem cells causes their permanent exit from the quiescent state, accelerates cell cycle re-entry in both stem cells and their progenitor myoblast cells (thus substantially expanding these two cell populations), and significantly decelerates terminal differentiation into myotubes; these effects of the depletion of a single pRB member deteriorate muscle fiber formation, slow muscle growth, and delay muscle repair [402]. Thus, even a partial decline in the abundance of pRBs alters the fate of muscle stem cells.

Among the cell cycle regulatory proteins that define the fate of quiescent adult stem cells are several cyclin-dependent kinase inhibitors. Their essential roles in maintaining the functionality of adult stem cells are discussed below.

As mentioned in Section 2.8.2, the interaction of cyclin-dependent kinase inhibitors p27 and p57 with the HSC70/cyclin D1 complex in the cytosol of hematopoietic stem cells prevents the nuclear import of this protein complex [301]. A depletion of both p27 and p57 in mice permits the nuclear import of the HSC70/cyclin D1 complex, thereby stimulating pRB hyperphosphorylation by the complex formed between cyclin D1 and cyclin-dependent kinases Cdk4/6 in the nucleus and impairing the ability of pRB to arrest the cell cycle of hematopoietic stem cells at the G_1/S checkpoint [301]. Due to these effects, mouse hematopoietic stem cells depleted of both p27 and p57 fail to maintain quiescence, are decreased in number, have low self-renewal activity, and exhibit a loss of the reconstitution ability in a transplantation assay [301]. Thus, a collective action of p27 and p57 is essential for sustaining the quiescence, self-renewal, and functionality of hematopoietic stem cells.

A depletion of the cyclin-dependent kinase 2 (Cdk2) inhibitor p21 in mice impairs the abilities of hematopoietic stem cells and neural stem cells to sustain quiescence, lowers their self-renewal potentials, reduces their reconstitution abilities in a transplantation assay, decreases their numbers, and ultimately causes an exhaustion of their pools [403,404]. Hence, p21 is indispensable for the maintenance of the quiescence, self-renewal ability, and fitness in both hematopoietic stem cells and neural stem cells.

Int. J. Mol. Sci. **2019**, *20*, 2158

3. Conclusions

In this review, we compared many metabolic, signal transduction, gene expression, epigenetic, stress survival, cell cycle regulation, and other traits of different types of mammalian and human adult stem cells. Our comparison indicates that adult stem cells have evolved an intricate network of cell-intrinsic mechanisms that control the establishment and preservation of these traits. These mechanisms regulate quiescence entry, maintenance, and exit in response to certain extrinsic pro-mitogenic and anti-mitogenic cues from the microenvironment of adult stem cells. One important challenge is to understand how the numerous cell-intrinsic mechanisms of quiescence entry, maintenance, and exit are integrated in space and time within mammalian and human adult stem cells. The other challenge is to decipher the hierarchical order and relative contributions of the different traits of adult stem cells in the ability of these cells to enter, sustain, and exit quiescence in response to specific cell-extrinsic factors within different tissue-specific microenvironments. Future work will also aim at understanding how the history of spatial and temporal changes in cell growth and division conditions is translated into the pattern of quiescence entry, maintenance, and exit in different types of adult stem cells. Since the impairment of a balance between the quiescence, proliferation, and differentiation of adult stem cells has been implicated in the pathophysiology of many diseases of old age, addressing these challenges in the future will increase our understanding of how this balance can be controlled to delay cellular and organismal aging and postpone the onset of aging-associated diseases.

Author Contributions: K.M., P.D., Y.M., D.M. and V.I.T. wrote the text. V.I.T. prepared the figures.

Funding: This research was supported by grants from the NSERC of Canada to Vladimir I. Titorenko. Karamat Mohammad and Paméla Dakik were supported by the Concordia University Graduate Fellowship Awards. Younes Medkour was supported by the Concordia University Public Scholars Program Award. Vladimir I. Titorenko is a Concordia University Research Fellow.

Acknowledgments: We are grateful to current and former members of the Titorenko laboratory for discussions.

Conflicts of Interest: The authors declare no conflict of interest.

References

1. Gray, J.V.; Petsko, G.A.; Johnston, G.C.; Ringe, D.; Singer, R.A.; Werner-Washburne, M. "Sleeping beauty": Quiescence in *Saccharomyces cerevisiae*. *Microbiol. Mol. Biol. Rev.* **2004**, *68*, 187–206. [CrossRef] [PubMed]
2. Cheung, T.H.; Rando, T.A. Molecular regulation of stem cell quiescence. *Nat. Rev. Mol. Cell. Biol.* **2013**, *14*, 329–340. [CrossRef] [PubMed]
3. Dhawan, J.; Laxman, S. Decoding the stem cell quiescence cycle—Lessons from yeast for regenerative biology. *J. Cell Sci.* **2015**, *128*, 4467–4474. [CrossRef] [PubMed]
4. Rumman, M.; Dhawan, J.; Kassem, M. Concise Review: Quiescence in Adult Stem Cells: Biological Significance and Relevance to Tissue Regeneration. *Stem Cells* **2015**, *33*, 2903–2912. [CrossRef]
5. Pardee, A.B. A restriction point for control of normal animal cell proliferation. *Proc. Natl. Acad. Sci. USA* **1974**, *71*, 1286–1290. [CrossRef] [PubMed]
6. Foster, D.A.; Yellen, P.; Xu, L.; Saqcena, M. Regulation of G1 Cell Cycle Progression: Distinguishing the Restriction Point from a Nutrient-Sensing Cell Growth Checkpoint(s). *Genes Cancer* **2010**, *1*, 1124–1131. [CrossRef]
7. Hartwell, L.H.; Culotti, J.; Pringle, J.R.; Reid, B.J. Genetic control of the cell division cycle in yeast. *Science* **1974**, *183*, 46–51. [CrossRef]
8. De Virgilio, C. The essence of yeast quiescence. *FEMS Microbiol. Rev.* **2012**, *36*, 306–339. [CrossRef]
9. Cameron, I.L.; Bols, N.C. Effect of cell population density on G_2 arrest in *Tetrahymena*. *J. Cell Biol.* **1975**, *67*, 518–522. [CrossRef]
10. Drewinko, B.; Yang, L.Y.; Barlogie, B.; Trujillo, J.M. Cultured human tumour cells may be arrested in all stages of the cycle during stationary phase: Demonstration of quiescent cells in G_1, S and G_2 phase. *Cell Tissue Kinet.* **1984**, *17*, 453–463. [CrossRef]
11. Costello, G.; Rodgers, L.; Beach, D. Fission yeast enters the stationary phase G_0 state from either mitotic G_1 or G_2. *Curr. Genet.* **1986**, *11*, 119–125. [CrossRef]

12. Baisch, H. Different quiescence states of three culture cell lines detected by acridine orange staining of cellular RNA. *Cytometry* **1988**, *9*, 325–331. [CrossRef] [PubMed]

13. Wei, W.; Nurse, P.; Broek, D. Yeast cells can enter a quiescent state through G_1, S, G_2, or M phase of the cell cycle. *Cancer Res.* **1993**, *53*, 1867–1870.

14. Takeo, K.; Tanaka, R.; Miyaji, M.; Nishimura, K. Unbudded G_2 as well as G_1 arrest in the stationary phase of the basidiomycetous yeast *Cryptococcus neoformans*. *FEMS Microbiol. Lett.* **1995**, *29*, 231–235.

15. Cooper, S. Reappraisal of serum starvation, the restriction point, G_0, and G_1 phase arrest points. *FASEB J.* **2003**, *17*, 333–340. [CrossRef]

16. Klosinska, M.M.; Crutchfield, C.A.; Bradley, P.H.; Rabinowitz, J.D.; Broach, J.R. Yeast cells can access distinct quiescent states. *Genes Dev.* **2011**, *25*, 336–349. [CrossRef]

17. Daignan-Fornier, B.; Sagot, I. Proliferation/quiescence: The controversial "aller-retour". *Cell Div.* **2011**, *6*, 10. [CrossRef] [PubMed]

18. Laporte, D.; Lebaudy, A.; Sahin, A.; Pinson, B.; Ceschin, J.; Daignan-Fornier, B.; Sagot, I. Metabolic status rather than cell cycle signals control quiescence entry and exit. *J. Cell Biol.* **2011**, *192*, 949–957. [CrossRef]

19. Roche, B.; Arcangioli, B.; Martienssen, R. Transcriptional reprogramming in cellular quiescence. *RNA Biol.* **2017**, *14*, 843–853. [CrossRef]

20. Rodgers, J.T.; King, K.Y.; Brett, J.O.; Cromie, M.J.; Charville, G.W.; Maguire, K.K.; Brunson, C.; Mastey, N.; Liu, L.; Tsai, C.R.; et al. mTORC1 controls the adaptive transition of quiescent stem cells from G_0 to G_{Alert}. *Nature* **2014**, *510*, 393–396. [CrossRef]

21. McGann, C.J.; Odelberg, S.J.; Keating, M.T. Mammalian myotube dedifferentiation induced by newt regeneration extract. *Proc. Natl. Acad. Sci. USA* **2001**, *98*, 13699–13704. [CrossRef]

22. Beauséjour, C.M.; Krtolica, A.; Galimi, F.; Narita, M.; Lowe, S.W.; Yaswen, P.; Campisi, J. Reversal of human cellular senescence: Roles of the p53 and p16 pathways. *EMBO J.* **2003**, *22*, 4212–4222. [CrossRef]

23. Fausto, N. Liver regeneration and repair: Hepatocytes, progenitor cells, and stem cells. *Hepatology* **2004**, *39*, 1477–1487. [CrossRef]

24. Pajcini, K.V.; Corbel, S.Y.; Sage, J.; Pomerantz, J.H.; Blau, H.M. Transient inactivation of Rb and ARF yields regenerative cells from postmitotic mammalian muscle. *Cell Stem Cell* **2010**, *7*, 198–213. [CrossRef]

25. Clevers, H. Stem cells. What is an adult stem cell? *Science* **2015**, *350*, 1319–1320. [CrossRef]

26. Coller, H.A.; Sang, L.; Roberts, J.M. A new description of cellular quiescence. *PLoS Biol.* **2006**, *4*, e83. [CrossRef]

27. Ochocki, J.D.; Simon, M.C. Nutrient-sensing pathways and metabolic regulation in stem cells. *J. Cell Biol.* **2013**, *203*, 23–33. [CrossRef] [PubMed]

28. Ito, K.; Suda, T. Metabolic requirements for the maintenance of self-renewing stem cells. *Nat. Rev. Mol. Cell Biol.* **2014**, *15*, 243–256. [CrossRef] [PubMed]

29. García-Prat, L.; Sousa-Victor, P.; Muñoz-Cánoves, P. Proteostatic and Metabolic Control of Stemness. *Cell Stem Cell* **2017**, *20*, 593–608. [CrossRef]

30. Kwon, J.S.; Everetts, N.J.; Wang, X.; Wang, W.; Della Croce, K.; Xing, J.; Yao, G. Controlling Depth of Cellular Quiescence by an Rb-E2F Network Switch. *Cell Rep.* **2017**, *20*, 3223–3235. [CrossRef] [PubMed]

31. Ren, R.; Ocampo, A.; Liu, G.H.; Izpisua Belmonte, J.C. Regulation of Stem Cell Aging by Metabolism and Epigenetics. *Cell Metab.* **2017**, *26*, 460–474. [CrossRef] [PubMed]

32. Bi, S.; Wang, H.; Kuang, W. Stem cell rejuvenation and the role of autophagy in age retardation by caloric restriction: An update. *Mech. Ageing Dev.* **2018**, *175*, 46–54. [CrossRef]

33. Lewis, D.B.; Gattie, G.T. The ecology of quiescent microbes. *ASM News* **1991**, *57*, 27–32.

34. Finkel, S.E. Long-term survival during stationary phase: Evolution and the GASP phenotype. *Nat. Rev. Microbiol.* **2006**, *4*, 113–120. [CrossRef] [PubMed]

35. Werner-Washburne, M.; Roy, S.; Davidson, G.S. Aging and the survival of quiescent and non-quiescent cells in yeast stationary-phase cultures. *Subcell. Biochem.* **2012**, *57*, 123–143. [PubMed]

36. Rittershaus, E.S.; Baek, S.H.; Sassetti, C.M. The normalcy of dormancy: Common themes in microbial quiescence. *Cell Host Microbe* **2013**, *13*, 643–651. [CrossRef] [PubMed]

37. Scheres, B. Stem-cell niches: Nursery rhymes across kingdoms. *Nat. Rev. Mol. Cell Biol.* **2007**, *8*, 345–354. [CrossRef] [PubMed]

38. Shim, J.; Mukherjee, T.; Banerjee, U. Direct sensing of systemic and nutritional signals by haematopoietic progenitors in *Drosophila*. *Nat. Cell Biol.* **2012**, *14*, 394–400. [CrossRef]

39. Heyman, J.; Kumpf, R.P.; De Veylder, L. A quiescent path to plant longevity. *Trends Cell Biol.* **2014**, *24*, 443–448. [CrossRef]

40. Seidel, H.S.; Kimble, J. Cell-cycle quiescence maintains *Caenorhabditis elegans* germline stem cells independent of GLP-1/Notch. *eLife* **2015**, *4*, e10832. [CrossRef]

41. Narbonne, P.; Gerhold, A.R.; Maddox, P.S.; Labbé, J.C. The C. *elegans* GSCs: A Powerful Model for In Vivo Study of Adult Stem Cell Regulation. *Int. J. Stem Cell Res. Ther.* **2016**, *3*, 044. [CrossRef]

42. Rovere, F.D.; Fattorini, L.; Ronzan, M.; Falasca, G.; Altamura, M.M. The quiescent center and the stem cell niche in the adventitious roots of *Arabidopsis thaliana*. *Plant Signal. Behav.* **2016**, *11*, e1176660. [CrossRef] [PubMed]

43. Schell, J.C.; Wisidagama, D.R.; Bensard, C.; Zhao, H.; Wei, P.; Tanner, J.; Flores, A.; Mohlman, J.; Sorensen, L.K.; Earl, C.S.; et al. Control of intestinal stem cell function and proliferation by mitochondrial pyruvate metabolism. *Nat. Cell Biol.* **2017**, *19*, 1027–1036. [CrossRef]

44. Velappan, Y.; Signorelli, S.; Considine, M.J. Cell cycle arrest in plants: What distinguishes quiescence, dormancy and differentiated G_1? *Ann. Bot.* **2017**, *120*, 495–509. [CrossRef] [PubMed]

45. Macedo, J.C.; Vaz, S.; Logarinho, E. Mitotic Dysfunction Associated with Aging Hallmarks. *Adv. Exp. Med. Biol.* **2017**, *1002*, 153–188.

46. Ahlqvist, K.J.; Suomalainen, A.; Hämäläinen, R.H. Stem cells, mitochondria and aging. *Biochim. Biophys. Acta* **2015**, *1847*, 1380–1386. [CrossRef] [PubMed]

47. Eridani, S.; Sgaramella, V.; Cova, L. Stem cells: From embryology to cellular therapy? An appraisal of the present state of art. *Cytotechnology* **2004**, *44*, 125–141. [CrossRef]

48. Mihaylova, M.M.; Sabatini, D.M.; Yilmaz, Ö.H. Dietary and metabolic control of stem cell function in physiology and cancer. *Cell Stem Cell* **2014**, *14*, 292–305. [CrossRef]

49. Ito, K.; Ito, K. Metabolism and the Control of Cell Fate Decisions and Stem Cell Renewal. *Annu. Rev. Cell Dev. Biol.* **2016**, *32*, 399–409. [CrossRef]

50. Santoro, A.; Vlachou, T.; Carminati, M.; Pelicci, P.G.; Mapelli, M. Molecular mechanisms of asymmetric divisions in mammary stem cells. *EMBO Rep.* **2016**, *17*, 1700–1720. [CrossRef]

51. Moore, D.L.; Jessberger, S. Creating Age Asymmetry: Consequences of Inheriting Damaged Goods in Mammalian Cells. *Trends Cell Biol.* **2017**, *27*, 82–92. [CrossRef] [PubMed]

52. Shen, Q.; Goderie, S.K.; Jin, L.; Karanth, N.; Sun, Y.; Abramova, N.; Vincent, P.; Pumiglia, K.; Temple, S. Endothelial cells stimulate self-renewal and expand neurogenesis of neural stem cells. *Science* **2004**, *304*, 1338–1340. [CrossRef]

53. Morrison, S.J.; Kimble, J. Asymmetric and symmetric stem-cell divisions in development and cancer. *Nature* **2006**, *441*, 1068–1074. [CrossRef]

54. Clayton, E.; Doupé, D.P.; Klein, A.M.; Winton, D.J.; Simons, B.D.; Jones, P.H. A single type of progenitor cell maintains normal epidermis. *Nature* **2007**, *446*, 185–189. [CrossRef]

55. Cicalese, A.; Bonizzi, G.; Pasi, C.E.; Faretta, M.; Ronzoni, S.; Giulini, B.; Brisken, C.; Minucci, S.; Di Fiore, P.P.; Pelicci, P.G. The tumor suppressor p53 regulates polarity of self-renewing divisions in mammary stem cells. *Cell* **2009**, *138*, 1083–1095. [CrossRef]

56. Zhang, Y.V.; Cheong, J.; Ciapurin, N.; McDermitt, D.J.; Tumbar, T. Distinct self-renewal and differentiation phases in the niche of infrequently dividing hair follicle stem cells. *Cell Stem Cell* **2009**, *5*, 267–278. [CrossRef]

57. Snippert, H.J.; van der Flier, L.G.; Sato, T.; van Es, J.H.; van den Born, M.; Kroon-Veenboer, C.; Barker, N.; Klein, A.M.; van Rheenen, J.; Simons, B.D.; et al. Intestinal crypt homeostasis results from neutral competition between symmetrically dividing Lgr5 stem cells. *Cell* **2010**, *143*, 134–144. [CrossRef]

58. O'Brien, L.E.; Soliman, S.S.; Li, X.; Bilder, D. Altered modes of stem cell division drive adaptive intestinal growth. *Cell* **2011**, *147*, 603–614. [CrossRef]

59. Shahriyari, L.; Komarova, N.L. Symmetric vs. asymmetric stem cell divisions: An adaptation against cancer? *PLoS ONE* **2013**, *8*, e76195. [CrossRef]

60. Katajisto, P.; Döhla, J.; Chaffer, C.L.; Pentinmikko, N.; Marjanovic, N.; Iqbal, S.; Zoncu, R.; Chen, W.; Weinberg, R.A.; Sabatini, D.M. Stem cells. Asymmetric apportioning of aged mitochondria between daughter cells is required for stemness. *Science* **2015**, *348*, 340–343. [CrossRef]

61. Ocampo, A.; Reddy, P.; Martinez-Redondo, P.; Platero-Luengo, A.; Hatanaka, F.; Hishida, T.; Li, M.; Lam, D.; Kurita, M.; Beyret, E.; et al. In vivo amelioration of age-associated hallmarks by partial reprogramming. *Cell* **2016**, *167*, 1719–1733. [CrossRef] [PubMed]

62. Wu, J.; Izpisua Belmonte, J.C. Stem Cells: A Renaissance in Human Biology Research. *Cell* **2016**, *165*, 1572–1585. [CrossRef] [PubMed]

63. Wu, J.; Ocampo, A.; Izpisua Belmonte, J.C. Cellular metabolism and induced pluripotency. *Cell* **2016**, *166*, 1371–1385. [CrossRef]

64. Ahmed, A.S.; Sheng, M.H.; Wasnik, S.; Baylink, D.J.; Lau, K.W. Effect of aging on stem cells. *World, J. Exp. Med.* **2017**, *7*, 1–10. [CrossRef] [PubMed]

65. García-Prat, L.; Sousa-Victor, P.; Muñoz-Cánoves, P. Functional dysregulation of stem cells during aging: A focus on skeletal muscle stem cells. *FEBS J.* **2013**, *280*, 4051–4062. [CrossRef]

66. Li, M.; Izpisua Belmonte, J.C. Genetic rejuvenation of old muscle. *Nature* **2014**, *506*, 304–305. [CrossRef]

67. Sousa-Victor, P.; Gutarra, S.; García-Prat, L.; Rodriguez-Ubreva, J.; Ortet, L.; Ruiz-Bonilla, V.; Jardí, M.; Ballestar, E.; González, S.; Serrano, A.L.; et al. Geriatric muscle stem cells switch reversible quiescence into senescence. *Nature* **2014**, *506*, 316–321. [CrossRef]

68. Goodell, M.A.; Rando, T.A. Stem cells and healthy aging. *Science* **2015**, *350*, 1199–1204. [CrossRef]

69. Akunuru, S.; Geiger, H. Aging, Clonality, and Rejuvenation of Hematopoietic Stem Cells. *Trends Mol. Med.* **2016**, *22*, 701–712. [CrossRef]

70. Almada, A.E.; Wagers, A.J. Molecular circuitry of stem cell fate in skeletal muscle regeneration, ageing and disease. *Nat. Rev. Mol. Cell Biol.* **2016**, *17*, 267–279. [CrossRef]

71. Chandel, N.S.; Jasper, H.; Ho, T.T.; Passegué, E. Metabolic regulation of stem cell function in tissue homeostasis and organismal ageing. *Nat. Cell Biol.* **2016**, *18*, 823–832. [CrossRef] [PubMed]

72. Ocampo, A.; Reddy, P.; Izpisua Belmonte, J.C. Anti-Aging Strategies Based on Cellular Reprogramming. *Trends Mol. Med.* **2016**, *22*, 725–738. [CrossRef] [PubMed]

73. Schultz, M.B.; Sinclair, D.A. When stem cells grow old: Phenotypes and mechanisms of stem cell aging. *Development* **2016**, *143*, 3–14. [CrossRef]

74. Soria-Valles, C.; López-Otín, C. iPSCs: On the Road to Reprogramming Aging. *Trends Mol. Med.* **2016**, *22*, 713–724. [CrossRef] [PubMed]

75. Zhang, H.; Ryu, D.; Wu, Y.; Gariani, K.; Wang, X.; Luan, P.; D'Amico, D.; Ropelle, E.R.; Lutolf, M.P.; Aebersold, R.; et al. NAD$^+$ repletion improves mitochondrial and stem cell function and enhances life span in mice. *Science* **2016**, *352*, 1436–1443. [CrossRef]

76. Artoni, F.; Kreipke, R.E.; Palmeira, O.; Dixon, C.; Goldberg, Z.; Ruohola-Baker, H. Loss of *foxo* rescues stem cell aging in *Drosophila* germ line. *eLife* **2017**, *6*, e27842. [CrossRef

77. Brunet, A.; Rando, T.A. Interaction between epigenetic and metabolism in aging stem cells. *Curr. Opin. Cell Biol.* **2017**, *45*, 1–7. [CrossRef]

78. García-Prat, L.; Muñoz-Cánoves, P. Aging, metabolism and stem cells: Spotlight on muscle stem cells. *Mol. Cell Endocrinol.* **2017**, *445*, 109–117. [CrossRef]

79. Revuelta, M.; Matheu, A. Autophagy in stem cell aging. *Aging Cell* **2017**, *16*, 912–915. [CrossRef] [PubMed]

80. Solanas, G.; Peixoto, F.O.; Perdiguero, E.; Jardí, M.; Ruiz-Bonilla, V.; Datta, D.; Symeonidi, A.; Castellanos, A.; Welz, P.S.; Caballero, J.M.; et al. Aged Stem Cells Reprogram Their Daily Rhythmic Functions to Adapt to Stress. *Cell* **2017**, *170*, 678–692. [CrossRef]

81. Keyes, B.E.; Fuchs, E. Stem cells: Aging and transcriptional fingerprints. *J. Cell Biol.* **2018**, *217*, 79–92. [CrossRef] [PubMed]

82. Mahmoudi, S.; Brunet, A. Bursts of Reprogramming: A Path to Extend Lifespan? *Cell* **2016**, *167*, 1672–1674. [CrossRef]

83. Meyer, K.; Yankner, B.A. Slowing Down Aging. *Cell Metab.* **2017**, *26*, 592–593. [CrossRef] [PubMed]

84. Zhang, Y.; Kim, M.S.; Jia, B.; Yan, J.; Zuniga-Hertz, J.P.; Han, C.; Cai, D. Hypothalamic stem cells control ageing speed partly through exosomal miRNAs. *Nature* **2017**, *548*, 52–57. [CrossRef]

85. Brack, A.S.; Conboy, M.J.; Roy, S.; Lee, M.; Kuo, C.J.; Keller, C.; Rando, T.A. Increased Wnt signaling during aging alters muscle stem cell fate and increases fibrosis. *Science* **2007**, *317*, 807–810. [CrossRef]

86. Chakkalakal, J.V.; Jones, K.M.; Basson, M.A.; Brack, A.S. The aged niche disrupts muscle stem cell quiescence. *Nature* **2012**, *490*, 355–360. [CrossRef]

87. Conboy, I.M.; Rando, T.A. Heterochronic parabiosis for the study of the effects of aging on stem cells and their niches. *Cell Cycle* **2012**, *11*, 2260–2267. [CrossRef]

88. Doles, J.; Storer, M.; Cozzuto, L.; Roma, G.; Keyes, W.M. Age-associated inflammation inhibits epidermal stem cell function. *Genes Dev.* **2012**, *26*, 2144–2153. [CrossRef] [PubMed]

89. Keyes, B.E.; Segal, J.P.; Heller, E.; Lien, W.H.; Chang, C.Y.; Guo, X.; Oristian, D.S.; Zheng, D.; Fuchs, E. Nfatc1 orchestrates aging in hair follicle stem cells. *Proc. Natl. Acad. Sci. USA* **2013**, *110*, E4950–E4959. [CrossRef] [PubMed]

90. Chen, C.C.; Murray, P.J.; Jiang, T.X.; Plikus, M.V.; Chang, Y.T.; Lee, O.K.; Widelitz, R.B.; Chuong, C.M. Regenerative hair waves in aging mice and extra-follicular modulators follistatin, dkk1, and sfrp4. *J. Investig. Dermatol.* **2014**, *134*, 2086–2096. [CrossRef] [PubMed]

91. So, W.K.; Cheung, T.H. Molecular Regulation of Cellular Quiescence: A Perspective from Adult Stem Cells and Its Niches. *Methods Mol. Biol.* **2018**, *1686*, 1–25. [PubMed]

92. Hsu, Y.C.; Fuchs, E. A family business: Stem cell progeny join the niche to regulate homeostasis. *Nat. Rev. Mol. Cell Biol.* **2012**, *13*, 103–114. [CrossRef] [PubMed]

93. Scadden, D.T. Nice neighborhood: Emerging concepts of the stem cell niche. *Cell* **2014**, *157*, 41–50. [CrossRef] [PubMed]

94. Khacho, M.; Slack, R.S. Mitochondrial activity in the regulation of stem cell self-renewal and differentiation. *Curr Opin Cell Biol.* **2017**, *49*, 1–8. [CrossRef] [PubMed]

95. Shyh-Chang, N.; Ng, H.H. The metabolic programming of stem cells. *Genes Dev.* **2017**, *31*, 336–346. [CrossRef]

96. Lisowski, P.; Kannan, P.; Mlody, B.; Prigione, A. Mitochondria and the dynamic control of stem cell homeostasis. *EMBO Rep.* **2018**, *19*, e45432. [CrossRef]

97. Wei, P.; Dove, K.K.; Bensard, C.; Schell, J.C.; Rutter, J. The Force Is Strong with This One: Metabolism (Over)powers Stem Cell Fate. *Trends Cell Biol.* **2018**, *28*, 551–559. [CrossRef]

98. Zhang, J.; Nuebel, E.; Daley, G.Q.; Koehler, C.M.; Teitell, M.A. Metabolic regulation in pluripotent stem cells during reprogramming and self-renewal. *Cell Stem Cell* **2012**, *11*, 589–595. [CrossRef]

99. Shyh-Chang, N.; Daley, G.Q.; Cantley, L.C. Stem cell metabolism in tissue development and aging. *Development* **2013**, *140*, 2535–2547. [CrossRef]

100. Folmes, C.D.; Nelson, T.J.; Martinez-Fernandez, A.; Arrell, D.K.; Lindor, J.Z.; Dzeja, P.P.; Ikeda, Y.; Perez-Terzic, C.; Terzic, A. Somatic oxidative bioenergetics transitions into pluripotency-dependent glycolysis to facilitate nuclear reprogramming. *Cell Metab.* **2011**, *14*, 264–271. [CrossRef] [PubMed]

101. Varum, S.; Rodrigues, A.S.; Moura, M.B.; Momcilovic, O.; Easley, C.A.; Ramalho-Santos, J.; Van Houten, B.; Schatten, G. Energy metabolism in human pluripotent stem cells and their differentiated counterparts. *PLoS ONE* **2011**, *6*, e20914. [CrossRef]

102. Stringari, C.; Edwards, R.A.; Pate, K.T.; Waterman, M.L.; Donovan, P.J.; Gratton, E. Metabolic trajectory of cellular differentiation in small intestine by Phasor Fluorescence Lifetime Microscopy of NADH. *Sci. Rep.* **2012**, *2*, 568. [CrossRef] [PubMed]

103. Flores, A.; Schell, J.; Krall, A.S.; Jelinek, D.; Miranda, M.; Grigorian, M.; Braas, D.; White, A.C.; Zhou, J.L.; Graham, N.A.; et al. Lactate dehydrogenase activity drives hair follicle stem cell activation. *Nat. Cell Biol.* **2017**, *19*, 1017–1026. [CrossRef]

104. St John, J.C.; Ramalho-Santos, J.; Gray, H.L.; Petrosko, P.; Rawe, V.Y.; Navara, C.S.; Simerly, C.R.; Schatten, G.P. The expression of mitochondrial DNA transcription factors during early cardiomyocyte in vitro differentiation from human embryonic stem cells. *Cloning Stem Cells* **2005**, *7*, 141–153. [CrossRef] [PubMed]

105. Prigione, A.; Fauler, B.; Lurz, R.; Lehrach, H.; Adjaye, J. The senescence-related mitochondrial/oxidative stress pathway is repressed in human induced pluripotent stem cells. *Stem Cells* **2010**, *28*, 721–733. [CrossRef]

106. Zhang, J.; Khvorostov, I.; Hong, J.S.; Oktay, Y.; Vergnes, L.; Nuebel, E.; Wahjudi, P.N.; Setoguchi, K.; Wang, G.; Do, A.; et al. UCP2 regulates energy metabolism and differentiation potential of human pluripotent stem cells. *EMBO J.* **2011**, *30*, 4860–4873. [CrossRef]

107. Schuijers, J.; Clevers, H. Adult mammalian stem cells: The role of Wnt, Lgr5 and R-spondins. *EMBO J.* **2012**, *31*, 2685–2696. [CrossRef] [PubMed]

108. Sánchez-Aragó, M.; García-Bermúdez, J.; Martínez-Reyes, I.; Santacatterina, F.; Cuezva, J.M. Degradation of IF1 controls energy metabolism during osteogenic differentiation of stem cells. *EMBO Rep.* **2013**, *14*, 638–644. [CrossRef] [PubMed]

109. Takubo, K.; Nagamatsu, G.; Kobayashi, C.I.; Nakamura-Ishizu, A.; Kobayashi, H.; Ikeda, E.; Goda, N.; Rahimi, Y.; Johnson, R.S.; Soga, T.; et al. Regulation of glycolysis by Pdk functions as a metabolic checkpoint for cell cycle quiescence in hematopoietic stem cells. *Cell Stem Cell* **2013**, *12*, 49–61. [CrossRef] [PubMed]

110. Maryanovich, M.; Zaltsman, Y.; Ruggiero, A.; Goldman, A.; Shachnai, L.; Zaidman, S.L.; Porat, Z.; Golan, K.; Lapidot, T.; Gross, A. An MTCH2 pathway repressing mitochondria metabolism regulates haematopoietic stem cell fate. *Nat. Commun.* **2015**, *6*, 7901. [CrossRef]

111. Wanet, A.; Arnould, T.; Najimi, M.; Renard, P. Connecting Mitochondria, Metabolism, and Stem Cell Fate. *Stem Cells Dev.* **2015**, *24*, 1957–1971. [CrossRef]

112. Prieto, J.; León, M.; Ponsoda, X.; Sendra, R.; Bort, R.; Ferrer-Lorente, R.; Raya, A.; López-García, C.; Torres, J. Early ERK1/2 activation promotes DRP1-dependent mitochondrial fission necessary for cell reprogramming. *Nat. Commun.* **2016**, *7*, 11124. [CrossRef]

113. Schell, J.C.; Rutter, J. Mitochondria link metabolism and epigenetics in haematopoiesis. *Nat. Cell Biol.* **2017**, *19*, 589–591. [CrossRef]

114. Eliasson, P.; Rehn, M.; Hammar, P.; Larsson, P.; Sirenko, O.; Flippin, L.A.; Cammenga, J.; Jönsson, J.I. Hypoxia mediates low cell-cycle activity and increases the proportion of long-term-reconstituting hematopoietic stem cells during in vitro culture. *Exp. Hematol.* **2010**, *38*, 301–310. [CrossRef]

115. Simsek, T.; Kocabas, F.; Zheng, J.; Deberardinis, R.J.; Mahmoud, A.I.; Olson, E.N.; Schneider, J.W.; Zhang, C.C.; Sadek, H.A. The distinct metabolic profile of hematopoietic stem cells reflects their location in a hypoxic niche. *Cell Stem Cell* **2010**, *7*, 380–390. [CrossRef]

116. Takubo, K.; Goda, N.; Yamada, W.; Iriuchishima, H.; Ikeda, E.; Kubota, Y.; Shima, H.; Johnson, R.S.; Hirao, A.; Suematsu, M.; et al. Regulation of the HIF-1alpha level is essential for hematopoietic stem cells. *Cell Stem Cell* **2010**, *7*, 391–402. [CrossRef]

117. Suda, T.; Takubo, K.; Semenza, G.L. Metabolic regulation of hematopoietic stem cells in the hypoxic niche. *Cell Stem Cell* **2011**, *9*, 298–310. [CrossRef]

118. Klimmeck, D.; Hansson, J.; Raffel, S.; Vakhrushev, S.Y.; Trumpp, A.; Krijgsveld, J. Proteomic cornerstones of hematopoietic stem cell differentiation: Distinct signatures of multipotent progenitors and myeloid committed cells. *Mol. Cell. Proteom.* **2012**, *11*, 286–302. [CrossRef]

119. Spencer, J.A.; Ferraro, F.; Roussakis, E.; Klein, A.; Wu, J.; Runners, J.M.; Zaher, W.; Mortensen, L.J.; Alt, C.; Turcotte, R.; et al. Direct measurement of local oxygen concentration in the bone marrow of live animals. *Nature* **2014**, *508*, 269–273. [CrossRef]

120. Zhang, C.C.; Sadek, H.A. Hypoxia and metabolic properties of hematopoietic stem cells. *Antioxid. Redox Signal.* **2014**, *20*, 1891–1901. [CrossRef]

121. Vannini, N.; Girotra, M.; Naveiras, O.; Nikitin, G.; Campos, V.; Giger, S.; Roch, A.; Auwerx, J.; Lutolf, M.P. Specification of haematopoietic stem cell fate via modulation of mitochondrial activity. *Nat. Commun.* **2016**, *7*, 13125. [CrossRef]

122. Pattappa, G.; Thorpe, S.D.; Jegard, N.C.; Heywood, H.K.; de Bruijn, J.D.; Lee, D.A. Continuous and uninterrupted oxygen tension influences the colony formation and oxidative metabolism of human mesenchymal stem cells. *Tissue Eng. Part C Methods* **2013**, *19*, 68–79. [CrossRef]

123. Stincone, A.; Prigione, A.; Cramer, T.; Wamelink, M.M.; Campbell, K.; Cheung, E.; Olin-Sandoval, V.; Grüning, N.M.; Krüger, A.; Tauqeer Alam, M.; et al. The return of metabolism: Biochemistry and physiology of the pentose phosphate pathway. *Biol. Rev. Camb. Philos. Soc.* **2015**, *90*, 927–963. [CrossRef]

124. Tsatmali, M.; Walcott, E.C.; Crossin, K.L. Newborn neurons acquire high levels of reactive oxygen species and increased mitochondrial proteins upon differentiation from progenitors. *Brain Res.* **2005**, *1040*, 137–150. [CrossRef] [PubMed]

125. Jang, Y.Y.; Sharkis, S.J. A low level of reactive oxygen species selects for primitive hematopoietic stem cells that may reside in the low-oxygenic niche. *Blood* **2007**, *110*, 3056–3063. [CrossRef]

126. Atashi, F.; Modarressi, A.; Pepper, M.S. The role of reactive oxygen species in mesenchymal stem cell adipogenic and osteogenic differentiation: A review. *Stem Cells Dev.* **2015**, *24*, 1150–1163. [CrossRef]

127. Ito, K.; Bernardi, R.; Morotti, A.; Matsuoka, S.; Saglio, G.; Ikeda, Y.; Rosenblatt, J.; Avigan, D.E.; Teruya-Feldstein, J.; Pandolfi, P.P. PML targeting eradicates quiescent leukaemia-initiating cells. *Nature* **2008**, *453*, 1072–1078. [CrossRef] [PubMed]

128. Ito, K.; Carracedo, A.; Weiss, D.; Arai, F.; Ala, U.; Avigan, D.E.; Schafer, Z.T.; Evans, R.M.; Suda, T.; Lee, C.H.; et al. A PML–PPAR-δ pathway for fatty acid oxidation regulates hematopoietic stem cell maintenance. *Nat. Med.* **2012**, *18*, 1350–1358. [CrossRef] [PubMed]

129. Ito, K.; Ito, K. Newly Identified Roles of PML in Stem Cell Biology. *Front. Oncol.* **2013**, *3*, 50. [CrossRef]

130. Ito, K.; Turcotte, R.; Cui, J.; Zimmerman, S.E.; Pinho, S.; Mizoguchi, T.; Arai, F.; Runnels, J.M.; Alt, C.; Teruya-Feldstein, J.; et al. Self-renewal of a purified Tie2⁺ hematopoietic stem cell population relies on mitochondrial clearance. *Science* **2016**, *354*, 1156–1160. [CrossRef] [PubMed]

131. Stoll, E.A.; Makin, R.; Sweet, I.R.; Trevelyan, A.J.; Miwa, S.; Horner, P.J.; Turnbull, D.M. Neural Stem Cells in the Adult Subventricular Zone Oxidize Fatty Acids to Produce Energy and Support Neurogenic Activity. *Stem Cells* **2015**, *33*, 2306–2319. [CrossRef]

132. Xie, Z.; Jones, A.; Deeney, J.T.; Hur, S.K.; Bankaitis, V.A. Inborn Errors of Long-Chain Fatty Acid β-Oxidation Link Neural Stem Cell Self-Renewal to Autism. *Cell Rep.* **2016**, *14*, 991–999. [CrossRef] [PubMed]

133. Ryall, J.G.; Dell'Orso, S.; Derfoul, A.; Juan, A.; Zare, H.; Feng, X.; Clermont, D.; Koulnis, M.; Gutierrez-Cruz, G.; Fulco, M.; et al. The NAD⁺-dependent SIRT1 deacetylase translates a metabolic switch into regulatory epigenetics in skeletal muscle stem cells. *Cell Stem Cell* **2015**, *16*, 171–183. [CrossRef]

134. Fukawa, T.; Yan-Jiang, B.C.; Min-Wen, J.C.; Jun-Hao, E.T.; Huang, D.; Qian, C.N.; Ong, P.; Li, Z.; Chen, S.; Mak, S.Y.; et al. Excessive fatty acid oxidation induces muscle atrophy in cancer cachexia. *Nat. Med.* **2016**, *22*, 666–671. [CrossRef]

135. Carracedo, A.; Cantley, L.C.; Pandolfi, P.P. Cancer metabolism: Fatty acid oxidation in the limelight. *Nat. Rev. Cancer* **2013**, *13*, 227–232. [CrossRef] [PubMed]

136. Knobloch, M.; Braun, S.M.; Zurkirchen, L.; von Schoultz, C.; Zamboni, N.; Araúzo-Bravo, M.J.; Kovacs, W.J.; Karalay, O.; Suter, U.; Machado, R.A.; et al. Metabolic control of adult neural stem cell activity by Fasn-dependent lipogenesis. *Nature* **2013**, *493*, 226–230. [CrossRef] [PubMed]

137. Beyaz, S.; Mana, M.D.; Roper, J.; Kedrin, D.; Saadatpour, A.; Hong, S.J.; Bauer-Rowe, K.E.; Xifaras, M.E.; Akkad, A.; Arias, E.; et al. High-fat diet enhances stemness and tumorigenicity of intestinal progenitors. *Nature* **2016**, *531*, 53–58. [CrossRef] [PubMed]

138. Murphy, M.P. How mitochondria produce reactive oxygen species. *Biochem. J.* **2009**, *417*, 1–13. [CrossRef]

139. Bigarella, C.L.; Liang, R.; Ghaffari, S. Stem cells and the impact of ROS signaling. *Development* **2014**, *141*, 4206–4218. [CrossRef]

140. Liang, R.; Ghaffari, S. Stem cells, redox signaling, and stem cell aging. *Antioxid. Redox Signal.* **2014**, *20*, 1902–1916. [CrossRef]

141. Cervantes, R.B.; Stringer, J.R.; Shao, C.; Tischfield, J.A.; Stambrook, P.J. Embryonic stem cells and somatic cells differ in mutation frequency and type. *Proc. Natl. Acad. Sci. USA* **2002**, *99*, 3586–3590. [CrossRef]

142. Kirby, D.M.; Rennie, K.J.; Smulders-Srinivasan, T.K.; Acin-Perez, R.; Whittington, M.; Enriquez, J.A.; Trevelyan, A.J.; Turnbull, D.M.; Lightowlers, R.N. Transmitochondrial embryonic stem cells containing pathogenic mtDNA mutations are compromised in neuronal differentiation. *Cell Prolif.* **2009**, *42*, 413–424. [CrossRef] [PubMed]

143. Ahlqvist, K.J.; Hämäläinen, R.H.; Yatsuga, S.; Uutela, M.; Terzioglu, M.; Götz, A.; Forsström, S.; Salven, P.; Angers-Loustau, A.; Kopra, O.H.; et al. Somatic progenitor cell vulnerability to mitochondrial DNA mutagenesis underlies progeroid phenotypes in Polg mutator mice. *Cell Metab.* **2012**, *15*, 100–109. [CrossRef] [PubMed]

144. Wahlestedt, M.; Ameur, A.; Moraghebi, R.; Norddahl, G.L.; Sten, G.; Woods, N.B.; Bryder, D. Somatic cells with a heavy mitochondrial DNA mutational load render induced pluripotent stem cells with distinct differentiation defects. *Stem Cells* **2014**, *32*, 1173–1182. [CrossRef]

145. Ma, H.; Folmes, C.D.; Wu, J.; Morey, R.; Mora-Castilla, S.; Ocampo, A.; Ma, L.; Poulton, J.; Wang, X.; Ahmed, R.; et al. Metabolic rescue in pluripotent cells from patients with mtDNA disease. *Nature* **2015**, *524*, 234–238. [CrossRef] [PubMed]

146. Armstrong, L.; Tilgner, K.; Saretzki, G.; Atkinson, S.P.; Stojkovic, M.; Moreno, R.; Przyborski, S.; Lako, M. Human induced pluripotent stem cell lines show stress defense mechanisms and mitochondrial regulation similar to those of human embryonic stem cells. *Stem Cells* **2010**, *28*, 661–673. [CrossRef] [PubMed]

147. Le Belle, J.E.; Orozco, N.M.; Paucar, A.A.; Saxe, J.P.; Mottahedeh, J.; Pyle, A.D.; Wu, H.; Kornblum, H.I. Proliferative neural stem cells have high endogenous ROS levels that regulate self-renewal and neurogenesis in a PI3K/Akt-dependant manner. *Cell Stem Cell* **2011**, *8*, 59–71. [CrossRef]

148. Morimoto, H.; Iwata, K.; Ogonuki, N.; Inoue, K.; Atsuo, O.; Kanatsu-Shinohara, M.; Morimoto, T.; Yabe-Nishimura, C.; Shinohara, T. ROS are required for mouse spermatogonial stem cell self-renewal. *Cell Stem Cell* **2013**, *12*, 774–786. [CrossRef]

149. Kohli, L.; Passed, E. Surviving changes: The metabolic journey of hematopoietic stem cells. *Trends Cell Biol.* **2014**, *24*, 479–487. [CrossRef] [PubMed]

150. Paul, M.K.; Bisht, B.; Darmawan, D.O.; Chiou, R.; Ha, V.L.; Wallace, W.D.; Chon, A.T.; Hegab, A.E.; Grogan, T.; Elashoff, D.A.; et al. Dynamic changes in intracellular ROS levels regulate airway basal stem cell homeostasis through Nrf2-dependent Notch signaling. *Cell Stem Cell* **2014**, *15*, 199–214. [CrossRef] [PubMed]

151. Bakker, S.T.; Passegué, E. Resilient and resourceful: Genome maintenance strategies in hematopoietic stem cells. *Exp. Hematol.* **2013**, *41*, 915–923. [CrossRef]

152. Adams, P.D.; Jasper, H.; Rudolph, K.L. Aging-Induced Stem Cell Mutations as Drivers for Disease and Cancer. *Cell Stem Cell* **2015**, *16*, 601–612. [CrossRef]

153. Maynard, S.; Swistowska, A.M.; Lee, J.W.; Liu, Y.; Liu, S.T.; Da Cruz, A.B.; Rao, M.; de Souza-Pinto, N.C.; Zeng, X.; Bohr, V.A. Human embryonic stem cells have enhanced repair of multiple forms of DNA damage. *Stem Cells* **2008**, *26*, 2266–2274. [CrossRef]

154. Saretzki, G.; Walter, T.; Atkinson, S.; Passos, J.F.; Bareth, B.; Keith, W.N.; Stewart, R.; Hoare, S.; Stojkovic, M.; Armstrong, L.; et al. Downregulation of multiple stress defense mechanisms during differentiation of human embryonic stem cells. *Stem Cells* **2008**, *26*, 455–464. [CrossRef]

155. Dannenmann, B.; Lehle, S.; Hildebrand, D.G.; Kübler, A.; Grondona, P.; Schmid, V.; Holzer, K.; Fröschl, M.; Essmann, F.; Rothfuss, O.; et al. High glutathione and glutathione peroxidase-2 levels mediate cell-type-specific DNA damage protection in human induced pluripotent stem cells. *Stem Cell Rep.* **2015**, *4*, 886–898. [CrossRef]

156. Miyamoto, K.; Miyamoto, T.; Kato, R.; Yoshimura, A.; Motoyama, N.; Suda, T. FoxO3a regulates hematopoietic homeostasis through a negative feedback pathway in conditions of stress or aging. *Blood* **2008**, *112*, 4485–4493. [CrossRef]

157. Renault, V.M.; Rafalski, V.A.; Morgan, A.A.; Salih, D.A.; Brett, J.O.; Webb, A.E.; Villeda, S.A.; Thekkat, P.U.; Guillerey, C.; Denko, N.C.; et al. FoxO3 regulates neural stem cell homeostasis. *Cell Stem Cell* **2009**, *5*, 527–539. [CrossRef]

158. Matsui, K.; Ezoe, S.; Oritani, K.; Shibata, M.; Tokunaga, M.; Fujita, N.; Tanimura, A.; Sudo, T.; Tanaka, H.; McBurney, M.W.; et al. NAD-dependent histone deacetylase, SIRT1, plays essential roles in the maintenance of hematopoietic stem cells. *Biochem. Biophys. Res. Commun.* **2012**, *418*, 811–817. [CrossRef]

159. Webb, A.E.; Pollina, E.A.; Vierbuchen, T.; Urbán, N.; Ucar, D.; Leeman, D.S.; Martynoga, B.; Sewak, M.; Rando, T.A.; Guillemot, F.; et al. FOXO3 shares common targets with ASCL1 genome-wide and inhibits ASCL1-dependent neurogenesis. *Cell Rep.* **2013**, *4*, 477–491. [CrossRef]

160. Rimmelé, P.; Bigarella, C.L.; Liang, R.; Izac, B.; Dieguez-Gonzalez, R.; Barbet, G.; Donovan, M.; Brugnara, C.; Blander, J.M.; Sinclair, D.A.; et al. Aging-like phenotype and defective lineage specification in SIRT1-deleted hematopoietic stem and progenitor cells. *Stem Cell Rep.* **2014**, *3*, 44–59. [CrossRef]

161. Mehta, A.; Zhao, J.L.; Sinha, N.; Marinov, G.K.; Mann, M.; Kowalczyk, M.S.; Galimidi, R.P.; Du, X.; Erikci, E.; Regev, A.; et al. The MicroRNA-132 and MicroRNA-212 Cluster Regulates Hematopoietic Stem Cell Maintenance and Survival with Age by Buffering FOXO3 Expression. *Immunity* **2015**, *42*, 1021–1032. [CrossRef]

162. Rimmelé, P.; Liang, R.; Bigarella, C.L.; Kocabas, F.; Xie, J.; Serasinghe, M.N.; Chipuk, J.; Sadek, H.; Zhang, C.C.; Ghaffari, S. Mitochondrial metabolism in hematopoietic stem cells requires functional FOXO3. *EMBO Rep.* **2015**, *16*, 1164–1176. [CrossRef] [PubMed]

163. Wakabayashi, N.; Itoh, K.; Wakabayashi, J.; Motohashi, H.; Noda, S.; Takahashi, S.; Imakado, S.; Kotsuji, T.; Otsuka, F.; Roop, D.R.; et al. Keap1-null mutation leads to postnatal lethality due to constitutive Nrf2 activation. *Nat. Genet.* **2003**, *35*, 238–245. [CrossRef] [PubMed]

164. Hochmuth, C.E.; Biteau, B.; Bohmann, D.; Jasper, H. Redox regulation by Keap1 and Nrf2 controls intestinal stem cell proliferation in Drosophila. *Cell Stem Cell* **2011**, *8*, 188–199. [CrossRef] [PubMed]

165. Mohrin, M.; Bourke, E.; Alexander, D.; Warr, M.R.; Barry-Holson, K.; Le Beau, M.M.; Morrison, C.G.; Passegué, E. Hematopoietic stem cell quiescence promotes error-prone DNA repair and mutagenesis. *Cell Stem Cell* **2010**, *7*, 174–185. [CrossRef] [PubMed]

166. Finkel, T.; Holbrook, N.J. Oxidants, oxidative stress and the biology of ageing. *Nature* **2000**, *408*, 239–247. [CrossRef] [PubMed]

167. D'Autréaux, B.; Toledano, M.B. ROS as signalling molecules: Mechanisms that generate specificity in ROS homeostasis. *Nat. Rev. Mol. Cell Biol.* **2007**, *8*, 813–824. [CrossRef]

168. Giorgio, M.; Trinei, M.; Migliaccio, E.; Pelicci, P.G. Hydrogen peroxide: A metabolic by-product or a common mediator of ageing signals? *Nat. Rev. Mol. Cell Biol.* **2007**, *8*, 722–728. [CrossRef] [PubMed]
169. Schieber, M.; Chandel, N.S. ROS function in redox signaling and oxidative stress. *Curr. Biol.* **2014**, *24*, R453–R462. [CrossRef]
170. Miyamoto, K.; Araki, K.Y.; Naka, K.; Arai, F.; Takubo, K.; Yamazaki, S.; Matsuoka, S.; Miyamoto, T.; Ito, K.; Ohmura, M.; et al. Foxo3a is essential for maintenance of the hematopoietic stem cell pool. *Cell Stem Cell* **2007**, *1*, 101–112. [CrossRef]
171. Tothova, Z.; Kollipara, R.; Huntly, B.J.; Lee, B.H.; Castrillon, D.H.; Cullen, D.E.; McDowell, E.P.; Lazo-Kallanian, S.; Williams, I.R.; Sears, C.; et al. FoxOs are critical mediators of hematopoietic stem cell resistance to physiologic oxidative stress. *Cell* **2007**, *128*, 325–339. [CrossRef] [PubMed]
172. Sykiotis, G.P.; Bohmann, D. Keap1/Nrf2 signaling regulates oxidative stress tolerance and lifespan in Drosophila. *Dev. Cell* **2008**, *14*, 76–85. [CrossRef]
173. Blanpain, C.; Mohrin, M.; Sotiropoulou, P.A.; Passegué, E. DNA-damage response in tissue-specific and cancer stem cells. *Cell Stem Cell* **2011**, *8*, 16–29. [CrossRef]
174. Lombard, D.B.; Chua, K.F.; Mostoslavsky, R.; Franco, S.; Gostissa, M.; Alt, F.W. DNA repair, genome stability, and aging. *Cell* **2005**, *120*, 497–512. [CrossRef]
175. Owusu-Ansah, E.; Banerjee, U. Reactive oxygen species prime Drosophila haematopoietic progenitors for differentiation. *Nature* **2009**, *461*, 537–541. [CrossRef]
176. Tormos, K.V.; Anso, E.; Hamanaka, R.B.; Eisenbart, J.; Joseph, J.; Kalyanaraman, B.; Chandel, N.S. Mitochondrial complex III ROS regulate adipocyte differentiation. *Cell Metab.* **2011**, *14*, 537–544. [CrossRef] [PubMed]
177. Lyublinskaya, O.G.; Borisov, Y.G.; Pugovkina, N.A.; Smirnova, I.S.; Obidina, J.V.; Ivanova, J.S.; Zenin, V.V.; Shatrova, A.N.; Borodkina, A.V.; Aksenov, N.D.; et al. Reactive Oxygen Species Are Required for Human Mesenchymal Stem Cells to Initiate Proliferation after the Quiescence Exit. *Oxid. Med. Cell. Longev.* **2015**, *2015*, 502105. [CrossRef]
178. Khacho, M.; Clark, A.; Svoboda, D.S.; Azzi, J.; MacLaurin, J.G.; Meghaizel, C.; Sesaki, H.; Lagace, D.C.; Germain, M.; Harper, M.E.; et al. Mitochondrial Dynamics Impacts Stem Cell Identity and Fate Decisions by Regulating a Nuclear Transcriptional Program. *Cell Stem Cell* **2016**, *19*, 232–247. [CrossRef] [PubMed]
179. Juntilla, M.M.; Patil, V.D.; Calamito, M.; Joshi, R.P.; Birnbaum, M.J.; Koretzky, G.A. AKT1 and AKT2 maintain hematopoietic stem cell function by regulating reactive oxygen species. *Blood* **2010**, *115*, 4030–4038. [CrossRef]
180. Biteau, B.; Jasper, H. EGF signaling regulates the proliferation of intestinal stem cells in Drosophila. *Development* **2011**, *138*, 1045–1055. [CrossRef]
181. Malinska, D.; Kudin, A.P.; Bejtka, M.; Kunz, W.S. Changes in mitochondrial reactive oxygen species synthesis during differentiation of skeletal muscle cells. *Mitochondrion* **2012**, *12*, 144–148. [CrossRef]
182. Ueda, T.; Nagamachi, A.; Takubo, K.; Yamasaki, N.; Matsui, H.; Kanai, A.; Nakata, Y.; Ikeda, K.; Konuma, T.; Oda, H.; et al. Fbxl10 overexpression in murine hematopoietic stem cells induces leukemia involving metabolic activation and upregulation of Nsg2. *Blood* **2015**, *125*, 3437–3446. [CrossRef]
183. Ito, K.; Hirao, A.; Arai, F.; Takubo, K.; Matsuoka, S.; Miyamoto, K.; Ohmura, M.; Naka, K.; Hosokawa, K.; Ikeda, Y.; et al. Reactive oxygen species act through p38 MAPK to limit the lifespan of hematopoietic stem cells. *Nat. Med.* **2006**, *12*, 446–451. [CrossRef]
184. Karigane, D.; Kobayashi, H.; Morikawa, T.; Ootomo, Y.; Sakai, M.; Nagamatsu, G.; Kubota, Y.; Goda, N.; Matsumoto, M.; Nishimura, E.K.; et al. p38α Activates Purine Metabolism to Initiate Hematopoietic Stem/Progenitor Cell Cycling in Response to Stress. *Cell Stem Cell* **2016**, *19*, 192–204. [CrossRef]
185. Kwon, B. p38α-mediated purine metabolism is linked to exit from quiescence of hematopoietic stem cells. *Stem Cell Investig.* **2016**, *3*, 69. [CrossRef]
186. Essers, M.A.G. Stressed-Out HSCs Turn Up p38α and Purine to Proliferate. *Cell Stem Cell* **2016**, *19*, 143–144. [CrossRef]
187. Wellen, K.E.; Hatzivassiliou, G.; Sachdeva, U.M.; Bui, T.V.; Cross, J.R.; Thompson, C.B. ATP-citrate lyase links cellular metabolism to histone acetylation. *Science* **2009**, *324*, 1076–1080. [CrossRef]
188. Sørensen, A.L.; Timoskainen, S.; West, F.D.; Vekterud, K.; Boquest, A.C.; Ahrlund-Richter, L.; Stice, S.L.; Collas, P. Lineage-specific promoter DNA methylation patterns segregate adult progenitor cell types. *Stem Cells Dev.* **2010**, *19*, 1257–1266. [CrossRef]

189. Yannarelli, G.; Pacienza, N.; Cuniberti, L.; Medin, J.; Davies, J.; Keating, A. Brief report: The potential role of epigenetics on multipotent cell differentiation capacity of mesenchymal stromal cells. *Stem Cells* **2013**, *31*, 215–220. [CrossRef]
190. Kaelin, W.G.; McKnight, S.L. Influence of metabolism on epigenetics and disease. *Cell* **2013**, *153*, 56–69. [CrossRef]
191. Hwang, I.Y.; Kwak, S.; Lee, S.; Kim, H.; Lee, S.E.; Kim, J.H.; Kim, Y.A.; Jeon, Y.K.; Chung, D.H.; Jin, X.; et al. Psat1-Dependent Fluctuations in α-Ketoglutarate Affect the Timing of ESC Differentiation. *Cell Metab.* **2016**, *24*, 494–501. [CrossRef] [PubMed]
192. Carey, B.W.; Finley, L.W.; Cross, J.R.; Allis, C.D.; Thompson, C.B. Intracellular α-ketoglutarate maintains the pluripotency of embryonic stem cells. *Nature* **2015**, *518*, 413–416. [CrossRef] [PubMed]
193. Moussaieff, A.; Rouleau, M.; Kitsberg, D.; Cohen, M.; Levy, G.; Barasch, D.; Nemirovski, A.; Shen-Orr, S.; Laevsky, I.; Amit, M.; et al. Glycolysis-mediated changes in acetyl-CoA and histone acetylation control the early differentiation of embryonic stem cells. *Cell Metab.* **2015**, *21*, 392–402. [CrossRef] [PubMed]
194. Ryall, J.G.; Cliff, T.; Dalton, S.; Sartorelli, V. Metabolic Reprogramming of Stem Cell Epigenetics. *Cell Stem Cell* **2015**, *17*, 651–662. [CrossRef] [PubMed]
195. TeSlaa, T.; Chaikovsky, A.C.; Lipchina, I.; Escobar, S.L.; Hochedlinger, K.; Huang, J.; Graeber, T.G.; Braas, D.; Teitell, M.A. α-Ketoglutarate Accelerates the Initial Differentiation of Primed Human Pluripotent Stem Cells. *Cell Metab.* **2016**, *24*, 485–493. [CrossRef] [PubMed]
196. Zhu, C.; Gao, Y.; Guo, H.; Xia, B.; Song, J.; Wu, X.; Zeng, H.; Kee, K.; Tang, F.; Yi, C. Single-Cell 5-Formylcytosine Landscapes of Mammalian Early Embryos and ESCs at Single-Base Resolution. *Cell Stem Cell* **2017**, *20*, 720–731. [CrossRef]
197. Chambers, S.M.; Shaw, C.A.; Gatza, C.; Fisk, C.J.; Donehower, L.A.; Goodell, M.A. Aging hematopoietic stem cells decline in function and exhibit epigenetic dysregulation. *PLoS Biol.* **2007**, *5*, e201. [CrossRef]
198. Noer, A.; Lindeman, L.C.; Collas, P. Histone H3 modifications associated with differentiation and long-term culture of mesenchymal adipose stem cells. *Stem Cells Dev.* **2009**, *18*, 725–736. [CrossRef]
199. Weishaupt, H.; Sigvardsson, M.; Attema, J.L. Epigenetic chromatin states uniquely define the developmental plasticity of murine hematopoietic stem cells. *Blood* **2010**, *115*, 247–256. [CrossRef]
200. Florian, M.C.; Dörr, K.; Niebel, A.; Daria, D.; Schrezenmeier, H.; Rojewski, M.; Filippi, M.D.; Hasenberg, A.; Gunzer, M.; Scharffetter-Kochanek, K.; et al. Cdc42 activity regulates hematopoietic stem cell aging and rejuvenation. *Cell Stem Cell* **2012**, *10*, 520–530. [CrossRef] [PubMed]
201. Beerman, I.; Bock, C.; Garrison, B.S.; Smith, Z.D.; Gu, H.; Meissner, A.; Rossi, D.J. Proliferation-dependent alterations of the DNA methylation landscape underlie hematopoietic stem cell aging. *Cell Stem Cell* **2013**, *12*, 413–425. [CrossRef]
202. Geiger, H.; de Haan, G.; Florian, M.C. The ageing haematopoietic stem cell compartment. *Nat. Rev. Immunol.* **2013**, *13*, 376–389.
203. Liu, L.; Cheung, T.H.; Charville, G.W.; Hurgo, B.M.; Leavitt, T.; Shih, J.; Brunet, A.; Rando, T.A. Chromatin modifications as determinants of muscle stem cell quiescence and chronological aging. *Cell Rep.* **2013**, *4*, 189–204. [CrossRef]
204. Benayoun, B.A.; Pollina, E.A.; Ucar, D.; Mahmoudi, S.; Karra, K.; Wong, E.D.; Devarajan, K.; Daugherty, A.C.; Kundaje, A.B.; Mancini, E.; et al. H3K4me3 breadth is linked to cell identity and transcriptional consistency. *Cell* **2014**, *158*, 673–688. [CrossRef]
205. Challen, G.A.; Sun, D.; Mayle, A.; Jeong, M.; Luo, M.; Rodriguez, B.; Mallaney, C.; Celik, H.; Yang, L.; Xia, Z.; et al. Dnmt3a and Dnmt3b have overlapping and distinct functions in hematopoietic stem cells. *Cell Stem Cell* **2014**, *15*, 350–364. [CrossRef]
206. Sun, D.; Luo, M.; Jeong, M.; Rodriguez, B.; Xia, Z.; Hannah, R.; Wang, H.; Le, T.; Faull, K.F.; Chen, R.; et al. Epigenomic profiling of young and aged HSCs reveals concerted changes during aging that reinforce self-renewal. *Cell Stem Cell* **2014**, *14*, 673–688. [CrossRef]
207. Beerman, I.; Rossi, D.J. Epigenetic Control of Stem Cell Potential during Homeostasis, Aging, and Disease. *Cell Stem Cell* **2015**, *16*, 613–625. [CrossRef]
208. Mayle, A.; Yang, L.; Rodriguez, B.; Zhou, T.; Chang, E.; Curry, C.V.; Challen, G.A.; Li, W.; Wheeler, D.; Rebel, V.I.; et al. Dnmt3a loss predisposes murine hematopoietic stem cells to malignant transformation. *Blood* **2015**, *125*, 629–638. [CrossRef]

209. Ugarte, F.; Sousae, R.; Cinquin, B.; Martin, E.W.; Krietsch, J.; Sanchez, G.; Inman, M.; Tsang, H.; Warr, M.; Passegué, E.; et al. Progressive Chromatin Condensation and H3K9 Methylation Regulate the Differentiation of Embryonic and Hematopoietic Stem Cells. *Stem Cell Rep.* **2015**, *5*, 728–740. [CrossRef]

210. Zhang, W.; Li, J.; Suzuki, K.; Qu, J.; Wang, P.; Zhou, J.; Liu, X.; Ren, R.; Xu, X.; Ocampo, A.; et al. Aging stem cells. A Werner syndrome stem cell model unveils heterochromatin alterations as a driver of human aging. *Science* **2015**, *348*, 1160–1163. [CrossRef]

211. Boonsanay, V.; Zhang, T.; Georgieva, A.; Kostin, S.; Qi, H.; Yuan, X.; Zhou, Y.; Braun, T. Regulation of Skeletal Muscle Stem Cell Quiescence by Suv4-20h1-Dependent Facultative Heterochromatin Formation. *Cell Stem Cell* **2016**, *18*, 229–242. [CrossRef] [PubMed]

212. Faralli, H.; Wang, C.; Nakka, K.; Benyoucef, A.; Sebastian, S.; Zhuang, L.; Chu, A.; Palii, C.G.; Liu, C.; Camellato, B.; et al. UTX demethylase activity is required for satellite cell-mediated muscle regeneration. *J. Clin. Investig.* **2016**, *126*, 1555–1565. [CrossRef] [PubMed]

213. Pagano, M.; Pepperkok, R.; Verde, F.; Ansorge, W.; Draetta, G. Cyclin A is required at two points in the human cell cycle. *EMBO J.* **1992**, *11*, 961–971. [CrossRef]

214. Blanpain, C.; Lowry, W.E.; Geoghegan, A.; Polak, L.; Fuchs, E. Self-renewal, multipotency, and the existence of two cell populations within an epithelial stem cell niche. *Cell* **2004**, *118*, 635–648. [CrossRef]

215. Fukada, S.; Uezumi, A.; Ikemoto, M.; Masuda, S.; Segawa, M.; Tanimura, N.; Yamamoto, H.; Miyagoe-Suzuki, Y.; Takeda, S. Molecular signature of quiescent satellite cells in adult skeletal muscle. *Stem Cells* **2007**, *25*, 2448–2459. [CrossRef]

216. Forsberg, E.C.; Passegué, E.; Prohaska, S.S.; Wagers, A.J.; Koeva, M.; Stuart, J.M.; Weissman, I.L. Molecular signatures of quiescent, mobilized and leukemia-initiating hematopoietic stem cells. *PLoS ONE* **2010**, *5*, e8785. [CrossRef]

217. Kamminga, L.M.; Bystrykh, L.V.; de Boer, A.; Houwer, S.; Douma, J.; Weersing, E.; Dontje, B.; de Haan, G. The Polycomb group gene Ezh2 prevents hematopoietic stem cell exhaustion. *Blood* **2006**, *107*, 2170–2179. [CrossRef] [PubMed]

218. Ezhkova, E.; Lien, W.H.; Stokes, N.; Pasolli, H.A.; Silva, J.M.; Fuchs, E. EZH1 and EZH2 cogovern histone H3K27 trimethylation and are essential for hair follicle homeostasis and wound repair. *Genes Dev.* **2011**, *25*, 485–498. [CrossRef]

219. Juan, A.H.; Derfoul, A.; Feng, X.; Ryall, J.G.; Dell'Orso, S.; Pasut, A.; Zare, H.; Simone, J.M.; Rudnicki, M.A.; Sartorelli, V. Polycomb EZH2 controls self-renewal and safeguards the transcriptional identity of skeletal muscle stem cells. *Genes Dev.* **2011**, *25*, 789–794. [CrossRef]

220. Hidalgo, I.; Herrera-Merchan, A.; Ligos, J.M.; Carramolino, L.; Nuñez, J.; Martinez, F.; Dominguez, O.; Torres, M.; Gonzalez, S. Ezh1 is required for hematopoietic stem cell maintenance and prevents senescence-like cell cycle arrest. *Cell Stem Cell* **2012**, *11*, 649–662. [CrossRef] [PubMed]

221. Locasale, J.W.; Cantley, L.C. Metabolic flux and the regulation of mammalian cell growth. *Cell Metab.* **2011**, *14*, 443–451. [CrossRef] [PubMed]

222. Libert, S.; Guarente, L. Metabolic and neuropsychiatric effects of calorie restriction and sirtuins. *Annu. Rev. Physiol.* **2013**, *75*, 669–684. [CrossRef] [PubMed]

223. Imai, S.; Guarente, L. NAD^+ and sirtuins in aging and disease. *Trends Cell Biol.* **2014**, *24*, 464–471. [CrossRef]

224. Yu, A.; Dang, W. Regulation of stem cell aging by SIRT1—Linking metabolic signaling to epigenetic modifications. *Mol. Cell. Endocrinol.* **2017**, *455*, 75–82. [CrossRef] [PubMed]

225. Rafalski, V.A.; Ho, P.P.; Brett, J.O.; Ucar, D.; Dugas, J.C.; Pollina, E.A.; Chow, L.M.; Ibrahim, A.; Baker, S.J.; Barres, B.A.; et al. Expansion of oligodendrocyte progenitor cells following SIRT1 inactivation in the adult brain. *Nat. Cell Biol.* **2013**, *15*, 614–624. [CrossRef]

226. Saharan, S.; Jhaveri, D.J.; Bartlett, P.F. SIRT1 regulates the neurogenic potential of neural precursors in the adult subventricular zone and hippocampus. *J. Neurosci. Res.* **2013**, *91*, 642–659. [CrossRef] [PubMed]

227. Zuccaro, E.; Arlotta, P. The quest for myelin in the adult brain. *Nat. Cell Biol.* **2013**, *15*, 572–575. [CrossRef] [PubMed]

228. Hu, B.; Guo, Y.; Chen, C.; Li, Q.; Niu, X.; Guo, S.; Zhang, A.; Wang, Y.; Deng, Z. Repression of SIRT1 promotes the differentiation of mouse induced pluripotent stem cells into neural stem cells. *Cell Mol. Neurobiol.* **2014**, *34*, 905–912. [CrossRef]

229. Fujita, Y.; Yamashita, T. Sirtuins in Neuroendocrine Regulation and Neurological Diseases. *Front. Neurosci.* **2018**, *12*, 778. [CrossRef]

230. Hisahara, S.; Chiba, S.; Matsumoto, H.; Tanno, M.; Yagi, H.; Shimohama, S.; Sato, M.; Horio, Y. Histone deacetylase SIRT1 modulates neuronal differentiation by its nuclear translocation. *Proc. Natl. Acad. Sci. USA* **2008**, *105*, 15599–15604. [CrossRef]

231. Ma, C.Y.; Yao, M.J.; Zhai, Q.W.; Jiao, J.W.; Yuan, X.B.; Poo, M.M. SIRT1 suppresses self-renewal of adult hippocampal neural stem cells. *Development* **2014**, *141*, 4697–4709. [CrossRef] [PubMed]

232. Fawal, M.A.; Davy, A. Impact of Metabolic Pathways and Epigenetics on Neural Stem Cells. *Epigenet. Insights* **2018**, *11*, 2516865718820946. [CrossRef]

233. Diaz-Ruiz, A.; Gonzalez-Freire, M.; Ferrucci, L.; Bernier, M.; Cabo, R. SIRT1 synchs satellite cell metabolism with stem cell fate. *Cell Stem Cell* **2015**, *16*, 103–104. [CrossRef]

234. Yun, J.; Johnson, J.L.; Hanigan, C.L.; Locasale, J.W. Interactions between epigenetics and metabolism in cancers. *Front. Oncol.* **2012**, *2*, 163. [CrossRef] [PubMed]

235. Gut, P.; Verdin, E. The nexus of chromatin regulation and intermediary metabolism. *Nature* **2013**, *502*, 489–498. [CrossRef] [PubMed]

236. Sharma, A.; Diecke, S.; Zhang, W.Y.; Lan, F.; He, C.; Mordwinkin, N.M.; Chua, K.F.; Wu, J.C. The role of SIRT6 protein in aging and reprogramming of human induced pluripotent stem cells. *J. Biol. Chem.* **2013**, *288*, 18439–18447. [CrossRef] [PubMed]

237. Rodgers, J.T.; Lerin, C.; Haas, W.; Gygi, S.P.; Spiegelman, B.M.; Puigserver, P. Nutrient control of glucose homeostasis through a complex of PGC-1α and SIRT1. *Nature* **2005**, *434*, 113–118. [CrossRef]

238. O'Brien, L.C.; Keeney, P.M.; Bennett, J.P. Differentiation of Human Neural Stem Cells into Motor Neurons Stimulates Mitochondrial Biogenesis and Decreases Glycolytic Flux. *Stem Cells Dev.* **2015**, *24*, 1984–1994. [CrossRef] [PubMed]

239. Guarente, L. The resurgence of NAD^+. *Science* **2016**, *352*, 1396–1397. [CrossRef] [PubMed]

240. Tang, A.H.; Rando, T.A. Induction of autophagy supports the bioenergetic demands of quiescent muscle stem cell activation. *EMBO J.* **2014**, *33*, 2782–2797. [CrossRef] [PubMed]

241. Wagers, A.J. How stem cells get "turned on". *EMBO J.* **2014**, *33*, 2743–2744. [CrossRef]

242. Brunet, A.; Sweeney, L.B.; Sturgill, J.F.; Chua, K.F.; Greer, P.L.; Lin, Y.; Tran, H.; Ross, S.E.; Mostoslavsky, R.; Cohen, H.Y.; et al. Stress-dependent regulation of FOXO transcription factors by the SIRT1 deacetylase. *Science* **2004**, *303*, 2011–2015. [CrossRef]

243. Kobayashi, Y.; Furukawa-Hibi, Y.; Chen, C.; Horio, Y.; Isobe, K.; Ikeda, K.; Motoyama, N. SIRT1 is critical regulator of FOXO-mediated transcription in response to oxidative stress. *Int. J. Mol. Med.* **2005**, *16*, 237–243. [CrossRef]

244. Yu, W.; Dittenhafer-Reed, K.E.; Denu, J.M. SIRT3 protein deacetylates isocitrate dehydrogenase 2 (IDH2) and regulates mitochondrial redox status. *J. Biol. Chem.* **2012**, *287*, 14078–14086. [CrossRef] [PubMed]

245. Brown, K.; Xie, S.; Qiu, X.; Mohrin, M.; Shin, J.; Liu, Y.; Zhang, D.; Scadden, D.T.; Chen, D. SIRT3 reverses aging-associated degeneration. *Cell Rep.* **2013**, *3*, 319–327. [CrossRef] [PubMed]

246. Qiu, X.; Brown, K.; Hirschey, M.D.; Verdin, E.; Chen, D. Calorie restriction reduces oxidative stress by SIRT3-mediated SOD2 activation. *Cell Metab.* **2010**, *12*, 662–667. [CrossRef] [PubMed]

247. Tao, R.; Coleman, M.C.; Pennington, J.D.; Ozden, O.; Park, S.H.; Jiang, H.; Kim, H.S.; Flynn, C.R.; Hill, S.; McDonald, W.H.; et al. Sirt3-mediated deacetylation of evolutionarily conserved lysine 122 regulates MnSOD activity in response to stress. *Mol. Cell* **2010**, *40*, 893–904. [CrossRef]

248. Someya, S.; Yu, W.; Hallows, W.C.; Xu, J.; Vann, J.M.; Leeuwenburgh, C.; Tanokura, M.; Denu, J.M.; Prolla, T.A. Sirt3 mediates reduction of oxidative damage and prevention of age-related hearing loss under caloric restriction. *Cell* **2010**, *143*, 802–812. [CrossRef]

249. Shackelford, D.B.; Shaw, R.J. The LKB1-AMPK pathway: Metabolism and growth control in tumour suppression. *Nat. Rev. Cancer* **2009**, *9*, 563–575. [PubMed]

250. Mihaylova, M.M.; Shaw, R.J. The AMPK signalling pathway coordinates cell growth, autophagy and metabolism. *Nat. Cell Biol.* **2011**, *13*, 1016–1023. [CrossRef]

251. Hardie, D.G.; Alessi, D.R. LKB1 and AMPK and the cancer-metabolism link—Ten years after. *BMC Biol.* **2013**, *11*, 36. [CrossRef]

252. Herzig, S.; Shaw, R.J. AMPK: Guardian of metabolism and mitochondrial homeostasis. *Nat. Rev. Mol. Cell Biol.* **2018**, *19*, 121–135. [CrossRef]

253. Gan, B.; Hu, J.; Jiang, S.; Liu, Y.; Sahin, E.; Zhuang, L.; Fletcher-Sananikone, E.; Colla, S.; Wang, Y.A.; Chin, L.; et al. Lkb1 regulates quiescence and metabolic homeostasis of haematopoietic stem cells. *Nature* **2010**, *468*, 701–704. [CrossRef]

254. Gurumurthy, S.; Xie, S.Z.; Alagesan, B.; Kim, J.; Yusuf, R.Z.; Saez, B.; Tzatsos, A.; Ozsolak, F.; Milos, P.; Ferrari, F.; et al. The Lkb1 metabolic sensor maintains haematopoietic stem cell survival. *Nature* **2010**, *468*, 659–663. [CrossRef]

255. Nakada, D.; Saunders, T.L.; Morrison, S.J. Lkb1 regulates cell cycle and energy metabolism in haematopoietic stem cells. *Nature* **2010**, *468*, 653–658. [CrossRef]

256. Durand, E.M.; Zon, L.I. Stem cells: The blood balance. *Nature* **2010**, *468*, 644–645. [CrossRef]

257. Mortensen, M.; Soilleux, E.J.; Djordjevic, G.; Tripp, R.; Lutteropp, M.; Sadighi-Akha, E.; Stranks, A.J.; Glanville, J.; Knight, S.; Jacobsen, S.E.; et al. The autophagy protein Atg7 is essential for hematopoietic stem cell maintenance. *J. Exp. Med.* **2011**, *208*, 455–467. [CrossRef] [PubMed]

258. Krock, B.; Skuli, N.; Simon, M.C. The tumor suppressor LKB1 emerges as a critical factor in hematopoietic stem cell biology. *Cell Metab.* **2011**, *13*, 8–10. [CrossRef]

259. Yilmaz, O.H.; Valdez, R.; Theisen, B.K.; Guo, W.; Ferguson, D.O.; Wu, H.; Morrison, S.J. Pten dependence distinguishes haematopoietic stem cells from leukaemia-initiating cells. *Nature* **2006**, *441*, 475–482. [CrossRef]

260. Zhang, J.; Grindley, J.C.; Yin, T.; Jayasinghe, S.; He, X.C.; Ross, J.T.; Haug, J.S.; Rupp, D.; Porter-Westpfahl, K.S.; Wiedemann, L.M.; et al. PTEN maintains haematopoietic stem cells and acts in lineage choice and leukaemia prevention. *Nature* **2006**, *441*, 518–522. [CrossRef]

261. Chen, C.; Liu, Y.; Liu, R.; Ikenoue, T.; Guan, K.L.; Liu, Y.; Zheng, P. TSC-mTOR maintains quiescence and function of hematopoietic stem cells by repressing mitochondrial biogenesis and reactive oxygen species. *J. Exp. Med.* **2008**, *205*, 2397–2408. [CrossRef]

262. Gan, B.; Sahin, E.; Jiang, S.; Sanchez-Aguilera, A.; Scott, K.L.; Chin, L.; Williams, D.A.; Kwiatkowski, D.J.; DePinho, R.A. mTORC1-dependent and -independent regulation of stem cell renewal, differentiation, and mobilization. *Proc. Natl. Acad. Sci. USA* **2008**, *105*, 19384–19389. [CrossRef] [PubMed]

263. Kharas, M.G.; Gritsman, K. Akt: A double-edged sword for hematopoietic stem cells. *Cell Cycle* **2010**, *9*, 1223–1224. [CrossRef]

264. Kharas, M.G.; Okabe, R.; Ganis, J.J.; Gozo, M.; Khandan, T.; Paktinat, M.; Gilliland, D.G.; Gritsman, K. Constitutively active AKT depletes hematopoietic stem cells and induces leukemia in mice. *Blood* **2010**, *115*, 1406–1415. [CrossRef]

265. Shorning, B.Y.; Zabkiewicz, J.; McCarthy, A.; Pearson, H.B.; Winton, D.J.; Sansom, O.J.; Ashworth, A.; Clarke, A.R. Lkb1 deficiency alters goblet and paneth cell differentiation in the small intestine. *PLoS ONE* **2009**, *4*, e4264. [CrossRef]

266. Yeung, T.M.; Chia, L.A.; Kosinski, C.M.; Kuo, C.J. Regulation of self-renewal and differentiation by the intestinal stem cell niche. *Cell. Mol. Life Sci.* **2011**, *68*, 2513–2523. [CrossRef] [PubMed]

267. Richmond, C.A.; Shah, M.S.; Carlone, D.L.; Breault, D.T. Factors regulating quiescent stem cells: Insights from the intestine and other self-renewing tissues. *J. Physiol.* **2016**, *594*, 4805–4813. [CrossRef]

268. Saxton, R.A.; Sabatini, D.M. mTOR Signaling in Growth, Metabolism, and Disease. *Cell* **2017**, *168*, 960–976. [CrossRef] [PubMed]

269. Kim, J.; Guan, K.L. mTOR as a central hub of nutrient signalling and cell growth. *Nat. Cell Biol.* **2019**, *21*, 63–71. [CrossRef] [PubMed]

270. Cunningham, J.T.; Rodgers, J.T.; Arlow, D.H.; Vazquez, F.; Mootha, V.K.; Puigserver, P. mTOR controls mitochondrial oxidative function through a YY1-PGC-1α transcriptional complex. *Nature* **2007**, *450*, 736–740. [CrossRef]

271. Morita, M.; Gravel, S.P.; Chénard, V.; Sikström, K.; Zheng, L.; Alain, T.; Gandin, V.; Avizonis, D.; Arguello, M.; Zakaria, C.; et al. mTORC1 controls mitochondrial activity and biogenesis through 4E-BP-dependent translational regulation. *Cell Metab.* **2013**, *18*, 698–711. [CrossRef]

272. Rosario, F.J.; Gupta, M.B.; Myatt, L.; Powell, T.L.; Glenn, J.P.; Cox, L.; Jansson, T. Mechanistic Target of Rapamycin Complex 1 Promotes the Expression of Genes Encoding Electron Transport Chain Proteins and Stimulates Oxidative Phosphorylation in Primary Human Trophoblast Cells by Regulating Mitochondrial Biogenesis. *Sci. Rep.* **2019**, *9*, 246. [CrossRef]

273. Zhou, J.; Shrikhande, G.; Xu, J.; McKay, R.M.; Burns, D.K.; Johnson, J.E.; Parada, L.F. Tsc1 mutant neural stem/progenitor cells exhibit migration deficits and give rise to subependymal lesions in the lateral ventricle. *Genes Dev.* **2011**, *25*, 1595–1600. [CrossRef]

274. Yilmaz, Ö.H.; Katajisto, P.; Lamming, D.W.; Gültekin, Y.; Bauer-Rowe, K.E.; Sengupta, S.; Birsoy, K.; Dursun, A.; Yilmaz, V.O.; Selig, M.; et al. mTORC1 in the Paneth cell niche couples intestinal stem-cell function to calorie intake. *Nature* **2012**, *486*, 490–495. [CrossRef]

275. Malam, Z.; Cohn, R.D. Stem cells on alert: Priming quiescent stem cells after remote injury. *Cell Stem Cell* **2014**, *15*, 7–8. [CrossRef]

276. Rodgers, J.T.; Schroeder, M.D.; Ma, C.; Rando, T.A. HGFA Is an Injury-Regulated Systemic Factor that Induces the Transition of Stem Cells into G(Alert). *Cell Rep.* **2017**, *19*, 479–486. [CrossRef] [PubMed]

277. Buszczak, M.; Signer, R.A.; Morrison, S.J. Cellular differences in protein synthesis regulate tissue homeostasis. *Cell* **2014**, *159*, 242–251. [CrossRef]

278. Vilchez, D.; Saez, I.; Dillin, A. The role of protein clearance mechanisms in organismal ageing and age-related diseases. *Nat. Commun.* **2014**, *5*, 5659. [CrossRef]

279. Vilchez, D.; Simic, M.S.; Dillin, A. Proteostasis and aging of stem cells. *Trends Cell Biol.* **2014**, *24*, 161–170. [CrossRef] [PubMed]

280. Brombin, A.; Joly, J.S.; Jamen, F. New tricks for an old dog: Ribosome biogenesis contributes to stem cell homeostasis. *Curr. Opin. Genet. Dev.* **2015**, *34*, 61–70. [CrossRef]

281. Moran-Crusio, K.; Reavie, L.B.; Aifantis, I. Regulation of hematopoietic stem cell fate by the ubiquitin proteasome system. *Trends Immunol.* **2012**, *33*, 357–363. [CrossRef]

282. Guan, J.L.; Simon, A.K.; Prescott, M.; Menendez, J.A.; Liu, F.; Wang, F.; Wang, C.; Wolvetang, E.; Vazquez-Martin, A.; Zhang, J. Autophagy in stem cells. *Autophagy* **2013**, *9*, 830–849. [CrossRef]

283. Strikoudis, A.; Guillamot, M.; Aifantis, I. Regulation of stem cell function by protein ubiquitylation. *EMBO Rep.* **2014**, *15*, 365–382. [CrossRef]

284. Werner, A.; Manford, A.G.; Rape, M. Ubiquitin-Dependent Regulation of Stem Cell Biology. *Trends Cell Biol.* **2017**, *27*, 568–579. [CrossRef]

285. Signer, R.A.; Magee, J.A.; Salic, A.; Morrison, S.J. Haematopoietic stem cells require a highly regulated protein synthesis rate. *Nature* **2014**, *509*, 49–54. [CrossRef]

286. Zhang, Y.; Duc, A.C.; Rao, S.; Sun, X.L.; Bilbee, A.N.; Rhodes, M.; Li, Q.; Kappes, D.J.; Rhodes, J.; Wiest, D.L. Control of hematopoietic stem cell emergence by antagonistic functions of ribosomal protein paralogs. *Dev. Cell* **2013**, *24*, 411–425. [CrossRef]

287. Cai, X.; Gao, L.; Teng, L.; Ge, J.; Oo, Z.M.; Kumar, A.R.; Gilliland, D.G.; Mason, P.J.; Tan, K.; Speck, N.A. Runx1 Deficiency Decreases Ribosome Biogenesis and Confers Stress Resistance to Hematopoietic Stem and Progenitor Cells. *Cell Stem Cell* **2015**, *17*, 165–177. [CrossRef]

288. Le Bouteiller, M.; Souilhol, C.; Beck-Cormier, S.; Stedman, A.; Burlen-Defranoux, O.; Vandormael-Pournin, S.; Bernex, F.; Cumano, A.; Cohen-Tannoudji, M. Notchless-dependent ribosome synthesis is required for the maintenance of adult hematopoietic stem cells. *J. Exp. Med.* **2013**, *210*, 2351–2369. [CrossRef]

289. Khajuria, R.K.; Munschauer, M.; Ulirsch, J.C.; Fiorini, C.; Ludwig, L.S.; McFarland, S.K.; Abdulhay, N.J.; Specht, H.; Keshishian, H.; Mani, D.R.; et al. Ribosome Levels Selectively Regulate Translation and Lineage Commitment in Human Hematopoiesis. *Cell* **2018**, *173*, 90–103. [CrossRef]

290. Hartman, N.W.; Lin, T.V.; Zhang, L.; Paquelet, G.E.; Feliciano, D.M.; Bordey, A. mTORC1 targets the translational repressor 4E-BP2, but not S6 kinase 1/2, to regulate neural stem cell self-renewal in vivo. *Cell Rep.* **2013**, *5*, 433–444. [CrossRef]

291. Liakath-Ali, K.; Mills, E.W.; Sequeira, I.; Lichtenberger, B.M.; Pisco, A.O.; Sipilä, K.H.; Mishra, A.; Yoshikawa, H.; Wu, C.C.; Ly, T.; et al. An evolutionarily conserved ribosome-rescue pathway maintains epidermal homeostasis. *Nature* **2018**, *556*, 376–380. [CrossRef] [PubMed]

292. Zismanov, V.; Chichkov, V.; Colangelo, V.; Jamet, S.; Wang, S.; Syme, A.; Koromilas, A.E.; Crist, C. Phosphorylation of eIF2α Is a Translational Control Mechanism Regulating Muscle Stem Cell Quiescence and Self-Renewal. *Cell Stem Cell* **2016**, *18*, 79–90. [CrossRef] [PubMed]

293. Cheung, T.H.; Quach, N.L.; Charville, G.W.; Liu, L.; Park, L.; Edalati, A.; Yoo, B.; Hoang, P.; Rando, T.A. Maintenance of muscle stem-cell quiescence by microRNA-489. *Nature* **2012**, *482*, 524–528. [CrossRef]

294. Crist, C.G.; Montarras, D.; Buckingham, M. Muscle satellite cells are primed for myogenesis but maintain quiescence with sequestration of Myf5 mRNA targeted by microRNA-31 in mRNP granules. *Cell Stem Cell* **2012**, *11*, 118–126. [CrossRef] [PubMed]

295. Sontag, E.M.; Samant, R.S.; Frydman, J. Mechanisms and Functions of Spatial Protein Quality Control. *Annu. Rev. Biochem.* **2017**, *86*, 97–122. [CrossRef] [PubMed]

296. Klaips, C.L.; Jayaraj, G.G.; Hartl, F.U. Pathways of cellular proteostasis in aging and disease. *J. Cell Biol.* **2018**, *217*, 51–63. [CrossRef] [PubMed]

297. Fan, G.C. Role of heat shock proteins in stem cell behavior. *Prog. Mol. Biol. Transl. Sci.* **2012**, *111*, 305–322. [PubMed]

298. Baharvand, H.; Fathi, A.; van Hoof, D.; Salekdeh, G.H. Concise review: Trends in stem cell proteomics. *Stem Cells* **2007**, *25*, 1888–1903. [CrossRef]

299. DeLany, J.P.; Floyd, Z.E.; Zvonic, S.; Smith, A.; Gravois, A.; Reiners, E.; Wu, X.; Kilroy, G.; Lefevre, M.; Gimble, J.M. Proteomic analysis of primary cultures of human adipose-derived stem cells: Modulation by adipogenesis. *Mol. Cell. Proteom.* **2005**, *4*, 731–740. [CrossRef]

300. Matsui, H.; Asou, H.; Inaba, T. Cytokines direct the regulation of Bim mRNA stability by heat-shock cognate protein 70. *Mol. Cell* **2007**, *25*, 99–112. [CrossRef]

301. Zou, P.; Yoshihara, H.; Hosokawa, K.; Tai, I.; Shinmyozu, K.; Tsukahara, F.; Maru, Y.; Nakayama, K.; Nakayama, K.I.; Suda, T. p57(Kip2) and p27(Kip1) cooperate to maintain hematopoietic stem cell quiescence through interactions with Hsc70. *Cell Stem Cell* **2011**, *9*, 247–261. [CrossRef] [PubMed]

302. Miharada, K.; Karlsson, G.; Rehn, M.; Rörby, E.; Siva, K.; Cammenga, J.; Karlsson, S. Cripto regulates hematopoietic stem cells as a hypoxic-niche-related factor through cell surface receptor GRP78. *Cell Stem Cell* **2011**, *9*, 330–344. [CrossRef]

303. Ribeil, J.A.; Zermati, Y.; Vandekerckhove, J.; Cathelin, S.; Kersual, J.; Dussiot, M.; Coulon, S.; Moura, I.C.; Zeuner, A.; Kirkegaard-Sørensen, T.; et al. Hsp70 regulates erythropoiesis by preventing caspase-3-mediated cleavage of GATA-1. *Nature* **2007**, *445*, 102–105. [CrossRef]

304. Tai-Nagara, I.; Matsuoka, S.; Ariga, H.; Suda, T. Mortalin and DJ-1 coordinately regulate hematopoietic stem cell function through the control of oxidative stress. *Blood* **2014**, *123*, 41–50. [CrossRef] [PubMed]

305. Vicart, P.; Caron, A.; Guicheney, P.; Li, Z.; Prévost, M.C.; Faure, A.; Chateau, D.; Chapon, F.; Tomé, F.; Dupret, J.M.; et al. A missense mutation in the αB-crystallin chaperone gene causes a desmin-related myopathy. *Nat. Genet.* **1998**, *20*, 92–95. [CrossRef]

306. Sanbe, A.; Osinska, H.; Saffitz, J.E.; Glabe, C.G.; Kayed, R.; Maloyan, A.; Robbins, J. Desmin-related cardiomyopathy in transgenic mice: A cardiac amyloidosis. *Proc. Natl. Acad. Sci. USA* **2004**, *101*, 10132–10136. [CrossRef] [PubMed]

307. Singh, B.N.; Rao, K.S.; Rao, C.M. Ubiquitin-proteasome-mediated degradation and synthesis of MyoD is modulated by alphaB-crystallin, a small heat shock protein, during muscle differentiation. *Biochim. Biophys. Acta* **2010**, *1803*, 288–299. [CrossRef] [PubMed]

308. Chen, J.; Shi, Z.D.; Ji, X.; Morales, J.; Zhang, J.; Kaur, N.; Wang, S. Enhanced osteogenesis of human mesenchymal stem cells by periodic heat shock in self-assembling peptide hydrogel. *Tissue Eng. Part A* **2013**, *19*, 716–728. [CrossRef] [PubMed]

309. Li, C.; Sunderic, K.; Nicoll, S.B.; Wang, S. Downregulation of Heat Shock Protein 70 Impairs Osteogenic and Chondrogenic Differentiation in Human Mesenchymal Stem Cells. *Sci. Rep.* **2018**, *8*, 553. [CrossRef] [PubMed]

310. Walter, P.; Ron, D. The unfolded protein response: From stress pathway to homeostatic regulation. *Science* **2011**, *334*, 1081–1086. [CrossRef] [PubMed]

311. Gardner, B.M.; Pincus, D.; Gotthardt, K.; Gallagher, C.M.; Walter, P. Endoplasmic reticulum stress sensing in the unfolded protein response. *Cold Spring Harb. Perspect. Biol.* **2013**, *5*, a013169. [CrossRef] [PubMed]

312. Melber, A.; Haynes, C.M. UPR^{mt} regulation and output: A stress response mediated by mitochondrial-nuclear communication. *Cell Res.* **2018**, *28*, 281–295. [CrossRef] [PubMed]

313. Shpilka, T.; Haynes, C.M. The mitochondrial UPR: Mechanisms, physiological functions and implications in ageing. *Nat. Rev. Mol. Cell Biol.* **2018**, *19*, 109–120. [CrossRef]

314. Araki, K.; Nagata, K. Protein folding and quality control in the ER. *Cold Spring Harb. Perspect. Biol.* **2011**, *3*, a007526. [CrossRef] [PubMed]

315. Malhotra, J.D.; Kaufman, R.J. ER stress and its functional link to mitochondria: Role in cell survival and death. *Cold Spring Harb. Perspect. Biol.* **2011**, *3*, a004424. [CrossRef]

316. Baker, M.J.; Tatsuta, T.; Langer, T. Quality control of mitochondrial proteostasis. *Cold Spring Harb. Perspect. Biol.* **2011**, *3*, a007559. [CrossRef] [PubMed]

317. Qureshi, M.A.; Haynes, C.M.; Pellegrino, M.W. The mitochondrial unfolded protein response: Signaling from the powerhouse. *J. Biol. Chem.* **2017**, *292*, 13500–13506. [CrossRef]

318. Rouault-Pierre, K.; Lopez-Onieva, L.; Foster, K.; Anjos-Afonso, F.; Lamrissi-Garcia, I.; Serrano-Sanchez, M.; Mitter, R.; Ivanovic, Z.; de Verneuil, H.; Gribben, J.; et al. HIF-2α protects human hematopoietic stem/progenitors and acute myeloid leukemic cells from apoptosis induced by endoplasmic reticulum stress. *Cell Stem Cell* **2013**, *13*, 549–563. [CrossRef]

319. Miharada, K.; Sigurdsson, V.; Karlsson, S. Dppa5 improves hematopoietic stem cell activity by reducing endoplasmic reticulum stress. *Cell Rep.* **2014**, *7*, 1381–1392. [CrossRef]

320. Sigurdsson, V.; Miharada, K. Regulation of unfolded protein response in hematopoietic stem cells. *Int. J. Hematol.* **2018**, *107*, 627–633. [CrossRef]

321. van Galen, P.; Kreso, A.; Mbong, N.; Kent, D.G.; Fitzmaurice, T.; Chambers, J.E.; Xie, S.; Laurenti, E.; Hermans, K.; Eppert, K.; et al. The unfolded protein response governs integrity of the haematopoietic stem-cell pool during stress. *Nature* **2014**, *510*, 268–272. [CrossRef]

322. Cui, K.; Coutts, M.; Stahl, J.; Sytkowski, A.J. Novel interaction between the transcription factor CHOP (GADD153) and the ribosomal protein FTE/S3a modulates erythropoiesis. *J. Biol. Chem.* **2000**, *275*, 7591–7596. [CrossRef]

323. Skalet, A.H.; Isler, J.A.; King, L.B.; Harding, H.P.; Ron, D.; Monroe, J.G. Rapid B cell receptor-induced unfolded protein response in non-secretory B cells correlates with pro- versus antiapoptotic cell fate. *J. Biol. Chem.* **2005**, *280*, 39762–39771. [CrossRef] [PubMed]

324. Zhang, K.; Wong, H.N.; Song, B.; Miller, C.N.; Scheuner, D.; Kaufman, R.J. The unfolded protein response sensor IRE1α is required at 2 distinct steps in B cell lymphopoiesis. *J. Clin. Investig.* **2005**, *115*, 268–281. [CrossRef]

325. Sigurdsson, V.; Takei, H.; Soboleva, S.; Radulovic, V.; Galeev, R.; Siva, K.; Leeb-Lundberg, L.M.; Iida, T.; Nittono, H.; Miharada, K. Bile Acids Protect Expanding Hematopoietic Stem Cells from Unfolded Protein Stress in Fetal Liver. *Cell Stem Cell* **2016**, *18*, 522–532. [CrossRef] [PubMed]

326. Xiong, G.; Hindi, S.M.; Mann, A.K.; Gallot, Y.S.; Bohnert, K.R.; Cavener, D.R.; Whittemore, S.R.; Kumar, A. The PERK arm of the unfolded protein response regulates satellite cell-mediated skeletal muscle regeneration. *eLife* **2017**, *6*, e22871. [CrossRef] [PubMed]

327. Mohrin, M.; Shin, J.; Liu, Y.; Brown, K.; Luo, H.; Xi, Y.; Haynes, C.M.; Chen, D. Stem cell aging. A mitochondrial UPR-mediated metabolic checkpoint regulates hematopoietic stem cell aging. *Science* **2015**, *347*, 1374–1377. [CrossRef] [PubMed]

328. Yau, R.; Rape, M. The increasing complexity of the ubiquitin code. *Nat. Cell Biol.* **2016**, *18*, 579–586. [CrossRef]

329. Gupta-Rossi, N.; Le Bail, O.; Gonen, H.; Brou, C.; Logeat, F.; Six, E.; Ciechanover, A.; Israël, A. Functional interaction between SEL-10, an F-box protein, and the nuclear form of activated Notch1 receptor. *J. Biol. Chem.* **2001**, *276*, 34371–34378. [CrossRef] [PubMed]

330. Koepp, D.M.; Schaefer, L.K.; Ye, X.; Keyomarsi, K.; Chu, C.; Harper, J.W.; Elledge, S.J. Phosphorylation-dependent ubiquitination of cyclin E by the SCFFbw7 ubiquitin ligase. *Science* **2001**, *294*, 173–177. [CrossRef] [PubMed]

331. Strohmaier, H.; Spruck, C.H.; Kaiser, P.; Won, K.A.; Sangfelt, O.; Reed, S.I. Human F-box protein hCdc4 targets cyclin E for proteolysis and is mutated in a breast cancer cell line. *Nature* **2001**, *413*, 316–322. [CrossRef]

332. Goh, E.L.; Zhu, T.; Leong, W.Y.; Lobie, P.E. c-Cbl is a negative regulator of GH-stimulated STAT5-mediated transcription. *Endocrinology* **2002**, *143*, 3590–3603. [CrossRef]

333. Jehn, B.M.; Dittert, I.; Beyer, S.; von der Mark, K.; Bielke, W. c-Cbl binding and ubiquitin-dependent lysosomal degradation of membrane-associated Notch1. *J. Biol. Chem.* **2002**, *277*, 8033–8040. [CrossRef]

334. Nakayama, K.; Nagahama, H.; Minamishima, Y.A.; Miyake, S.; Ishida, N.; Hatakeyama, S.; Kitagawa, M.; Iemura, S.; Natsume, T.; Nakayama, K.I. Skp2-mediated degradation of p27 regulates progression into mitosis. *Dev. Cell* **2004**, *6*, 661–672. [CrossRef]

335. Tsunematsu, R.; Nakayama, K.; Oike, Y.; Nishiyama, M.; Ishida, N.; Hatakeyama, S.; Bessho, Y.; Kageyama, R.; Suda, T.; Nakayama, K.I. Mouse Fbw7/Sel-10/Cdc4 is required for notch degradation during vascular development. *J. Biol. Chem.* **2004**, *279*, 9417–9423. [CrossRef]

336. Welcker, M.; Orian, A.; Jin, J.; Grim, J.E.; Harper, J.W.; Eisenman, R.N.; Clurman, B.E. The Fbw7 tumor suppressor regulates glycogen synthase kinase 3 phosphorylation-dependent c-Myc protein degradation. *Proc. Natl. Acad. Sci. USA* **2004**, *101*, 9085–9090. [CrossRef]

337. Yada, M.; Hatakeyama, S.; Kamura, T.; Nishiyama, M.; Tsunematsu, R.; Imaki, H.; Ishida, N.; Okumura, F.; Nakayama, K.; Nakayama, K.I. Phosphorylation-dependent degradation of c-Myc is mediated by the F-box protein Fbw7. *EMBO J.* **2004**, *23*, 2116–2125. [CrossRef]

338. Schmidt, M.H.H.; Dikic, I. The Cbl interactome and its functions. *Nat. Rev. Mol. Cell Biol.* **2005**, *6*, 907–918. [CrossRef]

339. Zeng, S.; Xu, Z.; Lipkowitz, S.; Longley, J.B. Regulation of stem cell factor receptor signaling by Cbl family proteins (Cbl-b/c-Cbl). *Blood* **2005**, *105*, 226–232. [CrossRef]

340. Naujokat, C.; Sarić, T. Concise review: Role and function of the ubiquitin-proteasome system in mammalian stem and progenitor cells. *Stem Cells* **2007**, *25*, 2408–2418. [CrossRef]

341. Reavie, L.; Della Gatta, G.; Crusio, K.; Aranda-Orgilles, B.; Buckley, S.M.; Thompson, B.; Lee, E.; Gao, J.; Bredemeyer, A.L.; Helmink, B.A.; et al. Regulation of hematopoietic stem cell differentiation by a single ubiquitin ligase-substrate complex. *Nat. Immunol.* **2010**, *11*, 207–215. [CrossRef] [PubMed]

342. Inuzuka, H.; Shaik, S.; Onoyama, I.; Gao, D.; Tseng, A.; Maser, R.S.; Zhai, B.; Wan, L.; Gutierrez, A.; Lau, A.W.; et al. SCF(FBW7) regulates cellular apoptosis by targeting MCL1 for ubiquitylation and destruction. *Nature* **2011**, *471*, 104–109. [CrossRef] [PubMed]

343. Rathinam, C.; Matesic, L.E.; Flavell, R.A. The E3 ligase Itch is a negative regulator of the homeostasis and function of hematopoietic stem cells. *Nat. Immunol.* **2011**, *12*, 399–407. [CrossRef] [PubMed]

344. Wertz, I.E.; Kusam, S.; Lam, C.; Okamoto, T.; Sandoval, W.; Anderson, D.J.; Helgason, E.; Ernst, J.A.; Eby, M.; Liu, J.; et al. Sensitivity to antitubulin chemotherapeutics is regulated by MCL1 and FBW7. *Nature* **2011**, *471*, 110–114. [CrossRef]

345. Takeishi, S.; Matsumoto, A.; Onoyama, I.; Naka, K.; Hirao, A.; Nakayama, K.I. Ablation of Fbxw7 eliminates leukemia-initiating cells by preventing quiescence. *Cancer Cell* **2013**, *23*, 347–361. [CrossRef] [PubMed]

346. King, B.; Boccalatte, F.; Moran-Crusio, K.; Wolf, E.; Wang, J.; Kayembe, C.; Lazaris, C.; Yu, X.; Aranda-Orgilles, B.; Lasorella, A.; et al. The ubiquitin ligase Huwe1 regulates the maintenance and lymphoid commitment of hematopoietic stem cells. *Nat. Immunol.* **2016**, *17*, 1312–1321. [CrossRef] [PubMed]

347. Matsuoka, S.; Oike, Y.; Onoyama, I.; Iwama, A.; Arai, F.; Takubo, K.; Mashimo, Y.; Oguro, H.; Nitta, E.; Ito, K.; et al. Fbxw7 acts as a critical fail-safe against premature loss of hematopoietic stem cells and development of T-ALL. *Genes Dev.* **2008**, *22*, 986–991. [CrossRef]

348. Rathinam, C.; Thien, C.B.; Langdon, W.Y.; Gu, H.; Flavell, R.A. The E3 ubiquitin ligase c-Cbl restricts development and functions of hematopoietic stem cells. *Genes Dev.* **2008**, *22*, 992–997. [CrossRef]

349. Thompson, B.J.; Jankovic, V.; Gao, J.; Buonamici, S.; Vest, A.; Lee, J.M.; Zavadil, J.; Nimer, S.D.; Aifantis, I. Control of hematopoietic stem cell quiescence by the E3 ubiquitin ligase Fbw7. *J. Exp. Med.* **2008**, *205*, 1395–1408. [CrossRef]

350. Rodriguez, S.; Wang, L.; Mumaw, C.; Srour, E.F.; Lo Celso, C.; Nakayama, K.; Carlesso, N. The SKP2 E3 ligase regulates basal homeostasis and stress-induced regeneration of HSCs. *Blood* **2011**, *117*, 6509–6519. [CrossRef]

351. Wang, J.; Han, F.; Wu, J.; Lee, S.W.; Chan, C.H.; Wu, C.Y.; Yang, W.L.; Gao, Y.; Zhang, X.; Jeong, Y.S.; et al. The role of Skp2 in hematopoietic stem cell quiescence, pool size, and self-renewal. *Blood* **2011**, *118*, 5429–5438. [CrossRef]

352. Simon, M.C.; Keith, B. The role of oxygen availability in embryonic development and stem cell function. *Nat. Rev. Mol. Cell Biol.* **2008**, *9*, 285–296. [CrossRef]

353. Kaelin, W.G., Jr. The von Hippel-Lindau tumour suppressor protein: O_2 sensing and cancer. *Nat. Rev. Cancer* **2008**, *8*, 865–873. [CrossRef] [PubMed]

354. Konishi, Y.; Stegmüller, J.; Matsuda, T.; Bonni, S.; Bonni, A. Cdh1-APC controls axonal growth and patterning in the mammalian brain. *Science* **2004**, *303*, 1026–1030. [CrossRef]

355. Lasorella, A.; Stegmüller, J.; Guardavaccaro, D.; Liu, G.; Carro, M.S.; Rothschild, G.; de la Torre-Ubieta, L.; Pagano, M.; Bonni, A.; Iavarone, A. Degradation of Id2 by the anaphase-promoting complex couples cell cycle exit and axonal growth. *Nature* **2006**, *442*, 471–474. [CrossRef]

356. Stegmüller, J.; Konishi, Y.; Huynh, M.A.; Yuan, Z.; Dibacco, S.; Bonni, A. Cell-intrinsic regulation of axonal morphogenesis by the Cdh1-APC target SnoN. *Neuron* **2006**, *50*, 389–400. [CrossRef]
357. Yang, Y.; Kim, A.H.; Yamada, T.; Wu, B.; Bilimoria, P.M.; Ikeuchi, Y.; de la Iglesia, N.; Shen, J.; Bonni, A. A Cdc20-APC ubiquitin signaling pathway regulates presynaptic differentiation. *Science* **2009**, *326*, 575–578. [CrossRef]
358. Huang, J.; Ikeuchi, Y.; Malumbres, M.; Bonni, A. A Cdh1-APC/FMRP Ubiquitin Signaling Link Drives mGluR-Dependent Synaptic Plasticity in the Mammalian Brain. *Neuron* **2015**, *86*, 726–739. [CrossRef] [PubMed]
359. Shao, R.; Liu, J.; Yan, G.; Zhang, J.; Han, Y.; Guo, J.; Xu, Z.; Yuan, Z.; Liu, J.; Malumbres, M.; et al. Cdh1 regulates craniofacial development via APC-dependent ubiquitination and activation of Goosecoid. *Cell Res.* **2016**, *26*, 699–712. [CrossRef]
360. Adhikary, S.; Marinoni, F.; Hock, A.; Hulleman, E.; Popov, N.; Beier, R.; Bernard, S.; Quarto, M.; Capra, M.; Goettig, S.; et al. The ubiquitin ligase HectH9 regulates transcriptional activation by Myc and is essential for tumor cell proliferation. *Cell* **2005**, *123*, 409–421. [CrossRef]
361. Zhong, Q.; Gao, W.; Du, F.; Wang, X. Mule/ARF-BP1, a BH3-only E3 ubiquitin ligase, catalyzes the polyubiquitination of Mcl-1 and regulates apoptosis. *Cell* **2005**, *121*, 1085–1095. [CrossRef]
362. Zhao, X.; Heng, J.I.; Guardavaccaro, D.; Jiang, R.; Pagano, M.; Guillemot, F.; Iavarone, A.; Lasorella, A. The HECT-domain ubiquitin ligase Huwe1 controls neural differentiation and proliferation by destabilizing the N-Myc oncoprotein. *Nat. Cell Biol.* **2008**, *10*, 643–653. [CrossRef]
363. Forget, A.; Bihannic, L.; Cigna, S.M.; Lefevre, C.; Remke, M.; Barnat, M.; Dodier, S.; Shirvani, H.; Mercier, A.; Mensah, A.; et al. Shh signaling protects Atoh1 from degradation mediated by the E3 ubiquitin ligase Huwe1 in neural precursors. *Dev. Cell* **2014**, *29*, 649–661. [CrossRef]
364. Urbán, N.; van den Berg, D.L.; Forget, A.; Andersen, J.; Demmers, J.A.; Hunt, C.; Ayrault, O.; Guillemot, F. Return to quiescence of mouse neural stem cells by degradation of a proactivation protein. *Science* **2016**, *353*, 292–295. [CrossRef]
365. Garcia-Prat, L.; Martinez-Vicente, M.; Muñoz-Cánoves, P. Autophagy: A decisive process for stemness. *Oncotarget* **2016**, *7*, 12286–12288. [CrossRef]
366. Leeman, D.S.; Hebestreit, K.; Ruetz, T.; Webb, A.E.; McKay, A.; Pollina, E.A.; Dulken, B.W.; Zhao, X.; Yeo, R.W.; Ho, T.T.; et al. Lysosome activation clears aggregates and enhances quiescent neural stem cell activation during aging. *Science* **2018**, *359*, 1277–1283. [CrossRef]
367. García-Prat, L.; Martínez-Vicente, M.; Perdiguero, E.; Ortet, L.; Rodríguez-Ubreva, J.; Rebollo, E.; Ruiz-Bonilla, V.; Gutarra, S.; Ballestar, E.; Serrano, A.L.; et al. Autophagy maintains stemness by preventing senescence. *Nature* **2016**, *529*, 37–42. [CrossRef]
368. Oliver, L.; Hue, E.; Priault, M.; Vallette, F.M. Basal autophagy decreased during the differentiation of human adult mesenchymal stem cells. *Stem Cells Dev.* **2012**, *21*, 2779–2788. [CrossRef]
369. Salemi, S.; Yousefi, S.; Constantinescu, M.A.; Fey, M.F.; Simon, H.U. Autophagy is required for self-renewal and differentiation of adult human stem cells. *Cell Res.* **2012**, *22*, 432–435. [CrossRef]
370. Eijkelenboom, A.; Burgering, B.M. FOXOs: Signalling integrators for homeostasis maintenance. *Nat. Rev. Mol. Cell Biol.* **2013**, *14*, 83–97. [CrossRef]
371. Warr, M.R.; Binnewies, M.; Flach, J.; Reynaud, D.; Garg, T.; Malhotra, R.; Debnath, J.; Passegué, E. FOXO3A directs a protective autophagy program in haematopoietic stem cells. *Nature* **2013**, *494*, 323–327. [CrossRef]
372. Liu, F.; Lee, J.Y.; Wei, H.; Tanabe, O.; Engel, J.D.; Morrison, S.J.; Guan, J.L. FIP200 is required for the cell-autonomous maintenance of fetal hematopoietic stem cells. *Blood* **2010**, *116*, 4806–4814. [CrossRef]
373. Mortensen, M.; Ferguson, D.J.; Edelmann, M.; Kessler, B.; Morten, K.J.; Komatsu, M.; Simon, A.K. Loss of autophagy in erythroid cells leads to defective removal of mitochondria and severe anemia in vivo. *Proc. Natl. Acad. Sci. USA* **2010**, *107*, 832–837. [CrossRef]
374. Mortensen, M.; Watson, A.S.; Simon, A.K. Lack of autophagy in the hematopoietic system leads to loss of hematopoietic stem cell function and dysregulated myeloid proliferation. *Autophagy* **2011**, *7*, 1069–1070. [CrossRef]
375. Ho, T.T.; Warr, M.R.; Adelman, E.R.; Lansinger, O.M.; Flach, J.; Verovskaya, E.V.; Figueroa, M.E.; Passegué, E. Autophagy maintains the metabolism and function of young and old stem cells. *Nature* **2017**, *543*, 205–210. [CrossRef] [PubMed]

376. Vázquez, P.; Arroba, A.I.; Cecconi, F.; de la Rosa, E.J.; Boya, P.; de Pablo, F. Atg5 and Ambra1 differentially modulate neurogenesis in neural stem cells. *Autophagy* **2012**, *8*, 187–199. [CrossRef] [PubMed]

377. Zhang, J.; Liu, J.; Huang, Y.; Chang, J.Y.; Liu, L.; McKeehan, W.L.; Martin, J.F.; Wang, F. FRS2α-mediated FGF signals suppress premature differentiation of cardiac stem cells through regulating autophagy activity. *Circ. Res.* **2012**, *110*, e29–e39. [CrossRef]

378. Zhang, J.; Liu, J.; Liu, L.; McKeehan, W.L.; Wang, F. The fibroblast growth factor signaling axis controls cardiac stem cell differentiation through regulating autophagy. *Autophagy* **2012**, *8*, 690–691. [CrossRef] [PubMed]

379. Wang, C.; Liang, C.C.; Bian, Z.C.; Zhu, Y.; Guan, J.L. FIP200 is required for maintenance and differentiation of postnatal neural stem cells. *Nat. Neurosci.* **2013**, *16*, 532–542. [CrossRef]

380. Fiacco, E.; Castagnetti, F.; Bianconi, V.; Madaro, L.; De Bardi, M.; Nazio, F.; D'Amico, A.; Bertini, E.; Cecconi, F.; Puri, P.L.; et al. Autophagy regulates satellite cell ability to regenerate normal and dystrophic muscles. *Cell Death Differ.* **2016**, *23*, 1839–1849. [CrossRef] [PubMed]

381. Fortini, P.; Ferretti, C.; Iorio, E.; Cagnin, M.; Garribba, L.; Pietraforte, D.; Falchi, M.; Pascucci, B.; Baccarini, S.; Morani, F.; et al. The fine tuning of metabolism, autophagy and differentiation during in vitro myogenesis. *Cell Death Dis.* **2016**, *7*, e2168. [CrossRef]

382. Pietras, E.M.; Warr, M.R.; Passegué, E. Cell cycle regulation in hematopoietic stem cells. *J. Cell Biol.* **2011**, *195*, 709–720. [CrossRef]

383. Sage, J. The retinoblastoma tumor suppressor and stem cell biology. *Genes Dev.* **2012**, *26*, 1409–1420. [CrossRef]

384. Matsumoto, A.; Nakayama, K.I. Role of key regulators of the cell cycle in maintenance of hematopoietic stem cells. *Biochim. Biophys. Acta* **2013**, *1830*, 2335–2344. [CrossRef] [PubMed]

385. Nakamura-Ishizu, A.; Takizawa, H.; Suda, T. The analysis, roles and regulation of quiescence in hematopoietic stem cells. *Development* **2014**, *141*, 4656–4666. [CrossRef] [PubMed]

386. Popov, B.; Petrov, N. pRb-E2F signaling in life of mesenchymal stem cells: Cell cycle, cell fate, and cell differentiation. *Genes Dis.* **2014**, *1*, 174–187. [CrossRef] [PubMed]

387. Dumont, N.A.; Wang, Y.X.; Rudnicki, M.A. Intrinsic and extrinsic mechanisms regulating satellite cell function. *Development* **2015**, *142*, 1572–1581. [CrossRef]

388. Kuerbitz, S.J.; Plunkett, B.S.; Walsh, W.V.; Kastan, M.B. Wild-type p53 is a cell cycle checkpoint determinant following irradiation. *Proc. Natl. Acad. Sci. USA* **1992**, *89*, 7491–7495. [CrossRef] [PubMed]

389. Harper, J.W.; Elledge, S.J.; Keyomarsi, K.; Dynlacht, B.; Tsai, L.H.; Zhang, P.; Dobrowolski, S.; Bai, C.; Connell-Crowley, L.; Swindell, E.; et al. Inhibition of cyclin-dependent kinases by p21. *Mol. Biol. Cell* **1995**, *6*, 387–400. [CrossRef]

390. Vousden, K.H.; Lane, D.P. p53 in health and disease. *Nat. Rev. Mol. Cell Biol.* **2007**, *8*, 275–283. [CrossRef] [PubMed]

391. TeKippe, M.; Harrison, D.E.; Chen, J. Expansion of hematopoietic stem cell phenotype and activity in Trp53-null mice. *Exp. Hematol.* **2003**, *31*, 521–527. [CrossRef]

392. Chen, J.; Ellison, F.M.; Keyvanfar, K.; Omokaro, S.O.; Desierto, M.J.; Eckhaus, M.A.; Young, N.S. Enrichment of hematopoietic stem cells with SLAM and LSK markers for the detection of hematopoietic stem cell function in normal and Trp53 null mice. *Exp. Hematol.* **2008**, *36*, 1236–1243. [CrossRef]

393. Akala, O.O.; Park, I.K.; Qian, D.; Pihalja, M.; Becker, M.W.; Clarke, M.F. Long-term haematopoietic reconstitution by Trp53-/-p16Ink4a-/-p19Arf-/- multipotent progenitors. *Nature* **2008**, *453*, 228–232. [CrossRef]

394. Liu, Y.; Elf, S.E.; Miyata, Y.; Sashida, G.; Liu, Y.; Huang, G.; Di Giandomenico, S.; Lee, J.M.; Deblasio, A.; Menendez, S.; et al. p53 regulates hematopoietic stem cell quiescence. *Cell Stem Cell* **2009**, *4*, 37–48. [CrossRef]

395. Asai, T.; Liu, Y.; Di Giandomenico, S.; Bae, N.; Ndiaye-Lobry, D.; Deblasio, A.; Menendez, S.; Antipin, Y.; Reva, B.; Wevrick, R.; et al. Necdin, a p53 target gene, regulates the quiescence and response to genotoxic stress of hematopoietic stem/progenitor cells. *Blood* **2012**, *120*, 1601–1612. [CrossRef]

396. Yamashita, M.; Nitta, E.; Suda, T. Regulation of hematopoietic stem cell integrity through p53 and its related factors. *Ann. N. Y. Acad. Sci.* **2016**, *1370*, 45–54. [CrossRef]

397. Meletis, K.; Wirta, V.; Hede, S.M.; Nistér, M.; Lundeberg, J.; Frisén, J. p53 suppresses the self-renewal of adult neural stem cells. *Development* **2006**, *133*, 363–369. [CrossRef]

398. Weinberg, R.A. The retinoblastoma protein and cell cycle control. *Cell* **1995**, *81*, 323–330. [CrossRef]

399. Giacinti, C.; Giordano, A. RB and cell cycle progression. *Oncogene* **2006**, *25*, 5220–5227. [CrossRef]

400. Harbour, J.W.; Dean, D.C. The Rb/E2F pathway: Expanding roles and emerging paradigms. *Genes Dev.* **2000**, *14*, 2393–2409.
401. Viatour, P.; Somervaille, T.C.; Venkatasubrahmanyam, S.; Kogan, S.; McLaughlin, M.E.; Weissman, I.L.; Butte, A.J.; Passegué, E.; Sage, J. Hematopoietic stem cell quiescence is maintained by compound contributions of the retinoblastoma gene family. *Cell Stem Cell* **2008**, *3*, 416–428. [PubMed]
402. Hosoyama, T.; Nishijo, K.; Prajapati, S.I.; Li, G.; Keller, C. Rb1 gene inactivation expands satellite cell and postnatal myoblast pools. *J. Biol. Chem.* **2011**, *286*, 19556–19564. [CrossRef] [PubMed]
403. Cheng, T.; Rodrigues, N.; Shen, H.; Yang, Y.; Dombkowski, D.; Sykes, M.; Scadden, D.T. Hematopoietic stem cell quiescence maintained by p21cip1/waf1. *Science* **2000**, *287*, 1804–1808. [CrossRef]
404. Kippin, T.E.; Martens, D.J.; van der Kooy, D. p21 loss compromises the relative quiescence of forebrain stem cell proliferation leading to exhaustion of their proliferation capacity. *Genes Dev.* **2005**, *19*, 756–767. [CrossRef]

International Journal of
Molecular Sciences

MDPI

Article

Disordered Expression of *shaggy*, the *Drosophila* Gene Encoding a Serine-Threonine Protein Kinase GSK3, Affects the Lifespan in a Transcript-, Stage-, and Tissue-Specific Manner

Mikhail V. Trostnikov [1,2], Natalia V. Roshina [1,2], Stepan V. Boldyrev [1,2], Ekaterina R. Veselkina [1], Andrey A. Zhuikov [1], Anna V. Krementsova [1,3] and Elena G. Pasyukova [1,*]

[1]   Institute of Molecular Genetics of RAS, Kurchatov Sq. 2, 123182 Moscow, Russia;
      mikhail.trostnikov@gmail.com (M.V.T.); nwumr@yandex.ru (N.V.R.); beibaraban34@gmail.com (S.V.B.);
      veselkinaer@gmail.com (E.R.V.); homkabrut@gmail.com (A.A.Z.); akrementsova@mail.ru (A.V.K.)
[2]   Vavilov Institute of General Genetics, Russian Academy of Sciences, Gubkin 3, 119991 Moscow, Russia
[3]   Emmanuel Institute of Biochemical Physics, Russian Academy of Sciences, Kosygin St. 4,
      119334 Moscow, Russia
[*]   Correspondence: egpas@rambler.ru

Received: 30 March 2019; Accepted: 2 May 2019; Published: 4 May 2019

Abstract: GSK3 (glycogen synthase kinase 3) is a conserved protein kinase governing numerous regulatory pathways. In *Drosophila melanogaster*, GSK3 is encoded by *shaggy* (*sgg*), which forms 17 annotated transcripts corresponding to 10 protein isoforms. Our goal was to demonstrate how differential *sgg* transcription affects lifespan, which GSK3 isoforms are important for the nervous system, and which changes in the nervous system accompany accelerated aging. Overexpression of three *sgg* transcripts affected the lifespan in a stage- and tissue-specific way: *sgg-RA* and *sgg-RO* affected the lifespan only when overexpressed in muscles and in embryos, respectively; the essential *sgg-RB* transcript affected lifespan when overexpressed in all tissues tested. In the nervous system, only *sgg-RB* overexpression affected lifespan, causing accelerated aging in a neuron-specific way, with the strongest effects in dopaminergic neurons and the weakest effects in GABAergic neurons. Pan-neuronal *sgg-RB* overexpression violated the properties of the nervous system, including the integrity of neuron bodies; the number, distribution, and structure of mitochondria; cytoskeletal characteristics; and synaptic activity. Such changes observed in young individuals indicated premature aging of their nervous system, which paralleled a decline in survival. Our findings demonstrated the key role of GSK3 in ensuring the link between the pathology of neurons and lifespan.

Keywords: GSK3 (glycogen synthase kinase 3); protein kinases; transcription; lifespan; the nervous system; *Drosophila melanogaster*

---

1. Introduction

GSK3 (glycogen synthase kinase 3) is an actively studied, highly conserved serine–threonine protein kinase that is primarily regulated by inhibition and is involved in multiple signaling pathways that regulate embryogenesis and differentiation; cell renewal, migration, and apoptosis; gene transcription; metabolism; and survival via more than 50 target proteins with phosphorylation sites for GSK3 [1,2]. In humans, there are two main forms of GSK3, α and β, encoded by the two highly homologous genes; in *Drosophila*, only one form exists, which is closer to GSK3β.

In *D. melanogaster*, GSK3 plays multiple roles in the development and function of the nervous system, starting from the early development stages. GSK3 directly phosphorylates aPKC (atypical protein kinase C) [3,4], a key component ensuring asymmetric neuroblast division during *D. melanogaster* early development. GSK3 participates in the development of neural precursor cells [5] and, via InR/TOR (Insulin Receptor/Target of Rapamycin) signaling, in the control of dendrite pruning during larval-pupal transition [6]. Moreover, GSK3 controls synapse formation, growth, and morphological structure by controlling the dynamics of the microtubule cytoskeleton [7–9] and regulates axonal stability and neural circuit integrity via the Wnt pathway [10]. GSK3 also regulates neurotransmitter release into NMJs (neuromuscular junctions) [8]. GSK3 directly phosphorylates microtubule-associated protein tau [11,12]. Overexpression of human tau in *D. melanogaster* disrupts axonal transport causing vesicle aggregation, and co-overexpression of constitutively active GSK3 enhances the effects [13]. Two functions of GSK3 specifically attracted our attention to this protein.

First, GSK3 participates in the control of asymmetric neuroblast division. Earlier, we demonstrated that functional changes in several genes involved in the regulation of this developmental process, namely, *aPKC* encoding the GSK3 target, aPKC, *escargot* and *inscutable*, affect the lifespan of adult flies [14,15]. Moreover, our previous studies suggest that changes in gene expression at the embryonic stage can be responsible for an increase in adult lifespan due to yet unknown mechanisms that provide this carry-over effect [16]. Therefore, we were interested in understanding whether GSK3 function, especially during the embryonic stage, also affects lifespan.

Second, GSK3 severely affects the functionality of the tau protein, which leads to the development of various age-dependent neurodegenerative diseases and significantly reduces life expectancy [11,17]. We were interested in understanding whether changes in GSK3 expression in the nervous system directly affect lifespan and, if yes, we aimed to identify which structural and functional properties of the nervous system accompany alterations in lifespan and rate of aging.

In *Drosophila melanogaster*, GSK3 is encoded by the gene, *shaggy* (*sgg*), which forms 17 annotated transcripts corresponding to 10 annotated protein isoforms (Available online: http://flybase.org/ reports/FBgn0003371). To address the questions raised, first, it was necessary to understand which transcripts and protein isoforms encoded by *sgg* are functional. Ten *sgg* transcripts and at least five Sgg proteins of different primary structure were initially demonstrated [18]. The major Sgg isoform that is detected at essentially all stages of the life cycle and in different tissues [18,19] is encoded by six transcripts annotated in FlyBase (Available online: http://flybase.org/reports/FBgn0003371), including *sgg-RB* (GenBank #AY122193.1). Only a few other *sgg* transcripts were reported as full-size cDNAs: *sgg-RA* (GenBank #X70863.1); *sgg-RD* (GenBank #BT133153.1); *sgg-RG* (GenBank #AY119664.1 and #BT050474.1); and *sgg-RO* (GenBank #BT072831.1). Analysis of the proteome revealed three Sgg isoforms, namely, Sgg-PA, Sgg-PB, and Sgg-PD, in the embryonic cell line, Kc167; no isoforms were found in fly tissues [20–22]. Two isoforms, Sgg-PA and Sgg-PB, corresponding to sgg39 and sgg10 isoforms of Ruel and coauthors [18], were detected in recent functional studies; however, the functionality of alterations in Sgg-PA expression was not revealed [9,23].

In this paper, we described the effects of the misexpression of four *sgg* transcripts (*sgg-RA*, *sgg-RB*, *sgg-RG*, and *sgg-RO*) on *Drosophila* lifespan and aging. Different *sgg* transcripts demonstrated stage- and tissue-specificity; for example, *sgg-RA* affected lifespan only when overexpressed in muscles, and *sgg-RO* affected lifespan only when overexpressed in embryos. The sex-specific increase in adult lifespan was observed due to increased *sgg-RO* expression in embryos. Altering *sgg-RO* expression in embryos was enough to affect adult lifespan, which suggested the existence of some carry-over mechanisms of an epigenetic or other nature. The essential *sgg-RB* transcript was functional in all tissues tested, and in the nervous system, only *sgg-RB* overexpression affected lifespan, causing severe shortening. If overexpression took place in middle-aged adult flies, detrimental effects were alleviated compared to those of lifelong overexpression. Overexpression of *sgg-RB* in different types of neurons accelerated the aging of flies in a neuron-specific manner, with the most pronounced effects in dopaminergic neurons and the least pronounced effects in GABAergic neurons. Pan-neuronal *sgg-RB* overexpression severely violated the structural and functional properties of the nervous system: The integrity of neuron bodies, the number, distribution, and structure of mitochondria, cytoskeletal properties in the presynaptic zones, and synaptic activity. Overall, our data demonstrated a wide variety of effects of GSK3 on survival and proved its causal role in the aging of the nervous system.

2. Results and Discussion

2.1. Differential Expression of Sgg Affects Lifespan

According to the annotation given in FlyBase (Available online: https://flybase.org/reports/FBgn0003371), *sgg*, like many other genes, can potentially form many transcripts. Questions about what set of transcripts exist in a living organism, whether each transcript has specific functions that are different from those of other transcripts—in particular, whether the transcripts are specific for a given stage of development, age, and tissue—remain open. Our primary goal was to understand what *sgg* transcripts that are functional in the nervous system affect lifespan. However, to obtain a broader view, we tested the effects of some *sgg* transcripts in a restricted set of other tissues and at different stages. Our choice of transcripts was based on their presence in GenBank as full-size cDNAs, which indicated that these variants are actually formed when the gene is expressed. Our choice of additional tissues was based on the fact that the fat body was repeatedly shown to play an important role in lifespan control [24–26], whereas the role of muscles, on the contrary, is not broadly known [27]. Finally, to test the functionality of *sgg* transcripts, we used a complex phenotypic trait—lifespan.

A transgenic line providing overexpression of the essential *sgg* transcript *sgg-RB*, sggB, constructed by [19], was obtained from the Bloomington *Drosophila* Stock Center, USA (Available online: http://flystocks.bio.indiana.edu/) together with the corresponding control line (Control B). To assess the functionality of *sgg-RA*, *sgg-RG*, and *sgg-RO*, we obtained three transgenic lines providing overexpression of these transcripts: sggA, sggG, and sggO; the line initially used for transformations was used as a control line (Control AGO). We also acquired two lines with transgenic constructs providing *sgg* knockdown (sggKD1 and sggKD2), together with the control lines (Control KD1 and Control KD2) provided by the same manufacturer (Available online: http://flystocks.bio.indiana.edu/Browse/TRiPtb.htm).

Overexpression of *sgg-RA*, *sgg-RB*, *sgg-RG*, or *sgg-RO* induced by GAL4 drivers should specifically increase the amount of the corresponding transcripts, while knockdown induced by GAL4 drivers should decrease the amount of all *sgg* transcripts. The level of decrease was expected to depend on the effectiveness of the knockdown provided by different lines. sggKD1 was expected to provide a stronger level of knockdown and sggKD2 the weaker level of knockdown, according to the manufacturer's description (https://bdsc.indiana.edu/stocks/rnai/rnai_n_z.html). A detailed description of all the lines is given in the Materials and Methods section.

To roughly characterize *sgg* effects on lifespan, we induced transcript-specific *sgg* overexpression and overall *sgg* knockdown in embryos (the D1 driver line) and in three tissues: The fat body (the D2 driver line), muscles (the D3 driver line), and the nervous system (the D4 driver line), using the GAL4-UAS binary system [28].

Overexpression of *sgg-RB* in embryos was lethal, and overexpression of *sgg-RA* and *sgg-RG* had no effect on the lifespan of both males and females (Table S1, Figure 1A,B). Overexpression of *sgg-RO* affected male and female lifespans in opposite directions: In males, lifespan was significantly increased, while in females, lifespan was decreased (Table S1, Figure 1A,B). In the second experiment aimed to verify this result, the effects were reproduced, although the level of the effects was somewhat lower in both males and females (Table S1, Figure 1C,D). As expected, the effect of the overall *sgg* knockdown was lethal when the stronger sggKD1 line was used, while the weaker knockdown decreased the lifespans of both males and females (Table S1, Figure 1E,F).

Overexpression of *sgg-RB* in the fat body was lethal, and overexpression of *sgg-RA*, *sgg-RG*, and *sgg-RO* had no effect on the lifespans of both males and females (Table S2, Figure 2A,B). Strong *sgg* knockdown in the fat body caused lethality, while weak knockdown decreased the male lifespan and did not affect the female lifespan (Table S2, Figure 2C,D). In females, the survival curve indicated a possible positive effect of the reduced *sgg* function on lifespan.

Overexpression of *sgg-RA* and *sgg-RB* in muscles decreased both male and female lifespans, and overexpression of *sgg-RG* and *sgg-RO* had no effect on the lifespans of both males and females (Table S3, Figure 3A–D). A negative effect of *sgg-RA* overexpression was confirmed in the second experiment (Table S3). Both strong and weak knockdown in muscles decreased male and female lifespans to varying degrees (Table S3, Figure 3E,F).

Overexpression of *sgg-RB* in the nervous system severely reduced male and female lifespans (Table S4, Figure 4A,B), and overexpression of *sgg-RA*, *sgg-RG*, and *sgg-RO* had no effect on the lifespan of both males and females (Table S4, Figure 4C,D). Strong *sgg* knockdown in the nervous system caused lethality, while weak knockdown decreased both male and female lifespans (Table S4, Figure 4E,F). Of note, weak knockdown in the nervous system affected the lifespan to a lesser degree than in muscles and in embryos, where the effect was most pronounced.

Overall, a set of lifespan assays produced several noteworthy results (Figure 5). First, the effects of *sgg* misexpression on lifespan were mostly detrimental. The most adverse results, that is, lethality, were observed when either overexpression or knockdown were induced in embryos or in the fat body. However, increased *sgg-RO* expression in embryos increased the adult lifespan in a sex-dependent manner. Detailed research aimed at analyzing the molecular mechanisms underlying this positive effect is currently under way. The effects of gene expression early in life on lifespan are of particular interest for us because we have previously shown that changes in gene expression in embryos can lead to a sex-specific increase in longevity [16]. The sex-dependent effects of alterations in gene transcription on lifespan have often been observed by others, although the nature of this dependence remains unclear [29].

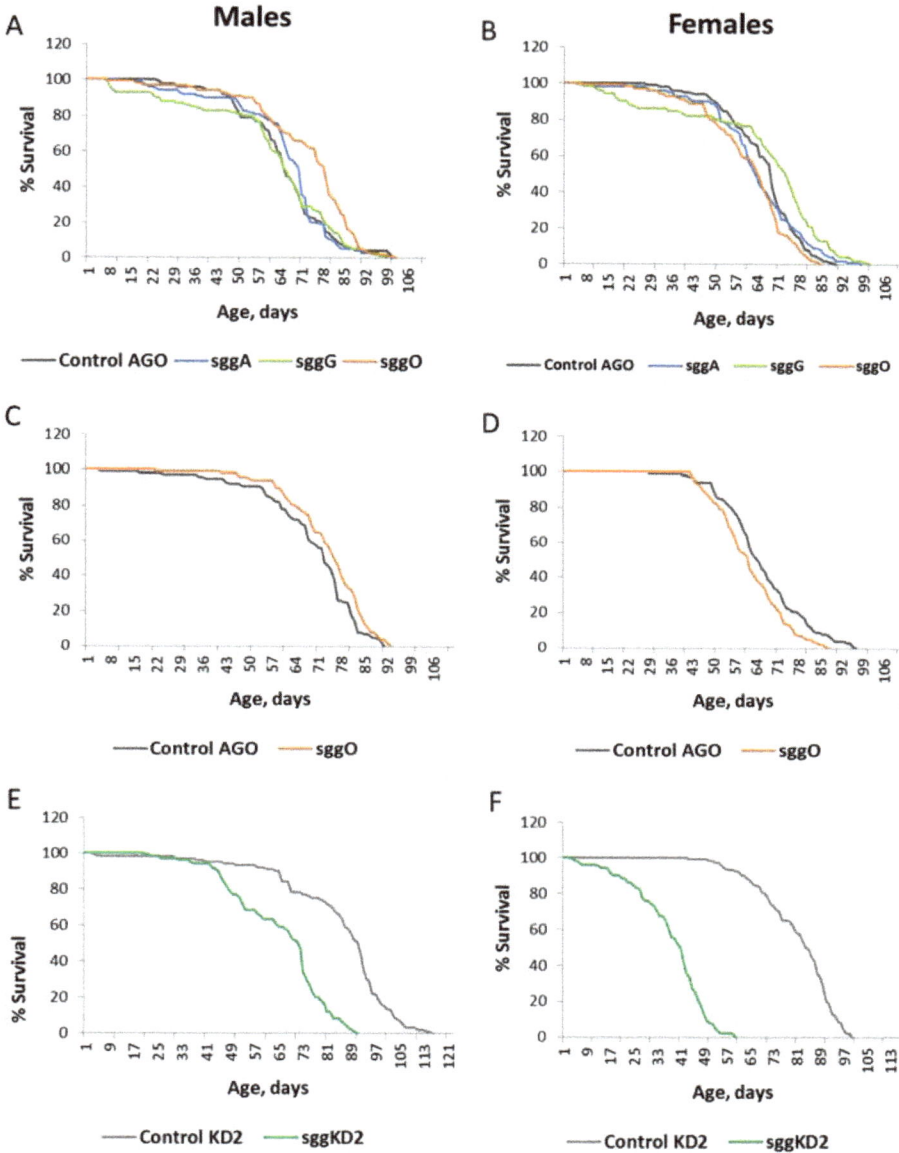

Figure 1. Effects of disordered *sgg* expression in embryos on male (**A,C,E**) and female (**B,D,F**) lifespans. Control AGO, sggA, sggG, sggO, Control KD2, and sggKD2 denote hybrid genotypes obtained as a result of crossing the corresponding lines with the driver line D1 inducing the expression of transgenic constructs in embryos. A full description of the genotypes is given in the Materials and Methods section.

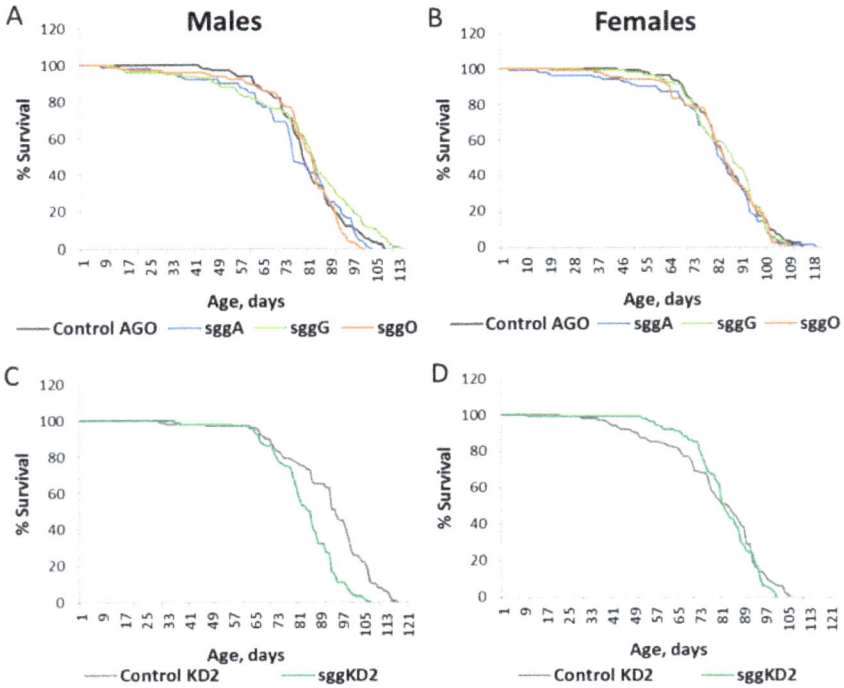

Figure 2. Effects of disordered *sgg* expression in the fat body on male (**A**,**C**) and female (**B**,**D**) lifespans. Control AGO, sggA, sggG, sggO, Control KD2, sggKD2, Control B, and sggA81T denote hybrid genotypes obtained as a result of crossing the corresponding lines with the driver line D2 inducing the expression of transgenic constructs in the fat body. A full description of genotypes is given in the Materials and Methods section. In panels **E** and **F**, the results of the two independent experiments are represented by solid and dotted lines.

Second, different transcripts had their own patterns of effects on lifespan. This fact confirmed that *sgg* functions in a cell-autonomous mode [19]. As expected, the previously well described [7,13,18,19] *sgg-RB* transcript affected lifespan when overexpressed in all tissues and at all stages tested, although the strength of the effect varied depending on the stage, age, and tissue. In the nervous system, *sgg-RB* appeared to be the only transcript with effects on lifespan. In embryos, of all minor transcripts tested, only the expression of *sgg-RO* affected the lifespan. As its effects were not detected in other tissues, we assume that this transcript may be specific for embryos and have a function that complements *sgg-RB* functions during development. Similarly, the *sgg-RA* transcript appeared to be functionally specific for muscles, where it also may complement *sgg-RB* functions. Transcripts that are specific for or preferred at certain developmental stages and particular tissues and organs have long been known (for example, [30–32]). The peculiarity of our data is that they allowed us to show in which tissues these transcripts are functional, judging by the integral phenotype used in our tests. Only one transcript, *sgg-RG*, did not show functionality in our tests, which, however, does not mean that it is generally deprived of any function, given the limited scope of our tests. Of note, we were able to overexpress individual transcripts, but knockdown reduced the expression of *sgg* transcripts all together; it would be interesting to evaluate the effects of decreased expression of individual transcripts on lifespan and other traits. We have already demonstrated that reduced transcription of a gene causes lifespan extension, whereas overexpression has detrimental effects [27].

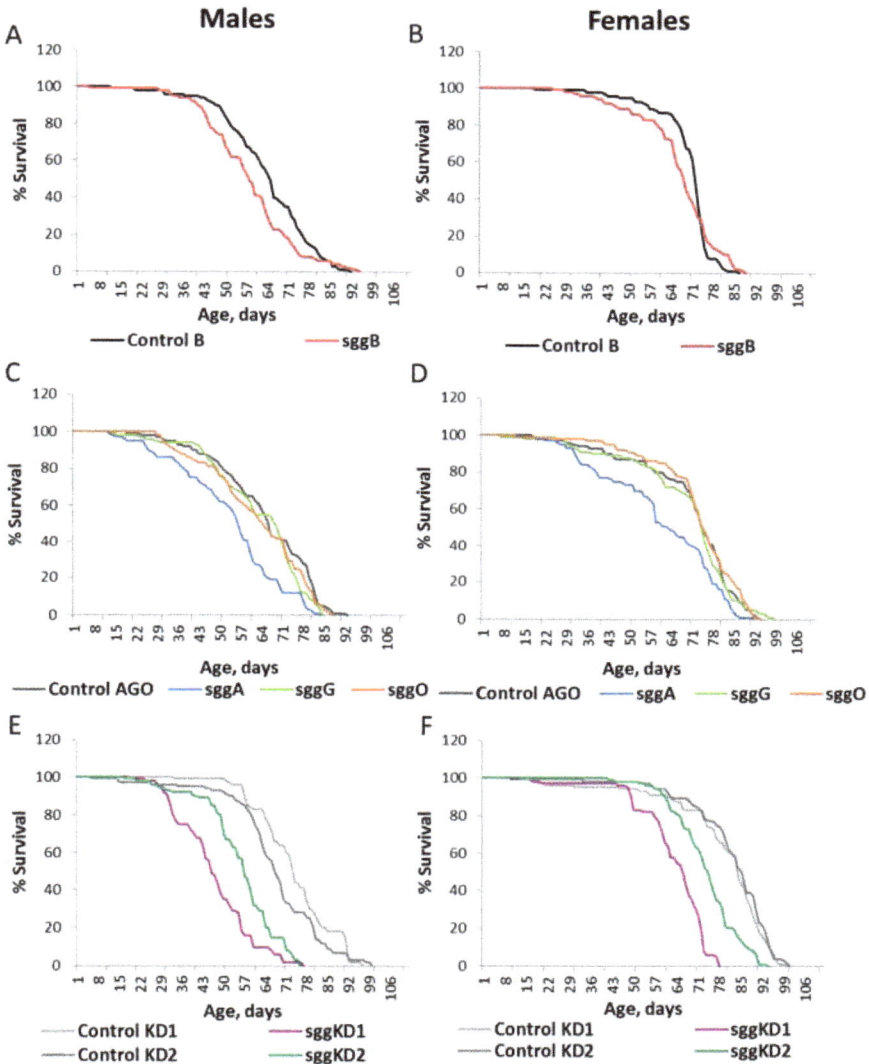

Figure 3. Effects of disordered *sgg* expression in muscles on male (**A**,**C**,**E**) and female (**B**,**D**,**F**) lifespans. Control B, sggB, Control AGO, sggA, sggG, sggO, Control KD1, sggKD1, Control KD2, and sggKD2 denote hybrid genotypes obtained as a result of crossing the corresponding lines with the driver line D3 inducing the expression of transgenic constructs in muscles. A full description of the genotypes is given in the Materials and Methods section.

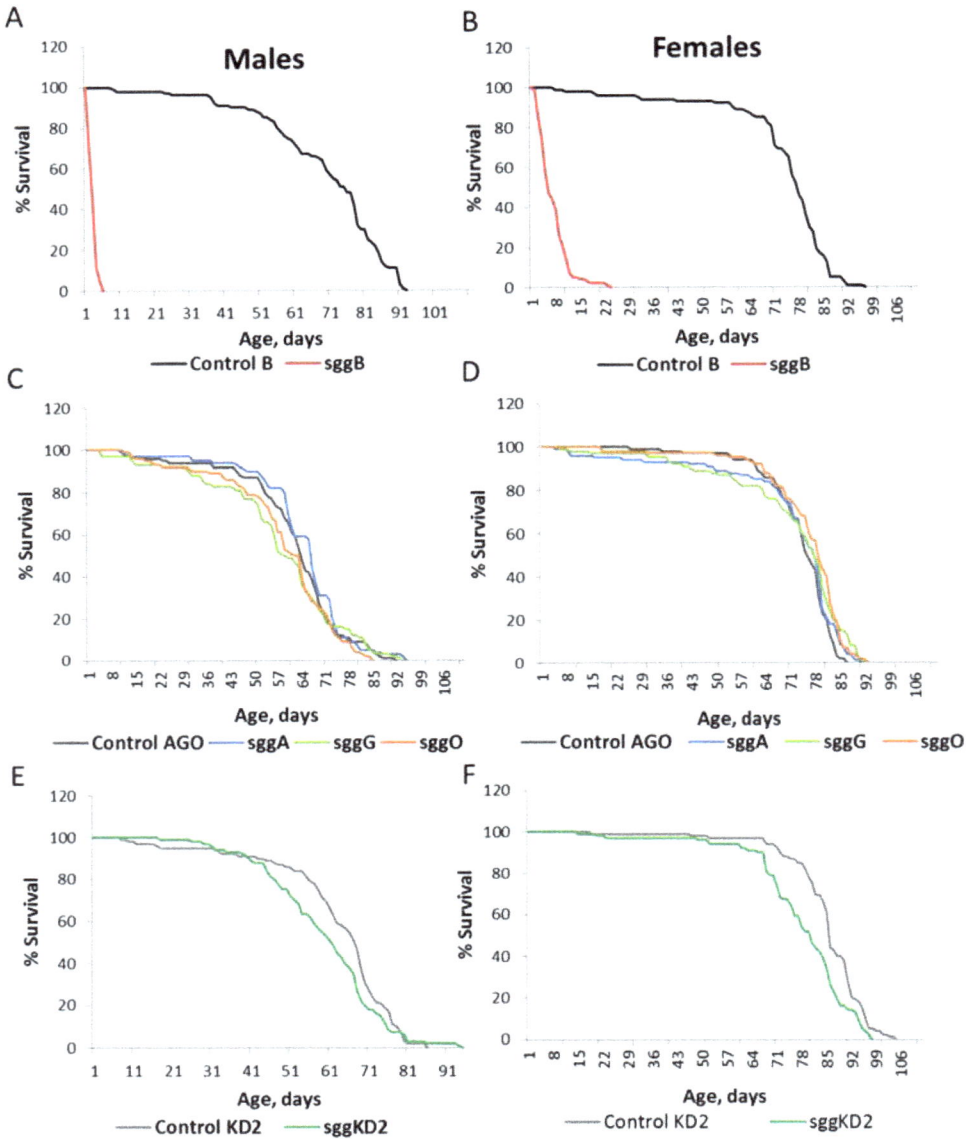

Figure 4. Effects of disordered *sgg* expression in the nervous system on male (**A,C,E**) and female (**B,D,F**) lifespans. Control B, sggB, Control AGO, sggA, sggG, sggO, Control KD2, and sggKD2 denote hybrid genotypes obtained as a result of crossing the corresponding lines with the driver line D4 inducing the lifelong expression of transgenic constructs in the nervous system. A full description of genotypes is given in the Materials and Methods section.

Tissues	Embryo (D1 driver)		Fat body (D2 driver)		Muscles (D3 driver)		Nervous system (D4 driver)	
Sex	♂	♀	♂	♀	♂	♀	♂	♀
sgg RA overexpression	-	-	-	-	V-	VV	-	-
sgg RB overexpression	L	L	L	L	V	V	VV	VV
sgg RG overexpression	-	-	-	-	-	-	-	-
sgg RO overexpression	ΛΛ	VV	-	-	-	-	-	-
Strong *sgg* knockdown	L	L	L	L	V	V	L	L
Weak *sgg* knockdown	V	V	V	-	V	V	V	V

Figure 5. Effects of disordered *sgg* expression in different tissues on male and female lifespans. Dark-red cells, L: lethal effects; red cells, V: lifespan significantly decreased; gray cells, -: no effect; green cells, Λ: lifespan significantly increased. The number of V or Λ characters in a cell denotes the number of independent experiments.

The fact that in the nervous system, only *sgg-RB* appeared to be functional gave us the necessary grounds for further analysis of the effects of *sgg* misexpression in the nervous system on the lifespan and structural and functional properties of neurons.

2.2. Overexpression of sgg-RB in the Nervous System Affects Lifespan in a Stage/Age- and Neuron-Specific Manner

We focused our efforts on the study of *sgg* overexpression in the nervous system because increased GSK3 function is thought to be associated with the development of several age-dependent pathological conditions, such as Alzheimer's disease and Parkinson's disease [12,33,34]. The increase in the number of *sgg-RB* transcripts and amount of GSK3 protein induced by the D4 panneuronal driver line was confirmed using real-time RT-qPCR and Western blotting, respectively ($p < 0.0001$ for comparisons between control and overexpressing individuals in both cases). Typical results of real-time RT-qPCR and Western blotting are illustrated in Figure S1.

Severe effects of *sgg-RB* overexpression in the nervous system on male and female lifespans were confirmed in the independent experiment, which was performed approximately a year after the first experiment (Table S4, Figure 6A,B). In both replicate experiments, the effects in males were more pronounced (Figure 4A,B and Figure 6A,B); however, flies of both sexes lived only several days.

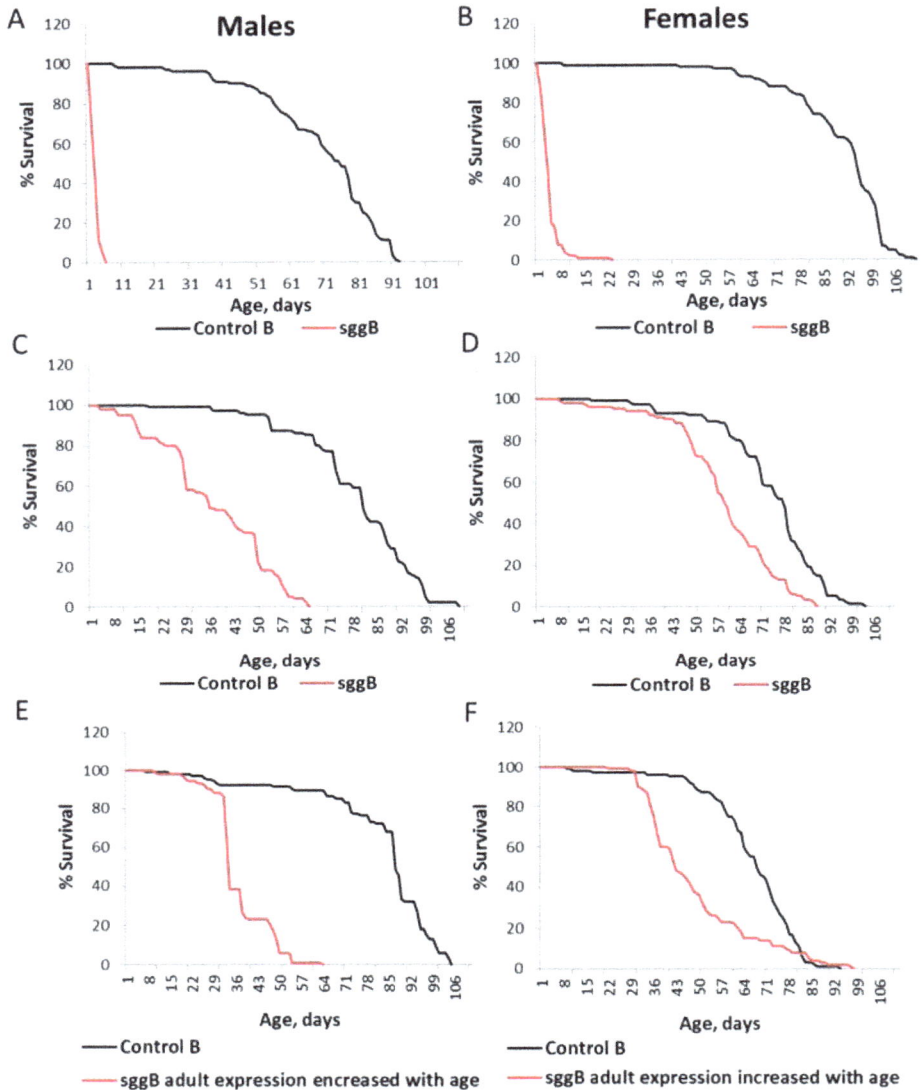

Figure 6. Effects of *sgg-RB* overexpression in the nervous system on male (**A,C,E**) and female (**B,D,F**) lifespans. Control B, sggB denote hybrid genotypes obtained as a result of crossing the corresponding lines with the driver line D4 inducing the lifelong expression of the transgenic construct in the nervous system (**A,B**), with the driver line D5 inducing the expression of the transgenic construct in the embryonic CNS (central nervous system) (**C,D**), and with the driver line D6 inducing the increasing expression of the transgenic construct in the adult nervous system with age (**E,F**). A full description of genotypes is given in the Materials and Methods section.

The *elav* driver (D4) used in the experiments described above provides lifelong expression in the nervous system, starting from the embryonic stage (see references at http://flybase.org/reports/FBti0002575.html).; therefore, we were not able to discriminate at which stage/age *sgg-RB* overexpression was significant for lifespan effects. To address this question, we used two additional drivers that induce the expression of transgenic constructs in the embryonic CNS (central nervous system) (D5; see references at http://flybase.org/reports/FBti0001255.html). and the increased with age expression of transgenic constructs in the adult nervous system (D6; see references at http://flybase.org/reports/FBti0040381.html). *sgg-RB* overexpression in the embryonic CNS decreased male and female lifespans, though to a lesser degree than lifelong panneuronal overexpression (Table S4, Figure 6C,D). This result indicated that normal *sgg-RB* expression in the CNS at the embryonic stage is important for lifespan determination; however, it does not fully account for the effects of lifelong panneuronal *sgg-RB* overexpression. According to the description given in FlyBase, the D6 driver should be switched on starting from the 50th day of life. However, survival curves obtained in our experiments (Table S4, Figure 6E,F) indicated that both control flies and flies with the *sgg-RB* transgene lived the same way until approximately the 30th day of life, but after that, the lifespans of both males and females with the *sgg-RB* transgene declined significantly faster compared to the control. These data clearly indicate that the D6 driver was switched on starting from the 30th day of life and that *sgg-RB* overexpression starting in the nervous system of middle-aged flies was sufficient to decrease their lifespans, though to a lesser degree than lifelong overexpression: The slope of survival curves in the latter case was much steeper both in males and females.

Overall, *sgg-RB* overexpression in the nervous system at different stages of development and aging negatively affected lifespan. Altering *sgg-RB* expression in the embryonic CNS was enough to affect the adult lifespan, which could be explained by epigenetic inheritance in cell lines or other yet unknown mechanisms. Based on our results, it is tempting to conclude that in the nervous system, embryonic *sgg-RB* overexpression is less detrimental than lifelong *sgg-RB* overexpression and *sgg-RB* overexpression in adults and that if overexpression took place in older adults, detrimental effects were alleviated. One of the theories of aging suggests that aging is caused by hyperfunction, that is, overactivity during adulthood of processes that are primarily important during development. Such hyperfunction can lead to hypertrophy-associated pathologies, which cause accelerated aging [35]. In the most general terms, smaller effects of overexpression in embryos would be in accordance with this theory. However, the strength of overexpression induced by different drivers in our experiments may be different and bias the comparisons. Experiments with a conditional tissue-specific transgene expression system using the elav-geneswitch driver could allow overexpression in larvae and adults at different ages and provide adequate comparison of the stage- and age-specific effects of altered *sgg-RB* expression. To compare the effects of overexpression in embryos and in adult flies, more sophisticated methods are needed, for example, combinations of GAL4 and GAL80 or split GAL4 drivers with different stage-specific patterns of expression, in accordance with the logic proposed in [36]. One cannot exclude the possibility that *sgg-RB* overexpression in old flies could become beneficial. Future experiments can shed light on these issues.

The *elav* driver (D4) provides panneuronal expression (see references at http://flybase.org/reports/FBti0002575.html); therefore, we were not able to discriminate which neurons are most sensitive to *sgg-RB* overexpression and are predominantly responsible for effects on lifespan. To address this question, we used several drivers (D7–D13) that induce the expression of transgenic constructs in neurons secreting different transmitters and in motor neurons. Two different drivers were used to induce *sgg-RB* overexpression in dopaminergic neurons, and in both cases, the effect was lethal. These neurons demonstrated maximum sensitivity to the level of *sgg-RB* expression. It remained unclear why in this case the panneuronal *sgg-RB* overexpression was not lethal; one mechanistic suggestion would be that in some neurons, *sgg-RB* overexpression had a positive effect on lifespan and that opposite effects counterbalanced each other. However, *sgg-RB* overexpression in all tested types of neurons decreased both male and female lifespans to varying degrees (Table S5, Figure S2). Overall, peptidergic neurons appeared to be highly sensitive; glutamate and cholinergic neurons, less sensitive; and motor and GABAergic neurons, the least sensitive to *sgg-RB* overexpression among the types of neurons tested, judging by the effect on lifespan. However, the level of the effect also depended on sex

(Figure S2). Other types of differently classified neurons might be worth testing in the future to better understand the specific impact of neurons on lifespan control provided by Sgg-RB.

Our results demonstrated that, despite the individual nuances, lifelong panneuronal *sgg-RB* overexpression would be an appropriate model system for further analyses of GSK impact on structural and functional properties of neurons. To better understand the association between the extent of lifespan reduction and alterations in neuronal characteristics, for further analyses, we also selected overexpression of *sgg-RB* in motor neurons as a second model system demonstrating moderate effects on lifespan. The choice of this additional model was also justified by the fact that we used larval NMJs (neuromuscular junctions) as a model synapse.

2.3. Overexpression of Sgg-RB in the Nervous System Affects Neuronal Structure and Function

We suggest that the lifespan reduction caused by aberrations in GSK3 expression in the nervous system is based on pathological changes in neurons. It was previously shown that *sgg* overexpression results in decreased numbers of synaptic boutons [7] and synapses [9], small synapses [8], and degeneration of presynaptic terminals and axons [10]. We decided to complement these results by some other structural and functional characteristics of the nervous system in flies with a dramatically decreased lifespan due to *sgg-RB* panneuronal overexpression and in flies with a moderately decreased lifespan due to *sgg-RB* overexpression in motor neurons.

2.3.1. Locomotion

First, we demonstrated that locomotion was highly impaired in 3- to 5-day-old males and females with panneuronal overexpression of *sgg-RB* (Figure 7A), indicating severe lesions in the functional status of the nervous system. These flies did not live longer than 10 days (males) or 15 days (females), and most of them hardly moved at all in the second half of life, so we were not able to characterize the age-dependent dynamics of their locomotion. In 3- to 5-day-old males and females with *sgg-RB* overexpression in motor neurons, locomotion was not changed compared to that of control flies, and in 20-day-old flies, locomotion was significantly increased compared to that of control flies (Figure 7B,C). In 40-day-old control flies, locomotion was still high, while in flies with *sgg-RB* overexpression in motor neurons, a substantial decline in locomotion was observed (Figure 7B,C). Of note, the mean lifespan of control males and females was 64 ± 3 and 68 ± 2 days, respectively (Table S4), and at the age of 40 days, the flies were not truly aging. The mean lifespan of males with *sgg-RB* overexpression in motor neurons was 35 ± 2 days (Table S4), and 40-day-old males were intensively aging, which was reflected in the rapid decline in locomotion that decreased below the control level. The mean lifespan of females with *sgg-RB* overexpression in motor neurons was 58 ± 2 days (Table S4), and the biological age of 40-day-old females was younger than in males, which was reflected in a smaller decline in locomotion that stayed at the control level. Overall, in our experiments, locomotion, as reported by [37], appeared to be a good marker of aging.

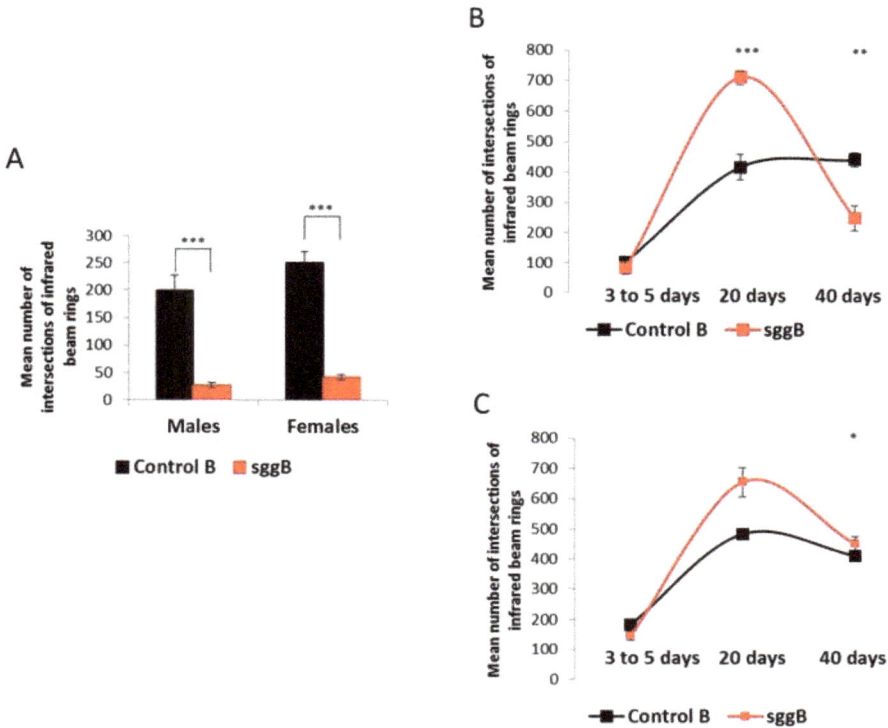

Figure 7. Effects of *sgg-RB* overexpression in the nervous system on locomotion of flies. Panneuronal *sgg-RB* overexpression in 1- to 3-day-old males and females (**A**). *sgg-RB* overexpression in motor neurons in males (**B**) and females (**C**). Control B, sggB denote hybrid genotypes obtained as a result of crossing the corresponding lines with the driver line D4 inducing the expression of transgenic constructs in the nervous system or the driver line D12 inducing the expression of transgenic constructs in motor neurons. A full description of genotypes is given in the Materials and Methods section. * denotes $p < 0.05$, ** denotes $p < 0.01$, *** denotes $p < 0.001$ determined by the Kruskal-Wallis test.

At the same time, an intriguing result of these experiments was that in 20-day-old flies with *sgg-RB* overexpression in motor neurons, the presumed decline in neuronal functions [7–10] and decreased lifespan were associated with elevated locomotion, indicating that the functional status of the nervous system necessary for ensuring motor functions was not only disturbed, but even improved. Simultaneous decreases in lifespan and increases in locomotion, both resulting from changes in the expression of a gene, have been reported previously [37]. It was suggested that mitochondrial function in neurons may change in such a way that the lifespan decreases due to the increased ROS (reactive oxygen species) production known to promote aging and shorten lifespan [38,39], whereas locomotion increases due to elevated mitochondrial activity and ATP levels. Indeed, the increase in mitochondrial function to a certain level should result in the increase of energy production accompanied by elevated levels of ROS. However, elevated ROS and ATP levels did not accompany a decrease in lifespan and an increase in locomotion described in [37], and the cellular mechanisms underlying an increase in locomotion remained obscure.

2.3.2. Mitochondria

The functionality of neurons critically depends on the energy supply, which is ensured by the work of mitochondria. Mitochondrial homeostasis, that is, precise control of mitochondrial number, integrity, and distribution, is especially critical to postmitotic cells, such as neurons. Violation of the fine-tuned interplay between mitochondrial function, energy metabolism, and neuronal activity is critical in the occurrence of various neurological pathologies [40]. We examined whether the number of mitochondria was affected by *sgg-RB* overexpression. For this purpose, we used larval NMJs, which are often used as a model system to study synapse development and function in *Drosophila* [41]. GSK3 is enriched in the presynaptic side of NMJs [7], so we focused on studying the presynaptic zone. Mitochondria are especially abundant at the presynaptic parts of axons. The number of GFP-labeled mitochondrial clusters was decreased in NMJs of individuals with panneuronal *sgg-RB* overexpression (Figure 8A,B) and *sgg-RB* overexpression in motor neurons (Figure 8C,D) to a similar extent. Thus, the decreased number of mitochondrial clusters was associated with both severe and moderate decreases in lifespan, but appeared to be not relevant to locomotion. Analysis of the brains of 3-day-old flies revealed swollen, dense mitochondria in all three individuals with pan-neuronal *sgg-RB* overexpression (Figure 9A–D). We failed to observe such mitochondria in control individuals (Figure 9E).

In individuals with *sgg-RB* overexpression, mitochondria were clustered predominantly in synaptic boutons, possibly to compensate for the overall decrease in number, whereas in control individuals, they were distributed more evenly along terminal ends of axons (Figure 8A,C). Mitochondria are produced in the neuronal soma and move to synapses, where they are needed for neuronal firing, along microtubules using kinesin motors; they can also move in the retrograde direction for reparation using dynein motors [42]. GSK3β is involved in the control of mitochondrial movement, both in increasing upon activation of the serotonin receptor [43] and in decreasing upon activation of the dopamine receptor [44,45]. GSK3β directly regulates dynein [46] and promotes anterograde movement [47]. We hypothesize that *sgg-RB* overexpression might affect mitochondrial movement and thus be responsible for disturbances in the distribution of mitochondria along axons.

2.3.3. Cytoskeleton

Not only mitochondrial movement, but also the overall functionality and architecture of neurons depend on the integrity of their cytoskeletons. We assessed the effects of *sgg-RB* overexpression in individuals with pan-neuronal *sgg-RB* overexpression and *sgg-RB* overexpression in motor neurons on the morphology of the microtubule network in NMJs, as visualized by antibodies to α-acetylated tubulin and to Futsch, a neuronal microtubule-associated protein [48]. Staining with antibodies to α-acetylated tubulin failed to reveal visual differences between the individuals with and without *sgg-RB* overexpression (Figure S3). Differences in the distribution of Futsch within the NMJs were found. In particular, in individuals with *sgg-RB* overexpression, only weak, diffuse Futsch staining in synaptic boutons of NMJs was observed compared to that in controls (Figure 10), indicating cytoskeletal abnormalities. In this case, the effect was more pronounced in individuals with pan-neuronal *sgg-RB* overexpression (Figure 10A) than in individuals with *sgg-RB* overexpression in motor neurons (Figure 10B), which correlates with the comparative effect on lifespan. It was shown that *sgg* interacts with Futsch, which functions downstream of *sgg* [7]; most likely, *sgg* controls synaptic growth through the direct phosphorylation of Futsch. Sgg is able to inhibit transcription factor AP-1 and the JNK (Jun-N-terminal kinase) cascade and, in this way, also affects synaptic growth [8]. Our data shed light on the cellular consequences of presumed increased phosphorylation of Futsch, which disappeared from synaptic boutons of NMJs, and demonstrated that Sgg-RB, when overexpressed, controlled the dynamics of the microtubule cytoskeleton.

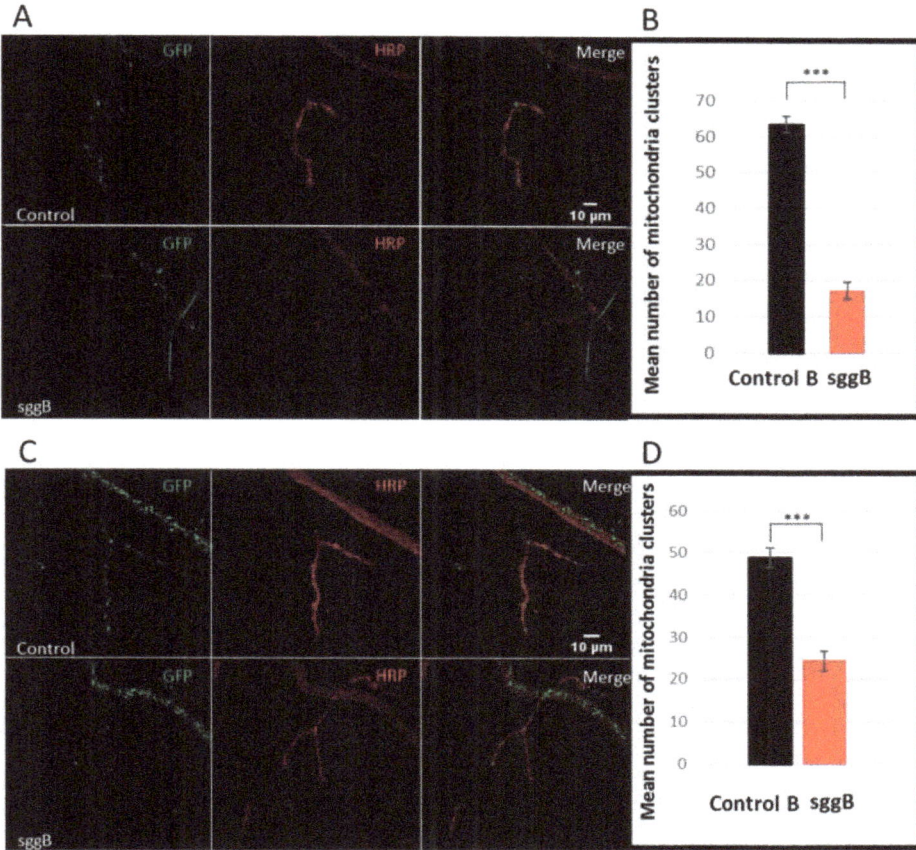

Figure 8. Mitochondrial clusters in NMJs of third-instar female larvae with panneuronal *sgg-RB* overexpression (**A**,**B**) and *sgg-RB* overexpression in motor neurons (**C**,**D**). Control B, sggB denote hybrid genotypes obtained as a result of crossing the corresponding lines with the driver line D4 inducing the expression of transgenic constructs in the nervous system or the driver line D12 inducing the expression of transgenic constructs in motor neurons. A full description of genotypes is given in the Materials and Methods section. Representative confocal images of NMJs (muscle 4, hemi-segment 34–) stained for mitochondria clusters (GFP, green) and neural membranes (HRP, horseradish peroxidase, red) (**A**,**C**). Bar = 10 μm. Quantification of the number of mitochondrial clusters (**B**,**D**). *** denotes $p < 0.001$ determined by the Kruskal-Wallis test.

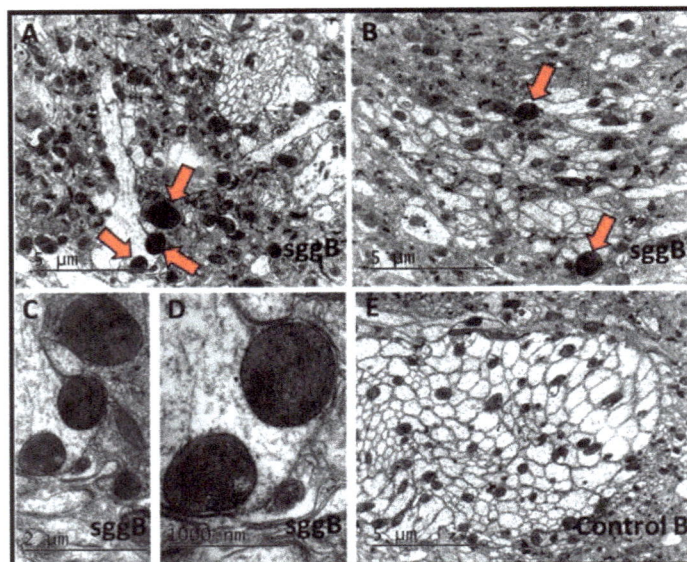

Figure 9. Fine structure of the adult brain in females with pan-neuronal *sgg-RB* overexpression. Control B, sggB denote hybrid genotypes obtained as a result of crossing the corresponding lines with the driver line D4 inducing the expression of transgenic constructs in the nervous system. A full description of genotypes is given in the Materials and Methods section. Three representative transmission electron microscopy images of the *Drosophila* inferior dorsofrontal protocerebrum at the age of 3 days (**A,B,E**); enlarged fragments of panel B (**C,D**). Mitochondria with violated structures are indicated by red arrows. Bar = 5 µm or 2 µm.

2.3.4. Active Synaptic Zones

Given that the number and size of synapses were shown to be affected by *sgg* function [7,8], we decided to evaluate the number of active synaptic zones in individuals with *sgg-RB* overexpression. The number of active zones in larval NMJs visualized by antibodies to BRP (Bruchpilot), a Drosophila homolog of vertebrate active zone protein ELKS [29], was significantly (approximately two-fold) decreased in individuals with pan-neuronal *sgg-RB* overexpression (Figure 11A,B) and *sgg-RB* overexpression in motor neurons (Figure 11C,D), indicating reduced synaptic activity. *sgg-RB* overexpression driven by both drivers had comparable effects on NMJs. These results supplemented the data on the role of Sgg in controlling the functionality of synapses, which demonstrated that it was involved in the regulation of neurotransmitter release [8]. Earlier, it was shown that the number of boutons was significantly decreased as a result of pan-neuronal GSK3 overexpression [7]. Importantly, although *sgg* overexpression has an inhibitory role on the quantity of boutons, it may not play an important role in bouton differentiation [7]. Overall, our data demonstrated that negative changes in lifespan caused by Sgg-RB overexpression in the nervous system were accompanied by negative changes in synaptic properties.

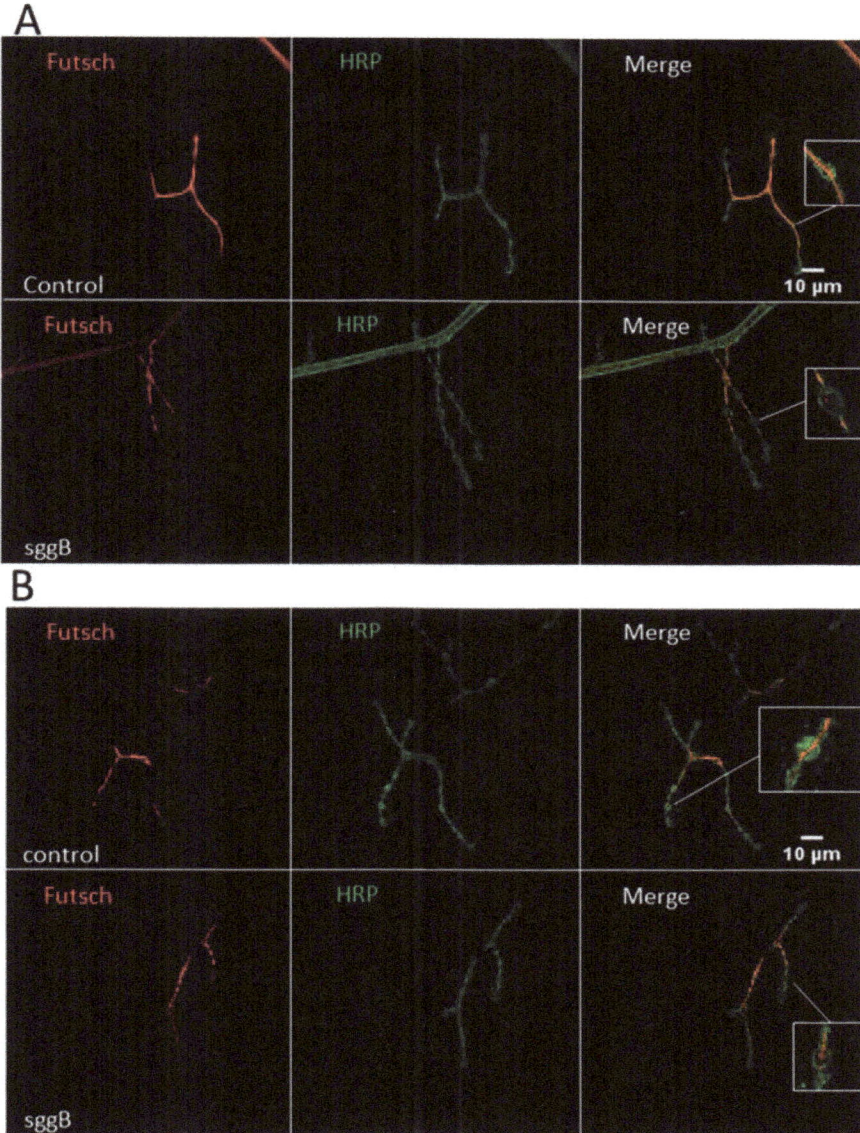

Figure 10. Distribution of Futsch in NMJs of third-instar female larvae with panneuronal *sgg-RB* overexpression (**A**) and *sgg-RB* overexpression in motor neurons (**B**). Control B, sggB denote hybrid genotypes obtained as a result of crossing the corresponding lines with the driver line D4 inducing the expression of transgenic constructs in the nervous system or the driver line D12 inducing the expression of transgenic constructs in motor neurons. A full description of genotypes is given in the Materials and Methods section. Representative confocal images of NMJs (muscle 4, hemi-segment 34–) stained for Futsch (red) and neural membranes (HRP, green). Bar = 10 μm.

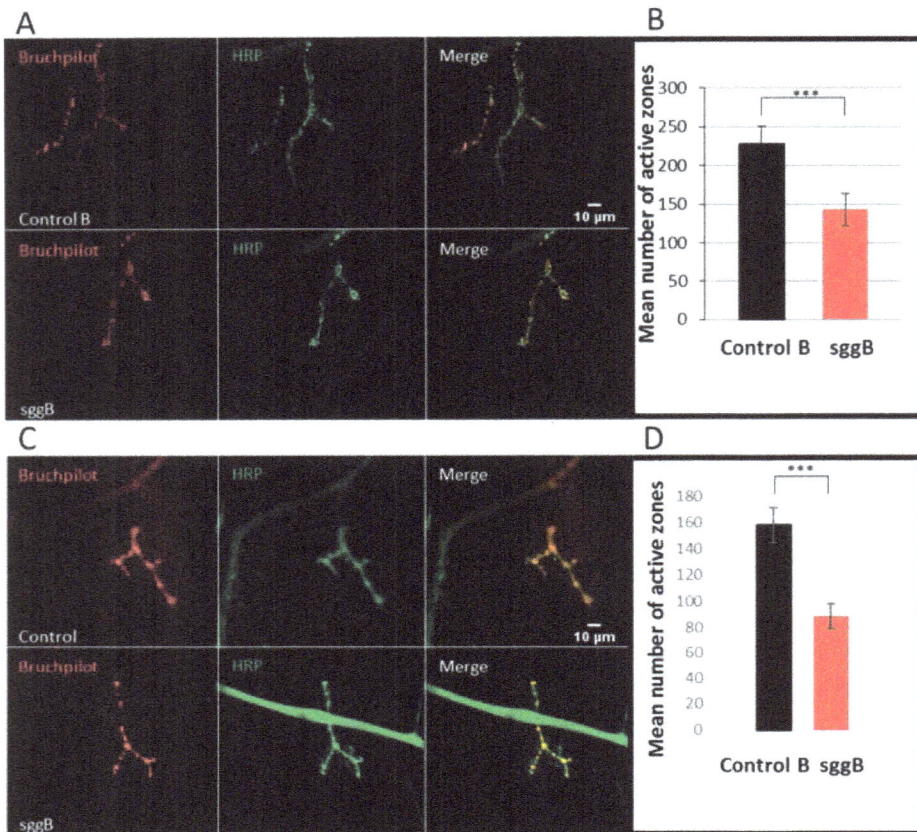

Figure 11. Active zones in NMJs of third-instar female larvae with panneuronal *sgg-RB* overexpression (**A,B**) and *sgg-RB* overexpression in motor neurons (**C,D**). Control B, sggB denote hybrid genotypes obtained as a result of crossing the corresponding lines with the driver line D4 inducing the expression of transgenic constructs in the nervous system or the driver line D12 inducing the expression of transgenic constructs in motor neurons. A full description of genotypes is given in the Materials and Methods section. Representative confocal images of NMJs (muscle 4, hemi-segment 34–) stained for active zones (BRP, red) and neural membranes (HRP, green) (**A,C**). Bar = 10 µm. Quantification of the number of active zones (**B,D**). *** denotes $p < 0.001$ determined by the Kruskal-Wallis test.

2.3.5. Neuronal Structure

We suggest that a significant decrease in the lifespan of individuals with pan-neuronal *sgg-RB* overexpression might be accompanied by pathological processes in the adult nervous system. Analysis of the brains of 3-day-old flies using transmission electron microscopy revealed a number of degenerative marks both in the bodies of the nerve cells and in the axonal structures in the neuropil region. The presence of "holes" surrounded by membranes and, in some cases, containing fragments of organelles apparently subjected to autophagy was demonstrated in the brains of all three individuals with pan-neuronal *sgg-RB* overexpression, whereas these structures were not found in the control brains (Figure 12A,B). These "holes" apparently represent destroyed areas of the neuron bodies and may indicate accelerated aging processes in the nervous system of individuals with pan-neuronal *sgg-RB* overexpression. In *Drosophila*, photosensitive structures, namely, rhabdomeres, are often used along with the brain to study neurodegenerative processes. We examined whether pathological changes occur in the rhabdomeres of individuals with pan-neuronal *sgg-RB* overexpression compared to those of controls. No changes were revealed in three control individuals, whereas in two out of three individuals with pan-neuronal *sgg-RB* overexpression, lacunae and irregular rhabdomere structures were found (Figure 12C,D), also indicating pathological processes in flies with pan-neuronal *sgg-RB* overexpression. Of note, in individuals with pan-neuronal *sgg-RB* overexpression, a high heterogeneity of external phenotypic manifestations was observed: For example, the wing shape varied fairly widely. In view of this fact, some difference in the level of degeneration of rhabdomeres was not surprising.

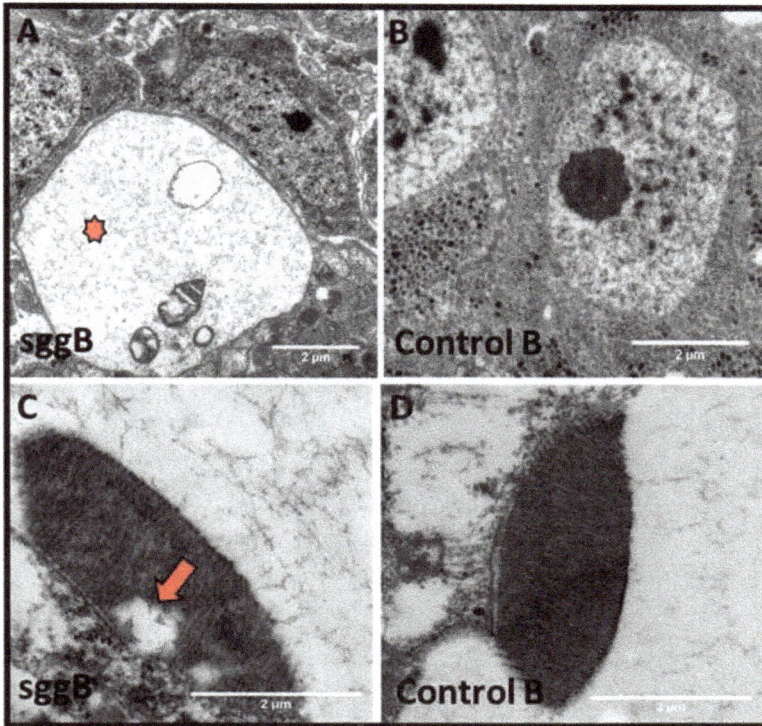

Figure 12. Signs of degeneration in the nervous system of females with pan-neuronal *sgg-RB* overexpression. Control B, sggB denote hybrid genotypes obtained as a result of crossing the corresponding lines with the driver line D4 inducing the expression of transgenic constructs in the nervous system. A full description of genotypes is given in the Materials and Methods section. Representative transmission electron microscopy images of the *Drosophila* brain (**A,B**) and rhabdomeres (**C,D**) at the age of 3 days. An example of a degeneration sign in the neuron body (**A**, indicated by the red star); rhabdomeres with violated structure (**C**, lacuna is indicated by the red arrow) compared to the respective controls (**B,D**). Bar = 1 μm or 2 μm.

Summing up the results presented in this chapter, we conclude that overexpression of the main Sgg-RB isoform in the nervous system had severe effects on its structural and functional properties. Overall, Sgg-RB overexpression in the nervous system affected the integrity of neuron bodies, the number, distribution, and structure of mitochondrial and cytoskeletal properties in the presynaptic zones, and synaptic activity. These changes indicated that Sgg-RB overexpression caused accelerated aging of the nervous system in young individuals. The accelerated aging led to a more or less sharp reduction in lifespan, depending on the cell types where Sgg-RB overexpression was induced. While pan-neuronal overexpression caused a severe lifespan decline and overexpression in motor neurons reduced the lifespan to a much lesser extent, and the effects of pan-neuronal overexpression and overexpression in motor neurons on the nervous system were mostly of similar scales. Thus, the rate of lifespan decline was not correlated with the rate of structural and functional changes in the nervous system. The only exception identified was a change in the structure of the cytoskeleton, namely, the Futsch protein distribution along the terminal parts of axons.

3. Materials and Methods

3.1. Fly Strains and Crosses

To provide *sgg* overexpression, several lines were used. The line providing *sgg-RB* overexpression w^{1118}; *P{w^{+mC} = UAS-sgg. B}MB5* (in short, sggB) with the transgenic construct encoding the normal PB form of GSK3 was obtained from the Bloomington *Drosophila* Stock Center (USA) (Available online: http://flystocks.bio.indiana.edu/); the w^{1118} (Control B) line without transgenic insertions was used as a control line [19].

To obtain lines providing *sgg-RA*, *sgg-RG*, and *sgg-RO* overexpression, cDNAs corresponding to these transcripts were cloned into the *pBID-UASC* vector (Addgene plasmid # 35200, Available online: https://www.addgene.org/35200/), which contains an attB site for *phi31* site-specific transformation of *Drosophila* embryos and UAS enhancer sequences [49]. Single clones with the unimpaired *sgg-RA*, *sgg-RG* and *sgg-RO* sequences and the line *y^1 M{vas-int.Dm}ZH-2A w*; M{3xP3-RFP.attP'}ZH-51C* with the second chromosome attP *phi31* integration site [50] were used for the transformation performed by BestGene, Inc. (Available online: https://www.thebestgene.com/HomePage.do). In the transgenic lines obtained, most of the auxiliary sequences were excised from the inserted transgenes due to induction of recombination between the *loxP* sites present in the *pBID-UASC* vector. One homozygous transgenic line of each genotype, namely, *y^1 M{vas-int.Dm}ZH-2A w*; M{3xP3-RFP.attP'}ZH-51C P{w^{+mC} = UAS-sgg.A}2M* (in short, sggA), *y^1 M{vas-int.Dm}ZH-2A w*; M{3xP3-RFP.attP'}ZH-51C P{w^{+mC} = UAS-sgg.G}1M* (in short, sggG), *y^1 M{vas-int.Dm}ZH-2A w*; and M{3xP3-RFP.attP'}ZH-51C P{w^{+mC} = UAS-sgg.O}1M* (in short, sggO), were used in further experiments. The line *y^1 M{vas-int.Dm}ZH-2A w*; M{3xP3-RFP.attP'}ZH-51C* (Control AGO) initially used for transformations was used as a control line with the same genetic background.

To provide *sgg* RNAi knockdown, two lines were obtained from the Bloomington *Drosophila* Stock Center: *y^1 sc* v^1; P{y$^{+t7.7}$ v$^{+t1.8}$ = TriP. HMS01751}attP40* (in short, sggKD1, with VALIUM20, hairpin size 21 bp, affects all *sgg* transcripts) and *y^1 v^1; P{y$^{+t7.7}$ v$^{+t1.8}$ = TriP. JF01255}attP2* (in short, sggKD2, with VALIUM1, hairpin size 400 bp, affects all *sgg* transcripts). *Y^1 v^1; P{y$^{+t7.7}$ = CaryP}attP40* (Control KD1) and *y^1 v^1; P{y$^{+t7.7}$=CaryP}attP2* (Control KD2) lines without transgenes providing RNAi were used as control lines for *sgg* RNAi knockdown, as suggested by the manufacturer (Available online: http://flystocks.bio.indiana.edu/Browse/TRiPtb.htm).

To induce *sgg* overexpression or knockdown, several driver lines were obtained from the Bloomington *Drosophila* Stock Center.

Y^1 w*; P{w$^{+mW.hs}$ = en2.4-GAL4}e22c; P{w^{+mC} = tGPH}4/TM3, Ser^1 (D1) was used to induce the expression of transgenic constructs in embryos.

W*; P{w^{+mC} = ppl-GAL4. P}2 (D2) was used to induce the expression of transgenic constructs in the fat body.

P{w^{+mC} = UAS-Dcr-2. D}1, w^{1118}; P{w^{+mC} = GAL4-Mef2. R}R1 (D3) was used to induce the expression of transgenic constructs in somatic muscle cells.

P{w$^{+mW.hs}$ = GawB}elavC155 w^{1118}; P{w^{+mC} = UAS-Dcr-2. D}2 (D4) was used to induce the lifelong expression of transgenic constructs in the nervous system.

W*; P{w$^{+mW.hs}$ = GawB}389 (D5) was used to induce the expression of transgenic constructs in the embryonic nervous system.

w^{1118}; *P{GawB}DJ695* (D6) was used to induce the increased expression of transgenic constructs in the adult nervous system with age.

w^{1118}; *P{w^{+mC} = Ddc-GAL4.L}Lmpt$^{4.36}$* (D7) and *w*; P{w[+mC] = UAS-mCD8::GFP. L}LL5/Cy;* and *P{w^{+mC} = ple-GAL4. F}3* (D8) were used to induce the expression of transgenic constructs in dopaminergic neurons.

$W*$; *P{w$^{+mW.hs}$ = GawB}386Y* (D9) was used to induce the expression of transgenic constructs in peptidergic neurons.

w^{1118}; *P{w$^{+mW.hs}$ = GawB}Vglut^{OK371}* (D10) was used to induce the expression of transgenic constructs in glutamatergic neurons.

$W*$; *P{w^{+mC} = ChAT-GAL4.7.4}19B* (D11) was used to induce the expression of transgenic constructs in cholinergic neurons.

$W*$; *P{GawB}D42* (D12) was used to induce the expression of transgenic constructs in motoneurons.

P{w^{+mC} = Gad1-GAL4.3.098}2/CyO (D13) was used to induce the expression of transgenic constructs in GABAergic neurons.

Additionally, the driver line w^{1118};*P{w^{+mC} = UAS-mito- HA-GFP.AP}2/CyO* (D14) was used to induce GFP expression in mitochondria.

D1, D3, and D4 driver lines proved to be effective in our previous work, according to the real-time RT-qPCR data [27].

To induce the expression of transgenic constructs, females of each of the driver lines were crossed to males of sggB, sggY214F, sggA81T, sggA, sggG, sggO, KD1, KD2, Control B, Control AGO, Control KD1, and Control KD2 lines. In all experiments, flies were kept at 25 °C on a standard medium of semolina, sugar, raisins, yeast, and agar with nipagin, propionic acid, and streptomycin.

3.2. Tests for Wolbachia

Prior to the experiments, all the lines were checked for the presence of *Wolbachia*, a *Drosophila* symbiont known to affect life history traits [51], via quantitative PCR (MiniOpticon real-time PCR detection system, Bio-Rad, Hercules, CA, USA) with primers for the 16S rRNA gene, 5′-CATACCTATTCGAAGGGATAG-3′ and 5′-AGCTTCGAGTGAAACCAATTC-3′ [52]. Negative results were obtained for all lines except sggA, D3, and D12. These lines were treated with tetracycline (0.25 mg/mL, [53]) for three generations followed by at least three generations of recovery before they were used in experiments.

3.3. Lifespan Assay

Lifespan was measured as described in [16]. Five virgin flies of the same genotype and sex, all collected on the same day from cultures with moderate density, were placed in replicate vials. Flies were transferred to vials with fresh food containing approximately 5 mL of standard medium without live yeast on the surface weekly. The number of dead flies was recorded daily. Experiments comparing fly lifespans were conducted simultaneously. Sample sizes were 50 to 100 flies per sex per genotype. The experiments that showed noteworthy results were repeated two to three times with an interval of approximately six months. The lifespan for each fly was estimated as the number of days alive from the day of eclosion to the day of death. Mean lifespans and survival curves were primarily used to characterize lifespan.

3.4. Locomotion Assay

Locomotion was measured as described in [16]. Flies were collected and maintained by the same procedures as for the lifespan assays, but without recording the deaths. Locomotion was measured at the same time each day in unmated males and females at age 3 to 5 days, 20 days and 40 days. Experiments comparing locomotion were conducted simultaneously. Sample sizes were 65 to 75 flies (13 to 15 vials) per sex per genotype. To measure locomotor activity, the vials were placed horizontally in a *Drosophila* Population Monitor (TriKinetics Inc, Waltham, MA, USA). Fly movement

303

along the walls or in the middle of the vial crossed the infrared beam rings along the length of the vial. Beam interruptions were detected, and totals were reported every five minutes to the host computer. Two measurements for five minutes were made for each vial. Locomotion was characterized as the mean number of beam interruptions per vial.

3.5. Real-Time RT-qPCR

Transcript levels were measured as described in [15,16]. Total RNA for real-time reverse transcription quantitative PCR (RT-qPCR) was extracted from batches of 30 whole bodies of 3- to 5-day-old males and females using TRIzol reagent (Invitrogen, Carlsbad, CA, USA) and DNase I (Sigma-Aldrich, St. Louis, MO, USA) according to the manufacturers' instructions.

First-strand cDNA was synthesized using SuperScript II Reverse Transcriptase (Invitrogen) with oligo(dT)$_{15}$ primers according to the manufacturer's instructions. Amounts of cDNA were determined by RT-qPCR using SYBR Green I in a MiniOpticon real-time PCR detection system (Bio-Rad).

Gdh and *Adh* housekeeping genes, characterized by relatively low expression comparable to *sgg* expression, were used as reference genes to normalize for differences in total cDNA between the samples. The forward and reverse primer sequences used were as follows: For *sggB*, shaggyPB1 5′-ATATACAGATCTTTTGTTTGGCAA-3′ and shaggyPB2 5′-AGGAGGAAGTTCTTGGACGA-3′; for *Gdh*, Gdh1 5′-TATGCCACCGAGCACCAGATTCC-3′ and Gdh2 5′-GGATGCCCTTCACCTTCTGCTT CTT-3′; for *Adh*, Adhd3: 5′-CGGCATCTAAGAAGTGATACTCCCAAAA-3′ and Adhr3: 5′-TGAGTGT GCATCGAATCAGCCTTATT-3′.

CFX Manager 3.1 software (Bio-Rad, 2012) was used to evaluate the relative gene expression. Inter-run calibrations were used for each panel of experiments since the experiments were conducted for several years. Three independent RNA extractions (biological replicates) per sex per genotype and three technical repeats for each RNA extraction were made.

3.6. Immunostaining and Microscopy

Immunostaining and microscopy were performed according to [27]. Male and female third-stage larvae and brains of 3- to 5-day-old unmated males and females were dissected in phosphate-buffered saline (PBS), fixed in 4% paraformaldehyde (Sigma-Aldrich, St. Louis, MO, USA) at room temperature for 20 min, and washed in PBS (3 × 15 min). For immunostaining, preparations were blocked in blocking buffer (BlockPRO, Visual Protein Biotechnology Corporation, Taiwan) for one hour at room temperature, incubated in primary antibodies (diluted in BlockPRO) overnight at 4 °C, washed in PBS (3 × 15 min), incubated in secondary antibodies (diluted in BlockPRO) for two hours, washed in PBS (3 × 15 min), and placed in a medium for immunofluorescence (VectaShield, Vector Labs, Burlingame, CA, USA). NMJs were analyzed in the fourth muscle of the third and fourth abdominal segments of larvae. A confocal laser scanning microscope (LSM 510, Zeiss, Oberkochen, Germany), ImageJ (http://rsb.info.nih.gov/ij/index.html) and LSM Image Browser (Zeiss) were used. Sample sizes were 10 to 15 specimens per genotype per experiment. To estimate bouton numbers and numbers of synaptic active zones, the same preparations were used. Mean numbers of type 1b boutons and satellite boutons were used to characterize NMJ morphology. The mean number of synaptic active zones was used to characterize synapse activity. Mean numbers of mitochondria and dopaminergic neurons were calculated.

The following primary antibodies were used: Mouse anti-Brp (mAb NC82, 1:200; Developmental Studies Hybridoma Bank (DSHB)) against Bruchpilot (BRP), a protein specific to active synaptic zones [54]; mouse anti-Futsch (mAb 22C10, 1:200; DSHB) against the microtubule-associated protein, Futsch [48]; mouse anti-α-acetylated tubulin (1:200; Santa Cruz Biotechnology, Dallas, TX, USA) against a tubulin isoform, a marker of microtubule networks [55]; Alexa Fluor 647-conjugated goat anti-HRP (1:400, Jackson ImmunoResearch), against Horseradish Peroxidase (HRP), a widely used marker of presynaptic membranes [7]. The secondary antibodies used were goat anti-mouse Cy3 conjugated (1:400, Jackson ImmunoResearch). Antibodies obtained from the DSHB were developed under the auspices of the NICHD and maintained by The University of Iowa, Department of Biology, Iowa City, IA 52242. Mitochondria were visualized with GFP using the D13 driver line. Dopaminergic neuron neurons were visualized with GFP using the D8 driver line.

3.7. Electron Microscopy

Three heads of 3- to 5-day-old unmated males and females of each genotype were dissected and fixed in 2.5% glutaraldehyde in cacodylate buffer at room temperature for 2 h. Samples were washed in sodium cacodylate buffer, fixed in 2% OsO_4 for 2 h at 4 °C, washed and, dehydrated by transferring through solutions of increasing alcohol concentration (30%, 50%, 70%, 90%, and 100%) for 15 min at room temperature followed by acetone for 15 min. Resin (Epon 812 with DDSA and MNA) infiltration was carried out according to the following protocol: Resin-acetone (1:3), resin-acetone (1:1), resin-acetone (3:1) at room temperature for 1 h, and resin alone for 12 h; and resin with a DMP 30 catalysator at 37 °C for 24 h and at 60 °C for 48 h. Hardened blocks were cut on the Leica UC7 (Leica) microtome into 70 nm slices, contrasted by uranyl acetate (30 min) and lead citrate (5 min) and then analyzed using a transmission electron microscope (JEM-1011, Jeol, Akishima, Tokyo, Japan).

3.8. Western Blotting

Approximately 50 heads of 3- to 5-day-old adults of each genotype were dissected and homogenized in 8 M urea solution. Equal amounts of samples from the supernatants were preincubated with sample buffer (deionized water, 0.5 M Tris-HCl, glycerol, 10% SDS, 0.5% bromphenol blue, DTT) for 5 min at 95 °C and separated in a 4–12% (*w/v*) acrylamide Bis/Tris SDS-PAGE gel using the vertical electrophoretic chamber Mini-Protean Tetra (Bio-Rad). Proteins were transferred from the gel to the PVDF membrane (Immobilon-P Membrane,Millipore, Burlington, MA, USA) using electroblotting (Mini Trans-Blot Modul, Bio-Rad), blocked in BlockPro blocking buffer (Visual Protein Biotechnology Corporation, Taiwan), and incubated with anti–GSK3 beta primary antibodies (1:300; ab18893, Abcam, Cambridge, Great Britain) for one hour. Bound antibodies were detected with goat anti–rabbit secondary antibodies conjugated with alkaline phosphatase (1:20000; A3687, Sigma). Prior to visualization, the membranes were incubated in the alkaline CDP buffer for 5 min and then in the Immun-Star AP-Substrate (Bio-Rad) for 7 min. After scanning, the relative intensity quantification of each band was evaluated with Image Lab software (Bio-Rad). Three independent protein extractions (biological replicates) per genotype were made.

3.9. Statistical Analyses

To compare control and mutant genotypes, Student's *t*-test and the nonparametric, distribution-free Kruskal-Wallis test were used for analyses of *sgg* transcript amounts; locomotion; and numbers of large boutons, satellite boutons, active zones, and mitochondria in NMJs. These two tests gave consistent results, so only the results of the Kruskal-Wallis test are reported here. Standard descriptive statistical analysis of lifespan [56,57] was performed to determine the mean lifespan and its accompanying variances, standard deviations, and standard errors; the median, minimum, and maximum lifespans; and the lifespans of the lower and upper quartiles and the 10 and 90 percentiles (Tables S1–S5). Survival curves were estimated using the Kaplan–Meier procedure. The nonparametric, distribution-free Mann-Whitney test and Kolmogorov-Smirnov test were used to evaluate the statistical significance of the difference between the survival curves.

4. Conclusions

Understanding the genetic and molecular basis of pathological changes of the nervous system underlying accelerated aging and shortening of lifespan are important for the development of evidence-based recommendations for life extension. GSK3 is an important enzyme governing numerous regulatory pathways and metabolic processes via interaction with InR/TOR, Wnt, JNK, and other signaling cascades [1,2]. GSK3 is specifically important in the nervous system, where

Int. J. Mol. Sci. **2019**, *20*, 2200

it is involved in the control of synaptic development, structure, and function due to regulation of axonal size, the microtubule cytoskeleton, and neurotransmitter release [7–10]. Here, we demonstrated GSK involvement in the regulation of mitochondria number, intracellular distribution and structure, and control of synaptic activity. We showed that GSK3 affected the axonal distribution of the microtubule-associated protein, Futsch, thus confirming the role of protein kinases in governing the architecture of the microtubule cytoskeleton. We also demonstrated that GSK3 overexpression caused signs of neurodegeneration in the brains of young individuals. All these findings certified that GSK3 misexpression is crucially important for the structural and functional integrity of the nervous system. The main GSK3 isoform is responsible for the normal operation of neurons.

It is generally clear that the structural and functional integrity of the nervous system, in turn, is important for ensuring a long life. Here, we showed that pathological changes in neurons caused by aberrations in GSK3 expression in the nervous system were paralleled by a rapid decline in survival and shortening lifespan. Our data also demonstrated that GSK3 misexpression in other tissues affected lifespan in a transcript-, stage-, and tissue-specific mode. We proved that some minor *shaggy* transcripts are functional and, moreover, may have the specificity required for a particular function. This result allowed us to lift the veil on the complexity of the gene structure and expression strategy. Most of the *Drosophila* genes have multiple annotated transcripts and proteins (https://flybase.org), and there are not many examples showing whether the alleged complexity of the structural organization of a gene is in line with reality and why such a sophisticated structure is needed. Of particular interest to us was the fact that the embryonic overexpression of one of the minor transcripts was enough to cause an increase in adult lifespan. This fact underscored the role of gene expression early in life for lifespan control and the existence of carry-over mechanisms of an epigenetic or other nature underlying this effect.

Altogether, our results demonstrated the key role of GSK3 in ensuring the link between the pathology of neurons and lifespan. The data presented in this article indicated how disentangling expression strategy and general gene biology might eventually provide a selective approach to the choice of potential drugs and therapies based on an understanding of the molecular mechanisms underlying the development of pathologies shortening lifespan. Obviously, treatment should be aimed at suppressing or enhancing the expression of certain gene transcripts in certain tissues. Our data allowed us to come closer to understanding the characteristics of *shaggy* gene expression, which are important for the search for potential drugs that specifically affect the pathologies of the nervous system and aging caused by the GSK3 malfunction.

Supplementary Materials: Supplementary Materials can be found at http://www.mdpi.com/1422-0067/20/9/2200/s1.

Author Contributions: Conceptualization, M.V.T., N.V.R., and E.G.P.; methodology, M.V.T., N.V.R., S.V.B., A.A.Z., A.V.K.; formal analysis, M.V.T., N.V.R., A.V.K., E.G.P.; investigation, M.V.T., N.V.R., S.V.B., E.R.V., A.A.Z.; writing—original draft preparation, M.V.T. and E.G.P.; writing—review and editing, M.V.T., N.V.R., S.V.B., E.R.V., A.A.Z., A.V.K., E.G.P.; visualization, M.V.T. and A.V.K.; supervision, E.G.P.; funding acquisition, M.V.T., N.V.R. and E.R.V.

Funding: This work was supported by the Russian State Budget project #AAAA-A19-119022590053-3 and by the Russian Foundation for Basic Research grants #14-04-01464-a for N.V.R., #18-34-00934-мол_a for E.R.V. and #19-34-80042-мол_эв_a for M.V.T.

Acknowledgments: We thank Tatiana Morozova for invaluable assistance in handling and transferring fly cultures, Svetlana Sarantseva and Ludmila Olenina for help in acquiring new methods and inspiring discussions, and Natalia Biserova, Alfiya Mustafina and Vladimir Popenko for tips on analyzing electron microphotographs. The equipment of the Shared Research Center of the Institute of Molecular Genetics of the Russian Academy of Sciences was used for this work. Stocks obtained from the Bloomington Drosophila Stock Center (NIH P40OD018537) were used in this study, and we express our sincere gratitude to the Center and its personnel for their continuous support. We also thank Addgene for providing the pBID-UASC plasmid and BestGene, Inc., for creating transgenic flies. We are grateful to American Journal Experts for English editing services.

Conflicts of Interest: The authors declare no conflict of interest.

References

1. McCubrey, J.A.; Rakus, D.; Gizak, A.; Steelman, L.S.; Abrams, S.L.; Lertpiriyapong, K.; Fitzgerald, T.L.; Yang, L.V.; Montalto, G.; Cervello, M.; et al. Effects of mutations in Wnt/β-catenin, hedgehog, Notch and PI3K pathways on GSK-3 activity-Diverse effects on cell growth, metabolism and cancer. *Biochim. Biophys. Acta Mol. Cell Res.* **2016**, *1863*, 2942–2976. [CrossRef]
2. Patel, P.; Woodgett, J.R. Glycogen Synthase Kinase 3: A Kinase for All Pathways? *Curr. Top. Dev. Biol.* **2017**, *123*, 277–302. [CrossRef]
3. Colosimo, P.F.; Liu, X.; Kaplan, N.A.; Tolwinski, N.S. GSK3beta affects apical-basal polarity and cell-cell adhesion by regulating aPKC levels. *Dev. Dyn.* **2010**, *239*, 115–125. [CrossRef] [PubMed]
4. Kaplan, N.A.; Colosimo, P.F.; Liu, X.; Tolwinski, N.S. Complex interactions between GSK3 and aPKC in *Drosophila* embryonic epithelial morphogenesis. *PLoS ONE* **2011**, *6*, e18616. [CrossRef] [PubMed]
5. Kanuka, H.; Kuranaga, E.; Takemoto, K.; Hiratou, T.; Okano, H.; Miura, M. *Drosophila* caspase transduces Shaggy/GSK-3beta kinase activity in neural precursor development. *EMBO J.* **2005**, *24*, 3793–3806. [CrossRef]
6. Wong, J.J.L.; Li, S.; Lim, E.K.H.; Wang, Y.; Wang, C.; Zhang, H.; Kirilly, D.; Wu, C.; Liou, Y.-C.; Wang, H.; et al. A Cullin1-based SCF E3 ubiquitin ligase targets the InR/PI3K/TOR pathway to regulate neuronal pruning. *PLoS Biol.* **2013**, *11*, e1001657. [CrossRef]
7. Franco, B.; Bogdanik, L.; Bobinnec, Y.; Debec, A.; Bockaert, J.; Parmentier, M.-L.; Grau, Y. Shaggy, the homolog of glycogen synthase kinase 3, controls neuromuscular junction growth in *Drosoph. J. Neurosci.* **2004**, *24*, 6573–6577. [CrossRef] [PubMed]
8. Franciscovich, A.L.; Mortimer, A.D.V.; Freeman, A.A.; Gu, J.; Sanyal, S. Overexpression screen in *Drosophila* identifies neuronal roles of GSK-3β *shaggy* as a regulator of AP-1-dependent developmental plasticity. *Genetics* **2008**, *180*, 2057–2071. [CrossRef]
9. Cuesto, G.; Jordán-Álvarez, S.; Enriquez-Barreto, L.; Ferrús, A.; Morales, M.; Acebes, Á. GSK3β inhibition promotes synaptogenesis in *Drosophila* and mammalian neurons. *PLoS ONE* **2015**, *10*, e0118475. [CrossRef] [PubMed]
10. Chiang, A.; Priya, R.; Ramaswami, M.; Vijayraghavan, K.; Rodrigues, V. Neuronal activity and Wnt signaling act through Gsk3-beta to regulate axonal integrity in mature *Drosophila* olfactory sensory neurons. *Development* **2009**, *136*, 1273–1282. [CrossRef]
11. Noble, W.; Hanger, D.P.; Miller, C.C.J.; Lovestone, S. The importance of tau phosphorylation for neurodegenerative diseases. *Front Neurol.* **2013**, *4*, 83. [CrossRef]
12. Ma, T. GSK3 in Alzheimer's disease: Mind the isoforms. *J. Alzheimers Dis.* **2014**, *39*, 707–710. [CrossRef] [PubMed]
13. Mudher, A.; Shepherd, D.; Newman, T.A.; Mildren, P.; Jukes, J.P.; Squire, A.; Mears, A.; Drummond, J.A.; Berg, S.; MacKay, D.; et al. GSK-3beta inhibition reverses axonal transport defects and behavioural phenotypes in *Drosoph. Mol. Psychiatry* **2004**, *9*, 522–530. [CrossRef]
14. Pasyukova, E.G.; Symonenko, A.V.; Roshina, N.V.; Trostnikov, M.V.; Veselkina, E.R.; Rybina, O.Y. Neuronal genes and developmental neuronal pathways in *Drosophila* life span control. In *Life Extension: Lessons from Drosophila*; Vaiserman, A.M., Moskalev, A.A., Pasyukova, E.G., Eds.; Springer International Publishing: Cham, Switzerland, 2015; pp. 3–37.
15. Symonenko, A.V.; Roshina, N.V.; Krementsova, A.V.; Pasyukova, E.G. Reduced neuronal transcription of *escargot*, the *Drosophila* gene tncoding a Snail-type transcription factor, promotes longevity. *Front. Genet.* **2018**, *9*, 151. [CrossRef]
16. Roshina, N.V.; Symonenko, A.V.; Krementsova, A.V.; Trostnikov, M.V.; Pasyukova, E.G. Embryonic expression of shuttle craft, a *Drosophila* gene involved in neuron development, is associated with adult lifespan. *Aging (Albany NY)* **2014**, *6*, 1076–1093. [CrossRef]
17. Hernandez, F.; Lucas, J.J.; Avila, J. GSK3 and tau: Two convergence points in Alzheimer's disease. *J. Alzheimers Dis.* **2013**, *33*, S141–S144. [CrossRef]
18. Ruel, L.; Bourouis, M.; Heitzler, P.; Pantesco, V.; Simpson, P. *Drosophila shaggy* kinase and rat glycogen synthase kinase-3 have conserved activities and act downstream of Notch. *Nature* **1993**, *362*, 557–560. [CrossRef]
19. Bourouis, M. Targeted increase in *shaggy* activity levels blocks wingless signaling. *Genesis* **2002**, *34*, 99–102. [CrossRef]

20. Bodenmiller, B.; Mueller, L.N.; Pedrioli, P.G.A.; Pflieger, D.; Jünger, M.A.; Eng, J.K.; Aebersold, R.; Tao, W.A. An integrated chemical, mass spectrometric and computational strategy for (quantitative) phosphoproteomics: Application to *Drosophila melanogaster* Kc167 cells. *Mol. Biosyst.* **2007**, *3*, 275–286. [CrossRef] [PubMed]

21. Brunner, E.; Ahrens, C.H.; Mohanty, S.; Baetschmann, H.; Loevenich, S.; Potthast, F.; Deutsch, E.W.; Panse, C.; de Lichtenberg, U.; Rinner, O.; et al. A high-quality catalog of the *Drosophila melanogaster* proteome. *Nat. Biotechnol.* **2007**, *25*, 576–583. [CrossRef]

22. Tress, M.L.; Bodenmiller, B.; Aebersold, R.; Valencia, A. Proteomics studies confirm the presence of alternative protein isoforms on a large scale. *Genome Biol.* **2008**, *9*, R162. [CrossRef] [PubMed]

23. Wolf, F.W.; Eddison, M.; Lee, S.; Cho, W.; Heberlein, U. GSK-3/Shaggy regulates olfactory habituation in *Drosophila*. *Proc. Natl. Acad. Sci. USA* **2007**, *104*, 4653–4657. [CrossRef]

24. Giannakou, M.E.; Goss, M.; Jünger, M.A.; Hafen, E.; Leevers, S.J.; Partridge, L. Long-lived Drosophila with overexpressed dFOXO in adult fat body. *Science* **2004**, *305*, 361. [CrossRef] [PubMed]

25. Bai, H.; Kang, P.; Tatar, M. *Drosophila* insulin-like peptide-6 (dilp6) expression from fat body extends lifespan and represses secretion of *Drosophila* insulin-like peptide-2 from the brain. *Aging Cell* **2012**, *11*, 978–985. [CrossRef] [PubMed]

26. Hoffmann, J.; Romey, R.; Fink, C.; Yong, L.; Roeder, T. Overexpression of Sir2 in the adult fat body is sufficient to extend lifespan of male and female *Drosophila*. *Aging (Albany NY)* **2013**, *5*, 315–327. [CrossRef]

27. Rybina, O.Y.; Sarantseva, S.V.; Veselkina, E.R.; Bolschakova, O.I.; Symonenko, A.V.; Krementsova, A.V.; Ryabova, E.V.; Roshina, N.V.; Pasyukova, E.G. Tissue-specific transcription of the neuronal gene *Lim3* affects *Drosophila melanogaster* lifespan and locomotion. *Biogerontology* **2017**, *18*, 739–757. [CrossRef]

28. Brand, A.H.; Perrimon, N. Targeted gene expression as a means of altering cell fates and generating dominant phenotypes. *Development* **1993**, *118*, 401–415. [PubMed]

29. Tower, J. Sex-specific gene expression and life span regulation. *Trends Endocrinol. Metab.* **2017**, *28*, 735–747. [CrossRef] [PubMed]

30. Vigoreaux, J.O.; Tobin, S.L. Stage-specific selection of alternative transcriptional initiation sites from the 5C actin gene of *Drosophila melanogaster*. *Genes Dev.* **1987**, *1*, 1161–1171. [CrossRef] [PubMed]

31. Uv, A.E.; Harrison, E.J.; Bray, S.J. Tissue-specific splicing and functions of the *Drosophila* transcription factor Grainyhead. *Mol. Cell. Biol.* **1997**, *17*, 6727–6735. [CrossRef]

32. Jones, L.; Richardson, H.; Saint, R. Tissue-specific regulation of *cyclin E* transcription during *Drosophila melanogaster* embryogenesis. *Development* **2000**, *127*, 4619–4630.

33. Golpich, M.; Amini, E.; Hemmati, F.; Ibrahim, N.M.; Rahmani, B.; Mohamed, Z.; Raymond, A.A.; Dargahi, L.; Ghasemi, R.; Ahmadiani, A. Glycogen synthase kinase-3 beta (GSK-3β) signaling: Implications for Parkinson's disease. *Pharmacol. Res.* **2015**, *97*, 16–26. [CrossRef]

34. Choi, H.; Koh, S.-H. Understanding the role of glycogen synthase kinase-3 in L-DOPA-induced dyskinesia in Parkinson's disease. *Expert. Opin. Drug Metab. Toxicol.* **2018**, *14*, 83–90. [CrossRef]

35. Blagosklonny, M.V. MTOR-driven quasi-programmed aging as a disposable soma theory: Blind watchmaker vs. intelligent designer. *Cell Cycle* **2013**, *12*, 1842–1847. [CrossRef]

36. Xie, T.; Ho, M.C.W.; Liu, Q.; Horiuchi, W.; Lin, C.-C.; Task, D.; Luan, H.; White, B.H.; Potter, C.J.; Wu, M.N. A genetic toolkit for dissecting dopamine circuit function in *Drosophila*. *Cell Rep.* **2018**, *23*, 652–665. [CrossRef]

37. Ridgel, A.L.; Ritzmann, R.E. Insights into age-related locomotor declines from studies of insects. *Ageing Res. Rev.* **2005**, *4*, 23–39. [CrossRef]

38. Cui, H.; Kong, Y.; Zhang, H. Oxidative stress, mitochondrial dysfunction, and aging. *J. Signal Transduct.* **2012**, *2012*, 646354. [CrossRef] [PubMed]

39. Orr, W.C.; Radyuk, S.N.; Sohal, R.S. Involvement of redox state in the aging of *Drosophila melanogaster*. *Antioxid. Redox Signal.* **2013**, *19*, 788–803. [CrossRef] [PubMed]

40. Kann, O.; Kovács, R. Mitochondria and neuronal activity. *Am. J. Physiol. Cell Physiol.* **2007**, *292*, C641–C657. [CrossRef]

41. Ruiz-Cañada, C.; Budnik, V. Introduction on the use of the *Drosophila* embryonic/larval neuromuscular junction as a model system to study synapse development and function, and a brief summary of pathfinding and target recognition. *Int. Rev. Neurobiol.* **2006**, *75*, 1–31. [CrossRef]

42. Course, M.M.; Wang, X. Transporting mitochondria in neurons. *F1000Res* **2016**, *18*, 5. [CrossRef]

43. Chen, S.; Owens, G.C.; Crossin, K.L.; Edelman, D.B. Serotonin stimulates mitochondrial transport in hippocampal neurons. *Mol. Cell. Neurosci.* **2007**, *36*, 472–483. [CrossRef] [PubMed]

44. Morfini, G.; Szebenyi, G.; Elluru, R.; Ratner, N.; Brady, S.T. Glycogen synthase kinase 3 phosphorylates kinesin light chains and negatively regulates kinesin-based motility. *EMBO J.* **2002**, *21*, 281–293. [CrossRef]

45. Chen, S.; Owens, G.C.; Edelman, D.B. Dopamine inhibits mitochondrial motility in hippocampal neurons. *PLoS ONE* **2008**, *3*, e2804. [CrossRef]

46. Gao, F.J.; Hebbar, S.; Gao, X.A.; Alexander, M.; Pandey, J.P.; Walla, M.D.; Cotham, W.E.; King, S.J.; Smith, D.S. GSK-3β phosphorylation of cytoplasmic dynein reduces Ndel1 binding to intermediate chains and alters dynein motility. *Traffic* **2015**, *16*, 941–961. [CrossRef]

47. Ogawa, F.; Murphy, L.C.; Malavasi, E.L.; O'Sullivan, S.T.; Torrance, H.S.; Porteous, D.J.; Millar, J.K. NDE1 and GSK3β associate with TRAK1 and regulate axonal mitochondrial motility: Identification of cyclic AMP as a novel modulator of axonal mitochondrial trafficking. *ACS Chem. Neurosci.* **2016**, *7*, 553–564. [CrossRef]

48. Roos, J.; Hummel, T.; Ng, N.; Klämbt, C.; Davis, G.W. *Drosophila* Futsch regulates synaptic microtubule organization and is necessary for synaptic growth. *Neuron* **2000**, *26*, 371–382. [CrossRef]

49. Wang, J.-W.; Beck, E.S.; McCabe, B.D. A modular toolset for recombination transgenesis and neurogenetic analysis of *Drosophila*. *PLoS ONE* **2012**, *7*, e42102. [CrossRef]

50. Bischof, J.; Maeda, R.K.; Hediger, M.; Karch, F.; Basler, K. An optimized transgenesis system for *Drosophila* using germ-line-specific phiC31 integrases. *Proc. Natl. Acad. Sci. USA* **2007**, *104*, 3312–3317. [CrossRef]

51. McGraw, E.A.; O'Neill, S.L. *Wolbachia pipientis*: Intracellular infection and pathogenesis in *Drosophila*. *Curr. Opin. Microbiol.* **2004**, *7*, 67–70. [CrossRef] [PubMed]

52. Werren, J.H.; Windsor, D.M. *Wolbachia* infection frequencies in insects: Evidence of a global equilibrium? *Proc. Biol. Sci.* **2000**, *267*, 1277–1285. [CrossRef] [PubMed]

53. Holden, P.R.; Jones, P.; Brookfield, J.F. Evidence for a *Wolbachia* symbiont in *Drosophila melanogaster*. *Genet. Res.* **1993**, *62*, 23–29.

54. Wagh, D.A.; Rasse, T.M.; Asan, E.; Hofbauer, A.; Schwenkert, I.; Dürrbeck, H.; Buchner, S.; Dabauvalle, M.-C.; Schmidt, M.; Qin, G.; et al. Bruchpilot, a protein with homology to ELKS/CAST, is required for structural integrity and function of synaptic active zones in *Drosophila*. *Neuron* **2006**, *49*, 833–844. [CrossRef]

55. Sherwood, N.T.; Sun, Q.; Xue, M.; Zhang, B.; Zinn, K. *Drosophila* spastin regulates synaptic microtubule networks and is required for normal motor function. *PLoS Biol.* **2004**, *2*, e429. [CrossRef]

56. Wilmoth, J.R.; Horiuchi, S. Rectangularization revisited: Variability of age at death within human populations. *Demography* **1999**, *36*, 475–495. [CrossRef] [PubMed]

57. Carey, J.R. *Longevity: The Biology and Demography of Life Span*; Princeton University Press: Princeton, NT, USA, 2003.

MDPI

St. Alban-Anlage 66

4052 Basel

Switzerland

Tel. +41 61 683 77 34

Fax +41 61 302 89 18

www.mdpi.com

International Journal of Molecular Sciences Editorial Office

E-mail: ijms@mdpi.com

www.mdpi.com/journal/ijms

www.ingramcontent.com/pod-product-compliance
Lightning Source LLC
Chambersburg PA
CBHW051714210326
41597CB00032B/5478